SOILS: THEIR PROPERTIES AND MANAGEMENT

Editors: P.E.V. Charman and B.W. Murphy

In association with the Department of
Land and Water Conservation, New South Wales

LAND & WATER
CONSERVATION

WITHDRAWN

OXFORD

OXFORD
UNIVERSITY PRESS

253 Normanby Road, South Melbourne, Australia

Oxford University Press is a department of the University of Oxford.
It furthers the University's objective of excellence in research, scholarship,
and education by publishing worldwide in

Oxford New York

Athens Auckland Bangkok Bogotá Buenos Aires Calcutta
Cape Town Chennai Dar es Salaam Delhi Florence Hong Kong Istanbul
Karachi Kuala Lumpur Madrid Melbourne Mexico City Mumbai
Nairobi Paris Port Moresby São Paulo Singapore Taipei Tokyo Toronto
Warsaw

with associated companies in Berlin Ibadan

OXFORD is a trade mark of Oxford University Press

in the UK and in certain other countries

© Soil Conservation Commission of New South Wales 1991, 2000
First published 1991
Reprinted 1992, 1993, 1994, 1998
Second edition published 2000

National Library of Australia
Cataloguing-in-Publication data:

Soils: their properties and management.

2nd ed.
Bibliographical.
Includes index.
ISBN 0 19 550994 3.

1. Soil conservation—New South Wales. 2. Soil management—New
South Wales. 3. Soils–New South Wales—Classification. 4. Soils—New
South Wales—Composition.
I. Charman, P.E.V. II. Murphy, B.W. III. Soil Conservation Service of
New South Wales.

631.49944

Edited by Eva Chan
Text designed by Derrick I. Stone Design
Cover designed by Modern Art Production Group
Typeset by Derrick I. Stone Design
Printed through Bookpac Production Services, Singapore

Contents

List of Figures

List of Tables

List of Plates

Contributors

Glen Atkinson, of the New South Wales Department of Land and Water Conservation, has experience in soil survey, soil mapping, and identification of soil limitations for land use.

Peter Charman, formerly of the Soil Conservation Service of New South Wales, is now retired. He has wide experience in soil conservation and soil classification, and in administering research and editing publications on land and soil management.

Greg Chapman, of the New South Wales Department of Land and Water Conservation, has been the leader of the Department's soil mapping program for several years and has experience in soil survey, soil mapping, and the identification of soil limitations for land use.

Phil Conacher, formerly of the New South Wales Department of Land and Water Conservation, has experience in the management and stabilisation of coastal areas in New South Wales.

Barry Craze, formerly of the New South Wales Department of Land and Water Conservation, has experience in laboratory procedures, including those for determining soil physical properties.

Bob Crouch, formerly of the New South Wales Department of Land and Water Conservation, has experience in soil mapping, soil classification, and has conducted research on erosion processes and hydrology.

Keith Edwards, formerly of the Soil Conservation Service of New South Wales, has experience in statistical analysis and was responsible for analysing data associated with long-term erosion experiments conducted by the Service.

David Eldridge, of the New South Wales Department of Land and Water Conservation, has conducted research on soil erosion and land stability in the rangelands of New South Wales.

Greg Elliott, formerly of the Soil Conservation Service of New South Wales, now works for the Australian Nuclear Science and Technology Organisation. He has conducted research on soil erosion using radioactive isotopes, and has experience in the area of soil chemistry as it relates to land rehabilitation.

Greg Fenton, of New South Wales Agriculture, is responsible for a state-wide program on the management of acid soils.

Guy Geeves, of the New South Wales Department of Land and Water Conservation, has conducted research on soil physical properties and land management.

Jonathon Gray, of the New South Wales Department of Land and Water Conservation, has experience in soil survey, soil mapping, and identification of soil limitations for land use.

David Greentree, of the New South Wales Department of Land and Water Conservation, has experience evaluating soil engineering properties for earthworks and in the site construction of earthworks.

Greg Hamilton, formerly of the Soil Conservation Service of New South Wales and now with the Western Australian Department of Agriculture, has conducted research on soil physical properties and land management.

Adrian Harte, of the New South Wales Department of Land and Water Conservation, has conducted research on the effects of conservation farming practices on soils and has worked as an advisory officer on land management, especially in relation to conservation farming.

Keith Helyar is a scientist with New South Wales Agriculture and is a leading authority on soil chemistry, especially soil phosphate and the processes and management of soil acidification under agricultural systems.

Richard Hicks, of the New South Wales Department of Land and Water Conservation, has experience in soil mapping and land evaluation, especially in relation to engineering properties and urban development.

Cathy Hird, formerly of the Soil Conservation Service of New South Wales, works as a private consultant and has experience in soil mapping, soil classification, and evaluating soils for urban land use.

Bill Johnston, of the New South Wales Department of Land and Water Conservation, has conducted research on revegetation and native vegetation in New South Wales.

John Lawrie, of the New South Wales Department of Land and Water Conservation, has experience in soil mapping and soil classification and worked as an advisory officer on land management, especially in relation to conservation farming.

John Leys, of the New South Wales Department of Land and Water Conservation, has conducted research on wind erosion, and has written several internationally recognised publications on this topic.

Dermot Mc Kane, of the New South Wales Department of Land and Water Conservation, has experience in soil survey, soil mapping, and identification of soil limitations for land use. He has worked in cooperation with CSIRO on an overview of soils of the Murray-Darling Basin.

Grant McTainsh, of the Faculty of Environmental Sciences at Griffith University, Queensland, has conducted research on wind erosion and land stability.

Mike Melville, of the School of Geography, University of New South Wales, is a leading authority on acid sulfate soils in New South Wales and has also worked overseas on such soils.

Rick Morse, formerly of the New South Wales Department of Land and Water Conservation, is currently a private consultant. He has experience in urban erosion issues, especially as they relate to streambank erosion and mass movement.

Phil Mulvey, managing director of Environmental Earth Sciences Pty Ltd, specialises in management of soil contamination and other environmental problems.

Brian Murphy works for the New South Wales Department of Land and Water Conservation. He has wide experience in soil mapping and soil classification, and in research and advisory roles on the effects of land management on soils.

Casey Murphy, of the New South Wales Department of Land and Water Conservation, has experience in soil survey, soil mapping, and identification of soil limitations for land use.

Ian Packer, of the New South Wales Department of Land and Water Conservation, has conducted research on the effects of conservation farming practices on soils, and has worked as an advisory officer on land management, especially in relation to conservation farming.

Ken Reynolds, of the New South Wales Department of Land and Water Conservation, has experience evaluating soil engineering properties for earthworks and has a long involvement with land rehabilitation work across the state.

Margaret Roper, formerly of the CSIRO Division of Plant Industry, has experience in research on soil biology and soil organic matter.

Col Rosewell works for the New South Wales Department of Land and Water Conservation. He has researched erosion processes and rates in New South Wales and has written several well recognised publications in this area.

Bill Semple, of the New South Wales Department of Land and Water Conservation, has conducted research on vegetation mapping and classification and has worked as an advisory officer on revegetation.

Bob Sonter, formerly of the Soil Conservation Service of New South Wales, has experience as an advisory officer in land management.

Roger Stanley, formerly of the New South Wales Department of Land and Water Conservation, has experience in the management and stabilisation of coastal areas in New South Wales.

Ian White, of the Centre for Resource and Environmental Studies at the Australian National University, is a leading international authority on acid sulfate soils, and has done extensive work in New South Wales and Queensland on such soils.

Andrew Wooldridge, of the New South Wales Department of Land and Water Conservation, has worked as an advisory officer on soil salinity in central western New South Wales.

Christoff Zierholz, of the New South Wales Department of Land and Water Conservation, has conducted research on soil erosion and land management.

Preface and Acknowledgments

Since the first publication of this book in 1991, soil conservation has moved on apace in Australia, highlighting the need for an update of this soils text and production of a second edition.

While some of the text remains virtually unchanged, it has all been reviewed and much has been rewritten. In a number of cases new authors have been involved in this process. The chapter structure has been retained but new sections have been added where appropriate. These have focused on aspects of soils relevant to new areas of environmental concern such as urban salinity, waste disposal and the exposure of acid sulfate soils in coastal regions.

The importance of soils in relation to the broader nature of environmental awareness has been acknowledged with a corresponding shift of emphasis in this edition. This change has been towards the relevance of soils to matters of water quality, catchment management and the ecological sustainability of land use practices. We have tried to put soils and soil conservation into the context of land management at its broadest level.

One important change in Australian soil science since 1991 has been the development of a new classification system for our soils, by Queenslander Ray Isbell. This system has been widely accepted by Australian soil scientists, and is featured in the relevant parts of this text.

The book retains its primary focus as a practically oriented soil science text, designed to give a sound understanding of soils for anyone concerned with land conservation. Although based on New South Wales soils and soil conservation experience, much of it has strong relevance to the rest of Australia. Following the wide use made of the first edition, it is hoped that students and teachers will again find this book useful in school and university courses into the new millennium.

As editors we wish to record our gratitude to the authors who were prepared to review their chapters and bring them up to date for this edition. We are particularly indebted to those new authors who rewrote some chapters and added a considerable amount of new material.

We also wish to thank Peter Fogarty, David Greentree, John Lawrie, Neil Rendell, Andrew Rawson, and Andrew Wooldridge, all of whom assisted with the review and assembly of additional material for this edition.

Our sincere thanks also go to Donna McKellar and Bernadette Mossman for all the word processing involved, and to Jill Henry of Oxford University Press for constant support and encouragement.

Peter E. V. Charman
Brian W. Murphy

The Nature of Soil

CHAPTER 1

B.W. Murphy

Soil is one of the world's most valuable assets and frequently it is the characteristics and fertility of this resource which determine a region's wealth. Not only is soil the main resource upon which agricultural production depends, but it is also used as a construction material and is the foundation material for many buildings and roads. Soil is also used as filter or sink for effluent and other waste material. Recreational facilities, including our National Parks, football fields, cricket pitches, beaches and ski fields, have soil as their base. Soils can also be used as indicators of the presence of mineral deposits. Therefore, we should all know something about our soils and how we can make best use of them in an environmentally acceptable way.

While soils are valuable they are also vulnerable, being subject to a wide range of natural and man-made processes which may cause them to be eroded or to lose their ability to perform the functions we expect of them.

Soils are developed by physical, chemical and biological processes, including the weathering of rock and the decay of vegetation. Soil materials include organic matter, clay, silt, sand and gravel, mixed together to form a natural medium in which most land plants grow.

Soil typically comprises several layers (or horizons) of soil material, more or less parallel to the earth's surface, which together form the soil profile. The type of soil at a particular site is dependent on the action of climate and organisms on parent material over long periods of time, and is also influenced by the topography. Soil can differ markedly from its parent material in morphology, properties and characteristics.

Soil means different things to different people and, depending on their occupation or interest, they see soils in a variety of ways.

To the pedologist, geomorphologist or geologist, soils are essentially the unconsolidated materials at the earth's surface, which have been derived from particular parent materials and subsequently influenced to a greater or lesser degree by the processes of weathering, new mineral formation, leaching, organic matter accumulation, erosion and deposition. Geologists often describe and map the 'regolith', which is described as 'the layer or mantle of loose, non-cohesive rock material, of whatever origin, that nearly everywhere forms the surface of the land and rests on bedrock'. Soils include those layers of regolith closest to the land surface and/or that regolith which has been subjected to the soil-forming processes discussed in this chapter.

The farmer or gardener sees the soil as a medium for plant growth and is mainly interested in obtaining the best possible plant growth from a particular soil.

Engineers or earthwork contractors see the soil as a construction material for dams or as a base for roads and road batters. They are mainly interested in the properties of the soil material which will contribute to a stable, long-lasting structure.

The builder, at least for small buildings, is interested in using the soil as a foundation. The soil should be stable and free from waterlogging and salt. The builder may also use soil in the form of clay as the basic material to make bricks.

The hydrologist is interested in how quickly, and to what degree, soils will absorb or shed rainfall and how the soil transmits that water across the landscape. The interaction between land, rainfall and vegetation gives us our water supplies, with soils playing a large part in determining both the quantity and quality of water in our streams, rivers and storages.

The conservationist is interested in soil because it is a fragile resource to be conserved, often subject to degradation when used beyond its capacity. Such practices as overgrazing, excessive cultivation, tree clearing and irrigation can lead to soil degradation. The most stable situation is achieved by maintaining a continuous cover of vegetation on the soil. However, this is not possible for many land uses, particularly considering the range of production systems necessary to meet the wide-ranging demand for commodities from the agricultural sector.

The environmentalist is interested in soils because particular soil types in combination with particular climates or geomorphological environments form unique and special ecosystems such as the alpine areas, the coastal forests, wetlands and native grasslands. The best management and care of these areas require knowledge and care of the soils on which they are formed.

Major forms of soil degradation include wind and water erosion, reduced fertility because of nutrient depletion, physical breakdown of soil structure, soil acidification, soil biological decline, salinisation and soil contamination. Unfortunately, soil degradation can also lead to the degradation of water resources, both in quality and quantity, and to the degradation of vegetation resources.

One of the main aims of environmental management should be to prevent the degradation of our soil resources or to reverse it where it has already occurred. To do this, an environmental manager is likely to consider all the previous points of view of soil at one time or another. This book outlines the important soil properties and in practical terms how our soils can be managed to best care for our total environment in a sustainable fashion.

1.1 Soil Formation

Many workers (Jenny, 1941; Corbett, 1969; FitzPatrick, 1971; Paton, 1978; Gray and Murphy, 1999) have recognised that soil formation needs to be discussed in terms of soil-forming factors. The soil-forming factors control the initial parent material weathering and the processes that in turn lead to soil development. In this book, soil formation is discussed from these two viewpoints.

1.1.1 Soil-forming Factors

Soil formation is traditionally considered to be dependent on five soil-forming factors described by Dokuchaev in 1883 (colour plate 1.1):
— parent material
— climate
— relief or topography
— organisms, including human activity
— time.

Various people have described and expanded on these factors, including Jenny (1941), Corbett (1969, 1972), FitzPatrick (1971), Birkeland (1974) and White (1987). Corbett (1972) describes how climate and parent material led to the formation of different soils in New South Wales. Given that each of the above factors can be a complex area of study in itself, it can be difficult to describe and account for the distribution of soils on the landscape.

More recently, these factors of soil formation have been challenged by Paton et al. (1995), who have suggested the use of the term 'lithospheric material' instead of parent material, and 'availability of water' instead of climate. These changes are suggested because of the broad interpretations on the original terms, and allow the importance of processes controlling parent material distribution to be given more emphasis than the traditional climatic control of soil distribution through the zonal concept of soils. They divide the major areas of lithospheric material on the basis of plate tectonics and recognise the following plate segments will have different lithospheric materials and processes of soil formation: plate centre (having largely granitic rocks); tensional margin (largely within the oceans); and compressional margins (many of the ancient and current areas of mountain development where rocks include basaltic and other less basic volcanic rocks).

One criticism of these ideas concerns the areas of Palaeozoic and Mesozoic plate margins that can occur within the current continental plates. For example the Lachlan Fold Belt, which covers much of central and southern New South Wales, is an example of an ancient compressional margin in the Palaeozoic Era which has the suggested characteristics of the margins, particularly the complexity of rock materials and soils, yet it is not considered in the Paton discussions. Nor does the theory account for the relatively extensive areas of Tertiary basalts within this state.

In the topography factor, Paton et al. (1995) emphasise that slope not only influences drainage characteristics of the soils but also affects the movement of soil down the slope under gravity as soil creep, and slope wash as erosion. Although no firm acceptance has been given to these ideas, and it appears they need further development and testing against experimental data, they do represent the initiation of some developing concepts of soil formation, which will hopefully lead to a better understanding of the distribution of soils on the earth's surface.

The major factors controlling the distribution of soils in the western part of New South Wales which is largely covered by the Murray-Darling Plains are described in Butler et al. (1983). The main parent materials are alluvial materials and aeolian materials. In eastern New South Wales, the distribution is described by Walker et al. (1983) and Hubble and Isbell (1983). The major variations in soils are associated with variations in parent material, especially variations in bedrock, although areas where climate or transported parent materials are important do occur in local areas such as on alluvial plains, along the coast and in the alpine areas. Chapters 8 and 9 describe the distribution of soils in New South Wales in more detail.

Parent Material

Soils form a cover over most of the earth's surface, often in association with weathered rock materials, volcanic ash, colluvium, alluvium, wind-blown deposits and accumulations of vegetative material. This overall mantle of relatively loose material overlying bedrock is called the regolith. The uppermost and vital layer that sustains humanity is the soil. However, it is necessary to recognise that the methods and procedures used to characterise soils are also applicable to characterise other regolith materials. Therefore the distinction between 'soil material' and 'other regolith material' can be difficult to make, although discussions of 'regolith materials' probably place more emphasis on the geological and geomorphological origins of materials, with less emphasis on the soil-forming processes to which they have been subjected. Generally regolith studies emphasise the origins of regolith materials and the potential for mineral exploration, while soil studies concentrate on the total physical and chemical properties of the regolith materials in relation to land management and environmental management.

Sometimes soils form directly from weathering rock (in situ), but many soils are formed from materials that have been transported and deposited. Whether rock or deposited materials, the parent material provides the starting point for soil formation. The main parent material groups are discussed in the following section, and also in FitzPatrick (1971).

Parent materials are affected to varying degrees by soil-forming processes, and an important principle to understand the types of soil that occur on the landscape is 'inheritance' as defined by Paton et al. (1995). Inheritance refers to the amount of the original parent material that is unaffected by soil-forming processes of weathering, transportation and so on (epimorphism). It is an assessment of the contribution that the original bedrock makes to the soil material at a site. This will vary greatly depending on the characteristics of the original bedrock, the degree of weathering of the bedrock, determined by the intensity of and length of time that weathering processes have affected the bedrock, and the types of transportation processes which the weathered bedrock material has undergone.

(a) Bedrock Parent Materials

A large proportion of soil parent material is bedrock. The mineral composition and grain size of the bedrock strongly influence the type of soil formed. A fundamental classification of bedrock parent material is based on the relative contents of silica and ferromagnesian minerals in the rocks (Corbett, 1969, 1972; FitzPatrick, 1971; Duff, 1993), and a brief summary is presented in Table 1.1. Siliceous (acidic) parent materials (granite, sandstone, rhyolite) are relatively high in silica and have high levels of quartz and potassium felspars. These rocks are sometimes termed felsic. More basic parent materials (basalt, gabbro) are high in the ferromagnesian minerals, such as olivine, amphibole, pyroxene and biotite mica. These are sometimes termed mafic. The mafic minerals have large quantities of bases and materials available for clay formation, and so the soils are usually clayey and often fertile. In contrast, the more siliceous parent materials are low in bases and clay-forming minerals, so the soils derived from these tend to be lower in clay content and less fertile.

The composition of sedimentary and metamorphic rocks varies widely, depending on the composition of the igneous rocks from which they were derived and the extent of weathering to which they have been subjected. In general, shales and schists tend to be slightly siliceous to intermediate, and sandstones tend to be siliceous. There are some exceptions to this general rule. For example, when sandstones consist of relatively unweathered rock fragments as in greywackes, the sandstones may be more intermediate in character. When shale-like rocks are composed of volcanic ash laid down in shallow water as might occur in tuffaceous rocks, these rocks may tend to be acidic to basic in character as the composition of volcanic ash varies widely (Duff, 1993).

The soils formed from limestones are dependent on the insoluble, highly basic residue remaining after weathering of the limestone. However, the soils in New South Wales that have formed from limestone tend to be those expected from rocks of intermediate to basic composition.

For a more detailed discussion of rock types in relation to soil formation, see FitzPatrick (1971), Corbett (1969, 1972), Birkeland (1974) and Gray and Murphy (1999). The links between the type of soils formed and the parent rock are discussed further in Paton et al. (1995), who distinguish quartz-rich and quartz-poor parent materials, and divide the clays produced by the quartz-poor parent materials into whether they are likely to form kaolin-rich or smectite-rich clays.

Table 1.1 Parent material classification (after FitzPatrick 1971)

I. Minerals

SILICEOUS (high silicon, low ferromagnesian content)	Quartz Potassium felspar	
	Muscovite mica (Similar to many clay minerals in sedimentary rocks)	Decreasing
INTERMEDIATE	Sodium felspars	resistance
	Calcium felspars	
	Biotite mica	to
	Amphibole (e.g. hornblende)	
	Pyroxene (e.g. augite)	weathering
BASIC (relatively low silicon, high ferromagnesian content)	Olivine	

II. Rock Types

	Igneous Rocks	Sedimentary and Metamorphic Rocks	Silicon Content	Ferromagnesian Content
SILICEOUS	Quartzite	Sandstones	>65%	<20%
	Granite/Rhyolite	Shale		
	Granodiorite/Dacite	Schist Slate		
INTERMEDIATE	Trachyte		55–65%	20–40%
	Andesite/Diorite	Limestones and calcareous sediments		
	Basalt/Gabbro			
BASIC	Peridotite/Dunite		45–55%	>40%

The distribution of rock types on the earth's surface is determined by geological history (Duff, 1993), and therefore, whenever soils are derived largely from bedrock, the geological history of an area can be a strong factor controlling the distribution of soils. For example, in New South Wales the soils derived from the ancient volcanic belts of andesitic volcanics and more recent basaltic volcanic rocks are quite different from the soils derived from the ancient marine sediments, terrestrial sandstones or siliceous granites that are now exposed. Paton et al. (1995) argue that as the distribution of lithospheric material at the earth's surface is controlled by plate tectonics, then ultimately plate tectonics has a major impact on the distribution of soils. Beckmann (1983) describes in broad terms how the geological and geomorphological history of Australia, in combination with climatic changes, has influenced the distribution of parent materials and the development of soils in Australia. He concludes many Australian soils have been formed on reworked older materials that have been redistributed by wind and water, as can be seen in the Western Plains and Riverina. The recent (Late Cainozoic) pattern of uplifts in the Eastern Highlands has led to the cutting of new land surfaces and wearing back of scarps. This has resulted in the exposure of new surfaces on the underlying bedrock, and bedrock geology then has a major influence on soil formation in these areas.

(b) Transported Materials

Where soil parent material is not bedrock, it is geological material that has been transported and deposited. The agents that transport materials about the earth's surface are summarised in Figure 1.1 and include mass movement under gravity (colluvial materials), water (alluvial materials), wind (aeolian materials) and ice (glacial materials). Sediments associated with lakes are also quite distinct, being a characteristic combination of alluvial and aeolian sediments. A comprehensive description of the kinds of transported materials that occur across the landscape is given in Reineck and Singh (1975), while Beattie (1972) and Butler et al. (1983) describe how soils vary on the riverine plains of New South Wales.

The source of transported materials, in combination with the processes of sorting and weathering during transport and deposition, influence the grain size and mineral composition of transported materials. These in turn influence the type of soil formed on them. The distribution of these transported parent materials across the landscape depends on the geological and geomorphological history of an area. Geological and geomorphological processes may cause very different parent materials to lie side by side in the landscape (see Figure 1.2).

The distribution of transported material on a landscape is often the result of a series or episodes of depositional events, which may be very recent or associated with previous periods of changing climate. Many of the more productive soils in New South Wales are formed on alluvium recently deposited along the floodplains of the major rivers. Vast areas may be covered by transported materials as seen in the alluvial and aeolian materials covering much of the Western Plains and the Riverina (Butler, 1956; McTainsh, 1985), or the sand dunes occurring along the coastline. Glacial depositional materials cover much of the central plains of North America and northern Europe, and vast aeolian deposits cover many parts of the world including the South Island of New Zealand on the Canterbury Plains, parts of South America and western Africa, and in northern China (Duff, 1993; Paton et al., 1995).

In coastal zones, the distribution of parent materials is controlled by a distinct set of geomorphological processes as described in Ward and McArthur (1983). The distribution of parent materials along the coastline can be attributed to the actions of waves and currents, and also to the deposition and movement of sediments associated with rivers and winds, and with the tectonic and climatic movements that can cause changes in sea level. Estuarine, marine, fluvial and terrestrial sedimentary environments can all occur along the coastline.

(c) Organic Materials

Organic material can be the major parent material for soils in cooler, wetter areas. The organic material from plant breakdown accumulates as peat and can build up to considerable depth. These soils in New South Wales cover only a small area and are largely confined to the alpine areas, or swamps or peat bogs in the more humid coastal areas and on the sandstone plateaux around Sydney.

Climate

Climate strongly influences soil-forming processes and is frequently the driving force behind many of them. The ratio of precipitation to evaporation influences the amount of water available for weathering, new mineral formation, leaching and the translocation of colloids, especially clay. The drier the climate, the less water available for these processes. Low temperatures also reduce weathering rates, but greatly increase rates of organic matter accumulation.

The water available for plant growth can also influence the rate of plant growth and hence the rate of organic matter accumulation and the activity of soil biota. The amount of plant growth can also affect surface cover levels and hence the amount of erosion that will occur under heavy rainfall.

Topography

Topography also influences the amount of water available for weathering, new mineral formation, leaching and the translocation of colloids (see Figure 1.3). Well-drained soils mainly occur upslope and there is generally an accumulation of water, weathering products and transported materials on lower slopes and in depressions. The net accumulation of water, bases and clay depends on the amount moving into and out of the profile. This is determined by the characteristics of the catchment affecting the profile site, the permeability of the subsoil and underlying layers, and the slope of the land at the site of the profile.

The topography can also influence the types of slope processes occurring at a site. The amount of soil

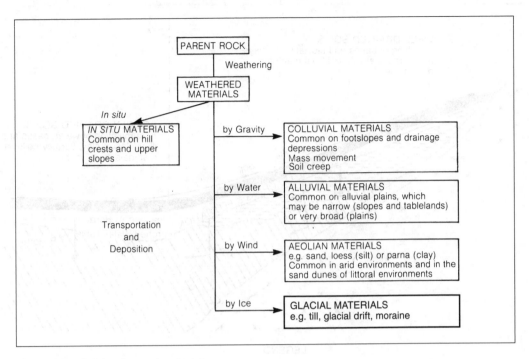

Figure 1.1 Origin of parent materials

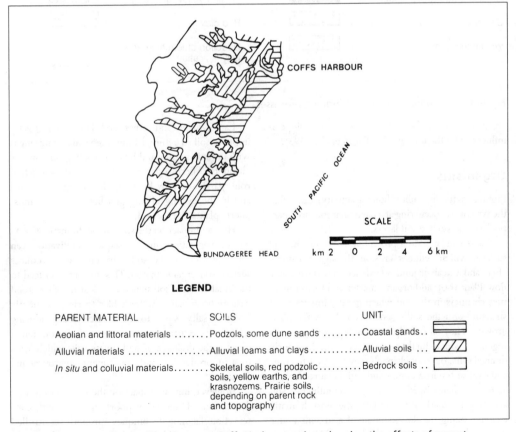

Figure 1.2 Generalised soil landscapes—Coffs Harbour region, showing the effects of parent materials on soil distribution (after Atkinson and Veness, 1981)

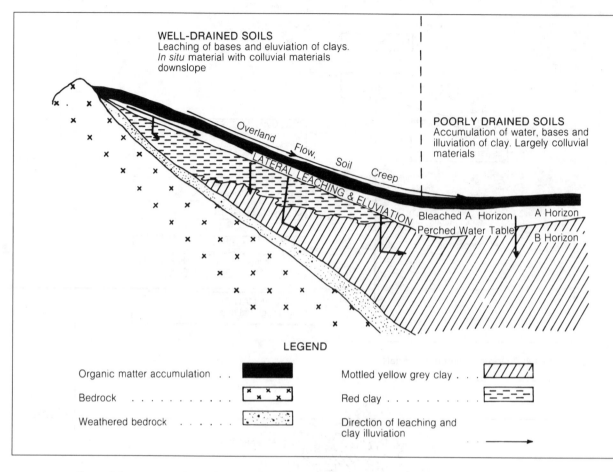

WELL-DRAINED SOILS
Leaching of bases and eluviation of clays.
In situ material with colluvial materials
downslope

POORLY DRAINED SOILS
Accumulation of water, bases and
illuviation of clay. Largely colluvial
materials

Overland Flow, Soil Creep

LATERAL LEACHING & ELUVIATION

Bleached A Horizon
Perched Water Table
A Horizon
B Horizon

LEGEND

Organic matter accumulation . .

Bedrock . . . .

Weathered bedrock . . . . . .

Mottled yellow grey clay . . .

Red clay . . . .

Direction of leaching and
clay illuviation

Figure 1.3 Topography and soil-forming processes

creep, and erosion and deposition along a slope is influenced by the topography (Paton et al., 1995).

Organisms

Biotic activity (including human activity) can affect the type of soil occurring at a particular location, particularly the surface soil layers.

Plants can provide a protective cover on the soil surface, which reduces erosion through a canopy effect and a near-ground retardance to surface water flow. Plant roots add organic matter and form miniature channels in the soil which greatly improve soil structure and the soil's general condition for plant growth. They also help in binding soil particles together, which further resists erosion. In poorly drained soils, old root channels may allow air to enter and oxidise the iron compounds, giving a red colour along the channels. Plants can greatly influence soil acidity, particularly at the surface, the most dramatic example being the development of acid soils under improved pastures in the tablelands of New South Wales (Williams, 1980).

Earthworms and other soil fauna can greatly improve soil structure, by ingesting soil and mixing it with organic matter and by constructing burrows or channels. Micro-organisms also play an important role, particularly by breaking down organic matter and in the cycling of nitrogen between the atmosphere, plants and the soil.

Human impact on the soil can be dramatic. Clearing of vegetation and subsequent cultivation can cause erosion of the soil. Overcultivation reduces surface soil organic matter. This, in turn, can lead to the breakdown of soil structure. Clearing of trees and irrigation can cause water tables to rise, frequently bringing salty water to the surface and producing saline soils. A particularly important environmental effect is the drainage of acid sulfate soils which leads to acid production and consequent environmental damage.

However, human impact on the soil is not always detrimental. Drainage of poorly drained land, conservation farming techniques, properly controlled irrigation and additions of nutrients and organic matter often increase soil productivity and stability.

Time

The final factor to be considered is time. Soil type is influenced by the amount of time that soil-forming processes have been occurring. Theoretically, at least, if a sufficient period of intense weathering has occurred, all soils should approach a similar state, regardless of parent material (Jenny, 1941; Chesworth, 1973; Birkeland, 1974). In practice, this rarely occurs, but may be approximately true on very old landscapes (Tertiary age or greater).

Soils of different ages are most evident on alluvial plains where they are often in close proximity to each other. Older soils are generally redder and have stronger horizon development.

1.1.2 Episodes of Landscape and Soil Development

In the northern hemisphere and in alpine areas, glaciation has been, or is, a major agent for transporting soil-forming materials. During the Pleistocene period culminating in the last Ice Age, about 10 000 years ago, large ice sheets stripped vast areas of accumulated soil (Paton, 1978). Thus, soil formation began again on fresh rock or fluvioglacial and aeolian deposits. This explains the relatively young soils found on the northern continents.

In Australia, most soils are relatively old, since there has been no stripping by glaciation except in small areas of the southern highlands and Tasmania. Some landscapes have been exposed to weathering since at least the late Tertiary period (about two to three million years ago). Australian soils are generally strongly weathered, and there are often complex arrays of different geological materials in any one area. They are, therefore, different from those found on glaciated areas of the northern hemisphere, where many concepts of soil formation were originally developed (see Paton, 1978, p. 110). Australian soils have not experienced the dramatic effects of continental glaciation, however periodic changes in geomorphological and pedogenic processes have occurred. Regional uplifting and changes in climate have led to distinct episodes of landscape development and soil formation (Butler, 1959; Beckmann, 1983; Walker and Butler, 1983).

The rejuvenation of the landscape by whatever means (uplift, glaciation, climatic change or vulcanism) does not necessarily lead to the total destruction of old soils and landscapes. Rejuvenation may only affect part of the landscape, causing the redistribution of weathering products (soils, clays and gravels), filling valleys, covering riverine plains and forming pediments. There need not be any input of fresh minerals, although vulcanism is often a source of fresh minerals for soil formation.

In the early- to mid-Tertiary (40 to 60 million years ago), the eastern highlands were uplifted as much as 600 m in places. This rejuvenated erosion, causing the old weathered surfaces of the highlands to be stripped, is thought to be responsible for the large volume of alluvial materials on the plains of the western-flowing rivers of New South Wales. Other more localised uplifts have instigated other episodes of erosion, with subsequent deposition in valleys.

The climate of New South Wales has varied considerably during the Tertiary and Quaternary periods (the last 60 million years). It is generally considered that laterite profiles formed in many areas when New South Wales had a hot-wet tropical climate. Although they have been largely stripped by subsequent erosion, remnants of these soils can be seen in many locations around Sydney (Burges and Beadle, 1952), for example, and Inverell (Corbett, 1969).

The term 'laterite' is not as widely used as it once was. The term 'ferricrete' is commonly used to describe material hardened by the enrichment of iron (Corbett, 1969; Hubble et al., 1983). The term 'laterite profile' is used to refer to profiles with a sequence of a ferruginous or iron-enriched zone (which may also be aluminium enriched), a mottled zone and a pallid zone (Hubble et al., 1983). It has been generally realised that ferricrete can arise in a number of ways besides intense weathering. Seepage waters can form l calised areas of ferricrete in less than ten years (Corbett, 1969; see also Hunt et al., 1977).

An important episode involved a cold phase during the late Quaternary (20 000 to 30 000 years ago), which rejuvenated extensive erosion and deposition in the eastern highlands. Many of today's soils in the eastern highlands may date from surfaces exposed and materials deposited at this time. Another less intense, regional episode of erosion and deposition may have been initiated by a climatic change about 1500 to 4000 years ago. This has led to a new series of small but significant alluvial deposits on the landscape and a new round of soil formation.

Arid periods in the interior of Australia have led to the development of sand dunes in south-western New South Wales and the transport and deposition of dust in parts of eastern Australia in the form of parna (Butler, 1956; Butler and Churchward, 1983). Although the thickness of this dust mantle appears to be very thin in places, it is likely that some amounts of dust from the arid interior are still being deposited on the eastern slopes and highlands.

More recently, clearing, grazing and cultivating by humans have led to a new episode of accelerated erosion and deposition which, although minor on a geological scale, has been highly significant locally.

Soil-forming Processes

Soil parent material, whether bedrock or deposited, is acted upon by various interacting processes to form soil. These are summarised by Birkeland (1974), White (1987) and Paton et al. (1995), and include the following:

Movement of Water through the Soil Profile

Although the movement of water through the soil profile is usually vertical, lateral movement becomes important on slopes. This process is associated with:
— leaching of soluble cations and anions, including calcium, magnesium, potassium, sodium, chloride, nitrate, sulphate and carbonate
— clay translocation, that is, the movement (clay eluviation) and deposition (clay illuviation) of suspended clay particles
— movement and accumulation of organic matter, silicon compounds and iron or aluminium oxides.

The speed of these processes is dependent on the ratio of precipitation to evaporation and on soil permeability. For example, highly permeable soils, such as sands, will be more readily leached by a given quantity of water than more impermeable soils such as clays.

The rate of operation of these processes is also dependent on topographic position. Weathering products and clays tend to be translocated from the better-drained soils, which generally occur upslope, to the less well-drained soils downslope (see Figure 1.3). However, the net rate of accumulation in the profiles is dependent on the amount of water moving in and out of the profile. It is possible for a soil to be saturated but, because of lateral flow, to have a net loss of weathering products and clay. This is the case with many bleached A_2 horizons above relatively impermeable B horizons.

Soil Drainage

Soil drainage, and particularly the amount of time a soil is wet or saturated, has very significant effects on the soil. In well-drained soils, which remain wet for only short periods, red and reddish brown colours dominate. Generally, the red colour may be attributed to oxidised iron. Imperfectly to poorly drained soils, which remain wet for at least several weeks and often all winter, have dull yellow and grey colours and are frequently mottled with patches of red, orange, white

and grey. These colours can be attributed to the varying degrees of hydration of iron and aluminium oxides. Very poorly drained soils, which are wet for most of the time, are usually very pale grey, bluish or olive-green; these are the gley colours. These colours indicate that ferric iron (Fe^{3+}) has been reduced to ferrous iron (Fe^{2+}) in the soil, but may also indicate that iron has been totally removed from the soil.

An important process in soils which undergo alternate periods of saturation and relative dryness is the segregation of iron and manganese within the soil. When the soil is saturated, there is a shortage of oxygen, and iron and manganese become reduced and mobile. When the soil dries out, the iron and manganese are redeposited as nodules or coatings. In extreme instances, hard, dense pans of iron and manganese nodules may form in the soil and be up to 30 to 50 cm thick.

Continued Weathering and New Mineral Formation

Weathering of primary minerals (those derived from the original rock) is a continuing process in soils and leads to the formation of secondary minerals (clays and oxides), which in turn may also be altered. Primary minerals which resist weathering and remain in the soil include quartz and, to a lesser extent, potassium felspar. However, even some of the more readily weathered minerals, such as hornblende, biotite and the plagioclases, may remain in the soil for a considerable time (Brewer, 1955). Buol et al. (1973) and Paton (1978) summarise the environments in which the new minerals, including clay minerals and the iron and aluminium oxides, form. The form of the secondary minerals produced is dependent on the relative concentrations of the weathering products, and the degree of weathering and leaching:
— least weathered: relatively high content of bases; aluminium, iron and silicon also present (montmorillonite, vermiculite or illite-like clays formed); the formation of montmorillonite requires the presence of relatively high levels of magnesium and low acidity
— moderately weathered: relatively low content of bases; aluminium and silicon dominant (kaolinite-like clays formed)
— highly weathered: negligible content of bases; relatively low silicon content; aluminium oxides or iron oxides dominant (bauxite, ferricrete formed).

New mineral formation also relates to the dissolution of iron and aluminium oxides and silicon compounds from primary minerals and clay minerals, with subsequent leaching from, or relocation and segregation in, the soil profile.

Solution, Precipitation and Accumulation of Soil Components

Mention has already been made of the solution and re-precipitation of iron and manganese compounds depending on their state of oxidation. Here the other important soil components are discussed: calcium carbonate, calcium sulphate (gypsum), and the more soluble salts, sodium chloride, sulphate and carbonate. The magnesium and potassium salts are also included within the discussion.

(i) Calcium carbonate accumulates in soils for two major reasons.

 (a) In arid and subhumid areas where leaching is relatively weak, the soluble constituents of soils are not removed from the profile. The soils are only wetted to a shallow depth (perhaps 1 m or less) before the moisture begins to evaporate. Any calcareous materials tend to accumulate at the depth of water penetration and the soil becomes dominated by calcium ions and, to a lesser extent, magnesium ions. The carbonate for the formation of calcium carbonate may be derived from carbon dioxide in the surface soil atmosphere, which is dissolved and leached down to the depth of water penetration, or alternatively it may be derived from the parent material.

 (b) The parent material may have a high content of calcium carbonate. Limestone, calcareous sediments or highly basic rocks are typical. In this case, calcareous soils may form, even in humid areas. However, given sufficient leaching time, the calcium carbonate may eventually be removed from the profile. The rates of removal of carbonate has been estimated by Jenny (1941), who observed that 200 to 250 years are required to remove carbonate from 1 m of a sandy soil in a humid climate.

(ii) Gypsum ($CaSO_4.2H_2O$) is deposited whenever calcium and sulphate reach high concentrations in soils. In basin conditions it can be differentially deposited out of sea water, being separated from other salts. The basin conditions allow the sequential precipitation of salts as the sea water recedes to the deepest part of a basin (Duff, 1993). However, gypsum is relatively soluble and it is readily removed from the soil if water is available. Hence gypsum is largely a feature of arid zone soils and those in the drier parts of the subhumid zone, where rainfall is not sufficient to remove it.

(iii) The soluble salts (mainly sodium chloride, sulphate and carbonate) affect both soil formation and soil use. Severely salt-affected soils, in a commercial sense, are largely non-productive. Although most common in arid areas, they also occur in subhumid and even humid areas.

The salt in a soil profile may owe its origin to either the release of sodium by weathering minerals such as sodic felspars, the presence of common salt in the soil parent material, as in marine sediments, or by atmospheric accessions (cyclic salt in rainfall or dust). Note that after lithification marine sediments do not always have free salt present, as it is compressed out of the rock during lithification (Gunn and Richardson, 1979; Isbell et al., 1983). The accumulation of salt is enhanced by seepage or rising water tables bringing salt from elsewhere in the landscape into the soil profile, where it is deposited if evaporation exceeds leaching and drainage.

Chloride is not often present in rock in any great quantities, and the weathering of rock minerals to clays is often associated with the formation of silicic acid and bicarbonates. Thus it is possible for sodium from the dissolution of rock materials to accumulate in clays to form soils with high sodium levels without chloride being present. The accumulation of sodium on the clays can adversely affect the physical characteristics of the soil. The clay is easily dispersed, the soil structure collapses when wet and soil pores are blocked, and dense impermeable horizons develop. Coarse prismatic and columnar B horizons often result from the influence of excessive exchangeable sodium. Soils so affected are termed sodic soils.

The surface soils may be affected, typically becoming hardsetting or having a coarse blocky structure.

Sodic soils may also form where soluble salts are leached from the soil leaving behind sodium on the soil exchange complex (on the clays).

Wetting and Drying

Wetting and drying is an important soil-forming process in those soils that shrink and expand substantially with changes in moisture content (expansive soils). The shrink-swell activity of the soil is an important consideration in the design of building and road foundations. Soils that strongly shrink on drying and swell when wet can develop distinctive features. These include surface cracking, gilgai formation (a surface microrelief of small mounds and hollows), and the development of a self-mulching surface soil (a highly pedal loose surface mulch).

Biological Activity

Biological soil-forming processes include:
—accumulation of organic matter, especially in surface soils

—development of soil structure and soil pores by plant roots, fungal hyphae, animal activity (such as earthworms) and decomposing organic matter; bioturbation can substantially affect soil properties by moving large quantities of soil materials within the soil profile, and laterally across the landscape (Paton et al., 1995)

—biological compounds possibly assisting water-aided movement of iron and aluminium in the profile; this process may be of importance in podzols

—numerous other biologically controlled reactions affecting nutrient cycling in the soil, particularly that of nitrogen

—human activities that act as a biological agent when cultivation and excessive grazing lead to erosion and reductions in organic matter and degradation of soil structure.

BIBLIOGRAPHY

Beattie, J. (1972), *Groundsurfaces of the Wagga Region, NSW*, Soil Publication No. 28, CSIRO, Australia.

Beckmann, G.G. (1983), 'Development of old landscapes and soils' in *Soils: An Australian Viewpoint*, Division of Soils, CSIRO, Melbourne; Academic Press, London.

Birkeland, P.W. (1974), *Pedology, Weathering and Geomorphology Research*, Oxford University Press, Melbourne.

Brewer, R. (1955), *Mineralogical Examination of a Yellow Podzolic Soil Formed on Granodiorite*, Soil Publication No. 5, CSIRO, Melbourne.

Buol, S.W., Hole, F.D. and McCracken, R.J. (1973), *Soil Genesis and Classification*, Iowa State University Press, Ames, USA.

Burges, A. and Beadle, N.C.W. (1952), 'The laterites of the Sydney District', *Australian Journal of Science* 14, 161–2.

Butler, B.E. (1956), 'Parna—an aeolian clay', *Australian Journal of Science* 18, 145–51.

Butler, B.E. (1959), *Periodic Phenomena in Landscape as a Basis for Soil Studies*, Soil Publication No. 14, CSIRO, Melbourne.

Butler, B.E. and Churchward, H.M. (1983), 'Aeolian processes' in *Soils: An Australian Viewpoint*, Division of Soils, CSIRO, Melbourne; Academic Press, London.

Butler, B.E., Blackburn, G. and Hubble, G.D. (1983), 'Murray-Darling Plains (VII)' in *Soils: An Australian Viewpoint*, Division of Soils, CSIRO, Melbourne; Academic Press, London.

Chesworth, W. (1973), 'The parent rock effect in the genesis of soil', *Geoderma* 10, 215–25.

Corbett, J.R. (1969), *The Living Soil: The Processes of Soil Formation*, Martindale Press, West Como, NSW.

Corbett, J.R. (1972), *Explanatory Notes—Soils Map of NSW*, NSW Department of Decentralisation and Development/School of Geography, University of NSW, Kensington.

Duff, D. (1993), *Holmes' Principles of Physical Geology*, Chapman and Hall, London, New York, Melbourne.

FitzPatrick, E.A. (1971), *Pedology: A Systematic Approach to Soil Science*, Oliver and Boyd, Edinburgh.

Gray, J. and Murphy, B.W. (1999), *Parent Material and Soils—A guide to the influence of parent material on soil distribution in Eastern Australia*, Technical Report No. 45, NSW Department of Land and Water Conservation, Sydney.

Gunn, R.H. and Richardson, D.P. (1979), 'The nature and possible origins of soluble salts in deeply weathered landscapes of eastern Australia', *Australian Journal of Soil Research* 17, 197–215.

Hubble, G.D. and Isbell, R.F (1983), 'Eastern Highlands (VI)' in *Soils: An Australian Viewpoint*, Division of Soils, CSIRO, Melbourne; Academic Press, London.

Hubble, G.D., Isbell, R.F. and Northcote, K.H. (1983), 'Features of Australian soils' in *Soils: An Australian Viewpoint*, Division of Soils, CSIRO, Melbourne; Academic Press, London.

Hunt, P.A., Mitchell, P.L.B. and Paton, T.R. (1977), 'Lateritic profiles and lateritic ironstones on the Hawkesbury Sandstone', *Geoderma* 19, 105–21.

Isbell, R.F., Reeve, R. and Hutton, J.T. (1983), 'Salt and sodicity' in *Soils: An Australian Viewpoint*, Division of Soils, CSIRO, Melbourne; Academic Press, London.

Jenny, H. (1941), *Factors of Soil Formation*, McGraw-Hill, New York.

McTainsh, G. (1985), 'Dust processes in Australia and West Africa', *Search* 16, 104–6.

Paton, T.R. (1978), *The Formation of Soil Material*, George Allen and Unwin, London.

Paton, T.R., Humphreys, G.S. and Mitchel, P.B. (1995), *Soils—A New Global View*, UCL Press, London.

Reineck, H.E. and Singh, I.B. (1975), *Depositional Sedimentary Environments*, Springer-Verlag, Berlin, Heidelberg, New York.

Walker, P.H. and Butler, B.E. (1983), 'Fluvial processes' in *Soils: An Australian Viewpoint*, Division of Soils, CSIRO, Melbourne; Academic Press, London.

Walker, P.H., Nicolls, K.D. and Gibbons, F.R. (1983), 'South-eastern Region and Tasmania (VIII)' in *Soils: An Australian Viewpoint*, Division of Soils, CSIRO, Melbourne; Academic Press, London.

Ward, W.T. and McArthur, W.M. (1983), 'Soil formation on coastal lands and the effects of sea level changes' in *Soils: An Australian Viewpoint*, Division of Soils, CSIRO, Melbourne; Academic Press, London.

White, R.E. (1987), *Introduction to the Principles and Practice of Soil Science*, Blackwell Scientific Publications, Oxford, Melbourne.

Williams, C.H. (1980), 'Soil acidification under clover pasture', *Australian Journal of Experimental Agriculture* 20, 561–7.

Forms of Erosion

CHAPTER 2

C.J. Rosewell, R.J. Crouch, R.J. Morse,
J.F. Leys, R.W. Hicks and R.J. Stanley

Erosion, the gradual wearing away of the earth's surface, varies from periods of relative inactivity to periods of intense activity. It has always taken place and always will. Over the long term, the surface of the earth is changing, with mountains rising, valleys being cut deeper and wider, the coastline receding or advancing. The physical pattern of the surface of the earth which we see today is not the result of some single cataclysmic sculpturing, but the result of changes so very slow that only after many centuries is the effect noticeable. Erosion is simply one aspect of this process of change. Human activities seldom slow down or halt the process, but frequently speed it up. The terms 'geological erosion' and 'natural erosion' are used to describe the erosion that occurs under natural environmental conditions. The long-term natural erosion rate is estimated at 1.5 to 1.7 t/ha/yr in mountainous lands and 0.1 to 7 t/ha/yr in undulating lands (Lal, 1990). Estimates of the lowering of the earth's surface by erosion are of the order of 30 mm/1000 years as a worldwide average (Duff, 1993). Rates may be higher for particular river systems with the estimate being 140 mm/1000 years for the Colorado River, 769 mm/1000 years for the Irrawaddy/Chindwin, and up to 3000 mm/1000 years for the Semani River in Albania (Duff, 1993).

An estimate for the rate of denudation in central New South Wales can be obtained from the occurrence of river gravels under basalt flows of known age on the top of Mount Panorama at Bathurst, and this gives an estimated rate of denudation of 8 to 9.5 mm/1000 years for the granite-dominated landscape in the vicinity of Bathurst (Murphy, 1985; Chan, 1998). Using basalt flows as markers, Chan (1998) has estimated that rates of geological weathering since the Eocene (41 million years) and Miocene (11 to 22 million years) periods vary in central western New South Wales from 0.4 mm/1000 years in an area 30 km west of Orange near Cudal, to 3.4 mm/1000 years south of Blayney at the prominent hill called Big Brother, to 15.4 mm/1000 years in proximity to the Belubula River between Blayney and Canowindra. The more rapid rate is associated with headwater erosion of a nick point associated with an old palaeosurface. This erosion is geologically controlled, as it was initiated by the formation of the Murray Basin to the west, which induced a lower

base level to which the Belubula River could erode. For the central west of New South Wales, these rates correspond to about 0.005 to 0.02 t/ha/yr, rates which are much lower than those generally observed under agricultural land use (see Chapter 3).

More rapid erosion rates of soil usually occur as a result of human activity and this is referred to as 'accelerated erosion' or just 'soil erosion'. Widespread agricultural activities generally result in a decrease in vegetative cover and are, therefore, a significant contributor to soil erosion. Accelerated soil erosion has historically caused disastrous consequences to the once-productive land of many nations. The 6100 million people (1999) on this planet rely on productive agriculture, so wise management of agricultural land is essential in meeting present world food needs and maintaining future soil productivity. Such management must incorporate farming systems that include adequate erosion control practices.

In some cases, the health and prosperity of communities can be threatened by severe erosion rates, as has happened at least in parts of Africa, Asia and South America where severe erosion has affected valuable agricultural land (Lal, 1990). It is estimated that accelerated erosion has irreversibly destroyed about 430 million hectares of arable land, which is about 30% of the land currently cultivated. Wolman (1985) has suggested that severe erosion losses have irreversibly destroyed many of the soils in the Middle East and Mediterranean lands, especially those on limestone.

In Australia, and particularly in New South Wales, erosion rates have been accelerated by agricultural activities, but forestry and urban activities have also increased erosion rates. Kaleski (1963) estimated that 10 million hectares in the Eastern and Central Divisions were affected by sheet erosion and nearly 9 million hectares were affected by gully erosion. He estimated that over 5 million hectares of the Western Division were affected by wind erosion. The report by the Soil Conservation Service of New South Wales (1988) gave similar values for sheet erosion, but this survey, which used a different methodology (Graham, 1989), estimated larger areas were affected by gully and wind erosion in the Eastern and Central Divisions being 16 million hectares and 12 million hectares respectively, and larger areas were affected by wind erosion in the Western Division (7.6 million hectares). A report by Stewart (1968) showed that the area affected by erosion in New South Wales had dropped in comparison to the area of erosion in 1948.

While the rates of erosion have not yet caused the severe soil degradation of other lands, it needs to remembered that on a world scale, agricultural activity has been going for only a relatively short time in Australia, and that some consideration needs to be given to what are acceptable rates of erosion and whether we are currently facing a long-term degradation of Australia's soil resources under current land management practices. Tolerable soil losses and potential effects of soil erosion on soil health are discussed in Chapter 3.

Soil erosion occurs wherever the soil surface is exposed to the agents of erosion such as water, wind or gravity. Whenever vegetation is cleared or lost due to climatic change, and the soil surface exposed, there are fewer plants to absorb the force of the wind, so wind erosion increases. When there is less vegetation to intercept the energy of falling raindrops and impede surface runoff flow, water erosion increases. With more surface runoff, streams and rivers become increasingly able to carry away large amounts of sediment. With less plant roots to bind soil on steep slopes, erosion due to mass movement increases.

Soil erosion usually occurs at much higher rates on land that is denuded and reshaped for urban development, roads and mining than on agricultural land. Although erosion is less harmful to these sites than to agricultural land, the eroded soil from both non-agricultural and agricultural sources often causes major problems downstream. There it may choke and pollute streams, block channels and rivers, cover roads and public utilities, and fill reservoirs and harbours. Thus, erosion may cause damage in three places: first, to the land from which the soil is removed; second, to the water that transports it; and third, to the site where it is deposited. Some sediment is deposited only temporarily. Subsequent storms or flood events, sometimes several years later, may re-entrain the sediment and move it further along the stream system (Trimble, 1975).

Erosion and sedimentation are major problems (ASCE, 1975). Soil erosion reduces the productivity of agricultural land. Sediment degrades water quality and may carry adsorbed polluting chemicals. Sediment is the world's greatest pollutant of surface waters (Robinson, 1971). Deposition of sediment on roads and public utilities, in stream channels, reservoirs, lakes, estuaries and harbours reduces their usefulness and requires costly removal (AWRC, 1969; Junor et al., 1979). About 20 000 million tonnes of sediment are transported to the oceans of the earth each year (Jansen and Painter, 1974).

Sediment sources include agricultural lands, construction sites, roadway embankments, cuts and drains, urban areas, disturbed forest lands and surface mines. Sediment sources may be classified according to the dominant type of erosion that affects them, such as sheet, rill, gully, tunnel, streambank or landslide. The erosion at any site may be due to the action of water, wind, mass movement or coastal processes, and these will be discussed in turn in the following sections.

Wind erosion is a major concern in many parts of the world. In Africa wind erosion is thought to be extending the southern boundary of the Sahara at the rate of 50 km/yr (Lal, 1990). It is especially a problem for soil degradation as the wind separates the finer fractions from the soil with most of the nutrients and organic matter, and leaves behind the coarser, less nutrient-rich fraction. In Australia wind erosion has been a major problem especially in semi-arid and arid sandy soils, such as in the mallee lands of south-eastern Australia and the mulga areas of south-western Queensland, but other areas of sandy soils and poor vegetative cover are frequently affected. The effects of wind erosion have been observed several times in the capital cities of south-eastern Australia as dust storms (Raupach et al., 1994). Dumsday (1973) has estimated that 1700 km^2 has been affected by wind erosion in New South Wales. Wind erosion has also affected the coastal sand dunes as described later in this chapter.

2.1 Water Erosion

2.1.1 General Water Erosion Processes

Soil erosion by water is a complex process being dependent on many, often interrelated, factors. The process is essentially one of detachment of particles from the soil, their transport and subsequent deposition. Eroded soil in transport is sediment, and sedimentation occurs when sediment is deposited. Sediment movement through a catchment is often intermittent, with detachment, transport and deposition of the eroded soil material recurring repeatedly. The rate of erosion depends on the climate, soil, topography, plant cover and land use.

The water erosion process usually begins when raindrops strike unprotected soil on the surface and detach soil particles. The potential ability of rainfall to cause erosion is referred to as its erosivity and is a function of the kinetic energy of the rain. When raindrops strike bare soil, practically all of the energy is consumed as work done against the soil surface in the disruption of soil aggregates, compaction of the soil surface and splash of soil particles into the air. Rain falling at an intensity of 50 mm/h for 30 minutes generates sufficient energy to raise the top 10 cm of soil to a height of 45 cm (Rosewell, 1985).

The impact of a single raindrop on a wet soil surface is like a bomb blast. Soil aggregates are disrupted and broken into constituent particles by the force of the impact. These particles are washed and packed into the soil surface pores, increasing the surface density and reducing the infiltration of water. As soon as a surface film of water develops, the splash of soil particles into the air reaches a maximum. The splashed particles may be transported considerable distances and on a sloping surface there is a net movement of soil downslope. The rate of soil detachment and net downslope movement by rainfall depend on soil properties, surface conditions, slope steepness and rainfall characteristics.

Detachment of soil particles is a function of the erosive forces of raindrop impact and flowing water, the susceptibility of the soil to detachment, the presence of material that reduces the magnitude of the eroding forces, and management of the soil that makes it less susceptible to erosion. Detachment occurs when the erosive forces of raindrop impact or flowing water exceed the soil's resistance to erosion.

Transport involves the detachment and suspension of sediment (entrainment) and its movement from its original location. Transport is a function of the forces of the transport agents, the transportability of the detached soil particles and the presence of material that hinders movement of sediment. Detached particles are transported by raindrop splash and flow. Deposition occurs when the sediment load of a given particle type exceeds the corresponding transport capacity of the flow (Foster, 1982). This process is known as sedimentation.

Either detachment or transport capacity may limit erosion and sediment load at a location on a slope (Meyer and Wischmeier, 1969). If the amount of sediment made available for transport by the detachment processes is less than the transport capacity, then the sediment load moving downslope will be the amount of detached sediment available for transport. If the available detached sediment exceeds the transport capacity, deposition occurs and the transport capacity controls the sediment load.

Most soil eroded by water is transported downslope by surface runoff, which is the portion of rainfall that becomes surface flow. Runoff does not occur, however, until the rainfall intensity exceeds the infiltration rate of the soil, and surface storage capacity of the land has been satisfied. Soils that have high infiltration rates or large surface storage capacity will, therefore, exhibit delayed runoff initiation and reduced runoff rates.

Once runoff begins, the quantity and size of material that it can transport depend on runoff velocity and turbulence, and these increase as the slope steepens and the flow increases. The larger and denser the disturbed material, the greater must be the flow velocity and turbulence to transport it. High rates of runoff can generally transport all rainfall-detached soil, detach soil itself by hydraulic shear and transport this material downslope.

There is considerable interaction between the two major erosion processes of raindrop impact and overland flow. Raindrop impact seals the soil surface and increases the rate of runoff. The impact of raindrops into runoff water increases its turbulence and the capacity of the flow to transport soil particles. Raindrop impact acting in conjunction with overland flow causes more erosion than when either acts alone.

Major Types of Water Erosion

Several forms of erosion can be identified based on the processes involved and the relative importance of the effects of detachment and transport by raindrop impact, detachment and transport by overland flow (stream power). Sheet and rill erosion refer to erosion that is dominated by raindrop impact and water flows of relatively low stream power. They are generally associated with upland areas where water flow is not highly concentrated. Gully erosion refers to erosion in areas where water flows become concentrated, especially in drainage depressions and on lower slopes. Raindrop impact is of lesser importance and stream power is moderate. Streambank erosion refers to erosion of the banks adjacent to large water flows such as in creeks and rivers, and raindrop impact has no effect.

Sheet Erosion

Sheet erosion is the removal of a fairly uniform layer of soil from the land surface by raindrop splash and/or runoff (see Plate 2.1). Soil eroded from upland slopes comes from:

—inter-rill areas (that part of the land surface between runoff channels)
—rills (the eroded portions of runoff channels that can be obliterated by subsequent tillage)
—gullies (eroded channels that are too large to be crossed by farm machinery).

The term 'sheet erosion' has often been used to include all erosion that can be obliterated by tillage (rill and inter-rill erosion), and unfortunately conjures up a picture of soil being removed uniformly in thin sheets. Runoff seldom occurs as smooth laminar flow and, in any case, smooth laminar flow only erodes at velocities higher than that found in runoff (Hudson, 1981). An example of an area affected by sheet erosion is shown in Hairsine et al. (1993), where two high intensity storms resulted in the sheet erosion of a cultivated area. The term 'sheet erosion' should be used to describe both movement by raindrop splash and transport of raindrop-detached soil by thin surface flows whose erosive capacity is increased by raindrop impact turbulence.

The detached soil particles are transported to the rills by the combined action of the thin inter-rill flow and raindrop impact. The flow alone can transport only the smallest particles, but raindrop impact entrains larger particles, significantly increasing the flow's transport capacity (Moss et al., 1979; Kinnell, 1990, 1994). Most transport to the rills is by inter-rill flow, while only a small fraction of the inter-rill sediment is splashed directly to the rills (Meyer et al., 1975b). On low slopes, delivery of inter-rill sedi-

Plate 2.1   Sheet erosion of this arable paddock has resulted in accumulation of sediment on the fence

ment to the rills may be limited by transport capacity. On steeper slopes, detachment limits delivery of inter-rill sediment (Meyer et al., 1975b).

Erosion on inter-rill areas is essentially independent of erosion in the rills, but erosion in the rills depends greatly on sediment inputs from the inter-rill area. If sediment inflow from the inter-rill areas exceeds the transport capacity of the flow, deposition occurs. If the sediment inflow is less than the transport capacity of flow in the rills, and if the flow's erosive forces exceed the resistance of the soil in the rills to detachment by flow, rill erosion occurs (Foster, 1982).

Most downslope movement of upland sediment is by flow in the rills. Even though excess transport capacity may exist in the inter-rill areas, this transport capacity does not add to the transport capacity of flow in the rills. Conversely, excess transport capacity in the rills is not available to transport sediment detached by raindrop impact on inter-rill areas (Foster, 1982).

Thus, sheet erosion is a relatively uniform removal of soil and is, therefore, not as visible as rill or gully erosion. On land where the soil is not protected by surface cover, soil is lost from most of the land surface by sheet erosion. The erosion rate is not greatly affected by the slope of the land or the location on the slope, since raindrop impact is relatively uniform over the area of land.

Rill Erosion

Rill erosion is erosion in numerous small channels which can be obliterated by normal tillage

(Houghton and Charman 1986) (see Plate 2.2). A depth of less than 300 mm may be used as a criterion to distinguish rills from gullies. Such distinction between rills and gullies is useful during erosion survey because the field surveyor is not required to use subjective judgment to decide if a rill is a gully. For studies of erosion processes, the flow characteristics rather than rill depth or size determine whether a set of channels are rills or gullies (Foster, 1982).

Rill erosion results primarily from soil detachment by concentrated runoff. It usually affects only a small proportion of the land surface, but is much more visible than inter-rill erosion. Rills may develop where runoff is concentrated due to topographic variations or tillage marks. However, such concentrated flow does not cause erosion until the flow's shear force exceeds the resistance of the soil. Therefore, concentrated runoff may flow for some distance downslope before rilling starts. Once rilling begins, it increases rapidly with greater flow accumulations, so rill erosion increases with the slope length. It also increases with slope steepness (see Plate 2.3). Flow in rills is characteristically narrow and incised in contrast to the broad and shallow flow where deposition occurs.

Almost all natural soil surfaces are subject to shear stress concentrations where flow occurs. If stress at these concentrations is greater than the soil's critical shear stress, erosion occurs. Shear stress is greatest at the bottom of the flow, which tends to incise the rill channels (Foster, 1982).

Many rills are initially formed by the upslope advance of gully-like headcuts where erosion is particularly intense. Rate of advance of headcuts and their dimensions are functions of slope, discharge, soil properties and incorporated material like crop residue (Meyer et al., 1975a).

After a headcut passes, other headcuts may form. The rills are eroded more deeply until a layer less susceptible to erosion restricts them. When this occurs, the rills widen and the erosion rate decreases (Foster, 1982). Other gully erosion processes such as undercutting and erosion of the side walls also occur in rills. Rill erosion is likely to occur if the incoming sediment load from the inter-rill areas is less than the flow's transport capacity. If the sediment load is greater, then deposition occurs.

Concentrated flow erosion is a form of rill erosion in topographic locations where overland flow is concentrated and erosion occurs to the base of the loose tilled layer and may be one or more metres wide (Foster, 1982). The term 'ephemeral gullies', or 'seasonally ephemeral gullies', has been used (USDA, 1984) as a description of an eroded channel in which ephemeral flows occur and that is much wider than a rill but shallow enough to be partially or completely filled by tillage.

Most erosion in these channels is from storms soon after secondary tillage operations or after seedbed preparation. After tillage, many soils reconsolidate and these channels become much less erodible over time during the growing season. Whereas the position of rills varies from year to year, concentrated flow erosion generally occurs in the same location each year. Concentrated flow areas become incised over several years, steepening adjacent overland flow slopes and accelerating nearby rill and inter-rill erosion.

Models of Sheet and Rill Erosion

Several models of sheet and rill erosion have been developed.

One approach is to define erosion risk as set out in Houghton and Charman (1986). This gives a qualitative assessment based on rainfall erosivity, slope and soil erodibility. It can be a useful management tool when more detailed quantitative data is not available.

Another is to use a quantitative but empirical model such as the Universal Soil Loss Equation or USLE (Wischmeier and Smith, 1978). The USLE and its successor, the Revised Universal Soil Loss Equation or RUSLE (Renard et al., 1997), are essentially based on the statistical analysis of a large volume of plot data. The RUSLE model has been adapted to New South Wales by Rosewell (1993). A number of models have added runoff estimation and overland flow transport processes to the USLE to provide improved prediction at the field and catchment scale. Examples are CREAMS (Knisel, 1980) and AGNPS (Young et al. 1989). Modern models are based on the processes of water erosion. These basic processes are the entrainment and transport of sediment based on raindrop impact, resistance of soil particles to detachment by flowing water, settling velocities of soil particles, stream power and other process-based factors controlling erosion rates. Examples of these models include WEPP (Laflen et al., 1991) and GUEST (Misra and Rose, 1989, 1996).

Gully Erosion

Gully erosion is the removal of soil by running water, resulting in the formation of channels sufficiently large that they disrupt normal farming operations and are too large to be filled during normal cultivation. This distinction between rills and gullies, and gullies and streams, is arbitrary. The most commonly accepted divisions are those of McDonald et al. (1984)—'gullies are deeper than 300 mm'—and the House of Representatives (1969)—'gullies are upland drainage ways, continuous or discontinuous,

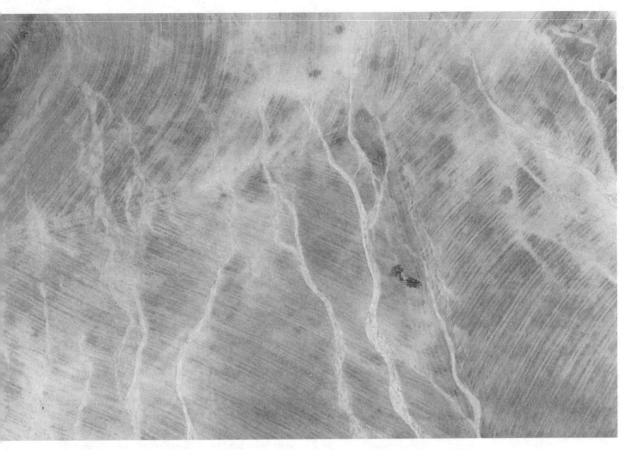

Plate 2.2   An aerial view of sheet and rill erosion of arable land after a storm

with steep sides and often with headward eroding scarps, usually conveying ephemeral runoff and draining areas smaller than 2.6 square kilometres' (see Plate 2.4).

Gully erosion involves several interacting processes depending on soil type, landform, land use and climate. Gullies initiate when equilibrium within a minor drainage line is upset by either increased discharge or decreased soil resistance to detachment and transport. Once the equilibrium is disturbed and gullies start forming, much more effort is required to regain stability than would have been needed to initially maintain a stable system.

Soil removal from gullies can be partitioned into gully head erosion, by which a gully lengthens, and which is caused primarily by flow over the gully head, and gully side erosion, which widens a gully and is caused by various combinations of diffuse flow over the side, seepage flow, flow along the gully and raindrop erosion. Ireland et al. (1939) described six main gully head types produced by different physical factors influencing drainage (Figure 2.1). In terms of erosion processes, four main types have been

Plate 2.3   Active rilling on a road batter

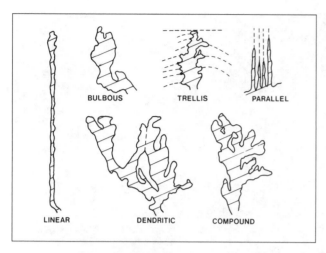

Figure 2.1 Gully head forms developed due to the influence of the drainage pattern (after Ireland et al., 1939)

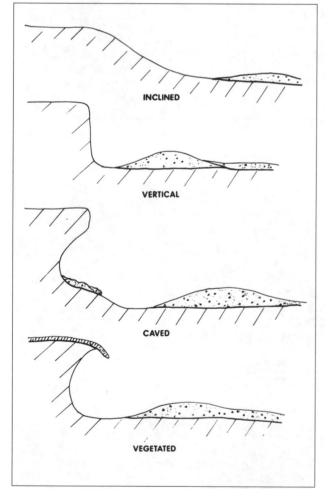

Figure 2.2 Gully head types dominated by different processes (after Ireland et al., 1939)

recorded (Figure 2.2), the different shapes evolving due to differing levels of resistance in the underlying soil strata.

In New South Wales, the inclined and caved heads are the most common. Inclined heads develop in the strongly aggregated clay soils with poor inter-aggregate cohesion. Under the influence of flowing water, the aggregates roll down the slope maintaining the inclined head (Figure 2.3). This form of head is usually associated with V-shaped gullies with very mobile sides. As the gullies cut deeper, soil removed from the foot of the sides causes parallel side retreat as the aggregates regain their angle of repose.

Caved heads formed by plunge pool and back trickle erosion develop where the topsoil is more resistant to erosion than is the subsoil (Figure 2.4). Retreat upslope is caused primarily by back trickle and splash saturating the vertical face. The saturated soil slumps into the gully producing a caved head. When the cave undermines the topsoil, it collapses into the gully moving the head upslope. The main action of the plunge pool is to cut the gully deeper and break up soil blocks from the head into transportable particles. Soils that disperse when saturated are predisposed to rapid erosion by these processes. Consequently, many soils in New South Wales that have dispersible B horizons are particularly prone to this form of gully erosion if the B horizon is exposed. Gully sides have been identified as major gully sediment sources (Blong et al., 1982). They are eroded under the influence of raindrops, runon and through-flow.

In addition to many of the factors that affect subsoil erodibility (Chapter 12), gully side stability is also affected by the following factors:
— *water table height:* a groundwater table higher than the gully floor reduces soil strength and

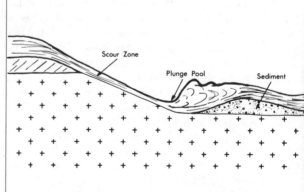

Figure 2.3 Major processes at an inclined gully head

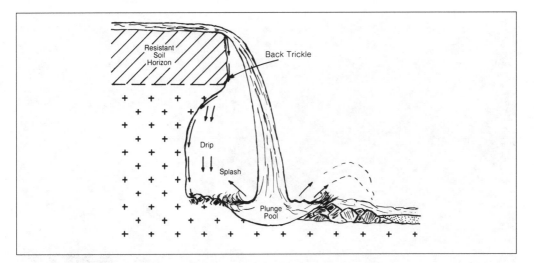

Figure 2.4   Major processes at a caved gully head

increases soil density; the extra load can lead to increased shear failure of the wetted gully wall

—*soil coherence:* subsoils with low wet coherence are prone to gully erosion

—*critical hydraulic gradients:* if the pressure exerted by throughflowing water exceeds the soil's shear strength, soil particles are more likely to be mobilised.

Most sides are exposed at the headcut as vertical soil faces that are then eroded under the influence of gravity and water until they are reduced to a stable angle. Gullies have been classified as 'V' or 'U' shaped depending on the stage of gully erosion, the relative resistance of the strata to erosion, and the dominant gully side erosion processes (Imeson and Kwaad, 1967). Other side shapes which evolve under the action of different processes include:

—*crenellated* by runon (Blong et al., 1982)

—*fluted* (or 'cathedral') by rilling (Veness, 1980)

—*tunnelling* by throughflow (Stocking, 1976).

Recognition of the dominant processes responsible for the evolution of a particular side form can help in indicating the treatment needed to regain stability.

Gullies evolved from rills by channel scouring or developed from isolated weaknesses are eroded along drainage lines, uphill from their point of initiation. If not continuous, with a channel lower down the slope to transport eroded material away, the eroded soil from a gully is deposited in a fan. This induces instability lower down the system, resulting in a series of discontinuous gullies that merge to form a continuous gully system (Figure 2.5). Within a gully, floor lowering occurs in a similar sequence of headcuts and zones of sediment accumulation.

Gully heads progressing up drainage lines have a continually decreasing catchment area, with a corresponding increase in the amount of water entering the system over the gully sides. Wherever the runoff water enters in a concentrated stream, branching will occur and lateral gullies will develop. Continuity and branching are commonly used to define gully erosion severity:

—discontinuous (minor)

—continuous (moderate)

—branching (severe).

Gully eroded areas are most effectively stabilised by gully filling and water diversion. This also removes the obstacle to movement around properties and the undesirable aesthetic quality of gullies. In cropping areas where filling brings the eroded land back into production, it is often the most practical control alternative. In grazing areas, the cost of filling is often not justified and the aim becomes one of preventing gully growth. Simple diversion of runoff water combined with the construction of a gully control structure to drown the gully head is a common practice. The removal of within-gully flow enables soil washed from sides to accumulate on the floor, reducing side slope and promoting stability. In more difficult situations where runoff cannot be diverted, the heads are protected by concrete, stone or earth flumes. These combine with grade stabilisation structures to reduce floor erosion and promote floor and side stability by encouraging vegetation establishment.

Tunnel Erosion

Tunnel erosion can be considered a special type of gully erosion and is sometimes called piping. It is the removal of subsurface soil by water while the surface

Plate 2.4 Serious gully erosion in the central tablelands

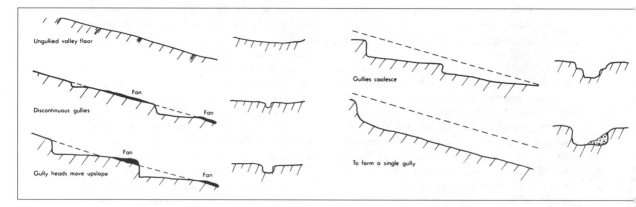

Figure 2.5 The evolution of a continuous gully system from discontinuous gullies (after Leopold, Wolman and Miller, 1964)

soil remains relatively intact. This produces long cavities beneath the soil surface, which enlarge until the surface soil is no longer supported and collapses forming circular holes from the cavity to the surface. If this process is not checked, further surface collapse converts the cavity into an open gully which continues to grow.

Erosion tunnels range in size from a few centimetres to several metres in diameter and occur under conditions ranging from equatorial rainforest to semi-arid rangeland (Crouch et al., 1986). They are formed by conditions that promote a lateral water flow of sufficient volume and velocity to detach and transport soil particles through subsurface channels.

Tunnel initiation can occur by water movement in soil cracks, saturation of A_2 or B horizon soil material, concentration of water to a point on a gully side or rodent activity (Figure 2.6). It is encouraged by any factor that promotes variation in soil permeability either within or between soil layers, and factors that promote soil detachment into transportable particles. The main requirements are water movement in defined paths and transportable soil (Crouch et al., 1986).

Dispersible soil is particularly susceptible to tunnel erosion because it breaks down readily into transportable particles which are moved at low threshold velocity. Clay movement and deposition block soil

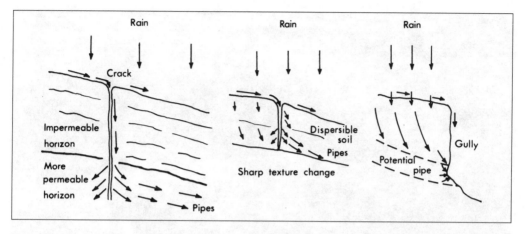

Figure 2.6 Three main mechanisms of tunnel formation — erosion tunnels may develop whenever there is enough water moving through soil to erode subsurface passages

pores, thus reducing the rate of water movement through the soil body and promoting flow along soil cracks, root holes and rodent burrows.

There are three main factors that affect tunnel initiation and development:
—water entry to the soil or a positive head of water
—water movement within the soil, or a hydraulic conductivity to allow water movement through the soil
—soil stability, or soil that is unstable to wetting and allows soil particles to be transported in the water flowing through the soil.

These are the factors that must be considered when developing erosion control techniques for tunnel-eroded areas and assessing the potential site susceptibility to tunnel erosion. For minimum tunnel erosion potential, water entry to the soil should occur evenly across the surface, water movement within soil should be even (and, if possible, vertical), and soil dispersibility should be low.

A soil with maximum tunnel erosion potential has water entering the soil in specific sites and moving laterally in defined paths through a dispersible soil. This is explained in detail by Crouch (1979).

Streambank Erosion

Streambank erosion is the removal of soil from streambanks by the direct action of stream flow, wind or wave action. It is associated with large water flows such as occur in creeks and rivers, and typically occurs during periods of high stream flow. It is one facet of the dynamic cycle between sediment deposition and sediment erosion occurring in a stream channel (see Plate 2.5). Erskine and Melville (1984) suggest that, at least in the sand bed channels of the Sydney Basin, large rainfall events are the prime

erosive agents, while small and moderate events may result in either erosion or deposition.

Erosion of sediment occurs in the stream channel when the stress applied by stream flow energy exceeds the resistance of the local material. A number of factors affect the rate at which sediment erosion may occur.

(a) Factors Increasing Fluvial Energy

(i) Any factor that can increase the overall discharge of a stream will increase the fluvial energy and predispose the banks to erosion. Figure 2.7 is a hypothetical hydrograph of a stream following a storm event. The overall shape of the curve and the height of the crest are dependent upon the intensity of the storm and the characteristics of the drainage basin. Drainage basin characteristics, which can be modified, affect the curve in the following way:

• if the curve has a steep slope and high crest, it can be assumed that the catchment is one with a large amount of direct runoff
• if the curve is fairly flat with a low crest, it is likely that there is substantial storage in the catchment, either on the surface or as groundwater; such storage tends to equalise flow throughout the discharge period. Therefore, in order to reduce streambank erosion, it is desirable to store water in the catchment for as long as possible by either maintaining catchment roughness (keeping maximum vegetative cover) or by installation of engineering structures such as detention basins. With very large storm events, the amount of water that may be stored becomes proportionately less,

Plate 2.5 Streambank erosion—showing undercutting on one side and sediment deposition on the other side of the stream

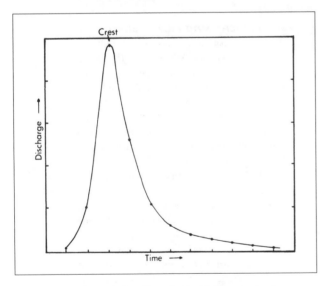

Figure 2.7 Hypothetical hydrograph of a stream following a storm event

and water-retarding attributes or structures become less effective.

(ii)   Fluvial energy acting on the outside bank of a stream is greatest when the ratio of the radius of the stream curvature to the stream width occurs in the range 1:2.0 to 1:3.5 (Hickin and Nanson, 1975). It can be very difficult to control streambank erosion and stabilise a section

of a stream without first changing the ratio beyond this critical stage.

(iii)   Increasing the slope of a stream by straightening it or dredging sediment may increase the stream power and flow energy applied to the bed and banks. It is essential to maintain the natural slope if modification of shape is necessary, otherwise streambank erosion may be initiated upstream.

(iv)   Artificial constriction of a flowline may increase flow energy at a point and predispose an otherwise stable area to extreme erosive forces. Several examples of this are evident in the state, particularly associated with the construction of road and rail facilities.

(v)   Obstacles to water flow may result in the deflection of channel flow towards the bank and locally increase fluvial energy. Such obstacles may originate from natural sources or from human activity.

(b) Factors Influencing Stability of Streambank Material

(i)   The size of the particles comprising streambanks affects the ability of streams to erode them. Sand, for example, is much more easily eroded than gravel, and silt is more easily eroded than sand. As stream velocity increases, so the fluvial energy and erosive power increases. The effect of velocity on particles of

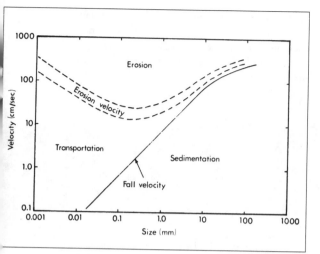

Figure 2.8   Curves of erosion and deposition for uniform material (after Hjulstrom, 1935)

varying sizes is demonstrated in Figure 2.8, which illustrates the minimum velocity necessary to move grains of a given size which are loose on the bed of a stream.

(ii)   Vegetative cover has a binding effect on what may otherwise be highly erodible materials. Table 2.1 (Langbein and Schumm, 1958) is data collected in several small drainage basins in the USA and indicates:

- the slowest rates of denudation occur when precipitation is high, giving rise to good plant growth and, hence, vegetative protection

- stream denudation is greatest where effective rainfall is too low to promote vegetative cover but high enough to provide excess runoff, for example in regions with about 250 to 380 mm annual rainfall.

Nanson and Hickin (1986) state that an added factor is the seasonal variation of stream flow.

Streams with little seasonal variation in flow are unlikely to have vegetative cover below water level and are more prone to streambank erosion below that level than streams with high seasonal variation.

(iii)   The dispersibility or erodibility of streambank material has a significant effect on the likelihood of streambank erosion occurring, as with other erosion forms (see Chapter 12).

The consequences of streambank erosion in New South Wales can be summarised in this way. Data are available which suggest that the annual suspended loads transported by Australian rivers are low by world standards (Holeman, 1968; Strakhov, 1967). Studies (Young, 1981) indicate that this is a result of the relatively old, pre-Quaternary landscape. Relief is low and stream channels have low slope gradients that reduce the capacity of streams to transport sediment. Further, the landscape contains many natural traps including alluvial fans, floodplains and swamps.

Streambank erosion data do not exist on a broad scale in New South Wales, but where it does occur on a local basis it can have severe consequences. Streambank erosion has resulted in substantial cost to the public and private purse. The outcome of continued damage caused by streambank erosion includes:

—high costs to stabilise river banks threatened by streambank erosion

—irreplaceable loss of prime agricultural land and abandonment of irrigation fields

—realignment of roads

—reconstruction of buildings

—repair or replacement of levees.

Movement of Nutrients and Chemicals on Sediments Moved by Water Erosion

Pollution in runoff, seepage or percolation resulting from land management activities has a major impact on water quality (Young et al., 1989). Sources of such pollution include soil erosion and sedimentation

Table 2.1   Rates of stream denudation (after Langbein and Schumm, 1958)

Effective Precipitation (mm)	Measured Mean Sediment Yield (t/km²)	Estimated Denudation (mm/1000 yrs)
250	240	90
250–380	270	100
380–760	190	70
760–1020	140	50
1020–1520	80	30

on rural and urban land and eroding streambanks, nutrients and organic materials from livestock wastes and agricultural land, and stormwater from urban areas. Nutrients such as phosphorus are strongly adsorbed onto clay particles and transported by soil erosion processes. Phosphorus is a major contributor to blue-green algae outbreaks in rivers and storages. An example of the nutrients that can be moved in soil particles washed from agricultural paddocks is given in Hairsine et al. (1993).

2.2 Wind Erosion

Wind erosion is the movement of soil particles by wind. It is of concern in coastal regions when sand is blown onto roads and public utilities, and when coastal landforms are destabilised. However, the largest areas affected by wind erosion in Australia are the inland dryland farming areas where the soil types are predominantly sandy and the rainfall is below 375 mm per annum. Since a significant proportion of our cereal grain production comes from these farming areas, it is imperative that wind erosion is controlled so as to allow long-term, stable and viable production. A large part of the slopes and plains of New South Wales are especially susceptible to wind erosion in drought times, even though the average rainfall is higher than 375 mm. The frequency of dust storms in New South Wales is presented in McTainsh et al. (1990).

Like water erosion, wind erosion is a natural process shaping the earth's surface, changing landforms and transporting geological materials from one place to another. Deflation is the process of lowering the land surface by removing soil particles by wind erosion. When previously stable soils are exposed to the force of wind by cultivation, deforestation, overgrazing or mining, then accelerated wind erosion can occur (Duff, 1993). These newly exposed soil materials have the capacity to produce large volumes of dust and wind-eroded materials as they have not been sorted by previous wind erosion events. They frequently have high proportions of materials susceptible to wind erosion.

2.2.1 Processes of Wind Erosion

General

Wind erosion occurs when the lift forces of the wind exceed the gravity and cohesion forces of the soil grains at the surface. Conditions when wind erosion is expected to be greatest are:

... when the soil is loose, dry and finely granulated; the soil surface is smooth and vegetative cover is sparse or absent; the susceptible area is sufficiently large; and the wind is strong and turbulent enough to move soil. (Lyles et al., 1985)

The processes of wind erosion, the factors controlling the rates of wind erosion, and the consequences of wind erosion are described in detail in a series of articles in a special issue of the *Australian Journal of Soil and Water Conservation* (1994, Volume 7, Number 3). Some key points that arise from this series of articles are the following:

(a) Sands and loamy sands are much more susceptible to wind erosion than sandy loams and loams, and generally clay loams and clays are relatively resistant to wind erosion (Leys et al., 1994; Leys, 1991a).

(b) Ground cover is essential to prevent wind erosion on susceptible soils, and the clearing and cultivation of these soils can lead to high rates of wind erosion (Leys and Heinjuis, 1992; Leys et al., 1994; Leys, 1991b).

(c) Rainfall conditions in the autumn before sowing can have a large effect on wind erosion because of their effect on weed growth, the timing and amount of cultivation, and the moisture content of the tilled soil (Leys et al., 1994; Neil et al., 1994).

(d) Wind erosion removes the finer fractions of the soil (particles < 0.90 mm), and it is this fraction which carries the majority of the nutrients, and nutrient- and water-holding capabilities of the soil. Therefore wind erosion has the capacity to severely reduce the fertility of soils (Leys and Heinjuis, 1992; Leys and McTainsh, 1994). For the paddock investigated in the above studies, the level of nitrogen is estimated as 16 times higher in the wind-blown dust than in the soil from which it was derived, and the cation exchange capacity and water-holding capacity of the soil was reduced by more than half. Miles and McTainsh (1994) showed that phosphorus levels as well as nitrogen levels were much higher in the wind-blown fraction compared to the original soil.

The distribution and causes of wind erosion in Australia are described in McTainsh et al. (1990) and in McTainsh and Leys (1993). An index of wind erosion for Australia is presented in Burgess et al. (1989).

Processes

The character of the wind, especially its velocity and turbulence, is the prime factor controlling the potential for wind erosion. The shape of the wind velocity profile will largely determine the capability of the wind to detach and lift soil particles from the surface. The shape of the velocity profile close to the ground is logarithmic with height (Figure 2.9) and is extremely important in determining sediment transport processes near the surface (Greeley and Iversen, 1985; Lyles et al., 1985).

The profile of wind velocity with height above the ground can be expressed as the following equation:

$$U_z = (U^*/K) [\log_e (z/z_0)]$$

where

U_z	=	the mean wind velocity at height z
U^*	=	the frictional velocity
K	=	a constant, 0.4
z_0	=	the roughness length or height above the ground where the wind velocity is zero

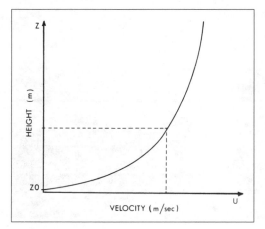

Figure 2.9   Wind velocity profile which is logarithmic with height

The frictional velocity (U*) is very important because it is directly proportional to the rate of increase of the wind velocity with the natural logarithm of the height. Therefore, U* is a measure of the velocity gradient of the wind and is strongly correlated with sediment transport. The only wind erosion model to be tested and calibrated in Australia is that of Shao et al. (1994).

Once erosion starts, the sediment is sorted according to its particle size, shape and weight.

There are three principal modes and related velocities that separate the soil into its different fractions (Figure 2.10):

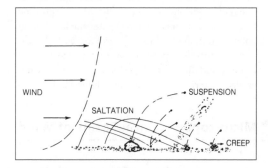

Figure 2.10   Diagram showing the three principal modes of aeolian transport

(a) The Creep Fraction

The 'creep' fraction (greater than 0.5 mm) moves at a very low mean velocity, and rolls and bumps its way across the unstable surface as a result of the impact of other faster-moving particles.

(b) The Saltation Fraction

The 'saltation' fraction (0.1–0.5 mm) moves at a moderate mean horizontal velocity by 'jumping' or 'bouncing' across the surface. Particles moved in this way usually bounce no higher than 50 cm and, when they return to the surface, impinge on other material. In this way they detach more particles, thereby perpetuating the saltation process. This fraction represents the greatest proportion, by weight, of soil moved in a wind erosion event, and often forms distinctive sand dunes of very fine sand. These are common on the downwind side of lakes that have variable levels of water, such as Lake George in southern New South Wales.

(c) The Suspension Fraction

The 'suspension' fraction (less than 0.1 mm) moves slightly slower than the mean surface wind and is carried by the wind horizontally over the surface. The volume of soil in suspension increases as the rate of saltation increases because of the continual disturbance of the soil surface caused by saltation (Bagnold,

1941). This fraction can move very large distances, and can form important geological and pedological deposits such as the loess deposits in northern China or the Canterbury Plains of New Zealand (Duff, 1993), and parna deposits in southern New South Wales (Butler and Churchwood 1983).

The symptoms of wind erosion are most commonly observed in the transport stage, such as in haze or dust storms (that is, primarily the suspension fraction), or the deposition stage, as the accumulation of soil particles on fence lines and across roads (that is, primarily the saltation and creep fractions)(see Plate 2.7).

Each fraction moves at a different velocity and in a different mode and, consequently, when deposition takes place it occurs at different rates and at different points within the landscape. Therefore, each depositional feature is representative of the size fraction involved and the rate of deposition. In this way wind-eroded materials are sorted into different size fractions.

2.2.2 Micro-topographical Forms of Wind Erosion

Close inspection of the surface soil usually reveals some movement of particles, for example the infilling of furrows on cultivated land and the accumulation on the leeward side of obstacles such as grass butts or stubble. In the case of sandy soils, a degree of sorting is also evident. In moderately severe cases of erosion, ripples of coarse sand will develop on the surface which in turn move across the surface, levelling the ground and leaving the entire area covered with a series of wave-like ripples. Finer-textured soils, such as loams, commonly have a 'swept' or 'baked' appearance following wind erosion.

When viewed in profile, the top few millimetres of a wind-eroded soil may be lighter-coloured than the soil it overlies. This effect has been attributed to the winnowing of the fine materials and organic matter (suspension fraction) from the exposed surface soil. It has been suggested that when this lighter layer of sand-size particles (saltation and creep fractions) is thick enough, it may, on some soils, become an 'armouring layer'. Soil loss ceases at this stage as there are no more wind-erodible size fractions available to the wind. The effect continues until the coarse sand layer is removed by a stronger wind event (see Plate 2.6).

Micro-topographical features which reflect the degree of wind erosion are:

(a) *smoothing of cultivation ridges*: where the tops of ridges have been rounded off and some accumulation is evident in furrows and other micro-depressions

(b) *wind sheeting, hummocking and ripple drift*: depending on the amount of vegetative residues present, wind sheeting and/or hummocking or ripple drift may occur. Wind sheeting is identified by a streaky appearance of the soil (especially on the leeward side of aggregates) or by a marked increase in the amount of coarse material at the surface.

Plate 2.6 Wind-eroded soil showing the winnowed surface layer consisting of coarse sand overlying uneroded soil

Hummocking is identified by accumulation of sandy material behind larger objects. Ripple drift is identified by a series of small wave-like ridges composed of coarse sand perpendicular to the erosive wind direction. Ripples are generally 1 to 2 cm high with wavelengths of 5 to 50 cm.

(c) *stripping and/or blowouts*: where significant deflation of the soil surface has occurred leaving a hard surface. In the case of blowouts, a major lowering of the surface has occurred.

These micro-topographical features may be produced in the duration of one windstorm event or over a period of time. These features may be obliterated by cultivation and thus are not permanent records of past erosional history, with the exception of large blowouts.

All the preceding micro-topographical features are a result of varying levels of saltation and deposition. If the saltation process is allowed to continue, sand ripples continue to build in height, forming sandy ridges and dunes.

2.3 Aeolian Landforms

Aeolian landforms are derived principally from two types of sediment deposits. Those from suspended sediments are called loess and those from saltation/deposition are called sand plains and sand dunes.

Loess landforms are rare in Australia although Pesci (1968) estimates that one tenth of the earth's land surface is covered by wind-blown dust ranging from 1 to 100 m thick. However, areas of eastern Australia are covered by parna, which is a wind-blown calcareous clay that has been transported in aggregates that are similar in size to loess (Butler, 1956). Parna has been most definitely identified in the Wagga region of New South Wales (Butler, 1956; Butler and Hutton, 1956; Beattie, 1972).

Sand plains and sand dunes occur in the drier regions of the world and cover a large proportion of inland Australia, especially in the south-west of New South Wales and along the coast. There are three major landform types:

—*sand shadows*: accumulations of sand found on the leeward side of obstacles such as bushes and boulders

—*sand drifts*: deposited on the leeward side of a gap between two obstacles

—*sand dunes*: ridges that feature deposition as well as transportation. There are three primary types of dunes:

1. longitudinal dunes, oriented parallel to the prevailing wind direction
2. transverse dunes, oriented perpendicular to the prevailing wind direction

3. parabolic dunes, which are U-shaped dunes pointing downwind.

Hack (1941) related dune type to wind strength, sand supply and amount of vegetation, and is a useful reference on the formation of dunes generally. Duff (1993) and Reineck and Singh (1975) describe the processes that lead to the formation of different kinds of sand dunes. An excellent description of aeolian landforms and the processes that form them can be found in Cooke et al. (1993).

Movement of Nutrients and Chemicals on Sediments Moved by Wind Erosion 2.2.4

The dust eroded from the soil is made up of minerals such as quartz, clays and organic matter. As a result, the eroded dust also carries soil nutrients and chemicals that may have been applied to the soil.

Studies in northern New South Wales (Leys and McTainsh, 1999) show that dust deposition to a riverine environment averaged 68 $g/m^2/yr$, 67% of which was deposited in dry periods. About 33% of the deposited dust was organic matter. Deposited dust was enriched in nutrients compared to the soils of the area, with an eight-fold increase for total N and organic matter, and about three-fold for total P. Total N deposition was about 1.244 $g/m^2/yr$ and total P about 0.285 $g/m^2/yr$. Predicted nutrient concentrations in a slow-flowing river (0.0039 $\mu g/L$ of N and 0.0008 $\mu g/L$ of P) are only a very small fraction of measured nutrient concentrations from rivers in the area. However, predictions of the concentrations in a still pond (3.41 $\mu g/L$ of N and 0.78 $\mu g/L$ of P) after one month of deposition are about 50% of measured nutrient concentrations. Measurements of the nutrient concentration of emitted dust from vehicles and cultivation, and deposited dust indicate that aeolian nutrient contributions are significant to riverine environments, especially during low-river flow conditions.

Dust can also transport herbicides and pesticides off-farm. Recent studies (Leys et al., 1999; Larney et al., 1999) have quantified: (1) the dust and endosulfan emissions caused by wind erosion, cultivation and vehicle movement on unsealed roads; and (2) the deposition of dust and endosulfan both on- and off-farm in the cotton-growing region of northern New South Wales over the summer of 1996/97. The major source of dust was wind erosion from roads, which was more significant than the cotton fields. This was because roads have more loose erodible sediment than cotton fields. Endosulfan (a pesticide used to control insects in cotton) transport was higher off roads than cotton fields because the erosion rate of the roads was higher.

Plate 2.7 Wind-eroded material burying a fence

Dust emission from a vehicle travelling at 80 km/h on an unsealed road (3.7 grams per metre travelled) was about double that for inter-row cultivation at 8 km/h (1.7 g/m). Unsealed roads were a greater source of endosulfan because of the greater frequency of vehicle movements compared to cultivation of fields, however endosulfan emissions are only a problem for the few days after aerial spraying.

Dust deposition was greatest near roads and decreased logarithmically with distance away from the road. Endosulfan source strength (combination of dust and vapour) was highest close to roads.

Dust, nutrients and endosulfan emissions can be reduced by restricting cultivation and vehicle movements on recently sprayed areas for a period of about three days after application and by leaving a buffer strip, in the order of 100 m, between sensitive areas and sprayed areas. The use of interception barriers such as windbreaks is a possible option for filtering dust from the air before it is transported off-farm. Increased aerial spraying precision could also reduce spray drift to roads, thereby further reducing endosulfan source strength. Therefore, nutrient and chemical transport from farms can be reduced by undertaking good land management practices as indicated by the cotton industry best management practice manual (Williams, 1997).

2.3 Mass Movement

Mass movement encompasses erosion processes in which gravity is the primary force acting to dislodge and transport land surface materials. It is a function of the gravitational stress acting on the land surface and the resistance of the surface soils and/or rock materials to dislodgement. When the gravitational stress exceeds this resistance, mass movement occurs. The occurrence of mass movement depends upon the interaction of various factors including landform, lithology, soil type, rainfall intensity and duration, drainage characteristics, vegetal cover and human intervention (Varnes, 1978; Selby, 1982; Duff, 1993).

Mass movement can broadly be categorised into two types: those in which colluvial material moves down steep slopes; and those in which soil material is subject to movement on slopes with a wider range of gradients (see Plate 2.8).

2.3.1 Causes of Mass Movement

Slope movement is seldom the result of a single factor; usually failure is the end result of activities and processes that have taken place over many years prior to the actual movement. The processes of rock and soil weathering, slope erosion and deposition, and man-made changes such as housing development and road construction can all contribute to slope failure, if the combination of landform, soil types, lithology, rock structure and climate is such that a hazard of mass movement exists.

Failures are significant if human life or property is at risk. In New South Wales, significant losses have occurred where houses and roads have been destroyed or severely damaged, particularly in the coastal districts between Newcastle and Wollongong.

In general, failures occur when the weight of the slope exceeds its restraining capability. This usually takes place following intense rainfall periods, when the slope weight has been increased dramatically by soil and rock saturation and/or a zone of weakness in the underlying material has been further weakened and lubricated by infiltrating water.

Disturbances to slope profiles that increase the weight factor (such as filling, buildings and so on) or remove their restraining capability (for example cutting into the slope toe for a road or house) will greatly increase the risk of failure unless slope-stabilising measures are employed. The principal stabilisation measure is to safely remove surface and subsurface water. This reduces the slope weight and reduces the risk of lubricating zones of weakness.

The most common triggering agent is the infiltration of water into the sloping land surface, which has the effect of both reducing the shear strength of the soil material and increasing the mass loading on the slope.

1. Factors leading to an increase in slope mass loading can be summarised as follows:
   (a) removal of lateral or underlying support
      (i) undercutting by water (for example, rivers and waves)
      (ii) weathering of weaker rock strata at the toe of the slope
      (iii) washing out of granular material by seepage
      (iv) man-made cuts and fills, excavations, draining of lakes or reservoirs
   (b) increased disturbing forces
      (i) natural accumulation of water, snow, talus
      (ii) man-made pressures (such as stockpiles of ore, tip-heaps, rubbish dumps or buildings)
   (c) transitory earth stresses
      (i) earthquakes
      (ii) continual passing of heavy traffic
   (d) increased internal pressure
      (i) build-up of pore-water pressure (such as in joints and cracks, especially in the tension crack zone at the rear of the slide).
2. Factors leading to a low shear strength or a reduction in the shear strength of the soil material can be summarised as follows:
   (a) materials which decrease in shear strength if water content increases such as clays, shale, mica, schist, talc, serpentine (for example, when the local water table is artificially increased in height by reservoir construction or leaking of water pipes, or as a result of stress release, vertical and/or horizontal, following slope modification)
   (b) materials which have low internal cohesion (such as consolidated clays, sands, porous organic matter)
   (c) bedrock formations which have faults, bedding planes, joints, foliation in schists, cleavage, brecciated zones and pre-existing shears
   (d) weathering changes
      (i) weathering reduces effective cohesion and, to a lesser extent, the angle of shearing resistance
      (ii) absorption of water leading to changes in the fabric of clays (such as loss of bonds between particles or the formation of fissures)
   (e) pore-water pressure increases as a result of: higher groundwater tables due to increased precipitation or reduced evapotranspiration; construction of water storages or dams; leaking of water pipes that have been installed near or above the site.

2.3.2 Classification of Forms of Mass Movement

There are various ways to classify mass movement including age, cause and degree of disruption of the displaced mass, but the preferred classification is based on the type of material—bedrock, debris and

Plate 2.8 Mass movement of a hillside slope

earth or soil materials—and the type of movement. The principal types of movement are defined as follows (see Varnes, 1978; Duff, 1993).

1. **Falls** are typically very rapid movements where the material in motion travels most of the distance through the air, such as in a rockfall.

2. **Topples** are movements dominated by a forward tilting motion about some pivot point below or low in the overturning mass.

3. **Lateral spreads** are movements dominated by lateral extension facilitated by fractures in the moving mass.

4. **Flows** involve material in motion moving downslope in a form similar to the behaviour of a viscous fluid, such that minor scarps within the moving material are usually not visible or are short-lived. Flows remain in contact with the ground and their movement varies from extremely slow to extremely rapid. The boundary between the moving and in-place material may be a sharp surface of differential movement or a zone of distributed shear. As well as subdividing flows on their type of material, water content is also commonly used as a criterion. The main type of flow relevant to soil conservation is earthflow.

5. **Slides** involve material in motion moving downslope as a mass, typically having distinct minor scarps within the moving material and in contact with the ground. Movement varies

from slow to rapid. Displacement is associated with finite shear failure along one or more surfaces, or within a relatively narrow zone, which are visible or may reasonably be inferred.

6. **Subsidence** whereby material displacement is predominantly vertically downwards, such as where an underground tunnel caves in causing surface settlement. This is a potential problem in old mining areas, especially if the old mines were relatively shallow.

Movement by a combination of these types constitutes a complex slope movement. Nomenclature of such complexes is based on the various component types and/or the different development stages of the movement. For example, a rock debris avalanche involves bedrock and debris material. Avalanche is a general term describing 'flow'-type movements that occur under a variety of moisture conditions, but are characterised by extremely rapid movement and a long and narrow track down the steep slopes on which they typically occur.

For further details of mass movement classification systems, as well as comprehensive coverage of its occurrence and environmental hazard, the reader should refer to Cooke and Doornkamp (1974), Varnes (1958, 1978) and Selby (1982).

Soil Factors   2.3.3

Slope failure usually occurs within the soil mantle rather than within unweathered bedrock, hence soil data provide a valuable means of analysing the physical behaviour of a slope, particularly where there are deep unconsolidated colluvial deposits. Soil investigations for landslip studies are divided essentially into two categories—field data and laboratory data. Field data may be generated by measurements and observations of the characteristics of the soil material as it exists in situ, but data may also need to be collected from field experiments or trials, especially those data relating to permeability and soil movement.

The following soil properties are of particular relevance to field classification in areas susceptible to landslip:
—texture of A and B horizons
—depth to bedrock
—nature of the parent material
—presence of colluvial detritus within the profile
—structure of A and B horizons
—profile and site drainage
—soil colour, especially presence of B horizon mottles
—special features, such as indurated horizons, gravel layers or evidence of the presence of a water table.

An important procedure in landslip analyses is the use of geophysics to identify soil and slope characteristics in areas susceptible to mass movement. Soil depth, the nature and thickness of parent material and bedrock lithology can be determined by interpretation of seismic travel-times using a signal enhancement seismograph. This technique avoids the depth limitation of soil augers and the difficulties involved with truck and trailer-mounted machinery on terrain associated with, or suspected of, mass movement.

Given the dependence of soil strength on soil moisture content, field investigations should also include studies of the soil water regime, looking at sources of water and the likely level of soil moisture through the year. Special attention should be given to the seasonal variation in water levels and the possibility of extra water being diverted onto the site by natural or man-made processes. If the investigations reveal potential for soil saturation to occur at a critical time or place within the soil material or regolith, this will greatly increase the chances of mass movement occurring.

Many laboratory procedures are available for evaluation of the engineering properties of soils. They generally relate in some way, or can be related, to soil strength, although they do not all necessarily measure this particular parameter (see Chapter 11). The laboratory tests commonly used to define the physical behaviour of a soil in relation to mass movement are:
—particle size distribution and Unified Soil Classification
—Atterberg limits (plastic and liquid limits, plasticity index)
—linear shrinkage
—Emerson Aggregate Test (dispersibility)
—shear strength (direct, confined, unconfined, triaxial, California Bearing Ratio)
—maximum compaction (for bulk density and optimum moisture content)
—cone penetrometer measurements
—field moisture content
—cation exchange capacity
—X-ray diffraction and X-ray fluorescence (for identification of clay mineralogy).

In most cases, it is not necessary to carry out all the laboratory tests listed here because many are time consuming or too costly for routine survey. Prior to each investigation, a thorough examination should be undertaken of the laboratory facilities and resources available, together with the scale and purpose of survey. Methods for cation exchange capacity and X-ray diffraction are impractical for routine analyses, although they provide useful evidence to support the clay mineralogy indicated by the Atterberg limits.

The scale of the survey is an important consideration in selection of laboratory tests; for example, triaxial shear tests are appropriate for intensive surveys on small sites at large scales, but are meaningless on a few samples over a large area. Therefore, prior to commencing a survey that involves landslip analysis, it is necessary to analyse carefully the practicality and predictive value of laboratory tests, particularly as the variability of properties is usually high in landslip terrain. This is to ensure that they are consistent with the requirements and purpose of the study.

2.4 Coastal Erosion

Coastal erosion is defined as the loss of sand, soil and rock material to the sea in response to wave action. A knowledge of the degree of erosion is important to the long-term management and protection of our beach areas.

2.4.1 Coastal Processes

The term 'coastal processes' refers to the forces active in shaping the coastal zone. These forcing functions are winds, waves, currents, tides and floods, and they influence the distribution of sediments in the coastal zone which shape the coastline (see Duff, 1993).

The primary force is wind. It generates waves as it blows across a waterbody, causes elevated ocean levels at the coast and transports loose sand. Sand blown inland from the beach can be trapped by vegetation or mechanical means to form coastal dunes. If not trapped, the sand may move further inland and result in permanent loss of sand from the beach.

Waves are the most obvious and direct cause of shoreline changes. Once material has been stirred up from the sea bottom by waves and becomes suspended in the water, it can be moved readily by any current that may be present. The combination of wave turbulence, rip currents and longshore currents moves sediments in large quantities. These sediments move onshore under the direct action of waves, offshore by rip currents and along the shore by longshore currents.

During floods, sediment is flushed from coastal streams into the marine environment. This sediment is typically silt, but marine sand that has been deposited by tidal and wave action, in the lower portion of the estuary, is also flushed into the offshore region. Between floods, wave and tidal action will move sand back into the estuary again, although sand flushed too far offshore by the flood will be permanently lost from the coastal system.

From this it is obvious that the sandy beach is only part of the total dynamic coastal system. This system extends from the back beach barrier, through the dunes, the berm, wash zone, surf zone, offshore bar and nearshore continental shelf out to a depth typically in the order of 30 to 50 m (Figure 2.11).

The general sequence of events in a dynamic coastal system is that storm waves remove significant quantities of sand from the beach berm to build a surf zone bar. Ensuing calm weather results in an onshore movement of this material to re-establish the berm. Such short-term beach fluctuations are usually much greater than any long-term trends of shoreline movement and, hence, act to mask those trends.

To gain a full understanding of the behaviour of a coastal system, a cyclic approach, known as the sediment cycle, has been developed in which all sediment losses and gains due to the forcing functions are identified (Figure 2.12). An example of the movement of sand associated with the cycle of sand movement on beaches is given in Quilty and Wearne (1975), who showed the effects of a cyclone and heavy seas on a beach, with the subsequent rebuilding of the beach when calmer weather returned. The study emphasised the importance of a frontal dune stabilised with vegetation to catch wind-blown sand, in minimising beach erosion.

A positive imbalance between the sediment gains and losses would result in an accreting shoreline. However, a negative imbalance, where the losses are greater than the gains, will result in a receding shoreline, which studies indicate is generally the case along the New South Wales coast.

Coastal Dunes and Erosion

2.4.2

The coastal system landward of the land/sea interface is generally characterised, in its natural state, by dunes. The dunes along the New South Wales coastline have been formed by the action of wind over tens of thousands of years. However, the extent of their formation has been governed by the amount of sand supplied to the beach by waves and currents. The varying supply of sand, as well as other factors (such as embayment size and orientation), has resulted in a number of different forms of dunes identifiable along the coastline.

Dune form and vegetation change, as one moves inland away from the active beach area, from a marine to a terrestrial environment. The beach berm is the flat area of the land/sea interface that is commonly recognised as the beach. Behind the beach is

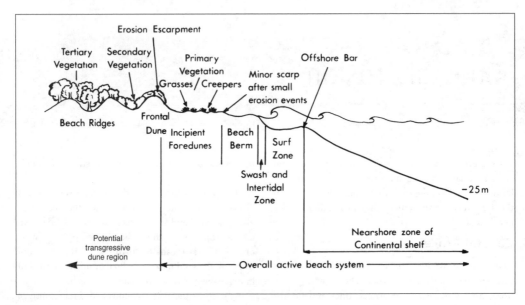

Figure 2.11 The coastal system

the series of dunes, the first being called the frontal dune. There may be a number of features on the seaward side of the frontal dune which indicate present-day processes. The presence of an erosion escarpment (a near-vertical face on the seaward side of the frontal dune) indicates the most landward extent of erosion due to past storm waves. Depending on the time lapse since the storm, this feature may be obscured by vegetation and/or an incipient dune. The incipient dune is a newly forming dune being supplied with wind-blown sand and may be stabilised with primary vegetation such as grasses and creepers (see Figure 18.1).

In many situations, especially in highly modified areas, the dune system may be completely covered by development and a seawall constructed on or between the frontal dune and the beach, thereby alienating sand in the back beach area from the coastal system, as at Bondi Beach. However, sometimes in these highly modified situations dunes can be artificially formed and vegetated to prevent the problems of wind-blown sand.

Coastal dunes are almost constantly under threat from wind erosion, and require a good vegetation cover to prevent strong winds removing sand from them. In areas where the vegetative cover is completely destroyed, vast areas of sand have become mobile to form transgressive dunes, such as happened at Stockton Beach and Kurnell, which fortunately have now been largely stabilised following major revegetation programs.

An example of beach erosion resulting from a cyclone is presented in Wearne and Quilty (1974). They showed that 12.3 m² of sand was lost from a 60 m transect in an unstable area or blowout during a major storm. This resulted in the lowering of the soil profile by 41 cm. In a transect along a natural dune area, the net loss of sand was 0.7 m² during the same storm. The study also showed that artificially constructed dunes, formed by brush or a fence, can accumulate sand and initiate the development of dunes in unstable situations. This study concluded that a frontal dune, stabilised by vegetation that can catch wind-blown sand, is essential for the long-term stability of a beach.

When vegetation is damaged or destroyed on the frontal dune, a 'blowout' generally occurs. A blowout is usually U-shaped and aligned to the major wind direction (south-east–north-west in New South Wales). Blowouts are formed through a lower section of dune which concentrates and funnels wind to cause increased wind velocities and further sand losses. With these losses, the blowout becomes deeper and, with the gradual collapse of the side walls, can grow to be a significant feature, particularly when several blowouts unite.

Sand moving inland through blowouts, or as transgressive dunes, can cause major problems and damage to roads, car parks, houses, public facilities and so on. The damage caused by wind-blown sand can be counted as a cost to the individual (house damage) or as a cost to the community and council (public build-

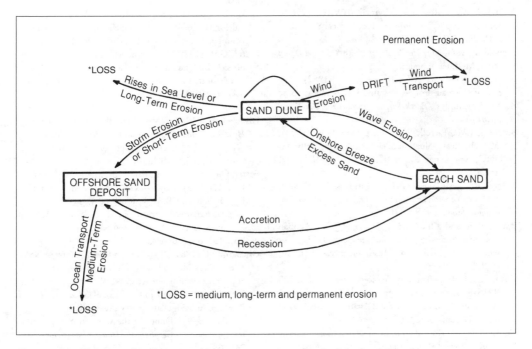

Figure 2.12 The coastal sediment cycle

ing and road damage). Although the area affected is small in relation to the area of the state as a whole, the social and economic impact of coastal sand drift is considerable. In earlier years, the most common practices causing coastal sand drift were the cutting of firewood, grazing of livestock, construction of buildings and roadways, uncontrolled bushfire and rabbits.

With population densities increasing in the major cities and towns along the New South Wales coastline, and new recreational pursuits evolving with modern technology, the threats to dunal vegetation are now coming from a different quarter.

The sheer numbers of people wishing to use a beach can result in trampling of vegetation underfoot, as they make their way from the car parks to the beaches. A further assault on the vegetative cover comes from the various forms of recreational vehicles such as four-wheel-drives, dune buggies and motorcycles. In recent years stabilisation of the frontal dunes of beaches has been largely successful, and programs such as Coastcare and Dunecare have stabilised many areas that were once unstable (see Chapter 18 for techniques of stabilisation).

BIBLIOGRAPHY

American Society of Civil Engineers (ASCE) (1975), *Sedimentation Engineering*, American Society of Civil Engineering, New York.

Anderson, M.G. and Kneal, P.E. (1980), 'Pore water pressure and stability conditions on a motorway embankment', *Earth Surface Processes and Landforms* 5, 37–46.

Australian Water Resources Council (AWRC) (1969), *Sediment Sampling in Australia*, Hydrological Series No. 3, Canberra, ACT.

Bagnold, R.A. (1941), *The Physics of Blown Sand and Desert Dunes*, Methuen, London.

Blong, R.J., Graham, O.P. and Veness, J.A. (1982), 'The role of sidewall process in gully development: some NSW examples', *Earth Surface Processes and Landforms* 7, 381–5.

Bradford, J.M., Farrell, D.A. and Larsen, W.E. (1973), 'Mathematical evaluation of factors affecting gully stability', *Soil Science Society of America: Proceedings* 37, 103–7.

Burgess, R.C., McTainsh, G.H. and Pitblado, J.R. (1989), 'An index of wind erosion in Australia', *Australian Geographical Studies* 27, 98–110.

Butler, B.E. and Churchward, H.M. (1983), 'Aeolian processes' in *Soils: An Australian Viewpoint*, Division of Soils, CSIRO Melbourne, Academic Press, London, 91–9.

Chan, R. (1998), 'Bathurst Geology Sheet 1:250 000 SI/55-8, Explanatory Notes, Geological Survey of New South Wales', Sydney and Australian Geological Survey Organisation, Canberra.

Cooke, R.U. and Doornkamp, J.C. (1974), *Geomorphology in Environmental Management*, Clarendon, Oxford.

Cooke, R.U., Warren, A. and Goudie, A.S. (1993), *Desert Geomorphology*, UCL Press, London.

Crouch, R.J. (1976), 'Field tunnel erosion—a review', *Journal of the Soil Conservation Service of NSW* 32, 98–111.

Crouch, R.J. (1979), 'The causes and processes of tunnel erosion in the Riverina', MSc (Agr) Thesis, University of New England.

Crouch, R.J. (1983), 'The role of tunnel erosion in gully head progression', *Journal of the Soil Conservation Service of NSW* 39, 148–155.

Crouch, R.J., McGarity, J.W. and Storrier, R.R. (1986), 'Tunnel formation processes in the Riverina area of NSW, Australia', *Earth Surface Processes and Landforms* 11, 157–68.

Dierickx, W. (1983), 'Hydraulic gradients near subsurface drains and soil erosion', *Transactions of the American Society of Agricultural Engineers* 26, 1409–12.

Duff, D. (1993), *Holmes' Principles of Physical Geology*, Chapman and Hall, London.

Dumsday, R.G. (1973), 'The economics of some soil conservation practices in the wheat belt of Northern New South Wales and Southern Queensland. A modelling approach', *Farm Management Bulletin No. 19*, University of New England, Armidale, New South Wales.

Elliott, G.L. (1976), *Some Properties of Tunnelling Soils in the Hunter Valley. I. Exchange Properties, Water Soluble Ions, Mechanical Analyses*, Tech. Bull. 15, Scone Research Centre, Soil Conservation Service of NSW, Sydney.

Erskine, W. and Melville, M.D. (1984), 'Sediment transport, supply and storage in sand-bed channels of the northern Sydney Basin', Proc. Drainage Basin and Sedimentation Conference, University of Newcastle, NSW.

Fentie, B., Coughlan, K.J. and Rose, C.W. (1997), 'Manual for Use of Program GUEST30'. Faculty of Environmental Sciences, Griffith University, Brisbane.

Foster, G.R. (1975), 'Mathematical simulation of upland erosion by fundamental erosion mechanics' in *Present and Prospective Technology for Predicting Sediment Yields and Sources*, ARS S-40, USDA Agricultural Research Service.

Foster, G.R. (1982), 'Modelling the erosion process', *Hydrologic Modelling of Small Watersheds*, ASAE Monograph No. 5, American Society of Agricultural Engineers, St Joseph, Michigan.

Foster, G.R., Young, R.A., Romkens, M.J.M. and Onstad, C.A.A. (1985), 'Processes of Soil Erosion by Water' in R.F. Follett and B.A. Stewart (eds), *Soil Erosion and Crop Productivity*, American Society of Agronomy, Crop Science Society of America and Soil Science of America, Madison, Wisconsin.

Graham, O.P. (1989), 'Land Degradation Survey of NSW 1987:1988: Methodology', Technical Report No. 7, Soil Conservation Service of NSW.

Greeley, R. and Iversen, J.D. (1985), *Wind as a Geological Process*, Cambridge University Press, UK.

Hack, J.R. (1941), 'Dunes of the western Navajo country', *American Geophysical Union Transactions* 31, 240–63.

Hairsine, P., Murphy, B., Packer, I. and Rosewell, C. (1993), 'Profile of erosion from a major storm in the south-east cropping zone', *Australian Journal of Soil and Water Conservation* 6(4), 50–55.

Hickin, E.J. and Nanson, G.C. (1975), 'The character of channel migration on the Beatton River, North East British Columbia, Canada', *Geological Society of America Bulletin* 86, 487–94.

Hjulstrom, F. (1935), 'Studies of the morphological activity of rivers as illustrated by the River Fyris', *University of Upsala Geological Institute Bulletin* 25, 221–527.

Holeman, J.N. (1968), 'The sediment yield of major rivers of the world', *Water Resources Research* 1, 737–47.

Houghton, P.D. and Charman, P.E.V. (1986), *Glossary of Terms used in Soil Conservation*, Soil Conservation Service of New South Wales, Sydney.

House of Representatives (1969), 91st Congress Committee on Public Works, *A Study of Streambank Erosion in the United States*, Report of the Chief of Engineers to the Secretary of the Army, Washington DC.

Hudson, N. (1981), *Soil Conservation*, Batsford Academic and Educational Ltd, London.

Imeson, A.C. and Kwaad, F.J.P.M. (1980), 'Gully types and gully prediction', *KNAG Geografisch Ti idschrift* XIV 5, 430–41.

Ireland, H.A., Sharp, C.F. and Eargle, D.H. (1939), 'Principles of Gully Erosion in the Piedmont of South Carolina', *USDA Technical Bulletin* 636, Washington DC.

Jansen, J.M.L. and Painter, R.B. (1974), 'Predicting sediment yield from climate and topography', *Journal of Hydrology* 21, 371–80.

Junor, R.S., Marston, D. and Donaldson, S.G. (1979), *A Situation Statement of Soil Erosion in the Lower Namoi Area*, Soil Conservation Service of NSW, Sydney.

Kaleski, L.G. (1963), 'The erosion survey of NSW', *Journal of the Soil Conservation Service of NSW* 19, 171–183.

Kinnell, P.I.A. (1990), 'The mechanics of raindrop induced flow transport', *Australian Journal of Soil Research* 28, 497–516.

Kinnell, P.I.A. (1994), 'The effect of predetached particles on erosion by shallow rain-impacted flow', *Australian Journal of Soil Research* 31, 127–142.

Knisel, W.G. (ed.) (1980), *CREAMS: A field scale model for chemicals, runoff, and erosion from agricultural management systems*, US Department of Agriculture, Conservation Research Report No. 26.

Lal, R. (1990), 'Soil erosion and land degradation: the global risks', in R. Lal and B.A. Stewart (eds.), *Advances in Soil Science* 11.

Laflen, J.M., Lane, L.J. and Foster, G.R. (1991), 'WEPP, a new generation of erosion prediction technology', *Journal of Soil and Water Conservation* 46, 34–38.

Langbein, W.B. and Schumm, S.A. (1958), 'Yield of sediment in relation to mean annual precipitation', *American Geophysical Union Transactions*. 39, 1076–84.

Larney, F.J., Leys, J.F., Müller, J. and McTainsh, G.H. (1998), 'Dust and endosulfan deposition in a cotton-growing area of northern New South Wales, Australia'. *Journal of Environmental Quality* 28, 692–701.

Leopold, L.G., Wolman, M.G. and Miller, J.P. (1964), *Fluvial Processes in Geomorphology*, Freeman, London.

Leys, J. and McTainsh, G. (1994), 'Soil loss and nutrient decline by wind erosion—cause for concern', *Australian Journal of Soil and Water Conservation* 7(3), 30–35.

Leys, J., Craven, P., Murphy, S., Clark, P. and Anderson, R. (1994), 'Integrated resource management of the mallee in south-western NSW', *Australian Journal of Soil and Water Conservation* 7(3), 10–19.

Leys, J.F. (1991a), 'Threshold friction velocities and soil flux rates of selected soils in south-west NSW, Australia' in *Aeolian Grain Transport 2. The erosional environment, Acta Mechanica*, Supplement 2, 103–112.

Leys, J.F. (1991b), 'The effect of prostrate vegetation cover on wind erosion', *Vegetatio* 91, 49–58.

Leys, J.F and Heinjus, D.J. (1992), 'Cover levels to control soil and nutrient loss from wind erosion in the South Australian Murray Mallee', *Proceedings of 7th International Soil Conservation Conference Sydney*.

Leys, J.F. and McTainsh, G.H. (1999), 'Dust and nutrient deposition to riverine environments of south-eastern Australia', *Zeitschrift fur Gemorphologie* (in press).

Lyles, L., Cole, G.W. and Hagen, L.J. (1985), 'Wind erosion: processes and prediction' in R.F. Follett and B.A. Stewart (eds), *Soil Erosion and Crop Productivity*, American Society of Agronomy, Crop Science Society of America and Soil Science of America, Madison, Wisconsin.

McDonald, R.C., Isbell, R.F., Speight, J.G., Walker, J. and Hopkins, M.S. (1984), *Australian Soil and Land Survey Field Handbook*, Inkata Press, Melbourne.

McTainsh, G.H. and Leys, J.F. (1993), 'Soil erosion by wind' in G.H. McTainsh and W.C. Boughton (eds), *Land Degradation Processes in Australia*, Longman, Melbourne, 188–233.

McTainsh, G.H., Lynch, A.W. and Burgess, R.C. (1990), 'Wind erosion in eastern Australia', *Australian Journal of Soil Research* 28, 323–39.

Meyer, L.D. and Wischmeier, W.H. (1969), 'Mathematical simulation of the process of soil erosion by water', *Transactions of the American Society of Agricultural Engineers* 12, 754–8, 762.

Meyer, L.D., Foster, G.R. and Nikolov, S. (1975a), 'Effect of flow rate and canopy on rill erosion', *Transactions of the American Society of Agricultural Engineers* 18, 905–11.

Meyer, L.D., Foster, G.R. and Romkens, M.J.M. (1975b), 'Source of soil eroded by water from upland slopes', in *Present and Prospective Technology for Predicting Sediment Yields and Sources*, ARS S-40, USDA Agricultural Research Service.

Miles, R. and McTainsh, G. (1994), 'Wind erosion and land management in the mulga lands of Queensland', *Australian Journal of Soil and Water Conservation* 7(3), 41–45.

Misra, R.K. and Rose, C.W. (1989), *Manual for the use of Program GUEST*, Division of Australian Environmental Studies, Griffith University, Nathan, Queensland.

Misra, R.K. and Rose, C.W. (1996), 'Application and sensitivity analysis of process-based erosion model GUEST', *European Journal of Soil Science* 47, 593–604.

Moss, A.J., Walker, P.H. and Hutka, J. (1979), 'Raindrop-simulated transportation in shallow water flows: an experimental study', *Sedimentary Geology* 22, 165–84.

Murphy, B.W. (1985), Regional distribution, soil formation and instability problems of soils in the Bathurst–Orange region of New South Wales', Master of Science in Agriculture thesis, University of Sydney.

Nanson, G.C. and Hickin, E.J. (1986), 'Statistical analysis of bank erosion and channel migration in Western Canada', *Geological Society of America Bulletin* 97, 497–504.

Neil, D., Yu, B. and Hesse, P. (1994), 'Wind, Rainfall, Dust—predicting wind erosion frequency', *Australian Journal of Soil and Water Conservation* 7(3), 36–40.

Pesci, M. (1968), 'Loess' in R.W. Fairbridge (ed.), *Encyclopedia of Geomorphology*, Reinhold, New York.

Quilty, J.A. and Wearne, A.H. (1975), 'Evaluation of a phase of extreme coastline erosion', *Journal of the Soil Conservation Service of NSW* 31, 179–192.

Raupach, M.R., McTainsh, G.H. and Leys, J.F. (1994), 'Estimates of dust mass in recent major dust storms', *Australian Journal of Soil and Water Conservation* 7(3), 20–24.

Reineck, H.E. and Singh I.B. (1975), *Depositional Sedimentary Environments*, Springer-Verlag, New York, Berlin.

Renard, K.G., Foster, G.R, Weesies, G.A. and McCool, D.K. (1997), 'Predicting Soil Erosion by Water: A Guide to Conservation Planning with the Revised Universal Soil Loss Equation (RUSLE)', USDA Agriculture Handbook No. 703, US Government Printer Office, Washington DC.

Robinson, A.R. (1971), 'Sediment, our greatest pollutant?', *Agricultural Engineering* 53, 406–8.

Rose, C.W., Coughlan, K.J., Ciesiolka, C.A.A. and Fentie, B. (1997), 'Program GUEST (Griffith University Erosion System Template)' in K.J. Coughlan and C.W. Rose (eds), *A New Soil Conservation Methodology and Application to Cropping Systems in Tropical Steeplands*, 34–58. ACIAR Report No. 40, ACIAR, Canberra.

Rosewell, C.J. (1985), 'Soil erosion on arable land' in P.E.V. Charman (ed.), *Conservation Farming*, Soil Conservation Service of NSW, Sydney.

Rosewell, C.J. (1993), 'SOILOSS—a program to assist in the selection of management practices to reduce erosion', Technical Handbook No. 11 (Second Edition), Department of Conservation and Land Management, Sydney.

Selby, M.J. (1982), *Hillslope Material and Processes*, Oxford University Press, Oxford.

Shao, Y., Raupach, M. and Short, D. (1994), 'Preliminary assessment of wind erosion patterns in the Murray-Darling Basin', *Australian Journal of Soil and Water Conservation* 7(3), 46–51.

Shao, Y., Raupach, M.R. and Leys, J.F. (1996), 'A model for predicting aeolian sand drift on scales from paddock to region', *Australian Journal of Soil Research* 34, 309–42.

Soil Conservation Service of NSW (1988), *Land Degradation Survey of NSW 1987–1988*, Soil Conservation Service of NSW, Sydney.

Stewart, J. (1968), 'Erosion survey of NSW Eastern and Central Divisions Re-assessment, *Journal of the Soil Conservation Service of NSW* 24, 139–54.

Stocking, M.A. (1976), 'Tunnel erosion', *Rhodesia Agricultural Journal* 73, 35–9.

Strakhov, N.M. (1967), *Principle of Lithogenesis*, Vol. 1, Plenum Publications, New York.

Trimble, S.W. (1975), 'Denudation studies: can we assume stream steady state?', *Science* 188, 1207–8.

United States Department of Agriculture (USDA) (1984), *Soil and Water Conservation and Education Progress and Needs*, USDA Soil Conservation Service, Washington DC.

Varnes, D.J. (1958), 'Landslide types and processes' in E. Eckel (ed.), *Landslides and Engineering Practice*, US Highway Research Board Special Report No. 29, Washington DC.

Varnes, D.J. (1978), 'Slope movement types and processes' in *Landslides: Analysis and Control*, Special Report No. 176, Transportation Research Board, National Academy of Sciences, Washington DC.

Veness, J.A. (1980), 'The role of fluting in gully extension', *Journal of the Soil Conservation Service of NSW* 36, 100–8.

Wearne, A.H. and Quilty, J.A. (1974), 'Frontal dune studies: the effects of cyclone Pam', *Journal of the Soil Conservation Service of NSW* 30, 201–12.

Williams, A. 1997, *Australian Cotton Industry Best Management Practices Manual*, Cotton Research and Development Corporation, Narrabri, NSW.

Wischmeier, W.H. and Smith, D. (1978), 'Predicting Rainfall Erosion Losses—A Guide to Conservation Planning', USDA Agricultural Handbook No. 537, US Government Printer Office, Washington DC.

Wolman, M.G. (1985), 'Soil erosion and crop productivity' in R.F. Follett and B.A. Stewart (eds.), *Soil Erosion and Crop Productivity*, American Society of Agronomy, Crop Science Society of America and Soil Science of America, Madison, Wisconsin.

Young, R.A., Onstad, C.A., Bosch, D.D. and Anderson, W.P. (1989), 'AGNPS: A nonpoint-source pollution model for evaluating agricultural watersheds', *Journal of Soil and Water Conservation* 44, 168–73.

Young, R.W. (1981), 'Denudation history of the south-central uplands of NSW', *Australian Geographer* 15, 77–8.

Soil Formation and Erosion Rates

CHAPTER 3

K. Edwards and C. Zierholz

Soil is a product of many processes. Weathering of rocks by a combination of physical, chemical and biotic processes results in the breakdown of the rock material and in the formation of new minerals. These products are the building blocks acted upon by soil-forming processes and lead to the development of soil at the site, or the products are transported to develop soil elsewhere (see Chapter 1; Paton, 1978; Selby, 1982). Under certain climatic conditions soils may also form from predominantly organic materials such as peat.

Weathering and erosion are part of the natural cycle of soil formation and recycling of materials at the earth's surface. However, the balance between current soil formation rates and erosion rates associated with current land uses can impact on soil quality at a site. Some land uses, by reducing surface cover, increasing runoff and reducing the resistance of soil to erosion, can lead to accelerated erosion.

How quickly does soil form and at what rates is it eroded? Without answers to these questions it is difficult to formulate policies and practices aimed at protecting the soil resource. This chapter presents available information concerning the rates of soil formation and erosion and discusses the implications for land management.

3.1 Soil Formation Rates

3.1.1 Soil-forming Materials

It has often been said that Australia has an old land surface, the implication of this being that the soils and underlying materials have been so strongly weathered that they are now infertile. This idea has to be qualified. While many areas of Australia have been exposed to weathering for millions of years, many of its soils and surfaces, particularly those in the eastern part, are quite young, in fact as young as many in other continents, including those which were widely affected by glaciation. Even in those parts of Australia where there has been severe and deep weathering, older surfaces have been dissected and progressively younger soils have been developing in the older weathered materials. Simultaneously, new surfaces have been developing on the sediments eroded from the old land surfaces and these sediments have in turn been reworked by the action of wind and water. (Beckmann, 1983)

Soil and parent material studies, mainly from southern and south-eastern Australia, with age relationships established by stratigraphy, correlation with former sea levels, or by radiocarbon dating, show that most differentiated soil profiles are formed on parent materials laid down as colluvial, alluvial or aeolian deposits during the Last (Russ-Wurm
30 000 to 40 000 years ago) Interglacial, or during an interstadial soon after it. (Mulcahy and Churchward, 1973)

Some of the thoughts advanced in these two quotations may be explained by referring to a general model of fluvial sediment differentiation in hillslope and riverine landscapes (Walker and Butler, 1983). The driving force in this form of landscape development is the erosion of material from hillslopes by rainfall and runoff, and the subsequent deposition of sediment further downslope when the gradient has decreased sufficiently. In the zone of net erosion, soils are mostly developed on material resistant to erosion or too coarse to be transported by rain splash and flow of water. This may occur on weathered bedrock or on previously deposited material (see Figure 3.1a). In the zone of net deposition, soils are developed on colluvium and sediment. Coarser materials are found further upslope and finer sediments downslope.

In riverine systems, alluvial material of varying size is deposited during flooding on levee banks and floodplains (see Figure 3.1b). Sediments generally become finer with increasing distance from the river channel.

Rates of erosion and sedimentation on hillslopes depend on factors such as continental uplift (which increases slopes and hence erosion rates) and by palaeoclimatic changes. In addition to fluvial sediments, aeolian deposits such as dunes and dust mantles are widespread in Australia (Butler and Churchward, 1983). These materials can also form base materials for, or contribute to, soil profile development, as is seen in the south-western slopes of New South Wales in the vicinity of Wagga Wagga and Junee where parna has been a major soil-forming material.

In summary, soils form on a range of materials including fluvial and aeolian sediments and weathered bedrock. The degree of soil profile differentiation is determined by the rate of soil-forming processes (see Paton, 1978) and the duration of periods of stability of the particular landscape. If the landscape is unstable, little profile differentiation will occur as erosion processes rework the materials continuously. However, if the landscape is stable and soil-forming processes occur at a comparatively rapid rate, profile differentiation will occur.

3.1.2 Soil Development and Weathering Rates

What does all this mean in relation to soil erosion and soil formation rates in Australia?

Two processes of soil formation need to be considered in evaluating the balance between the rate of soil formation and soil erosion (Hall et al., 1985; Nowak et al., 1985). A third can be added on the basis of the geomorphological processes described above. These are considered following.

(a) The rate of soil formation in unconsolidated materials such as existing soil materials, loose sands, recently deposited alluvium, or weathered bedrock (saprolite). Included are the formation of topsoil from subsoil, and the development of specific types of soil horizons such as organic layers, clay B horizons, cemented layers, and bleached horizons.

(b) The rate of soil formation from hard, relatively unweathered rock materials. In the long term this is the rate which is most critical in areas where soils overlay bedrock in the vicinity of the soil surface.

(c) The preceding discussion (Figures 3.1 and 3.2) has shown that in riverine systems soils can also be developed as a result of the mobilisation and deposition of eroded soil materials. This can occur as a consequence of wind or water erosion.

McCormack and Young (1981) also identified two factors that need to be considered when finding a balance between erosion rates and rates of soil formation. They are the formation of the A horizon from the existing soil material, and the weathering of parent rock into a favourable root zone (see Figure 3.2).

Soil Formation on Unconsolidated Materials

Few data are available on rates of soil development on unconsolidated materials in Australia. Most of the available data are from studies of alluvium using radiocarbon dates. Walker and Coventry (1976) have shown that an organic profile develops to its maximum extent in alluvium within 1000 years, followed by B horizon development to a maximum within

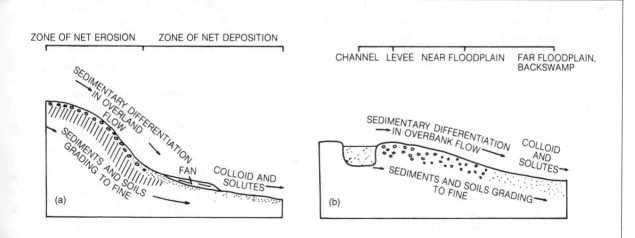

Figure 3.1 Fluvial sediment differentiation in hillslope and riverine landscapes (after Walker and Butler, 1983)

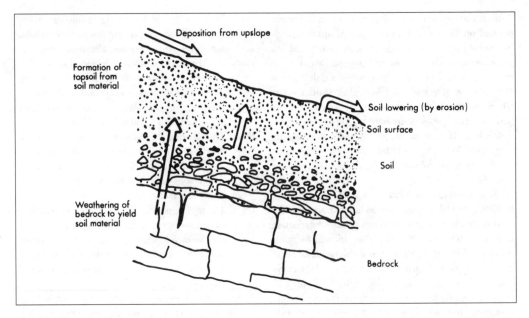

Deposition from upslope

Formation of
topsoil from
soil material

Soil lowering (by erosion)

Soil surface

Soil

Weathering of
bedrock to yield
soil material

Bedrock

Figure 3.2   Some of the processes operating on the soil profile—the balance of these needs to be considered

30 000 years (that is, 30 mm per 1000 years). Little and Ward (1981) observed soil profile differentiation to a depth of 165 cm in alluvium dated at 750 000 years which indicates an average rate of profile development of about 2 mm in 1000 years. Walker (1980) estimates that rates of the order of 30 mm per 1000 years (that is, about 0.4 t/ha/yr) are appropriate for development of soils with strongly developed texture profiles. He considers that these rates would apply to soils developed on alluvium, but that the rates of formation on bedrock would be considerably lower.

The time for particular kinds of soil horizons to develop in unconsolidated materials in North America has been estimated by Hall et al. (1985). The development of an A_1 horizon and an organically enriched horizon is estimated to take 24 to 30 years; a clayey B horizon 2000 to 2350 years; a cemented layer more than 2000 years; and a well-structured B horizon 250 to 550 years. These rates are more rapid than those estimated for Australia.

Other rates of development of particular horizons are estimated by Jenny (1941). Using glacial moraines, Jenny showed that lowering of pH, nitrogen accumulation and slight clay formation can occur within 50 to 60 years on newly deposited moraine materials. The time for an iron-rich (podsolised) B horizon to develop on sand deposits is estimated at 1000 to 1500 years under a forest. Based on the leaching of carbonates from dune systems, Jenny estimated that calcium carbonate can be leached from a sandy profile within 250 to 350 years.

These rates of soil formation are very dependent on the presence of water and leaching. The rates of

soil formation will be slower in drier areas than in areas receiving higher rainfall. In some special cases, soil formation can proceed at a rapid rate. For example, rapid profile development has been observed on coastal sand dunes after mining for heavy minerals. Paton et al. (1976) noted formation of podzol profiles within five years and development of bleached A_2 material (to 45 cm) within 10 years.

It is thought that the rate of development of topsoil can be increased by additional aeration, incorporation of organic matter and mixing of the upper soil layers by cultivation. Bennett (1939) suggested that topsoil can be developed at the rate of approximately 1 mm per year under such conditions in the USA. The accuracy of this estimate is unknown, but it has formed the basis for the frequently adopted figure of 11.2 t/ha/yr as a maximum allowable soil loss for deep soils of the USA (McCormack and Young, 1981). The concept of whether any soil loss is acceptable is very much open to question and, even if some loss is acceptable, the figures which are quoted are thought to be an order of magnitude too high (Stocking, 1978).

Soil Formation on Unweathered Bedrock

The yield of completely weathered material from rock is extremely variable and depends on rock type, climate and vegetation. There are at least three methods available for measuring this: by weighing prepared rock tablets; by using a micro-erosion meter; or by studying historical monuments. Estimated rates vary through two orders of magnitude, 2 to 200 mm per 1000 years. Values at the higher end of the range were measured on readily soluble limestone (Saunders and Young, 1983). Weathering rates ranging from close to zero to 200 mm of weathering per 1000 years quoted by Selby (1982) may be more appropriate estimates for general Australian conditions.

Jenny (1941) recorded some work by Hilgar who observed that some sedimentary rocks can weather quite quickly. Hilgar noted that 10 to 20 mm fragments of a particular sandstone weathered quite quickly under wetting and drying, and broke down almost completely within 20 years. Corresponding fragments of limestone had only slightly changed in size. However, Hilgar also noted that the chemical composition of the sandstone had barely changed, while that of the limestone had changed considerably. He concluded that the sandstone showed rapid physical weathering but slow chemical weathering, and the limestone slow physical weathering but relatively rapid chemical weathering. Other work recorded by Jenny included estimates of weathering rates using tombstones, which showed that 250 to 500 years was required to weather limestone to a depth of 25 mm, and in the same period, a clay slate barely showed any weathering effects at all. Other work on the weathering of rock materials used in buildings showed that some sandstones were strongly weathered in 100 to 200 years, while others may take 700 to 900 years to be strongly weathered. He did note that a porphyry was only slightly weathered at the surface after 400 years. Jenny maintained that the presence of moisture is a critical factor in the rates of weathering, and that in very dry climates, weathering rates are much slower than in wetter climates.

Beckmann and Coventry (1987) have suggested that soil formation rates by weathering and other processes may be so low as to be virtually unmeasurable, especially in the more arid areas.

An estimate of soil formation rates can be obtained from the general rates of denudation of the landscape over geological time periods. These estimates would indicate that the general rates of denudation are 10 mm/1000 years (see Chapter 2). Assuming that soil materials or saprolite are not accumulating to any great depth, as is the case for many of the more resistant rock types, and that soil formation rates are not accelerated by the bedrock being closer to the surface, then this long-term denudation rate can be assumed to be a long-term soil formation rate. The second assumption that soil formation rates are not increased when the bedrock is closer to the surface may not be valid, but it still appears that soil formation from bedrock is usually a very slow process.

The general conclusion is that the rates of soil formation from bedrock are very slow, and for some

rock types such as hard igneous rocks, extremely slow. Therefore the capacity for bedrock to provide new soil material is very limited in the short term, and that for the very long-term maintenance of soil profiles, erosion rates need to be correspondingly low.

Soil Loss Tolerance Values

To maintain soil quality or productivity in the long term, there has to be a balance between soil formation rates and soil loss rates. Otherwise, there is a long-term degradation of the soil resource, a process that has affected areas in the Mediterranean and Middle East which have a long history of agriculture (Lowdermilk,1953; Wolman,1985; Lal 1990). The Soil Loss Tolerance Value (or T-value) has been used in the USA to predict the level of soil loss that can be tolerated before productivity is affected. The T-value is defined as:

the maximum level of soil erosion that will permit a high level of crop productivity to be maintained economically and indefinitely. (Wischmeier and Smith,1978)

In the USA the T-value is in the range of 2–11 t/ha/yr, the range of values being based on the depth of soil favourable for plant growth and on the assumption that an A horizon can develop in a permeable, medium-textured material in well-managed cropland at about this rate (Hall et al., 1985). The rate of 11 t/ha/yr is approximately equal to 1 mm/yr. Compliance with the accepted T-value is tested using the Universal Soil Loss Equation.

However, while the rate of soil formation of an A horizon from an unconsolidated material might be expected to be 2–11 t/ha/yr, the rate of soil formation from consolidated parent materials such as hard igneous rocks, or dense shales, is much less (Smith and Stamey, 1965; Nowak et al., 1985), and is often much less than 1 t/ha/yr (see previous section). As well, it is the rate of soil formation at the interface of subsoil and consolidated parent material that is the controlling factor in soil renewal and the maintenance of overall soil depth with erosion (Smith and Stamey, 1965). The rate of 2–11 t/ha/yr used as a T-value is likely to be an order of magnitude higher than the soil formation rates from consolidated parent materials (Stocking, 1978).

The T-value can be expected to vary considerably between soils depending on whether the 'new' or 'replacement' soil is required to develop from unconsolidated alluvium, a well-structured subsoil, a dense hard subsoil, saprolite, or a relatively hard unweathered bedrock. A soil on a very deep alluvium, or aeolian deposit, will have a very different T-value from that of a shallow soil on a hard consolidated parent material, or a soil with a shallow surface soil over a dense, coarsely structured subsoil.

The rate of productivity decline of a soil with soil loss has been termed soil vulnerability (Pierce et al., 1983). Thus soils which have only a shallow depth of soil favourable for plant growth can be termed vulnerable soils.

There is little information on T-values or vulnerability of soils to erosion for the soils of New South Wales. Information is required on the depth of soil materials that are favourable for plant growth, and whether the soil material favourable for plant growth overlies alluvium, aeolian deposits, saprolite or colluvial materials. Estimates are required for the rates of change such as:

(i) alluvium to topsoil; rates of topsoil formation will be more rapid for younger alluvial materials than for older alluvial materials, especially those that have been affected by salinity or sodicity

(ii) rock to saprolite

(iii) saprolite to soil

(iv) rock to subsoil

(v) subsoil to topsoil.

Vegetation growth and moisture can increase the rates of topsoil development. In some cases the additions of soil ameliorants such as gypsum, agricultural lime and fertilisers can greatly increase the rates of topsoil formation. Waterponding in combination with plant growth has been used to increase the rate of development of soil favourable for plant growth from dense hard saline subsoils that have been exposed by erosion in semi-arid areas of New South Wales (Cunningham 1987; Rhodes and Ringrose-Voase 1987).

Further information on likely T-values for the soils of this state will be of assistance in the allocation of resources for improving land management, and for assessing the likely impact of land management on the long-term viability of soil resources.

3.2 Soil Loss Rates

3.2.1 Erosion and Soil Loss

While the areal extent of many forms of land degradation has been documented and is being monitored, the actual rates of degradation are largely undocumented. Olive and Walker (1982) provide a list of studies with published erosion rates. They acknowledged that it was not an exhaustive list, but also indicated there was a lack of data from large areas of forest and agricultural land in Australia. Since that publication, more data have been obtained and previously collected data have emerged. In spite of this, there is still a dearth of soil loss data.

Erosion is a continuing geomorphological process involved in landscape development as a smoothing or levelling of the earth's surface by the relocation of weathered material. Geological or natural erosion (that erosion due only to the forces of nature) needs to be distinguished from accelerated erosion (resulting from human activities). The same processes operate in each case and the distinction is often only a matter of degree and rate. It should be remembered that natural erosion events can also be catastrophic, but that such events occur only during the natural coincidence of specific conditions which rarely happens in the time scale humans can witness directly. It is possible for some land use changes to have an equivalent effect on erosion rates as major climatic changes or changes of base levels have had in geological history.

For the purposes of discussion, it is necessary to differentiate between the terms *soil erosion*, *soil loss* and *sediment yield*. These terms are frequently interchanged yet each has its own meaning. Soil erosion operates in conjunction with mass movement and solution to denude the earth's surface of weathered material, with the relative importance of each process at any point depending on the particular environment (Kirkby, 1980). Soil loss defines the net removal of soil from a surface by a combination of erosion processes. Sediment yield represents the amount of eroded soil delivered to a point or reference as a result of erosion, deposition and transport processes.

Wind and water are the two major agents of erosion in Australia. Discussion in this chapter focuses on reported erosion rates by water (or undifferentiated combinations of wind and water). It must be recognised that wind can be the dominant form of erosion in some parts of Australia. Wind erosion has been less researched and, consequently, fewer data are available than for water erosion (see McTainsh and Leys, 1993).

During a typical erosion event, soil particles are removed by raindrop detachment and/or runoff. The measured soil loss is the amount of eroded material moved off the particular area. The amount of soil lost is, however, less than the amount of soil eroded because some of the detached material is subsequently deposited within the area. The observed sediment yield is the amount of material moving past a given point of interest. This may be a measuring device in a stream for large, catchment-size experiments, or a collection trough for a plot scale experiment. The measured sediment yield is dependent on the gross amount of erosion in a catchment and on the efficiency of transport of eroded material out of the catchment.

Material may be redeposited within the paddock from which it was eroded owing to filtering by vegetation or litter and settling in depressions. Sediment may also be deposited where changes in slope or obstacles cause the velocity of the runoff water to decrease. Such features result in a decrease in the ability of the runoff to transport sediment and, hence, deposition occurs.

At least four different methods can be used to estimate the rate of loss of material from the land (Young, 1969). The measurement of the suspended and dissolved material transported by rivers and the measurement of sediment accumulated in reservoirs use sediment yield to derive soil loss. The method of direct measurement of soil loss uses techniques to investigate surface processes on slopes, including rates of soil creep, surface wash and landslides. This is achieved by using plot scale experiments and either simulating erosion events artificially or monitoring natural erosion events. A fourth method is the comparison of the ages of soil materials. A number of approaches have been developed which utilise different dating techniques based on natural or artificial radionucleotides and mineral/magnetic tracers. This approach may give short-term residence times of sediments (for example, thermoluminescence) or long-term averages to determine periods of landform change (for example, radiocarbon dating).

This chapter concentrates on contemporary water erosion rates using data from plots and small catchments. Reference to yields from larger catchments is also made. Most of the evidence available to us in Australia utilises the measurement of soil loss from plots and hillslopes, although the other methods also have been used.

Rates of soil loss and sediment yield have been reported in a range of units. For ease of comparison the reported rates are presented here in t/ha unless otherwise specified. To convert the reported fig-ures to depth of soil in millimetres, the figure in t/ha may be divided by 14 (assuming a soil bulk density of $1.4 \ g/cm^3$).

Care must be taken in considering the presented values as they imply a uniform rate of removal both temporally and spatially. Erosion events have been observed to be highly variable in magnitude, space and time. For a given event, most of the erosion tends to occur in a limited proportion of the total reference area or in a small fraction of the overall observation time. Generally speaking, most soil loss occurs in relatively few but major events; that is, most erosion events contribute negligible amounts of sediment and a few contribute the majority (Edwards, 1980).

This variability in temporal patterns of soil loss means that extrapolating soil loss from short periods of record is prone to error. For example, the meteorological conditions experienced during a period of study may have been atypical of the area and hence soil losses recorded during that period will not be indicative of the 'real' (long-term) conditions at that site. Similarly, because of the high spatial variability in soil loss, great care must be exercised when extrapolating data to other sites.

A number of factors have been identified as controlling soil erosion. These factors include rainfall erosivity, soil erodibility, slope length and steepness, and management practices imposed on the land (Wischmeier and Smith, 1978). The latter factor mainly exerts its control through modifying the soil surface condition and amount and type of protective material at the ground surface, particularly in agricultural and forestry environments. The level of ground cover and the amount of soil disturbance are most important factors in determining changes in the amount of erosion experienced at a site. Rainfall erosivity determines absolute levels of soil loss and is seen to vary widely throughout New South Wales (see Figure 3.3). Erosivity increases to the north and also increases from inland areas to coastal areas.

Rates of removal of soil material may be expressed as depth of soil (for example, mm), volume of soil per unit area (m^3/km^2), or weight per unit area $(t/ha \ or \ g/m^2)$.

If a bulk density of material of $1.4 \ g/cm^3$ is assumed, then:

1 mm depth	=	14 tonnes per hectare
1 mm	=	1000 cubic metres per square kilometre
1 m^3/km^2	=	0.014 tonnes per hectare
1.4 t/km^2	=	1.4 tonnes per square kilometre

Figure 3.3 Isoerodent map of New South Wales in SI units (MJ. mm)/ha. h.y (Rosewell and Turner 1992)

3.2.2 Plot Studies

Plots have been widely used by soil conservationists and geomorphologists to assess rates of surface soil movement. Enclosed plots, that collect all the runoff and sediment loss from a defined area, and open plots, with no defined boundaries, have been used for this purpose. Open plots use collection devices such as Gerlach troughs that collect the sediment from the upslope area (see de Ploey and Gabriels, 1980, for a discussion on sediment measurement). The usefulness of closed plots may be limited by their length being too short to allow generation of overland flow in some instances. Open plots have an undefined and variable catchment area, which can make interpretation of results difficult.

Bare Fallow Plots

On bare fallow plots, soil loss is measured from areas of soil that are exposed to natural rainfall and maintained free of vegetation and in a cultivated condition. This is generally thought to represent the greatest erosion hazard. Measurements from such plots have been made by the Soil Conservation Service in New South Wales (Armstrong, 1990; Rosewell, 1992) and by CSIRO (Kinnell, 1983) in the Australian Capital Territory at Ginninderra. The plots were of 0.01 ha in area with a slope length of 41

m. Observed losses generally ranged from 30 t/ha/yr to 90 t/ha/yr (see Table 3.1).

Cropped Plots

The potential levels of soil loss from bare fallow plots are expected to be reduced by the presence of vegetation cover on the plots. A major study conducted in New South Wales measured soil losses from closed plots of 41 m length for periods of up to 33 years (Edwards, 1987).

The plots were either permanent native pasture used in grazing trials, improved pasture or sown with crops. The sites studied were at Cowra, Gunnedah, Inverell, Wagga Wagga and Wellington, all in the wheat-belt to the west of the Great Dividing Range, as well as at Scone in the Hunter Valley.

When the mean annual soil losses from all the cropping plots are considered together (2040 plot years of data), the mean soil loss per plot was 2.40 t/ha with a standard deviation of 3.25 t/ha (see Table 3.2). The median value was 1.06 t/ha which is significantly less than the mean, indicating, as suggested earlier, that more soil loss resulted from fewer events. There was a significant difference between the losses from the summer cropping plots and those from the winter cropping plots, the mean value from the former being about five times greater than from

Table 3.1 Soil loss from bare fallow plots (t/ha)

Location	Soil	Slope (%)	Years	Annual soil loss (t/ha)
Cowra	Dr2.22, Non-calcic brown	8	3.5	31.3
Ginninderra	Dy, Yellow podzolic	4	2.0	44.0
Gunnedah	Ug5.13, Black earth	8	5.7	87
Inverell	Uf6.21, Chocolate soil	9–11	8.0	51.4
Wagga Wagga	Dr2.32, Red-brown earth	8	2.5	59.6

Table 3.2 Annual soil loss from cropping plots (t/ha)

	Winter Crops	Summer Crops	All
NUMBER OF PLOTS	78	12	90
MEAN	1.52	8.11	2.40
STD DEVIATION	1.66	4.98	3.25
MEDIAN	0.97	9.58	1.06

Plate 3.1 A rainfall simulator used for plot studies of runoff and soil loss under controlled conditions

the latter. The median value for summer crops was about 10 times greater than for the winter cropping plots. The cropping effect was confounded with the sites as summer crops were only grown at Gunnedah and Inverell while winter crops were grown at Cowra, Wellington and Wagga Wagga, as well as at Gunnedah and Inverell.

Pasture Plots

Mean annual soil losses from pasture plots were considerably lower than those from the cropping plots and, within the pasture plots, differences were also apparent. Based on 1700 plot years of data from 80 plots, the mean annual soil loss from the pasture plots was 0.24 t/ha, but this was highly variable with a standard deviation of 0.40 t/ha. Median annual soil loss from the plots was 0.05 t/ha and this figure is thought to give a more reliable indication of the 'normal' rate of soil loss and again shows the skewed distribution of data with few large events affecting the average (but not the median) value. The maximum mean annual loss recorded was 1.87 t/ha and the minimum mean value was zero.

These pasture plots can be separated into three different classes: the first class consists of those plots that had experienced at least one cultivation (at the time of establishment or re-establishment); the second class being those plots that were virtually undisturbed by either cultivation or grazing (that is the nil or light grazing plots from the grazing treatments); and the third being those plots from the grazing experiments that fell into the moderate or heavy grazing category. The plots with minimal disturbance differed significantly from those that had received a cultivation, but neither group differed significantly from the moderately disturbed group, as shown in Table 3.3. Again, as for the collective analysis, for each group of plots the median value is less than the mean indicating a dominant effect of larger events.

Table 3.3 Annual soil loss from pasture plots (t/ha)

	Cultivated	Moderate Disturbance	Minimal Disturbance	All
NUMBER OF PLOTS	33	28	18	79
MEAN	0.37	0.22	0.05	0.24
STD DEVIATION	0.51	0.34	0.06	0.40
MEDIAN	0.10	0.03	0.02	0.05

Other Quantitative Plot Studies

In a study comparing rates of surface processes in tropical and temperate climates on soils formed from sandstone and granite parent materials, Williams (1973) found no significant difference in rates between the different lithologies or climates. Steel collection trays 45 cm long on unbounded plots were used for recording sediment movement (see Table 3.4).

Soil losses in two small pasture catchments in the Nogoa watershed, in central Queensland, were measured over a three-year period with annual soil losses ranging from 0.01 to 21.1 t/ha with a median value of 0.46 t/ha (Ciesiolka, pers. comm.) Rates of loss recorded in individual, modified, 1 metre-wide Gerlach troughs were influenced by geomorphic factors and pasture cover. For example, where ground cover was 10% or less, the median annual soil loss was increased to 10.9 t/ha.

Further north in canefields, losses of the order of 380 t/ha were reported over one wet season while, in more recent trials, losses ranged from 70 to 150 t/ha (Prove, 1984). Lower rates have been achieved through the use of differing cultural practices, for example less than 5 t/ha (no burn and no cultivation) to less than 23 t/ha (no cultivation plus burn) in one season.

Humphreys and Mitchell (1983) assessed a range of different soil surface processes at a number of natural sites on sandstone hillslopes in the Sydney Basin. Rainwash was determined using catch trays with no plot boundaries. Relatively high rates of soil movement were recorded at Blackheath which were found to be associated with lyrebird activity (see Table 3.5). It was shown that such faunal activity can result in significant downslope movement of soil.

Pilgrim et al. (1986) measured rates varying between 0.1 and 0.4 t/ha/yr on woodlands on deep weathered granite in a 400 mm annual rainfall area in south-west Western Australia. Four unbounded plots were used for measurement over a period of 6.2 years and their mean annual soil loss was 0.3 t/ha/yr. Slope angles of the plots ranged from 1.0 to 13.0 degrees.

In a study of rehabilitation of coal mine sites in the Hunter Valley of New South Wales, Elliott and Dight (1986) found a range in soil loss from 0.4 to 3.0 t/ha/yr for revegetated plots. On areas of native pasture, losses were 0.9 to 1.0 t/ha/yr, while on bare overburden, rates averaged 11.8 t/ha/yr. Plots were 2 m² in area and the observations were made over a five-year period.

At the Nabarlek mine site in the Northern Territory, Duggan (pers. comm.) reports soil losses ranging from 20 to 102 t/ha/yr on embankments over a period of one wet season. Plots were 2 m² in area on a slope of approximately 17 degrees. The lowest rates were recorded from rock mulched plots and the highest from untreated plots.

Using five plots, each 2 m² in area, Elliott et al. (1984) recorded mean annual soil losses over a 3.5-year period between 0.015 and 16.03 t/ha/yr. The

Table 3.4 Soil loss due to slopewash (after Williams, 1973)

Location	Slope (degrees)	Slope Length (m)	Years	Soil Loss* (t/ha/yr)
GRANITE, NT	1–3	60–291	2	0.76 (0.17–2.37)
SANDSTONE, NT	1–15	27–121	2	0.78 (0.27–1.41)
GRANITE, NSW	2–14	25–226	3	0.76 (0.02–4.07)
SANDSTONE, NSW	3–25	18–290	3	1.44 (0.17–4.32)
* Average value is given; range in brackets				

Table 3.5   Soil movement on Sydney Basin sandstone hillslopes (after Humphreys and Mitchell, 1983)

Location	Soil	Slope (degrees)	Slope Length (m)	Years	Soil Loss* (t/ha/yr)
CATTAI	Dy	4.4–7.5	11–38	2.7	0.04 (0.02–0.08)
	Uc				0.25
CORDEAUX	Uc	1–4.5	63–100	2.7	0.32 (0.04–0.68)
BLACKHEATH	Uc	3–26	5–30	2.0	0.81 (0.19–29.66)
LIDSDALE	Dy	0–10	5–180	2.9	0.86 (0.07–6.97)

* Average value is given; range in brackets

plots were located in both pasture and cultivated areas in the Hunter Valley of New South Wales.

Based on observations covering one year, Blong et al. (1982) found sediment yields following bushfires equivalent to 2.4 to 8.0 t/ha on closed plots 4 m long by 2 m wide. The site, near Narrabeen Lagoon, Sydney, had duplex soils with deep sandy topsoils and the authors stipulated that the drought conditions prevalent during the observation period may have resulted in lower than potential erosion.

Following bushfires in the Royal National Park south of Sydney, Atkinson (1984a) reported soil losses of 2.2 to 7.5 t/ha for one event and losses of 22 to 40 t/ha for another event. Closed plots of 9 m² located on shallow sandy soils (Uc5.22) on slopes of 6–13% were used in this study. Over a 12-month period, total soil losses ranged from 39.6 to 64.2 t/ha (Atkinson, 1984b).

In a different type of study using historical information and making estimates of soil loss based on field inspection, rates of 21.7 t/ha/yr over a 50-year period were inferred (Ring, 1982). These rates are from a 90 ha area (6% mean slope) of former cropping country near Wagga Wagga.

Rates of soil loss up to 209 t/ha/yr were reported for arid rangeland by Fanning (1994). Rates of soil loss were determined using erosion pins on a severely eroded surface in a small (19 km²) catchment in western New South Wales over a 10-year period. The highest rate was measured on rilled surfaces, 59.5 t/ha/yr on flat surfaces and 30.6 t/ha/yr on vegetated hummocky surfaces.

Use of Radioactive Isotopes

Caesium-137 is a product of atmospheric nuclear testing. It becomes attached to clay and silt particles. Any detectable changes in caesium concentration in the soil beyond that expected due to radioactive decay may be attributed to erosion or deposition (Ritchie and McHenry, 1990). Elliott et al. (1984) demonstrated that for soils in which net erosion is occurring, the changes in content of caesium-137 in

the original soil and sediment are highly correlated with that expected for the observed soil loss.

By measuring concentrations of caesium-137 in soils in southern central Queensland, Reece and Campbell (1984) inferred that clearing trees and shrubs from pasture lands increased erosion by 8.2 t/ha/yr over that from uncleared, ungrazed country. Grazing of uncleared areas led to increased losses of 12.4 t/ha/yr in the same period. The loss from the uncleared, ungrazed country is unknown. These rates appear relatively high for those commonly expected for grazing lands.

A survey of erosion in New South Wales was based on measurements of the caesium-137 technique (Elliott et al., 1997). The survey measured caesium-137 levels along selected toposequences across a range of soils and land uses, and showed that the caesium-137 reference value was related to average annual rainfall. The estimated net soil losses from slopes under different land uses and in different regions from Elliott et al. are summarised below. These values are based on net soil losses from a slope, and do not give the movement of soil within the slope. Values of soil loss at some particular points are called spot soil losses.

(i)   potato growing on north coast
      11.6 to 41.2 t/ha/yr

(ii) vineyard in Hunter Valley
      9.0 to 13.8 t/ha/yr

(iii) wheat/pasture rotation on north-west slopes
      1.3 to 8.4 t/ha/yr; spots to 17 t/ha/yr

(iv) wheat/pasture on far north-west slopes (North Star)
      2.6 to 10.3 t/ha/yr; spots to 27.1 t/ha/yr

(v)   wheat/pasture on central west slopes
      0.87 to 1.24 t/ha/yr; spots to 2.4 to 6.2 t/ha/yr

(vi) wheat/pasture south-west slopes
      0 t/ha/yr; spots to 24 t/ha/yr

(vii) grazing central plains (Cobar)
      7.3 t/ha/yr; spots to 43 t/ha/yr

(viii) grazing southern tablelands
        0.05 to 0.54 t/ha/yr; spots to 5.13 t/ha/yr
(ix)   grazing central west slopes
        0.15 t/ha/yr
(x)   grazing Hunter Valley
        0.19 to 1.51 t/ha/yr; spots to 12 t/ha/yr
(xi)   grazing north coast
        0.07 to 0.13 t/ha/yr
(xii)   grazing north-west slopes
        0.51 to 1.20 t/ha/yr
(xiii) grazing north coast
        0.48 t/ha/yr

These values are based on a few selected slope transects and so there are some limitations in using them to predict erosion rates under the various land uses in these areas. However, the values are in general agreement with the figures presented earlier, using other techniques to measure erosion.

3.2.4 Occasional Observations

Occasional observations of soil loss tend to provide overestimates of long-term soil loss rates. Such observations are generally made subsequent to a major erosion event and the lower levels of losses in the preceding and subsequent periods tend to remain unknown. The following paragraphs provide a number of such observations and it should be considered that these estimates provide soil loss rates in the upper range of that expected.

Over a period of three months on potato lands in the Donnybrook area, Western Australia, McFarlane (undated) recorded losses of 140 to 686 t/ha (10 to 49 mm). These lands were cultivated up and down slopes ranging from 7.8% to 20.0% (mean 13.4%) and slope lengths ranging from 50 to 120 m. Soil losses of 350 t/ha were recorded in cultivated land at Inverell, New South Wales, following a series of storms (Harte et al., 1984). This estimate was obtained by measuring cross-sections of rilled ground. McFarlane and Ryder (1986) determined losses for cultivated paddocks as being from 32 to 350 t/ha for a single storm falling on freshly seeded soils in Western Australia using a similar technique. From deposition in alluvial fans, losses of 133 to 307 t/ha were calculated for the same storm for different sites.

The sorting effect the erosion process can have on the distribution of sediment is emphasised by the results of Marschke (1986), who observed coarser materials transported by saltation and fine sediments carried as suspended load. Observations were made using contour bay plots of approximately 1 ha. Soil loss under a stubble-burning tillage practice in one event was 26.5 t/ha, of which the suspended load was only 1.59 t/ha. Under zero tillage, total loss was 0.6 t/ha for the same event, with a suspended load

of 0.03 t/ha. The suspended load was determined by grab samples at a flume while the bedload was assessed by deposition in the contour bank channel. At Hay in semi-arid New South Wales, soil loss recorded from one event was 0.46 t/ha from an uneroded site while at an adjacent scalded site, soil loss for the same event was 8.65 t/ha (Alchin, 1983).

Erosion following a major storm at Cowra in central western New South Wales was characterised by Hairsine et al. (1993). During the storm 81 mm of rain fell in 45 minutes, giving an average intensity of 108 mm/h, but the maximum intensity during the storm was 360 mm/h. The rainfall erosivity of the storm was 3430 MJmm/(ha.h), compared to the long-term average annual rainfall erosivity of 1120 MJmm/(ha.h) for Cowra. This was a very infrequent storm based on existing rainfall data.

The paddock where the erosion was measured had a long-term tillage trial with traditional tillage (TT), reduced tillage (RT) and direct drill (DD) treatments. The TT and RT treatments had been recently cultivated and were in a loose, easily detached condition with soil particles being easily picked up by flow (entrainment). The amount of erosion was estimated using transects which measured the width and depth of rills. In areas of sheet erosion, the depth of sheet erosion was estimated by measuring the height of remnant pedestals.

Erosion rates for the loosely tilled soils were 342 t/ha and 364 t/ha for the TT and RT treatment respectively, and for the untilled DD treatment the erosion rate was 65 t/ha.

Conclusions from this study were the following.

(i)   The very high erosion rates from a single event show that individual erosion events can be a major cause of erosion of soils. The long-term average annual soil losses estimated using the SOILOSS (Rosewell, 1993) were 46, 42 and 28 t/ha for the TT, RT and DD treatments respectively, values much less than the erosion from this single storm. If a highly erosive storm occurs when the soil is in an erosion-susceptible condition, high erosion losses can be expected.

(ii)   Differences between uneroded soils and soil material deposited within the paddock can be substantial, with much of the clay, silt and very fine sand being removed from the paddock in the erosion process.

(iii)   Estimates of nutrients lost from the paddock in the finer soil fractions were 290 kg/ha, 315 kg/ha and 42 kg/ha of nitrogen for the TT, RT and DD treatments respectively; 15.8 kg/ha, 12.1 kg/ha and 3.5 kg/ha of phosphorus for the treatments respectively; and 2.8 t/ha and

2.6 t/ha of soil carbon for the TT and RT treatments respectively. Apart from these nutrients being worth in the vicinity of $230/ha in 1993 prices for the TT and RT treatments, these are nutrients that have been mobilised and can be delivered to streams and water bodies within the catchment. Although all the nutrients may not be delivered directly to streams because of deposition on lower slopes and within drainage depressions, the mobilisation of these nutrients represents a threat to water quality within the catchment.

A further study mapped the development of rills on a cultivated paddock following an extended period of low intensity rainfall (Murphy and Flewin, 1993). Soil loss was estimated by measuring the volume of rills and the bulk density of the soil along transects across the eroded area. A total of 199 mm of rain fell over a period of 80 days, with maximum intensity 20 mm/h, and the rainfall erosivity was 191 MJmm/(ha.h). Total soil loss was 48 m³/ha or 78 t/ha.

Conclusions from this study were the following.

(i)   The rates of erosion seemed to be very high for the relatively low rainfall erosivity. Low infiltration rates caused by surface sealing and soil saturation resulted in high runoff rates that contributed considerably to the measured high rates of erosion. The rates of erosion were also attributed to the soil being in a loose cultivated condition for much of the period, which greatly increased soil erodibility.

(ii) The measured erosion was much higher than the long-term annual average erosion rate estimated using SOILOSS (Rosewell, 1993). As Rosewell points out with regard to the USLE or SOILOSS program:

.. the rainfall factor does not adequately express hydrology, especially antecedent moisture conditions, as they affect peak rate and runoff. Similarly the erodibility of a soil will vary widely with different moisture, tillage, biological and chemical factors.

The USLE and SOILOSS programs can be used to predict long-term average soil loss for different areas, but they cannot predict erosion for individual, short-term rainfall or storm periods.

(iii) The need for conservation farming practices to control erosion of cropping land was demonstrated here, as erosion was significant in lengths as short as 20 to 30 m, and some rills extended to the foot of the erosion bank installed at the top of the area on which the erosion was measured. Based on this data, erosion control banks alone are inadequate to

control erosion to a sufficient degree to prevent long-term soil loss under cultivation.

(iv) Soils that have low infiltration rates are likely to have more frequent erosion events because of the increased occurrence of runoff events.

Catchment Studies   3.2.5

Olive and Walker (1982) provide a summary of suspended sediment yield data for Australian rivers, and the reader is referred to that summary for details. Of the 38 studies reported therein, 26 have sediment yields of less than 1 t/ha, nine have yields of 1 to 2 t/ha, and the remaining three have yields in excess of 2 t/ha. Mostly the catchments are in excess of 1000 ha, with more than half in excess of 5000 ha and one quarter in excess of 50 000 ha.

On much smaller catchments in the eastern Darling Downs of Queensland, the effect of alternative stubble management practices on soil loss has been demonstrated at two sites (Freebairn and Wockner, 1986a, b). On a black earth at Greenmount, mean annual soil loss over an eight-year period ranged from 2.1 t/ha/yr for zero tillage to 61.0 t/ha/yr when bare fallows were maintained prior to sowing of the crop. Stubble mulching reduced soil loss to 5.3 t/ha/yr, stubble incorporation to 17.9 t/ha/yr, and summer cropping with a winter fallow 22.3 t/ha/yr. At a second site, Greenwood, over a six-year period losses were 31.6 t/ha/yr (bare fallow), 19.8 t/ha/yr (summer crop), 7.8 t/ha/yr (stubble incorporated), 3.9 t/ha/yr (stubble mulch) and 1.8 t/ha/yr (zero tillage). This site had a grey clay soil. Catchment areas for these studies were confined between contour banks and ranged in size from 0.7 to 1.4 ha. Total soil movement was estimated using the sum of coarse sediment deposited in the contour channels and the extrapolated load of suspended sediment discharged from the catchment. The calculations indicated that 80–90% of the sediment was deposited in the channel.

An extension of this work to Capella, central Queensland, shows conflicting results. Sallaway (pers. comm.) obtained data from field-sized areas (10 to 16 ha) planted to three crop types using three different tillage practices. Data for four cropping seasons (1982–86) include a very wet autumn in 1983 which generated increased runoff and soil loss for the zero tillage treatment (see Table 3.6). The results for a more 'normal' year are shown in Table 3.7.

Mean annual soil losses recorded using sediment traps in pineapple fields in Nambour, south-eastern Queensland, ranged from 7.0 to 36.4 t/ha (Capelin and Truong, 1985). These results were obtained over a four-year period on red podzolic soils on lands with slopes of 11–17%.

Table 3.6 Soil loss (t/ha), Emerald, Queensland, 1982–86

Crop	Zero Tillage	Reduced Tillage	Conventional
Sunflower	4.66	3.61	9.92
Sorghum	6.81	2.89	5.85
Wheat	1.11	2.84	4.35

Table 3.7 Soil loss (t/ha), Emerald, Queensland, 1985–86

Crop	Zero Tillage	Reduced Tillage	Conventional
Sunflower	1.30	4.75	8.82
Sorghum	4.29	6.29	16.13
Wheat	0.00	0.35	2.53

On rural catchments used for grazing, in the New England region of New South Wales, suspended sediment yields ranging from 0.02 to 0.12 t/ha were measured in a full year (Field, 1984). The coarse sediment component was observed to be insignificant. The catchments ranged in size from 764 to 1364 ha.

On a rural catchment at Collie, in Western Australia (94 ha, cleared and sown to improved pasture), Abawi and Stokes (1982) estimated annual soil losses to be between 1.5 and 2.0 t/ha over a four-year period. Coarse sediment load was estimated at 1.3 t/ha/yr.

Hartley et al. (1984) investigated land use effects on sediment yield in a study in the Adelaide Hills, South Australia. Suspended sediment yields for one year were estimated to be 2.4 t/ha for a 410 ha catchment used for intensive horticulture, 0.5 t/ha for an 830 ha urbanised catchment, and 0.2 t/ha for a 500 ha catchment covered with virgin scrub. In this study, the authors acknowledge that the recorded yields were less than the total movement of soil within the catchment because eroded material larger than 5 microns would have been redeposited within the catchment.

The effect of disturbance associated with mining activities may be gauged using data from the Northern Territory. Reported sediment yields were estimated as 24.6 t/ha/yr in the first year following the initial disturbance, declining to 0.6 t/ha/yr in the fourth year (Duggan, pers. comm.) This study catchment was 22 ha in area. The soil loss figures from the fifth year onwards were found to be similar to those from larger (1600 ha) natural catchments with annual losses of 0.4 t/ha/yr which are thought to represent natural, background levels.

Mean annual sediment losses from small catchments at Wagga Wagga, New South Wales, varied from 0.02 t/ha/yr on a 7.5 ha catchment treated with improved pasture and soil conservation structures, to 1.65 t/ha/yr on a 7.3 ha catchment with native pasture (Adamson, 1974). The reduced sediment loss is attributed to the improvement of ground cover and the removal of gullies as a source of sediment. Soils were yellow solodic soils (Dy2.42) and mean catchment slope was 12.5% for the treated catchment and 10.6% for the untreated catchment.

From two small catchments in the Nogoa catchment in central Queensland, soil losses ranged from 1.6 to 14.2 t/ha in one catchment (9.6 ha), to 10 to 113 t/ha in the other (12 ha) over a three-year period (Ciesiolka, pers. comm.) In the 12 ha catchment, it was estimated that at least 94% of the sediment was derived from a gully and the remainder from hillslopes. The contribution from hillslopes was much greater in the 9.6 ha catchment and was estimated to be between 31% and 80% of the annual load.

Similarly, sediment yield from two catchments in the Hunter Valley of New South Wales contrasted in the relative contribution of coarse sediment load and suspended sediment load. In a comparable six-year period, coarse sediment load from one catchment was 9.0 t/ha/yr and suspended load was 2.0 t/ha/yr while on the second catchment, coarse sediment load was 0.4 t/ha/yr and suspended sediment load was 0.7 t/ha/yr (Lang, 1984). Both catchments were 40 to 50 ha in area and were used for grazing sheep.

Costin (1980) reported soil loss from an 88 ha catchment in the Australian Capital Territory ranging from 0.004 to 0.376 t/ha over a four-year period, with an average of 0.191 t/ha. Most of the soil loss was in fine suspension with little as coarse sediment load or coarse floating debris. These yields were from a catchment with improved pastures sown on soils described as red and yellow podzolic soils with a slope of 3.7%.

3.2.6 Sediment Surveys of Reservoirs

Water storage structures act as sediment traps and provide a record of erosion over time. Using the estimated volume of sediment in the structure, the loss of soil per unit area of the contributing catchment can be calculated. Two major problems are associated with this method. They are the unknown amount of sediment that is flushed through the system and not trapped in the reservoir (trapping efficiency), and the contribution of sediments resulting from shore erosion due to wave action.

Abrahams (1972) reported sediment yields based on rates of accumulation of sediment in 14 storages in eastern Australia. These volumetric rates have been converted to a gravimetric basis and summarised by Olive and Walker (1982) using a specific gravity of 2.5, but a lower value appears more appropriate. Consequently, values in the range of 0.20 to 3.78 t/ha/yr are indicated. As noted by these authors, the values are likely to underestimate the real values. Most of the coarser sediments would have been trapped by the storage, but only an unknown proportion of the suspended load would have been trapped.

Wasson and Galloway (1984) showed that rates of erosion in the Umberumberka Creek catchment near Broken Hill, in western New South Wales, have declined during this century. From the time of construction of a reservoir in 1915 to the early 1940s, the average erosion rate in the 42 000 ha catchment was estimated to be 5.2 t/ha/yr. In the subsequent period to 1982, the level of erosion reduced to 2.0 t/ha/yr. Climatic factors are thought to have played a major part in reducing the rate of erosion.

Bishop (1984) indicates much lower erosion rates for the Warragamba Dam catchment that forms Sydney's major water supply. An overall rate of 0.04 t/ha/yr is estimated based on the rate of sedimentation of the reservoir over three years. The total area of that catchment is 784 000 ha. In a smaller sub-catchment of 45 000 ha within that area, a rate of 0.32 t/ha/yr was estimated which shows the extent of local variability.

By studying the volume of sediment deposited in farm dams in small rural catchments in south-eastern New South Wales, Neil and Fogarty (1991) found sediment yields in the range from 0.08 to 0.76 t/ha/yr in catchments ranging from about 10 to 1000 ha. A survey of town water supply dams in New South Wales showed such structures have trapped an average of 1.58 t/ha/yr (Morse and Outhet, 1986). This survey covered 12 dams with an average catchment area of 22 400 ha.

3.2.7 Erosion Rates Following Bushfires

A significant amount of data has been compiled on the rates of soil erosion following fire, overseas and in Australia. In an attempt to quantify the effects of fire on runoff and erosion, Zierholz (1997) compiled a comparison of 51 post-fire erosion studies of various scales. This comparison expressed erosion rate as a percentage increase over unburned (background) erosion rate. The comparison was then used to identify variables thought responsible for the magnitude of erosion response observed at the different scales of measurement and is summarised in Table 3.8.

This summary illustrates the variability in erosion rates following fire. A number of factors were considered to contribute to the observed variation. These factors include fire intensity, rainfall intensity, geomorphology and the dominant erosion processes observed. Fire intensity is important because it determines the amount of soil cover burned and the heating of the soil. The combined effects of these processes expose the soil to erosive forces, decrease infiltration and reduce surface roughness. Rainfall intensity and timing is important as it determines the scale and magnitude of the erosion event and when it occurs during the recovery phase. Generally, the sooner after the fire the event occurs, the greater the erosion potential. Landscape factors such as soil type, vegetation, terrain and drainage patterns are also important as they control the soil erodibility and type of erosion process, as well as the transport routing for sediments.

It should be noted that the effects of fire on soil erosion do not affect soil erodibility as cultivation does. The key role fire has in accelerating erosion is in the removal of soil cover and the alteration of soil hydraulic characteristics. Comparing observations of erosion response at plot scale indicates that cultivated soil is much more prone to erosion while burned but otherwise undisturbed areas are relatively resistant. At catchment scale, however, widespread high intensity bushfires can have the effect of leaving large areas exposed and capable of generating significantly greater volumes and rates of runoff. Cultivation tends to be discontinuous along paddock and property boundaries and leaves some sort of riparian buffer zone where sediment may be trapped.

3.2.8 Rates of Wind Erosion

The rates of wind erosion are less well documented because of the difficulties of measuring wind erosion (McTainsh and Leys, 1993; Miles and McTainsh, 1994). However, the measures of wind erosion made using wind erosion tunnels give fluxes of over 100 g/m/s (Leys et al., 1994). This shows the large potential for wind erosion to erode soil and indicates

Table 3.8 Factors associated with trends in erosion response following fires

	Response (% increase in Sediment Yield)							
	Not significant (0–10%)		Low to medium (10–500%)		Medium to high (500–10 000%)		Catastrophic (> 10 000%)	
Scale of study	Plot	Catchment	Plot	Catchment	Plot	Catchment	Plot	Catchment
Fire intensity	low-intensity control burns	low or medium patchy burns	low to hot control burns	control burn, patchy	moderate to intense control burns, wildfires	wildfire	annual burning	wildfires, complete burning of catchm
Rain or Runoff observed	low to medium	average	up to 70 mm/h simulated rainfall	average	high rain intensities	medium events	14.2% runoff —no intensity given	large rain events
Geomorphology/ Environment	slopes to 25%, woodland forests	various	woodland shrubland medium slopes	mountain Fynbos, medium slopes	medium to steep slopes, chaparral, forests	mountain Fynbos, medium slopes	hardwood forest, chaparral woodland	steep, mountain terrain, forested/ chaparral
Type of Erosion	lack of runoff	lack of continuity	sheet erosion	sheet erosion	sheet and rill erosion	sheet, rill and gully erosion		gully eros debris flo
Examples	Gilmour and Cheney, 1968;Versfeld, 1981	Biswell and Schults, 1976	Imeson et al., 1992; Lavee et al., 1995	Scott and van Wyk, 1992	Atkinson, 1984b; Debano et al., 1979	Scott and van Wyk, 1990	Adams et al., 1947; Copley et al., 1944	Good, 197 Brown, 19 USDA, 19!

that soil may be moved from paddocks in the order of tonnes/ha per hour under strong wind conditions on a susceptible soil (Leys and Heinjus, 1992).

However, actual measures of wind erosion are usually much lower because natural wind velocities are usually less and more variable, as are the soil surface conditions when the winds are blowing. Fluxes measured in wind conditions on a paddock varied from 15.2 kg/m/week (total flux up to height of 2.3 m) to 4.2 kg/m/week (dust fraction less than 90 mm), depending on the height and particle sizes measured (Leys and McTainsh, 1994). This gave an erosion rate of 4.2 tonnes of dust leaving the paddock in a period of one week.

Measures of wind erosion on mulga lands in south-western Queensland give erosion losses of 2.8 to 5.2 t/ha/yr on hard mulga soils with a crust, to 89.5 t/ha/yr on soft mulga soils (Miles and McTainsh, 1994).

A major concern with wind erosion is that regardless of actual wind erosion rates, wind erosion removes the finer soil particles as dust, and so removes much of the nutrient store in the soil. The removal of this finer fraction also reduces the capacity of the soil to store nutrients and to store water (Leys and Heinjus, 1992; Leys and McTainsh, 1994).

Wind erosion may also have important environmental consequences because of dust and visibility problems, and the potential for chemicals such as pesticides to be transported on the eroded dust particles.

3.3 Conclusions

This chapter has presented a range of studies that have reported rates of soil formation and erosion. The data collated here are by no means a complete list of the available literature but a fair cross-section of the research that has been carried out in this field.

When summarising the available data on soil formation and erosion rates, a number of considerations need to be made. A broad range of methods has been used by different researchers, so the results are not directly comparable. Results have been obtained over different periods of time and, as highlighted earlier, short periods of records can lead to biased results. Also, the catchment areas on which observations were made varied enormously, from 2 m² to hundreds of thousands of hectares. The relative importance of the various types of erosion processes tends to change depending on the scale of observation. Overall sediment yield per unit area tends to decrease as catchment size increases because the number of sediment sinks also increases. In short, the sediment transported out of a large catchment is less than the amount of soil eroded within the catchment. At the lower end of the scale, soil loss rates may increase with plot/hillslope size, as an increase in catchment size at that scale provides greater opportunity for concentrated flow. Small plots tend to be transport limited (sheet erosion only) while large plots have increased capacity for detachment and transport of sediment (rill erosion).

General conclusions to be reached from the presented data are:

(i)   in Australia, soil loss rates generally increase from south to north and from inland to coastal regions as a result of changes in rainfall erosivity

(ii) land use and management practices have a major effect on soil loss; any practice that reduces the amount of protective ground cover, lowers infiltration rates and loosens the soil and so the energy for entrainment of soil particles, increases the risk of high soil loss

(iii) observed soil loss per unit area tends to decrease with increasing catchment size

(iv) the estimated values for mean annual soil loss tend to be dominated by major events; that is, a small number of erosion events account for most of the erosion, while a few major events

can cause major erosion losses if erosive rain falls on an erosion-susceptible soil

(v)   losses from individual events can be many times the mean annual loss and can reach high levels, of the order of 300 to 700 t/ha

(vi) losses from undisturbed forested catchments are likely to be in the range 0 to 1 t/ha/yr; bushfires and other disturbances can lead to accelerated erosion rates, of the order of 10 to 50 t/ha/yr, especially during the period before any regeneration occurs

(vii) under pasture, mean annual losses are of the order of up to 1 t/ha/yr for non-gullied plots

(viii) under cropping, erosion rates are of the order of 1 to 50 t/ha/yr, but are more likely to be at the lower end of this range; winter cropping in the southern parts of New South Wales will yield the lowest values, while summer cropping in the north will yield the highest

(ix) the growth of crops such as sugar cane in tropical and subtropical areas can lead to losses of 100 to 500 t/ha/yr

(x)   bare fallow conditions in temperate areas can be expected to yield soil losses of the order of 50 to 100 t/ha/yr

(xi) soil formation rates are poorly defined, but are almost certainly less than 1 t/ha/yr and more likely to be less than 0.5 t/ha/yr, and the formation of soil materials from consolidated bedrock material is likely to be much less than 0.5 t/ha/yr; soil loss rates from all but the most conservative land uses (for example, good pasture) are likely to exceed those rates of formation and lead to a diminution of the soil resource.

BIBLIOGRAPHY

Abawi, G.Y. and Stokes, R.A. (1982), *Wights Catchment Sediment Study 1977–1981*, Water Resources Technical Report No. 100, Sydney.

Abrahams, A.D. (1972), 'Drainage densities and sediment yields in eastern Australia', *Australian Geographical Studies* 10, 19–41.

Adams, F., Ewing, P.A. and Huberty, M.R. (1947), 'Hydrologic aspects of burning brush and woodland-grass ranges in California', California Division of Forestry.

Adamson, C.M. (1974), 'Effects of soil conservation treatment on runoff and sediment loss from a catchment in south-western New South Wales, Australia', Proceedings Paris Symposium, *Effects of man on the interface of the hydrological cycle with the physical environment*, IAHS-AISH Publication No. 113, 3–14.

Alchin, B.M. (1983), 'Runoff and soil loss on a duplex soil in semi-arid New South Wales', *Journal of Soil Conservation Service of NSW* 39, 176–87.

Armstrong, J.L. (1990), 'Runoff and soil loss from bare fallow plots at Inverell, New South Wales', *Australian Journal of Soil Research* 28, 659–675.

Atkinson, G. (1984a), 'Soil erosion following wildfire in a sandstone catchment', National Soils Conference, ASSSI, Brisbane.

Atkinson, G. (1984b), 'Erosion damage following bushfires', *Journal of Soil Conservation Service of NSW* 40, 4–9.

Beckmann, G.G. (1983), 'Development of old landscapes and soils' in *Soils: An Australian Viewpoint*, Division of Soils, CSIRO, Melbourne; Academic Press, London, 51–72.

Beckmann, G.G. and Coventry, R.J. (1987), 'Soil erosion losses: squandered withdrawals from a diminishing account', *Search* 18, 21–26.

Bennett, H.H. (1939), *Soil Conservation*, McGraw-Hill, New York.

Bishop, P. (1984), 'Modern and ancient rates of erosion of central eastern NSW and their implications' in R.J. Loughran, *Drainage Basin Erosion and Sedimentation*, University of Newcastle, Vol. 1, 35–42.

Biswell, H.H. and Schultz, A.M. (1976), 'Surface runoff and erosion as related to prescribed burning', *Journal of Forestry* 55, 372–4.

Blong, R.J., Riley, S.J. and Crozier, P.J. (1982), 'Sediment yield from runoff plots following bushfire near Narrabeen Lagoon, NSW', *Search* 13, 36–8.

Brown, J.A.H. (1972), 'Hydrologic effects of a bushfire in a catchment in south-eastern New South Wales', *Journal of Hydrology* 15, 77–96.

Butler, B.E. and Churchward, H.M. (1983), 'Aeolian processes' in *Soils: An Australian Viewpoint*, Division of Soils, CSIRO, Melbourne; Academic Press, London, 91–9.

Capelin, M.A. and Truong, P.N. (1985), 'Soil erosion within pineapple fields in south-east Queensland', *Proceedings Fourth Australian Soil Conservation Conference*, Maroochydore, Queensland.

Copley, T.L., Forrest, L.A., McCall, A.G. and Bell, F.G. (1944), 'Investigations in erosion control and reclamation of eroded land at the central Piedmont Conservation Experiment Station, Statesville, North Carolina, 1930–1940', US Department of Agriculture, Soil Conservation Service Technical Bulletin 873.

Costin, A.B. (1980), 'Runoff and soil loss and nutrient losses from an improved pasture at Ginninderra, Southern Tablelands, New South Wales', *Australian Journal of Agricultural Research* 31, 533–46.

Cunningham, G.M. (1987), 'Reclamation of scalded land in western New South Wales', *Journal of the Soil Conservation Service of NSW* 43, 52–60.

Debano, L.F., Rice, R.M. and Conrad, C.E. (1979), 'Soil heating in chaparral fires: effects on soil properties, plant nutrients, erosion and runoff', Research Paper PSW-145, Pacific Southwest Forest and Range Experimental Station, Forest Service, US Department of Agriculture, Berkeley, California.

Edwards, K. (1980), 'Runoff and soil loss in the wheatbelt of New South Wales', *Proceedings Agricultural Engineering Conference, Geelong*, The Institution of Engineers, Australia, Barton, ACT, 94–8.

Edwards, K. (1987), *Runoff and Soil Loss Studies in New South Wales*, Technical Handbook No. 7, Soil Conservation Service of NSW, Sydney.

Elliott, G.L. and Dight, D.C.C. (1986), *An Evaluation of the Surface Stability of Rehabilitated Overburden in the Upper Hunter Valley, NSW*, Report to NSW Coal Association.

Elliott, G.L., Campbell, B.L. and Loughran, R.J. (1984), 'Correlation of erosion and erodibility assessments using caesium 137', *Journal of Soil Conservation Service of NSW* 40, 24–9.

Elliott, G.L, Loughran, R.J., Packer, I., Maliszewski, L.T., Curtis, S.J., Saynor, M.J., Morris, C.D. and Epis, R.B. (1997), 'A National Reconnaissance Survey of Soil Erosion New South Wales', Report prepared for the Australian National Landcare Program, Department of Primary Industries and Energy, Project Number 1989–90: No 8, University of Newcastle, NSW.

Fanning, P. (1994), 'Long-term contemporary erosion rates in an arid rangelands environment in western New South Wales, Australia', *Journal of Arid Environments* 28, 173–187.

Field, J.B. (1984), 'Erosion in a catchment in New England, NSW' in R.J. Loughran, *Drainage Basin Erosion and Sedimentation*, University of Newcastle, Vol. 2, 43–58.

Freebairn, D.M. and Wockner, G.H. (1986a), 'A study of soil erosion on vertosols of the eastern Darling Downs, Queensland. I. Effects of surface conditions on soil movement within contour bay catchments', *Australian Journal of Soil Research* 24, 135–58.

Freebairn, D.M. and Wockner, G.H. (1986b), 'A study of soil erosion on vertosols of the eastern Darling Downs, Queensland. II. The effect of soil, rainfall and flow conditions on suspended sediment losses', *Australian Journal of Soil Research* 24, 159–72.

Gilmour, D.A. and Cheney, N.P. (1968), 'Experimental prescribed burn in radiata pine', *Australian Forestry* 32, 171–8.

Good, R.B. (1973), 'A preliminary assessment of erosion following the wildfires in Kosciusko National Park, NSW in 1973', *Journal of Soil Conservation Service of NSW* 29, 191–9.

Hairsine, P., Murphy, B., Packer, I. and Rosewell, C. (1993), 'Profile of erosion from a major storm in the South-east Cropping Zone', *Australian Journal of Soil and Water Conservation* 6(4), 50–5.

Hall, G.F., Logan, T.J. and Young, K.K. (1985), 'Criteria for determining tolerable erosion rates', in R.F. Follett and B.A. Stewart (eds), *Soil Erosion and Produc-*

tivity, Americal Society of Agronomy Inc., Crop Science Society of America Inc., Soil Science Society of America Inc., Madison, Wisconsin, USA.

Harte, A.J., Enright, N.F. and Watt, L.A. (1984), 'Soil loss measuring simplified', *Journal of Soil Conservation Service of NSW* 24, 62–3.

Hartley, R.E., Maschmedt, D.J. and Chittleborough, D.J. (1984), 'Land management: key to water quality control', *Water* 11, 18–21.

Humphreys, G.S. and Mitchell, P.B. (1983), 'A preliminary assessment of the role of bioturbation and rainwash on sandstone hillslopes in the Sydney Basin' in R.W. Young and G.C. Nanson (eds), *Aspects of Australian Sandstone Landscapes, Australian and New Zealand Geomorphology*, Group Special Publication No. 1, University of Wollongong, 66–80.

Imeson, A.C., Verstraten, J.M., Van Mulligan, E.J. and Sevink, J. (1992), 'The effects of fire on runoff and water repellency on infiltration and runoff under Mediterranean type forest', *Catena* 19, 345–61.

Jenny, Hans (1941), *Factors of Soil Formation*, McGraw-Hill, New York, London.

Kinnell, P.I.A. (1983), 'The effect of kinetic energy of excess rainfall on soil loss from non-vegetated plots', *Australian Journal of Soil Research* 21, 445–53.

Kirkby, M.J. (1980), 'The problem' in M.J. Kirkby and R.P.C. Morgan (eds), *Soil Erosion*, John Wiley and Sons Ltd, Chichester, UK.

Lal, R. (1990), 'Soil erosion and land degradation: the global risks', in R. Lal and B.A. Stewart (eds), Soil Degradation, special issue of *Advances in Soil Science 11*, Springer-Verlag, New York and Berlin.

Lang, R.D. (1984), 'Temporal variations in catchment erosion and implications for estimating erosion rates' in R.J. Loughran, *Drainage Basin Erosion and Sedimentation*, University of Newcastle, Vol. 1, 43–50.

Lavee, H., Kutiel, P., Segev, M. and Benyamini, Y. (1995), 'Effect of surface roughness on runoff and erosion in a Mediterranean type ecosystem: the role of fire', *Geomorphology* 11, 227–34.

Leys, J. and McTainsh, G. (1994), 'Soil loss and nutrient decline by wind erosion—cause for concern', *Australian Journal of Soil and Water Conservation* 7(3), 30–5.

Leys, J.F. and Heinjus, D.J. (1992), 'Cover levels to control soil and nutrient loss from wind erosion in the South Australian Murray Mallee', *Proceedings of 7th International Soil Conservation Conference Sydney*.

Little, I.P. and Ward, W.T. (1981), 'Chemical and mineralogical trends in a chronosequence developed on alluvium in eastern Victoria, Australia', *Geoderma* 25, 173–88.

Lowdermilk, W.C. (1953), 'Conquest of the land through seven thousand years', Agricultural Information Bulletin 99, US Department of Agriculture, Washington DC.

McCormack, D.E. and Young, K.K. (1981), 'Technical and societal implications of soil loss tolerance' in R.P.C. Morgan (ed.), *Soil Conservation: Problems and Prospects*, John Wiley and Sons Ltd, Chichester, 365–76.

McFarlane, D.J. (undated), *Water Erosion on Potato Land during the 1983 Growing Season Donnybrook*, Soil Conservation Branch, Division of Resource Management Technical Report No. 26, WA Department Agriculture.

McFarlane, D.J. and Clinnick, P.F. (1984), 'Annual rainfall erosion index for Australia', *Erosion Research Newsletter*, No. 9, Darwin, NT.

McFarlane, D.J. and Ryder, A.T. (1986), *Report on Water Erosion in the Irishtown-Wongamine Area North of Northam during June, 1986*, Soil Conservation Branch, Division of Resource Management, Internal Report 1317185, WA Department Agriculture.

McTainsh, G.H. and Leys, J.F. (1993), 'Soil erosion by wind' in G.H. McTainsh and W.C. Boughton (eds), *Land Degradation Processes in Australia*, Longman, Melbourne, 188–233.

Marschke, G.W. (1986), 'Surface hydrology and erodibility of a black earth under four tillage practices', *Proceedings Conference Agricultural Engineering, Adelaide*, The Institution of Engineers, Australia.

Miles, R, and McTainsh, G. (1994), 'Wind erosion and land management in the mulga lands of Queensland', *Australian Journal of Soil and Water Conservation*, 7(3), 41–5.

Morse, R.J. and Outhet, D.N. (1986), 'Sediment management on a total catchment basis', *Journal of Soil Conservation Service of NSW* 42, 11–14.

Mulcahy, M.J. and Churchward, H.M. (1973), 'Quaternary environments and soils in Australia', *Soil Science* 116, 156–69.

Murphy, B.W. and Flewin, T.C. (1993), 'Rill erosion on a structurally degraded sandy loam surface soil', *Australian Journal of Soil Research* 31, 419–36.

Neil, D. and Fogarty, P.J. (1991), 'Land use and sediment yield on the Southern Tablelands of New South Wales', *Australian Journal of Soil and Water Conservation* 4(2), 33–9.

Nowak, P.J., Timmons, J.F, Carlson, J., and Miles, R. (1985), 'Economic and social perspectives on T values relative to soil erosion and crop productivity', in R.F. Follett and B.A. Stewart (eds), *Soil Erosion and Productivity*, Americal Society of Agronomy Inc., Crop Science Society of America Inc., Soil Science Society of America Inc., Madison, Wisconsin, USA.

Olive, L.J. and Walker, P.H. (1982), 'Processes in overland flow: erosion and production of suspended material' in E.M. O'Loughlin and P. Cullen (eds), *Prediction in Water Quality*, Australian Academy of Science, Canberra, 87–119.

Paton, T.R. (1978), *The Formation of Soil Material,* George Allen and Unwin, London.

Paton, T.R., Mitchell, P.B., Adamson, D., Buchanan, R.A., Fox, M.D. and Bowman, G. (1976), 'Speed of podzolization', *Nature* 260, 601–2.

Pierce, F.T., Larson, W.E., Dowdy, R.H., and Graham, W.A.P. (1983), 'Productivity of soils: assessing long-term changes due to erosion, *Journal of Soil and Water Conservation* 38(1), 39–44.

Pilgrim, A.T., Puvaneswaran, P. and Conacher, A.J.

(1986), 'Factors affecting natural rates of slope development', *Catena* 13, 169–80.

de Ploey, J. and Gabriels, D. (1980), 'Measuring soil loss from experimental studies' in M.J. Kirkby and R.P.C. Morgan (eds), *Soil Erosion*, John Wiley and Sons Ltd, Chichester, UK, 63–108.

Prove, B.G. (1984), 'Soil erosion and conservation in the sugar canelands of the wet tropical coast', *Erosion Research Newsletter*, No. 10, Darwin, NT.

Reece, R.H. and Campbell, B.L. (1984), 'The use of Cs-137 for determining soil erosion differences in a disturbed and non-disturbed semi-arid ecosystem', *Proceedings Second International Rangelands Congress*, Adelaide.

Ring, P.J. (1982), 'Soil erosion, rehabilitation and conservation: a comparative agriculture engineering/economic case study', *Proceedings Conference Agricultural Engineering*, Armidale, 197–202.

Ritchie, J.C. and McHenry, J.R. (1990), 'Application of radioactive fallout caesium-137 for measuring erosion and sediment accumulation rates and patterns—a review', *Journal of Environmental Quality* 19, 215–33.

Rosewell, C.J. (1992), 'The erodibility of five soils in New South Wales' in G.J. Hamilton, K.M. Knowles and R. Attwater (eds), *Proceedings 5th Australian Soil Conservation Conference*, Vol. 3, Department of Agriculture, Perth, 112–15.

Rosewell, C.J. (1993), 'SOILOSS—a program to assist in the selection of management practices to reduce erosion', Technical Handbook No. 11 (Second Edition), Department of Conservation and Land Management, Sydney.

Saunders, I. and Young, A. (1983), 'Rates of surface processes on slopes, slope retreat and denudation', *Earth Surface Processes and Landforms* 8, 473–501.

Scott, D.F. and Van Wyk, D.B. (1990), 'The effects of wildfire on soil wettability and hydrological behaviour of an afforested catchment', *Journal of Hydrology* 121, 239–56.

Scott, D.F. and Van Wyk, D.B. (1992), 'The effects of fire on soil water repellency, catchment sediment yields and streamflow' in B.W. van Wilgen, D.M. Richardson, F.J. Kruger and van Hensbergen H.J. (eds), *Fire in South African Mountain Fynbos, 1992*, Springer-Verlag, Berlin, 216–39.

Selby, M.J. (1982), *Hillslope Materials and Processes*, Oxford University Press, Oxford.

Smith, R.M. and Stamey, W.L. (1965), 'Determining the range of tolerable erosion', *Soil Science* 100, 414–24.

Stocking, M. (1978), 'A dilemma for soil conservation', *Area Institute of British Geographers* 10, 306–8.

USDA (1954), 'Fire-flood sequence on the San Dimas Experimental Forest Station', USDA California Forest and Range Experimental Station, Forest Service Technical Paper No. 6.

Versfeld, D.B. (1981), 'Overland flow on small plots at the Jonkershoek Forestry Research Station', *South African Forestry Journal* 119, 35–40.

Walker, P.H. (1980), 'Soil morphology, genesis and classification in Australia', in *Proceedings National Soils Conference, ASSSI*, Sydney.

Walker. P.H. and Butler, B.E. (1983), 'Fluvial processes' in *Soils: An Australian Viewpoint*, Division of Soils, CSIRO, Melbourne; Academic Press, London, 83–90.

Walker, P.H. and Coventry, R.J. (1976), 'Soil profile development in some alluvial deposits of eastern New South Wales', *Australian Journal of Soil Research* 14, 305–17.

Wasson, R.J. and Galloway, R.W. (1984), 'Erosion rates near Broken Hill before and after European settlement' in R.J. Loughran, *Drainage Basin Erosion and Sedimentation*, University of Newcastle, Vol. 1, 213–20.

Williams, M.A.J. (1973), 'The efficacy of creep and slopewash in tropical and temperate Australia', *Australian Geographical Studies* 11, 628.

Wischmeier, W.H. and Smith, D.D. (1978), *Predicting Rainfall Erosion Losses—A Guide to Conservation Planning*, Agriculture Handbook No. 537, US Department of Agriculture.

Wolman, M.G. (1985), 'Soil erosion and crop productivity: a world-wide perspective', in R.F. Follett and B.A. Stewart (eds), *Soil Erosion and Productivity*, American Society of Agronomy Inc., Crop Science Society of America Inc., Soil Science Society of America Inc., Madison, Wisconsin, USA.

Young, A. (1969), 'Present rate of land erosion', *Nature* 224, 851–2.

Zierholz, C. (1997), 'The effect of fire on runoff and erosion in Royal National Park, New South Wales', MSc Thesis, School for Research and Environmental Management, Australian National University.

Other Forms of Soil Degradation

CHAPTER 4

P.E.V. Charman

Apart from soil erosion, there are a number of other forms of soil degradation which are of concern to land managers and soil conservationists in New South Wales. The term 'soil degradation' may be generally defined as the decline in soil quality caused through its improper use by humans (Houghton and Charman, 1986). Soil quality is an assessment of the capability of the soil to carry out specific functions. A more formal definition is:

Soil quality is the capacity of a specific kind of soil to function, within natural or managed ecosystem boundaries, to sustain plant and animal productivity, maintain or enhance water and air quality, and support human health and habitation. (Soil Science Society of America, 1995)

Doran (1996) and Larson and Pierce (1994) identified the key functions of soil as being:

(i) to provide a medium for plant growth and biological activity; soils play a key role in the storage and cycling of elements required by biological systems, decomposing organic wastes and detoxifying some hazardous materials

(ii) to regulate and partition water flow and storage in the environment; the ability of the soil to store and transmit water is a major factor regulating water availability to plants and the transport of environmental pollutants to surface and groundwater

(iii) to serve as an environmental buffer in the formation and destruction of environmentally hazardous compounds.

Soil degradation impairs or reduces the capability of soil to carry out these functions. Degradation of soil quality has generally been brought about by agricultural use of the soil, as well as by urbanisation, deforestation and mining.

The best agricultural soils are those which are deep, fertile and resistant to erosion. They take in rainfall easily without sealing at the surface, but retain enough of it to keep plants growing during dry spells and alive in droughts. Their structure is such that air and plant roots can penetrate easily and they can be cultivated under a wide range of moisture conditions.

These desirable properties in soils evolve according to the conditions of their formation and the rocks from which they have developed. Such properties depend on the proportions of clay, silt and sand particles in the soil and the way they are bound together to give the soil structure. The amount of organic matter, usually from the breakdown of plants and animal residues, is of considerable importance too. This has a vital role in providing nutrients for plant growth, as well as chemical substances that bind the soil particles together into aggregates. A wide range of organisms in soils serves to break down organic matter for these purposes, and also creates channels and cavities through which air and water can move. This is discussed further in Chapter 14.

While soil erosion mainly occurs during more extreme climatic events, other forms of soil degradation are more insidious, occurring slowly and gradually over time. Consequently, their economic effects are often more difficult to assess directly. However, many soil scientists and soil conservationists believe that processes like soil acidification and soil structure decline will, in the long term, lead to more serious overall effects on productivity and the environment if they are not checked. This is apart from their role as precursors of erosion, which itself leads to long-term losses of productive potential and adverse environmental effects.

The main forms of soil degradation of current concern are soil salinisation, soil fertility decline including nutrient decline and soil acidification, soil structural degradation, development of acid sulfate soils, and soil contamination. These are described in the following sections. The reader is also referred to Chapters 10 and 14 where the soil properties affected are described in more detail.

4.1 Soil Salinisation

Salinisation results from the accumulation of free salts in part of a landscape to an extent that causes degradation of the vegetation and/or soils. The occurrence and effects of high soil salinity are discussed in Chapter 13.

Soil salinity is of major concern in two important situations—one in irrigated soils and one in non-irrigated (dryland) soils. It is also a problem in some urban areas.

4.1.1 Irrigation Salinity

This form of salinity problem relates to the increasing build-up of salts, mainly sodium chloride, in soils used for irrigation along the Murray and Murrumbidgee Rivers. Such salinisation results from the repeated use of saline river water for irrigation; a cycle has developed in which the salinity of both the soils and rivers, particularly the Murray, is increasing. Rising water tables also mobilise salt that is stored deeper in the soil profile, bringing it to within the root zone of plants, and even to the surface. The problems of Murray River salinity have interstate implications, as the river is a major resource for New South Wales, Victoria and South Australia. Salinisation is increasingly a problem affecting other inland river systems too.

Most rivers naturally contain some salt, but usually at a very low level. However, the level of salt in a river depends on the nature and geology of the catchment and soil landscapes from which the water originates. It also depends on the relative amounts of surface and subsurface water that make up the flow. The use of the river water for irrigation, however, introduces another complicating factor. The slightly saline water is removed from the river and applied to growing crops across large areas of alluvial soils on either side. If more water is applied than the crops need, water tables tend to rise, bringing salts from deeper in the soil profile into the root zone, which affects crop growth. Surplus water passes through the soil, picking up more salts, and finds its way back to the river by natural drainage. Thus, there is a slight increase in the salinity of the river, and the effect may be multiplied again and again as the water is reused for further irrigation downstream. If less water is applied than the crops need, the salts in the irrigation water tend to be retained in the soil. Salinity thus builds up with repeated water applications until both crops and soils are adversely affected by the high salt levels and elevated water tables.

The problems of rising water tables and salinisation of soils are endemic in irrigation schemes worldwide. To a very large extent they are inevitable because extra water is applied to soils, which, with the vagaries of climate, must frequently coincide with rain to produce a situation where there is more water in the soil than plants can transpire and the atmosphere evaporate. Under such conditions, accessions to the water table occur. The older the irrigation scheme, the closer the water tables rise to the soil surface. Once they rise to within 1 to 2 m depth, the soil near the surface remains moist for lengthy periods between irrigations, and water and salts move from the water table to the surface under the influence of evaporation and capillary rise.

The processes involved are somewhat more complex than this simplified account would suggest, with the types of crops and soils involved and the methods of irrigation used being critical factors determining whether or not salt levels will have serious effects on productivity. However, the dilemma remains—how to avoid an increase in the salinity level of the river water without increasing that of the soil. Both have adverse consequences in the long term, and no easy solution has yet been found. Irrigation efficiency, drainage and the close matching of irrigation practices and water needs of both crops and soils are likely to be important factors in any answer to the problem.

Dryland Salinity 4.1.2

Generally, the cause of dryland salinity is attributed to changes in land use which affect water movement through the landscape. A typical situation occurs following the clearing of trees from hillslopes, which reduces transpiration and allows an increase in rainfall intake beyond the root zone and a rise in water tables lower down the slope. Increased subsurface seepage dissolves salts in the soil and, with lateral flow through the landscape, moves from hillslopes to valley floors. Rising water tables also bring salt to the surface from deeper in the profile, mobilising salt reserves that were previously stable. Salty water surfaces in patches depending on the geomorphology and topography of the site. The salt becomes concentrated by evaporation at these locations and the normal vegetation is killed (see Table 13.1). Stock tend to be attracted to the salt and congregate at the site. Their trampling results in further deterioration and extension of the bare patch until it becomes an erosion hazard as well as an area of low or nil productivity (see Plate 4.1).

Other forms of dryland salinisation include saline scalds, which develop when topsoil is removed to expose a subsoil that is naturally high in salts. Although these are partly natural features, in that they are mentioned in historical records as being present at the time of European settlement, their incidence has increased with grazing pressures caused by domestic livestock and plague populations of rabbits (Hamilton and Lang, 1978).

In the drier semi-arid parts of the state, many soils are naturally saline below the surface soil, and wind erosion, by removing topsoil, causes the formation of large areas of scalding. This is most common in western New South Wales.

The occurrence of areas of salt-affected land is somewhat cyclic in pattern, depending on the balance between wetter and drier seasons. However, the last few years has seen a large increase in the area affected, as salinity has begun to move further up the slopes in some areas and relatively large areas have been subjected to salinisation. A further problem of saline areas on lower slopes and in drainage depressions is the influence of these saline areas on water quality. Runoff coming from these areas can have high salinity levels, which has a large potential to influence water quality in creeks and rivers. A brief discussion on the distribution of secondary salinisation in New South Wales is presented in Chapter 13.

Urban Salinity

4.1.3

In urban situations saline soils present a high erosion hazard, have low permeability, and usually only support a sparse vegetation cover. The build-up of salts can also be a threat to building foundations and brickwork as the salt can cause the cement and bricks to crumble if salt levels become high. Urban salinity is associated with high water tables which may result from overclearing, especially the removal of trees, excessive watering, disruption of natural drainage lines or overflow of septic tanks and sullage pits. The types of problems caused by urban salinity include: damage to foundations and brickwork; rising damp; unhealthy grass, shrubs and trees; waterlogged ground; corrosion of water, gas and sewerage pipes; pot-holing of roads; and failure of septic tanks.

For these reasons, it is important that saline areas, or potentially saline areas, are identified at the planning stage so that appropriate measures can be applied as development proceeds. These would incorporate vegetation that is both salt tolerant and has minimal water needs, minimisation of watering of parks and gardens, and a good drainage system to deal with all storm water.

Plate 4.1　　An aerial view of serious dryland salinisation on the southern tablelands of New South Wales (see 'Dryland Salinity', p. 61)

4.2 Soil Fertility Decline

Two main categories of degradation will be considered here—the decline in the overall chemical fertility of soils and the more specific processes of soil acidification.

4.2.1 Chemical Fertility Decline

The main causes of chemical soil fertility decline are product removal from the land, lack of replacement by land management practices, erosion and leaching. It has been occurring in our soils since Europeans first started farming here, in particular growing crops. In better-watered areas near the coast the decline has not been so marked, since there the quantity and reliability of rainfall have allowed soil organic matter levels to be better maintained. On grazing lands it has also been easier to counteract the loss of humus, thanks to returns by grazing animals and less cultivation being involved than for cropping.

Organic matter has always been the focus of the intrinsic fertility of our soils, and across the great cropping lands west of the divide the story has been one of organic matter loss due to repeated cultivation and, consequently, oxidation of this important resource. Under a tillage regime, and with an unreliable climate, it has been difficult to maintain the levels of organic matter in the soil, particularly where no attempts have been made to re-incorporate plant residues after the crop has been harvested. Hence levels of 3% or 4% have dropped to levels below 1%, in some cases over just a few years. Fertility levels have only been kept up by the addition of artificial fertilisers.

Superphosphate has traditionally been the main fertiliser used, following the recognition in the early 1900s that many Australian soils were deficient in the element phosphorus, a major and essential plant nutrient. Together with the introduction of subterranean clover, the 'sub and super' revolution of the thirties, forties and fifties saw superphosphate fertiliser elevated in status to become an essential part of ley farming, as well as of purely cropping or grazing enterprises. This brought about many improvements in farm productivity and checked the losses in soil organic matter that had resulted from exploitative agricultural practices.

But the dominance of superphosphate and the ley farming of the times left a legacy of their own, in terms of shallow-rooted pastures, increased soil acidity, new weed invasions and nutrient imbalance in many soils. The other aspects of chemical fertility, apart from phosphorus fertilisation and nitrogen from sub-clover, tend to have been neglected. Potassium, calcium and sulfur, and in some cases the trace elements also, were left in the wings rather than on centre stage, although the growing of canola has refocused interest in sulfur as a fertiliser.

As well as phosphorus, the other two major nutrient elements essential to plant growth are nitrogen and potassium. The intrinsic nitrogen status of soils in their virgin state is linked to the level of organic matter in them. Most organic nitrogen ultimately comes from the atmosphere and is assimilated into soils by micro-organisms, many associated with legumes. From this organic state it may be transformed by further soil microbial activity into inorganic forms which can be taken up directly by plants. It is because of this mobility of nitrogen in soils, being constantly cycled between atmosphere, plants and soil, that the link with organic matter is so important. Organic matter levels are often closely linked with total nitrogen levels in soil, but not necessarily with forms of nitrogen that are readily available to plants. Constant cultivation of soils with only moderate fertility results in oxidation of their organic matter and release of the stored nitrogen to the growing crop, or possibly to loss by leaching or erosion. Decline in fertility can be rapid and replacement of nitrogen, by fertiliser addition or growth of legumes, becomes vital to sustainable cropping.

Potassium is the other major element of importance, particularly in a cropping situation, and must be added to soils in regular amounts to avoid undue nutrient depletion. Most of our soils are, however, reasonably well provided with potassium, and deficiencies usually only show up where crop production is intensive, as in the horticultural industry.

Calcium and sulfur are important minor elements that are essential for plant growth, not only in cropping situations but also on grazing lands. Calcium has a two-fold importance—as a constituent of most farm produce, and as a chemical requirement for keeping soils at a satisfactory pH level. Hence the importance of lime, the main source of added calcium, in modern agriculture. In cropping situations,

Table 4.1 Nutrients removed by various forms of production (kg) (after Hyland, 1995)

	Nitrogen (kg)	Phosphorus (kg)	Potassium (kg)	Sulfur (kg)	Calcium (kg)
One fat lamb	2.3	0.2	0.1	0.2	0.4
One wool fleece	0.7	trace	0.1	trace	trace
1000 litres milk	6.0	1.0	1.4	0.6	1.2
One tonne legume hay	30	3.0	25	2.0	9.0
One tonne cereal hay	20	2.0	18	1.4	0.6
One tonne wheat	23	3.0	4.0	1.5	0.4
One tonne canola	41	7.0	9.0	10	4.0

sulfur is likely to be required as a fertiliser addition, particularly for high demand crops such as canola. It is also of benefit to many pasture lands, where sulfur-fortified superphosphate may be used.

Table 4.1 gives data from Hyland (1995) and shows the approximate amounts of these main elements removed in various farm products. More detailed information on soil nutrient depletion is to be found in Chapter 13.

4.2.2 Soil Acidification

The majority of soils in the eastern half of New South Wales are naturally slightly acid, at least in the top few centimetres. Generally, the wetter the climate, the more acid they are likely to be. The natural acidity level of these surface soils probably ranges somewhere on the acid side of neutral, typical values for pH in water (pHw) being between 5.5 and 7.0. The majority of plants normally grow best in this sort of range, provided other soil chemical and physical conditions are satisfactory. The soil acidity problem appears to have occurred when farming practice has caused the pHw to drop below 5.5, or pH in calcium chloride (pHca) below 4.7. It is these soils which are generally referred to as 'acid soils'. At pHw less than 5.5 or pHca less than 4.7, aluminium toxicity and certain nutrient deficiencies can severely affect the growth of some plants, among them canola and lucerne and some sensitive wheat varieties (see Colour plate 4.3).

• However, there are some soils which are naturally very acidic and have a very low buffering capacity to soil acidification processes. These soils are often the more sandy soils formed on sandstones and siliceous granites in New South Wales.

The problem of very acidic soils was first recognised by Williams (1980), and in a program of soil conservation work in the catchment of Pejar Dam on the southern tablelands of New South Wales (Cumming, 1982). Many paddocks with a long history of regular superphosphate application on improved clover-based pastures had a soil pHw of 5.5 or less. This was markedly affecting pasture growth to the extent of creating an erosion hazard due to depleted pasture cover. The former Soil Conservation Service then developed a liming strategy, as part of the overall soil conservation program, to raise the pHw level of affected soils to between 6.0 and 6.5. In this way, normal pasture growth is restored, full advantage can be taken of other fertilisers and the erosion hazard is minimised.

The cause of the increased acidity is complex, being generally associated with nitrate leaching and the build-up of organic matter from pasture improvement involving legumes and superphosphate application. It occurs more commonly in sandy soils and in climatic environments where leaching is more effective. More details on the processes of soil acidification are presented in Chapter 13.

Fortunately, soil acidification is a slow process. Many years of farming practices and superphosphate application may go by before the level of acidity becomes serious. When it does, manganese and aluminium in the soil can become toxic to plants, and calcium and molybdenum may become unavailable to them.

These chemical effects can result in establishment failure, increase in plant disease and poor plant growth. Because of this, the plants' response to normal fertilisers is restricted and overall productivity is reduced. Stock grazing such pastures may also suffer, possibly in response to changing calcium/phosphorus ratios or manganese levels.

The solution to preventing acidification is based on reducing nitrate leaching and/or adding agricultural lime or some other liming agent. Depending on method of application, fineness of lime product, soil type and subsequent rainfall, this can rapidly restore the pH in the soil to a satisfactory level (see Colour plate 13.2).

4.3 Soil Structural Degradation

Declining soil structure as a result of the agricultural use of land is yet another form of soil degradation, although frequently it goes unrecognised. There are two broad categories within this form of degradation—those caused by cultivation and those caused by irrigation.

4.3.1 Soil Structure and Cultivation

Under most forms of cultivation, the structural condition of the soil tends to deteriorate. Many of our traditional cultivation techniques, developed under European climatic conditions, have a generally destructive effect on soil aggregates. When structure is destroyed, soil surfaces seal, soils become compacted, water infiltration and aeration are reduced, plant germination and growth are retarded and erosion hazard increases. These effects can involve topsoils and/or subsoils.

Many of the soils of the wheat-belt that have been cultivated for many years have probably been affected. Such soils have depended largely on organic matter to sustain their physical fertility and stability. Those soils that are generally light textured are subject to rapid degradation when cultivated. Soil organic matter and aggregation decline dramatically during the first year of cultivation, and the soils become subject to water and wind erosion. A number of years of pasture (five to 10 years) is required to reverse this structural decline.

Other soils that also suffer from soil structure problems are the grey and brown clays associated with the riverine plains of the Murray-Darling Basin. Many of these tend to have a surface sealing problem as a result of their clayey texture and the dispersive effect of sodium associated with the clay. Under native pasture, these features are of little concern but, once cultivated, problems of germination and poor plant growth arise in locations where the sodium effects are at a maximum. Treatment with gypsum has been proposed for such problem areas. The economics of its application are likely to dictate that only seriously affected sites are worth treating.

Another form of soil structure degradation that has come to light in recent years is hardpan formation, whereby a hard layer forms at some depth in the soil and restricts the penetration of plant roots and limits water penetration into the subsoil. Hardpans

are natural features of some soils, but can also be formed by years of cultivation at the same depth. These are called plough-pans, and they occur in many parts of the wheat-belt where soils have been consistently cultivated. A number of deep ripping and subsoiling implements developed recently have demonstrated the value of loosening the soil to a depth greater than that of normal tillage. The technique is seen to be of particular value on certain soil types that are cropped using conservation tillage methods. The deep ripping would normally be carried out every few years.

Soil Structure and Irrigation 4.3.2

The problem of soil structural breakdown under irrigation is directly attributable to the effect of wetting on the more clayey soils of the irrigation areas. Two main types of breakdown are recognised—that due to slaking, or the natural collapse of soil aggregates in water, and that due to dispersion as a result of sodium on the soil clays. The type of irrigation also affects the degree of breakdown. Flood irrigation is generally more harmful than spray or trickle irrigation because the latter methods wet the soil more slowly, reducing slaking. With furrow irrigation, soil structural stability is affected mainly in the furrows, but not in the slowly wetted rows. In the case of spray irrigation, the advantages of slow wetting may be counteracted by the disruptive effect of the droplets hitting the soil. Of the four types—flood, furrow, spray and trickle irrigation—the latter is likely to be the least damaging to soil structure. In these situations of irrigated clay soils, gypsum can be used to improve soil structure where it can be shown to be economic.

Apart from soils whose surface conditions are prone to breakdown when abruptly saturated in furrow or flood irrigation applications, the magnitude of the problem has been worsened by laser levelling of irrigation soils. Such levelling is undertaken to achieve two objectives: agronomic operational efficiency and improved irrigation efficiency. In the process of such levelling, however, subsoils with sodic properties are often exposed. These are even less stable and less permeable than their surface soils. The result is smooth irrigation bays with highly variable infiltration properties that are agronomically undesirable and which depress overall production.

The problem of degraded subsoil structure has appeared in a number of irrigation areas. It arises through clayey subsoils not draining quickly enough for the necessary cultural operations to be undertaken without causing them to deform. In consequence, as soon as the surface is dry, these operations are undertaken and the still wet subsoils are compressed, losing much of their macroporosity. This makes them even less well drained, anaerobic and denser, and plant roots fail to penetrate them. The solution to this problem has been, so far, to undertake a crop rotation using a combination of irrigated and dryland cropping. During the dryland cropping (usually wheat), the natural shrinking and cracking that occur on drying of these soils go a long way to rehabilitating any structural deformation they may have suffered during an irrigation phase.

The basic cause of soil degradation in irrigation areas lies in the limitations of the soil resources and the landscape, together with the inappropriate irrigation techniques and soil management/agronomic practices. Many irrigation soils have the following features:

(i)   are lacking in stability and are susceptible to dispersion on wetting, a problem made worse by regular cultivation which reduces organic matter levels, and the rapid wetting of soils under irrigation

(ii)   are fine textured and drain poorly

(iii)   are liable to plastic deformation when worked or driven over in a moist state (thus worsening their poor water transmission and aeration properties)

(iv)   occur in generally flat topography making drainage disposal a problem

(v)   are underlain by highly saline soils over large areas

Plate 4.2   Severe land and soil degradation

(vi) some irrigation channels are located on higher ground in the riverine plain and these higher areas happen to be the location of prior streams and old levees; they can have coarser textured soils and this can result in high flows into the water table.

The consequences of such poor soil conditions and inappropriate irrigation practices are that less water infiltrates with each irrigation, creating excessive tailwater and a need for more frequent applications. Compounded with poorly draining subsoils, these extra irrigations ensure that soils remain wet for longer and much of the water in the subsoils is not used by plants. Despite poor drainage, this excess water is eventually added to the water table and waterlogging and salinisation may follow.

4.4 Acid Sulfate Soils

Another form of soil acidification is due to the disturbance of acid sulfate soils. These occur along much of the coastline of New South Wales, in low-lying areas such as those occupied by mangrove swamps and salt marshes, as well as tidal lakes and estuaries.

The processes and formation of acid sulfate soils are discussed more fully in Chapter 13. As soils have developed over thousands of years in these locations, strongly influenced by the alternation of tidal incursion and estuarine flooding by river water, continual waterlogging of subsoils has resulted in anaerobic conditions at depth and the formation of sulfide layers. Typically these layers are dominated by iron sulfide formed from the weathering of the ancient sediments in the absence of oxygen. Under natural conditions such sulfide layers are not allowed to oxidise significantly, owing to continued waterlogging, although in severe drought conditions some oxidation may occur and sulfuric acid is produced. However, the amounts released are kept in balance by subsequent floods or tidal inflows, which neutralise and disperse acid which might get into streams.

The more serious problem occurs when these areas are drained for agriculture or other forms of development. Excavation or drainage exposes the iron sulfide layers to the air, and oxidation takes place with the formation of sulfuric acid and consequent pollution of nearby coastal streams. The amount of acid produced may well be more than the streams can cope with naturally, and the acid then adversely affects plants, fish and other organisms. The excavated spoil from drained areas may be left exposed and continues to oxidise to the detriment of any attempts to grow vegetation or crops in the area. Rainfall tends to leach the acid from such spoil areas into streams and the process may continue over many years if no treatment action is taken. If the sulfide layers are clayey in character, oxidation can be seriously prolonged, whereas the acid production from sandy materials tends to be short-lived in comparison.

The primary effects of this acid on the soil involve a drastic lowering of pH and the solution of iron and aluminium (and in some cases heavy metals) to levels which allow only very limited plant growth. Acid-tolerant plants may grow but are likely to have adverse effects on grazing animals, particularly if the stock are also drinking acid water. The water draining from these soils is very acid (pH in the range 2 to 4), and as it reaches waterways and streams its effects on aquatic life and ecology can be devastating, with severe fish kills and destruction of crustaceans and smaller aquatic fauna. Water plants and habitat also are likely to be affected owing to excess aluminium and precipitated iron compounds accumulating in stream water.

When such coastal areas are being drained for urban development the potential for acid sulfate soils threatens installations made of concrete, iron and steel involving drainage pipes and building foundations. The use of acid landfill may also affect areas proposed for landscaping into gardens or lawns.

Management options for dealing with the problem of acid sulfate soils initially involve recognition of their potential and avoidance of development of areas where there is a serious threat. Maps are available showing where the problem is most likely to occur, and a number of field indicators are available to assist recognition of affected local areas. Liming, the use of shallow drains and maintaining a cover of water over such land may also have their place in specific locations, subject to inspection by personnel with appropriate expertise and experience.

4.5 Soil Contamination

Under increasingly intensive land use which characterises many areas in New South Wales, it is inevitable that in some situations contamination of the soil will occur. This typically involves the tainting of soils with man-made chemicals which may be organic or inorganic in form, or even the build-up of excessive natural products in the soil due to human mismanagement. Soils may become contaminated because of direct intentional application of hazardous materials to the soil, inadvertent applications in conjunction with other materials such as fertilisers and fungicides, spillages, deposition of contaminated sediments from water erosion or wind erosion, or as seepages through the soil and rock strata.

Heavy metals may be inadvertently added to the soil in fertilisers, the most widely recognised being cadmium in phosphate fertilisers. The amount of cadmium in phosphate fertilisers varies with the source of the rock used to make the fertilisers. Phosphogypsum produced from the manufacture of phosphate fertilisers may also contain significant levels of cadmium. Fluorine may also be added to soil in gypsum. Although substances can be added to soils in fertilisers and ameliorants, the amounts added are usually sufficiently low as to not be of immediate concern. However, the fact that these substances can be added to soils in fertilisers and ameliorants means that the amounts being added should be monitored, and action taken before any problems arise. Given the more stringent controls on permitted levels of heavy metals in foodstuffs, this has become more critical. This is especially the case in acid soils and where leafy or high vegetable matter crops (root or tuber crops such as potatoes) are grown for human consumption (Incitec, 1997).

Very high levels of soil contamination are a result of the direct addition of relatively large amounts of chemicals and other contaminants such as heavy metals or hazardous waste to soils in the course of human activity. Inorganic contaminants are usually associated with mining activities, as metals and other non-organic toxicants released during processing are often more highly available than those accumulated during normal weathering processes. Elements such as chromium, copper, manganese, nickel, cobalt, lead, zinc and arsenic may be of concern from mine wastes.

Other mine wastes which may have high levels of sulfides or cyanides are also of concern.

Lead levels in soils are usually higher in urban areas than rural areas, because of additions to the soil from car emissions and industrial processes. Mercury levels in soils are also likely to be higher in urban areas, although the use of mercury in some agricultural fungicides has led to some build-up in certain agricultural areas (Tiller, 1992).

Complex substances such as fuels, oils, pesticides, herbicides, drenches and dips are the main sources of contamination that are organically based. In many cases oils have an adverse physical effect on soils rather than direct chemical toxicity. Organochlorine compounds, widely used in the past as insecticides, are now restricted owing to their toxicity to larger organisms. Likewise termiticides such as aldrin, dieldrin and chlordane are currently banned for this purpose. Current pesticides, often organophosphates, break down more readily in soil and do not present such a residual toxicity problem. Further details on these matters are to be found in Chapter 13.

Another form of soil contamination that may become a problem is the possible disposal on land of treated sewage. Without adequate processing this is known to be a source of many viral and bacterial infections, as well as of heavy metal toxicity via plants grown in soils used for disposal of this material. Much research is being done on this problem because of the environmental desirability of using soils for the disposal of effluent wastes. Chapter 21 covers this subject in more detail in its urban context.

The contamination of soil is not the only problem, as once a soil is contaminated, it may result in the contamination of water bodies as seepage from these contaminated soils flows into lakes, rivers and groundwater reserves.

It is necessary to remember that heavy metals may be unusually high in soils that are associated with particular geological deposits. For example, lead levels are relatively high in the vicinity of Broken Hill, and chromium and copper levels may be high in soil associated with ultrabasic rock. In fact soil analyses are sometimes used in mineral exploration to look for mineral deposits. Thus relatively high levels of heavy metals are not always a sign of soil pollution resulting from human activity.

BIBLIOGRAPHY

Cumming, R.W. (1982), 'pH survey of acid soils in the Pejar area', *Journal of the Soil Conservation Service of NSW* 38, 13–18.

Doran, J.W. (1996), 'Soil health and global sustainability' in *Proceedings of an International Symposium, Soil Quality is in the Hands of the Land Manager*, held April 1996 at Ballarat, University of Ballarat, Victoria.

Hamilton, G.J. and Lang, R.D. (1978), 'Reclamation and control of dryland salt affected soils', *Journal of the Soil Conservation Service of NSW*, 34, 28–36.

Hazelton, P.A and Koppi, A.J. (eds) (1995), 'Soil Technology—Applied Soil Science—A Course of Lectures', Australian Society of Soil Science Inc. NSW Branch, and Department of Agricultural Chemistry and Soil Science, University of Sydney.

Houghton, P.D. and Charman, P.E.V. (1986), *Glossary of Terms Used in Soil Conservation*, Soil Conservation Service of NSW, Sydney.

Hyland, M. (1995), 'Soil nutrient depletion', *Australian Journal of Soil and Water Conservation* 8(1), 28–30.

Incitec (1997), 'Heavy metals in fertilisers and agriculture', Agri-topic No. 142, March 1997, Incitec Ltd, Brisbane.

Larson, W.E. and Pierce, F.J. (1994), 'The dynamics of soil quality as a measure of sustainable management' in J.D. Doran, D.C. Coleman, D.F. Bezdicek, and B.A. Stewart (eds), *Defining Soil Quality for a Sustainable Environment*, Soil Science Society of America Special Publication No. 35, Madison, Wisconsin.

Taylor, S. (1993), *Dryland Salinity—Introductory Extension Notes*, Second Edition, Department of Land and Water Conservation, Sydney.

Tiller, K.G. (1992), 'Urban soil contamination in Australia', *Australian Journal of Soil Research* 30, 937–57.

Williams, C.H. (1980), 'Soil acidification under clover pasture', *Australian Journal of Experimental Agriculture and Animal Husbandry* 20, 561–7.

The Soil Profile

CHAPTER 5

B.W. Murphy and C.L. Murphy

Classifying requires the grouping of individuals. What, then, constitutes a soil individual? For practical purposes, the soil individual is the soil profile, a vertical section of soil extending from the surface to the parent material or some specified depth.

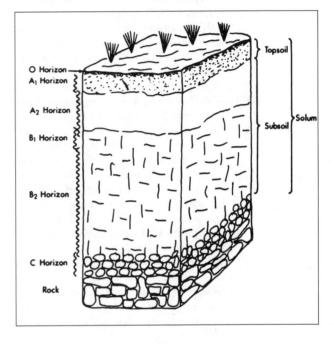

Figure 5.1   An idealised soil profile

5.1 Soil Profile Morphology

The soil profile is a column of soil extending downwards from the soil surface through all its horizons to parent material, other substrate material or to a specified depth. It may be exposed in a pit, cutting or gully, or by a coring machine. In practice, the soil profile is frequently considered as a column of soil with ends of 900 cm² in area (Northcote, 1979) and extending to either a depth of 1 m, or to parent material or other underlying materials unrelated to the soil profile. The latter situation may occur with buried soils or where deposited materials overlie bedrock.

The soil profile is generally composed of three major layers, designated A, B and C horizons, which occur approximately parallel to the land surface. The A and B horizons are layers that have been formed by weathering and soil development, and comprise the solum. The C horizon is partially weathered parent material. A surface organic horizon (O or P) and/or a subsolum (D) horizon may also occur.

The boundaries between successive soil horizons are specified by their width and shape. The following descriptions of the horizons are based on McDonald et al. (1990). An idealised profile is shown in Figure 5.1.

P HORIZON is a layer dominated by organic material in varying stages of decomposition which have accumulated under water or in areas of excessive wetness such as in swamps.

O HORIZON is a surface layer of plant materials in varying stages of decomposition not significantly mixed with the mineral soil. Often it is not present or only poorly developed in agricultural soils, but is fairly common in forest soils. When highly developed, it can be divided into two parts, O_1 and O_2.

O_1 HORIZON is the surface layer of undecomposed plant materials and other organic debris.

O_2 HORIZON is a layer consisting of partly decomposed organic debris. The original form of the organic debris cannot be recognised.

A HORIZON is the top layer of mineral soil. It can be divided into two main parts, A_1 and A_2.

A_1 HORIZON is the surface soil and is generally referred to as topsoil. It has an accumulation of organic matter, a darker colour and maximum biological activity relative to other horizons. This is usually the most useful part of the soil for plant growth and revegetation. It is typically from 5 to 30 cm thick. This horizon may be further divided into subhorizons if desired: for example, A_{11} and A_{12}.

A_2 HORIZON is a layer of soil paler in colour than the A_1 and B horizons, poorer in structure and lower in organic matter. It is often the zone of maximum leaching, clay translocation and weathering. It is characterised by losses of clay minerals, bases, sesquioxides and organic matter. When these processes are particularly strong, this horizon is white or grey, and is known as a bleached horizon. This is often an indication of impeded soil drainage. The A_2 horizon is typically from 5 to 70 cm thick, however it is not present in all soil profiles. These horizons generally have poorer structure and lower organic matter than the horizons above, and they are more dense and have lower infiltration rates. They can be a high erosion risk when exposed.

A_3 HORIZON is a transitional horizon between the A and B but is dominated by the overlying A_1 or A_2 horizons.

B HORIZON is the layer of soil below the A horizon. It is usually finer in texture (more clayey), denser and stronger in colour. Once exposed, it is usually a poor medium for plant growth. Thickness ranges from 10 cm to over 2 m. It can be divided into two main parts, B_1 and B_2.

B_1 HORIZON is a transitional horizon dominated by properties characteristic of the underlying B_2 horizon.

B_2 HORIZON is a horizon of maximum development, owing to concentration of clay minerals, iron, aluminium and/or translocated organic material. Structure and consistence are gener-

Table 5.1   Soil horizon designations (after McDonald et al., 1990)

b	buried horizon
c	accumulation of concretions of nodules of iron, aluminium or manganese
d	densipan, very fine sandy earthy pan
e	conspicuous bleached horizon
f	faunal accumulation such as earthworm casts
g	strongly gleyed horizon, indicative of permanent or periodic waterlogging
h	accumulation of amorphous organic matter-aluminium complexes, common in the B horizon of poorly drained podzols (podosols)
j	sporadic bleached horizon
k	horizon with an accumulation of carbonates
m	strongly cemented or indurated horizon
p	ploughed horizon or layer
q	horizon with accumulation of secondary silica
r	layers of weathered rock
s	horizon with accumulation of sesquioxide-aluminium, often present as B horizons of free-draining podzols (podosols)
t	accumulation of silicate clay
w	colour and/or structure development in B horizon with little sesquioxide-organic matter complexes present
y	accumulation of calcium sulfate (gypsum)
z	accumulation of salts more soluble than calcium sulfate (gypsum) and calcium carbonate.

ally unlike that of the A and C horizons, and colour is typically stronger. This horizon may be further divided into subhorizons based on colour, structure, texture or other morphological features. These subhorizons are called B_{21}, B_{22} and B_{23}.

B_3 HORIZON is a transitional horizon between the overlying B horizon and the subsolum material but whose properties are dominated by those of the overlying B horizon.

C HORIZON consists of the weathered, consolidated or unconsolidated layers of parent material below the B horizon. These are rarely affected by biological soil-forming processes.

The C horizon is recognised by its lack of soil formation activity and development, and by remnants of geologic features such as depositional layers or 'ghost' rock structure. Its thickness is variable.

D HORIZON consists of layers below the soil profile that differ markedly from the parent material and from the soil profile in character and pedological organisation. Included are buried soils, depositional layers and bedrock which has not influenced the soil profile, as may occur when aeolian materials are deposited on bedrock.

5.2 Soil Description in the Field

Accurate, reliable descriptions of soils in the field require a degree of training and experience. Detailed guidelines for the field examination of soils are given in the *Australian Soil and Land Survey Field Handbook* (McDonald et al., 1990) and the *Soil Data System—Site and Profile Information Handbook* (Abraham and Abraham, 1992). It is recommended that soils information captured in the field be recorded in a systematic manner such as on any one of the data cards that are now available. The New South Wales Soil Data System has a number of soil data cards that can be used for various purposes and levels of information collected. A generalised introduction to examination of soils in the field is given in this chapter.

Choice of Site 5.2.1

Sites for soil profile description are chosen for a number of reasons. Usually a site is representative of one or more of the following:

—a landform element
—a vegetation type
—a land use type
—a particular rock type or geological formation
—an area of soils upon which a trial or research project is conducted
—a particular type of soil
—a particular pattern on a remote sensing image such as a satellite image or an airborne gamma ray image (radiometrics).

The choice of site, particularly in soil mapping, may be facilitated by the use of aerial photographs, geological maps, geomorphological maps, vegetation maps, satellite imagery, specialised remote sensing images such as airborne gamma ray spectrometry or electromagnetic induction or, if available, soil maps.

Site Description 5.2.2

The site description is necessary to establish the place of the soil profile to be described in the landscape. This enables statements to be made about what the soil profile is representative of, whether a particular landform, vegetation, land use, rock type or geological formation.

While vital if the soil profile is to be used in soil mapping, a good site description is also valuable information if the soil profile description is being used for the characterisation of a research site, a reference profile for a particular soil or a demonstration. For research sites, a good soil profile description is essential, otherwise the representativeness of the research work and the likely extent of the extrapolation of the research results remain unknown.

The site description comprises information on the landform element, the landform pattern, vegetation, lithology, geological formation, land use, profile drainage, erosion, microrelief and evidence of salting. At the site, the condition of the surface soil is recorded, whether self-mulching, hardsetting, crusted or other special condition.

Preparation of the Soil Profile for Examination

The soil profile can best be observed in a vertical section of soil exposed in a pit or cutting dug for the purpose. The site of the pit should be chosen very carefully, taking into account such factors as ground cover, degree of erosion or deposition, surface drainage, proximity to trees and stumps, and so on—in fact, anything likely to detract from the soil as being representative of the area being considered. The site should normally be chosen only after preliminary exploratory test-hole boring. Once selected, the location of the soil site should be accurately marked on a map or aerial photograph.

As far as possible, the pit should be dug so that the sun shines on one side of the pit for as long as possible, especially at the time of sampling. The depth of a soil profile pit will vary with the soil body and sampling requirements. In deep soils, 2 m is usually adequate. The main consideration in these soils is that the B_2 horizon should be penetrated. Sampling of layers below the floor of the pit can be taken using an auger. If using an existing trench or cutting it may be necessary to cut it back at least 30 cm to ensure that the profile is not contaminated by slump, wash or lateral sediment movement (see Plate 5.1).

Where it is not possible to dig a pit, a spade and auger may be used. Use the spade to cut down and bring out the top 30 to 45 cm of soil, placing this on the ground. Then, auger out the soil, beginning at the bottom of the spade hole and placing each boring in order by the soil taken with the spade, so that measurements can be made and the profile viewed as a whole. Continue augering to the required depth, as for the pit.

The use of mechanical soil augers can be very cost effective as it ensures the maximum use of expensive field time. Augers that take 15 cm diameter cores are almost ideal, although augers that take smaller cores are also useful. A photo of a mechanical auger taking a 15 cm core is shown in Plate 5.2. The 15 cm core is large enough for reliable field description, although some soil structural units are still too large to be readily observed with this core. There is usually ample soil available for both field description and sampling for laboratory analyses.

Describing the Profile

When the face of the pit or cutting has been cleaned down, or when the auger borings are complete, the profile is ready to be described. One of two approaches may be taken in describing the soil profile.

(a)   Mark out the observable changes in colour, structure, texture or fabric and describe the soil above each boundary change. If there are

Plate 5.1   A soil profile prepared for description in a soil pit

Plate 5.2 A soil profile prepared for description after mechanical augering

no observable changes in layers more than about 10 cm thick, it may be necessary to take the second approach.

(b)   Take a metre rule or tape and mark the soil profile (trench face or augered profile) into 10 cm portions starting with the surface of the predominantly mineral soil as 0 cm and working downwards. Surface accumulations of organic matter are measured upwards from their junction with the mineral soil. When this is done, examine each 10 cm portion to determine whether or not a visible soil change occurs between the 10 cm markings and, where it does, mark this also. Now the process of describing the profile by 10 cm and part 10 cm portions may be started.

A photo of a profile is often a valuable asset in preparing reports or presenting information in presentations. For a small effort in time, the profile photo collects a lot of visual information that usually cannot be gathered again without considerable cost and effort.

Soil Properties Described in the Examination of a Profile

5.2.5

The following soil properties should be recorded for each layer or horizon. The determination of these properties is described in three basic references: *Australian Soil and Land Survey Field Handbook* (McDonald et al., 1990), *Soil Data System—Site and Profile Handbook* (Abraham and Abraham, 1992), and *A Factual Key for the Recognition of Australian Soils* (Northcote, 1979).

(a) Texture

Texture gives an estimate of the relative amounts of clay, silt and sand in a soil, and is based on the behaviour of the soil when worked at different moisture contents. Texture is usually carried out on the less than 2 mm fraction of the soil. There are six field texture groups identified (Northcote, 1979). There is only an approximate relationship between texture determined in the field and particle size analysis done in the laboratory, as other factors than the relative amounts of clay, silt and sand affect the behaviour of soil when determining texture.

Texture Group	Approx. Clay Content
1. Sands	less than 5%
2. Sandy Loams	10–15%
3. Loams	20–25%
4. Clay Loams	30–35%
5. Light Clays	35–40%
6. Heavy Clays	greater than 45%

Each of these texture groups is further subdivided for the purpose of field identification. The full procedure is set out in the references just cited, and is based on the ribboning behaviour of moist soil worked in the hand (see Plate 5.3). Considerable experience is necessary to obtain consistent results with the method described.

The six main texture groups should be apparent as follows:

Sands have very little or no coherence and cannot be rolled into a stable ball (bolus). Individual sand grains adhere to the fingers.

Sandy loams have some coherence, can be rolled into a stable ball and will form a ribbon 15–25 mm long. Sand grains can be felt during manipulation.

Loams will form a ribbon about 25 mm long. The soil ball is easy to manipulate and has a smooth spongy feel with no obvious sandiness.

Clay loams will form a ribbon 25–50 mm long. The soil ball is becoming plastic, capable of being moulded into a stable shape.

Light clays will form a ribbon 50–70 mm long. Plastic behaviour is evident in the soil ball, which has a smooth feel with some resistance to ribboning.

Heavy clays will form a ribbon more than 75 mm long. The soil ball is smooth and very plastic, with moderate–strong resistance to ribboning.

The texture groups may be further subdivided as described in the preceding references.

Texture, when determined in the field, is also influenced by organic matter (particularly at high levels), clay types, soil cations and the degree of soil structural development. Highly structured soil, high in free iron, can texture substantially lighter than would be expected from their clay contents. These soils are called subplastic.

(b) Texture Changes with Depth—the Texture Profile

Sampling of the profile should be frequent enough to detect the requisite texture changes, which are

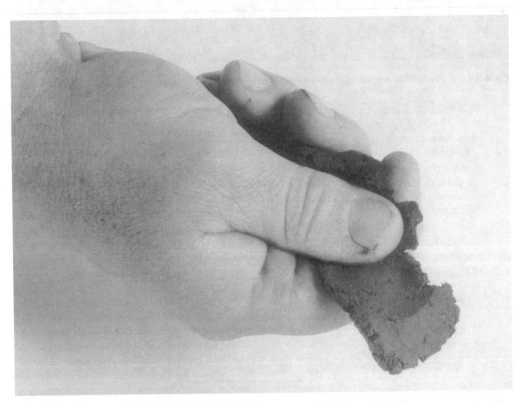

Plate 5.3 The field ribboning procedure used for assessing soil texture

defined in the Factual Key (Northcote, 1979) as follows:

Division	Range of Texture Down the Solum
Uniform (U)	Within one main texture group or the span thereof
Gradational (G)	Exceeds one main texture group or the span thereof (gradual increase in clay content)
Duplex (D)	Texture contrast between A and B horizons at least 1.5 main texture groups, or the span thereof, occurring over 10 cm or less

However, as the Factual Key (Northcote, 1979) is being gradually replaced by the new Australian Soil Classification (Isbell, 1996) (see Chapter 6), it is necessary to also include the criteria for texture contrast soils in this new classification which differs slightly from that in the Factual Key. A major division in the new classification is for soils with a 'clear and abrupt textural B horizon'. This type of B horizon must have twice as much clay as the A horizon above if the A horizon has less than 20% clay; or if the A horizon has more than 20% clay, then the B horizon must have an increase of clay of at least 20% relative to the A horizon. Under this criterion, some of the soils defined as duplex soils in the Factual Key will not have a 'clear and abrupt textural B horizon'. Therefore not all duplex soils in the Factual Key will automatically be in the soil orders in the Australian Soil Classification requiring a clear and abrupt textural B horizon.

(c) Description of Organic Soils

Although technically speaking organic soils do not have textural names they can be assessed by examining the degree of distinctiveness of the plant remains.

Fibric Peat

Fibrous peat, undecomposed or weakly decomposed plant remains which are readily identifiable

Hemic Peat

Semi-fibrous peat, moderately to well decomposed organic material, plant remains difficult to identify

Sapric Peat

Humified peat, strongly to completely decomposed organic matter

Sandy Peat

Organic material sandy to touch

Loamy Peat

Obvious mineral particle content, smooth to touch

Clayey Peat

Obvious fine mineral particle content, sticky when wet and weakly coherent.

(d) Colour

Soil colour is the most obvious property of the soil. While not important in itself, it can indicate much about a soil's history and likely behaviour. For example, well-drained soils tend to be red and poorly drained soils tend to be pale yellow or grey, often with orange mottles. Dark soils frequently, but not always, contain higher levels of organic matter.

Soil colour is usually described in terms of common colours such as red, orange, yellow, brown, black or grey. Bluish or greenish greys may occur in waterlogged soils and are referred to as gley colours. Soil colour charts such as those of Munsell Colour Company (1975) and Fujihara Industry Company (1967) should be used for more scientific or objective assessments of soil colour.

It is important that colours are determined on the moist soil (except where otherwise required) with a Munsell soil colour chart or its equivalent. This is the standard chart for soil colour determination. Soil colour is expressed as a letter/number code combining hue (spectral colour, normally in terms of red and yellow), value (dark and light) and chroma (intensity of colouration).

For example, in the colour code 5YR 4/6,
   5YR is the hue
   4 is the colour value
   6 is the chroma rating.

Each page of the standard soil colour chart represents one level of hue which is shown at the top of the page (5YR, 10R and so on). Changes in value occur down the page and changes in chroma across the page. Thus, light colours are at the top of each page and dark ones at the bottom, dull colours on the left of each page and bright colours on the right. For the Factual Key, the combinations of value and chroma are grouped into five categories. These are set out in a diagram on page 11 of the Factual Key and, for convenience, the boundaries can be drawn in on each page of the colour chart if desired. The categories are called value/chroma ratings (V/C 1–5) and are important criteria for classifying soils using the Factual Key. Standard viewing conditions are overcast sky or open shade, sidelit at 45 degrees. The soil aggregate should be viewed through the holes in the page using the appropriate achromatic mask.

(e) Mottling

Soil material is either whole-coloured or mottled. Mottles are blotches of colour different from the main soil colour. Usually at least 10% of the soil should be affected before a soil is described as mottled. Estimation charts are included in the standard soil colour charts for this purpose. The abundance, size, contrast colour and distinctiveness of mottles can also be described (McDonald et al., 1990; Abraham and Abraham, 1992). Mottles are most commonly associated with poorly drained soils that become seasonally wet, but are sometimes associated with weathering.

(f) Structure (Pedality)

The term 'structure' refers to the arrangement of soil particles, and if this includes natural aggregates these are called peds. Thus, a soil may be described as pedal or apedal depending on the presence or absence of peds. Examples of pedal soils include those with blocky structured B horizons, typical of many red podzolic soils, and those with naturally self-mulching A horizons, such as in the black cracking clays. Examples of apedal soils are loose incoherent sands, B horizons of red earths or massive clays in which the subsoil layer appears as a solid mass without any marked aggregation.

Peds are described as smooth or rough-faced. The former are relatively dense, tough, easily defined structural units in which the ped surfaces are almost shiny. The latter are less dense and characteristically porous and friable—they lack any lustre and are not easily distinguished.

Soil structure is described by type or shape and size of ped with the common types of soil structure being shown in Figure 5.2. Different soil structures are often indicators of properties of soil. Lenticular structure is common in soils with a high shrink-swell potential while columnar structure often indicates hard dense subsoils which are sodic. Polyhedral structure often indicates highly aggregated soils with high free iron contents.

(g) Fabric

The term 'fabric' is used to describe the appearance of the soil material when viewed under a ×10 hand lens. Four categories of earthy, sandy, rough-ped and smooth-ped fabric are used, the first two generally applying to the apedal condition and the second two to pedal soils. These categories are defined by Northcote (1979) as:

Earthy Fabric

Soil material is coherent with pores but few if any peds; soil particles are coated with clays and oxides arranged around the pores

Sandy Fabric

Soil material is coherent with few if any peds and consists of closely packed sand grains

Rough-ped Fabric

Peds are evident and more than 50% are rough-faced with porous, lacklustre surfaces; they are usually porous and friable

Smooth-ped Fabric

Peds are evident and more than 50% are smooth-faced with almost shiny surfaces; they are relatively dense, tough, easily defined structural units

(h) Coherence

This term describes the degree to which soil material is held together at different moisture levels. At least two-thirds has to remain united for a soil to be described as coherent, at the given moisture level.

(i) Soil pH

Soil pH is a measure of the acidity and alkalinity of the soil. A soil is referred to as being acidic if the pH is less than about 6.5; neutral if the pH is between 6.5 and 8.0; or alkaline if the pH is greater than 8.0.

Soil pH can be determined in the field using Raupach indicator (Raupach and Tucker, 1959) or Universal indicator (see Colour plate 5.5).

Soil reaction trends, or how the soil pH changes down the profile, are used in the Factual Key to classify soils. Four reaction trends, shown in Figure 5.3, are recognised. For many soils, the pH in the lower B horizon is diagnostic.

(j) Consistence

This refers to the resistance to deformation of soil material at different moisture contents. It can be assessed by the force required to crumble an air-dry, 2 cm lump of soil between thumb and finger as follows:

(i)   very small force required
(ii)   small but significant force required
(iii)   moderate force required
(iv)   strong force required
(v)   very strong force required, beyond power of thumb and finger.

(k) Pans

Pans are hardened and/or cemented layers in the soil which present considerable resistance to the manual boring of an auger hole under normal conditions. Their hardness would not generally be greatly affected by changes in moisture content. The common types of pans are caused by cementation due to calcium carbonate, silica, sesquioxides, humus or

(a) Platy

fine, < 2 mm thickness
medium, 2 to 5 mm thickness
coarse, > 5 mm thickness

(b) Lenticular

fine, < 10 mm thickness
medium, 10 to 50 mm thickness
coarse, > 50 mm thickness

(c) Prismatic (flat top)

fine, < 20 mm thickness
medium, x = 20 to 50 mm thickness
coarse, x > 50 mm thickness

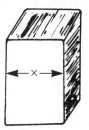

(d) Columnar (domed top)

fine, < 20 mm thickness
medium, x = 20 to 50 mm thickness
coarse, x > 50 mm thickness

(e) Angular blocky,   Subangular blocky,
(sharp edges)   (rounded edges)

fine, < 20 mm thickness
medium, x = 20 to 50 mm thickness
coarse, x > 50 mm thickness

(f) Polyhedral (many–sided)

fine, < 2 mm thickness
medium, 2 to 5 mm thickness
coarse, > 5 mm thickness

(g) Granular Round

fine, < 2 mm thickness
medium, 2 to 5 mm thickness
coarse, > 5 mm thickness

NB

The size names of fine, medium and coarse are frequently not given, only the size is recorded as in McDonald et al. (1990)

Figure 5.2   Soil structure names

iron oxides. Clay pans are layers of very low porosity clay. Pans usually have low permeability and frequently restrict root growth.

(l) Segregations and Concretions

These are discrete mineral accumulations in the soil that occur as a result of the concentration of some constituent, usually by biological or chemical means.

Segregations tend to be soft, concretions hard. Types include calcium carbonate, gypsum, manganese compounds, iron compounds and worm casts. They vary widely in size, shape, hardness and colour.

(m) Coarse Fragments

These are particles, not included under Texture, which are greater than 2 mm in diameter. Most

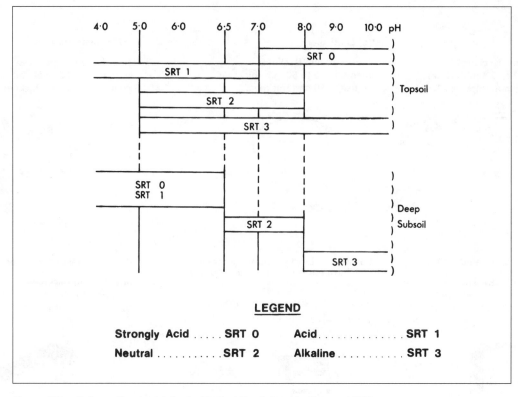

Figure 5.3   Soil reaction trends in the Factual Key (after Northcote, 1979)

commonly they are rock fragments such as quartz, shale or basalt floaters.

Field Soil Tests

Contents of a Soil Testing Kit
The contents of a typical soil testing kit are as follows (see also Plate 5.4):

Reagents: distilled water, 5% silver nitrate solution, 5% barium chloride solution, barium sulphate powder, hydrochloric acid (1_N), hydrogen peroxide (6%), Raupach indicator and/or Universal indicator (note that a colour chart and a special grade of barium sulphate are necessary for use with these indicators).

Apparatus: test tubes, evaporating dish, petri dish, beaker, spatula, corks, spotting tile, dropper bottles for indicator and acids, hand lens (×10), knife or trowel, spade, mattock, geological pick, measuring tape, soil auger, filter paper (Whatman No. 42), clinometer, compass, data cards, pencil or pen.

Soil Tests
(i)   *Soil Texture*: Referred to earlier in this chapter and basic references include McDonald et al. (1984) and Northcote (1979).

(ii)   *Emerson Aggregate Test*: A beaker should be partly filled with distilled water and a natural clod of soil (1 cm diameter) carefully placed in the

water. *Do not stir or shake*. Appearance of a 'cloud' of clay surrounding the clod indicates a degree of dispersibility. Quick disintegration of the particle indicates high erodibility. The test should also be repeated in groundwater if this is available. (See Chapter 10 for a more detailed description of the test.)

(iii)   *Soil Colour*: Refer to Colour in the previous section, and Munsell Colour Company (1975) and Fujihara Industry Company (1967).

(iv)   *Soil pH*: *Either* mix a small amount of soil with a few drops of Raupach indicator, and make into a paste on the spotting tile. Sprinkle the paste with barium sulphate powder (special grade) and compare colour formed with chart after three minutes (Raupach and Tucker, 1959); *or* place about 1 cm depth of soil in a test tube and the same depth of barium sulphate powder. Fill the test tube about one third full of distilled water, add a few drops of Universal indicator and shake thoroughly for one minute. Allow to settle and compare colour with chart. If cloudy or poorly coloured, add more barium sulphate and indicator, and reshake.

(v)   *Calcium Carbonate*: Effervescence visible on the application of a few drops of hydrochloric acid.

Plate 5.4 The contents of a field kit used for soil description

(N) indicates presence of calcium carbonate in soil.

(vi) *Soluble Carbonates in Seepages*: Add barium chloride solution—a white precipitate that clears on the addition of hydrochloric acid indicates presence of soluble carbonates.

(vii) *Soluble Chlorides in Seepages*: Add silver nitrate solution—a white precipitate gradually turning purple indicates presence of chlorides (probably sodium chloride). Presence of carbonates or chlorides in seepages may indicate dispersible soils. To determine presence of soluble salts in soils, an extract of the soil can be made by shaking the soil thoroughly with distilled water, filtering and treating the filtrate as before.

(viii) *Soluble Sulphates in Seepages*: Add barium chloride solution—a white precipitate that does not clear in hydrochloric acid indicates presence of sulphates.

(ix) *Organic Matter*: Add hydrogen peroxide (H_2O_2)—effervescence usually indicates organic matter. H_2O_2 also reacts with concretions of manganese compounds but these are more likely to be formed in the B horizon. A positive reaction to H_2O_2 in the surface soil is most likely organic matter, but a positive reaction in the subsoil is most likely to be manganese.

Interpretation of Soil Morphological Data

5.2.7

A number of interpretations have to be made from the soil profile data before classification of the soil can be achieved.

(a) The degree of pedological organisation has to be assessed. Evidence of this is to be found in all soils in which chemical, biological or physical changes have taken place. It reflects soil-forming processes and may be revealed as horizons, colour changes, texture changes, mottling, concretions of lime or iron compound, structural development and consistence changes.

(b) Horizons, where appropriate, should be assigned to the layers previously described. Of particular relevance is the presence of an A_2 horizon which is lighter than the A_1 horizon above and the B horizon below. If present, the A_2 horizon should be assessed as bleached or unbleached. A bleached horizon is very light in colour and has a high value and low chroma (typically V/C = 3), which must be determined on *dry soil*. 'Sporadically bleached' and 'conspicuously bleached' are terms used to describe the amount of bleaching present. Other horizons, as per Table 5.1, may be designated.

(c) The texture profile as defined in the Factual Key and the new Australian Soil Classification should be determined as described earlier in this chapter.

(d) It may be necessary to decide if sufficient mottling is present for a soil to be classified as mottled. Normally this is about 10%.

(e) Soil pH—the degree of acidity or alkalinity in a soil—is required in most soil classifications and for identifying potential limitations of a soil.

(f) The presence of sesquioxide/aluminium and humus/aluminium horizons is required for the new Australian Soil Classification.

(g) Laboratory analysis is required for the classification of some groups in the new Australian Soil Classification; for example, tests for free iron, base status, exchangeable sodium percentage and water pH values are sometimes required (Isbell, 1996).

Other interpretations of the data will be required before final classification of a soil is made, but these will depend on the classification system used.

BIBLIOGRAPHY

Abraham, S.M. and Abraham, N.A. (eds) (1992), *Soil Data System—Site and Profile Information Handbook*, Department of Conservation and Land Management, Sydney.

Fujihara Industry Company (1967), *Standard Soil Colour Charts*, Revised Edition, Fujihara Industry Company, Tokyo.

Isbell, R.F. (1996). *The Australian Soil Classification*, CSIRO, Melbourne.

McDonald, R.C., Isbell, R.F., Speight, J.G., Walker, J. and Hopkins, M.S. (1990), *Australian Soil and Land Survey Field Handbook*, Inkata Press, Melbourne.

Munsell Colour Company (1975), *Munsell Soil Colour Charts*, Munsell Colour, Macbeth Division of Kollmorgen Corporation, Maryland.

Northcote, K.H. (1979), *A Factual Key for the Recognition of Australian Soils*, Rellim Technical Publications, Glenside, SA.

Raupach, M. and Tucker, B.M. (1959), 'The field determination of soil reaction', *Journal of the Australian Imstitute of Agricultural Science* 25, 129–33.

Systems of Soil Classification

CHAPTER 6

B.W. Murphy and C.L. Murphy

Classification is the grouping of things according to their similarities and differences. It summarises information and eases communication between people. Usually a classification implies a whole collection of attributes in addition to those on which the classification is made. In the case of soils, they may be grouped according to the way they were formed, their parent material, their morphology or soil properties relevant to a particular purpose.

Soil classification is useful because it:

—groups soils into categories which provide a basis for communication

—summarises information so that many statements about one particular soil are likely to apply to other soils in the same group

—provides a basis for the scientific and logical evaluation of soils

—provides a basis for categorising experience with soils so that information about a group of soils can be built upon.

Naturally, no system of soil classification is going to satisfy everyone interested in soils. A soil conservationist, a vegetable grower, a civil engineer and an environmentalist would have different criteria to determine the most useful soil classification. It is also unlikely that one classification will completely fulfil the needs of one individual. This particularly applies to anyone concerned with sustainable land management, who is interested in soils from a number of points of view.

Any group of objects may be classified in a variety of ways. People may be grouped according to nationality, sex, education, age or occupation. Each is valid depending on the purpose of the classification. Likewise, soils have been classified in a large number of ways (see Table 6.1), but two broad types of classification can be distinguished—general purpose and specific purpose.

General purpose classifications are those based on a collection of different and often quite unrelated features or attributes. They are designed with no specific purpose in mind, but aim to provide a useful summary of available information that may prove useful in a wide variety of contexts. Most of the common systems of soil classification are of this type.

Specific purpose classifications are those based on a narrow range of specified soil properties. Often these are related to particular land use problems such as are associated with acid soils, sodic soils, saline soils, expansive soils or acid sulfate soils. Such systems are usually used when investigating particular problems of soil management.

6.1 Soil Classification Systems in Australia

The first soil classification system was based on soil genesis and developed by the Russians. It was subsequently modified and improved by the Americans to the great soil group system, before being introduced into Australia by Prescott in the 1930s (Prescott, 1931). Prescott emphasised the role of climate and vegetation in soil formation and also recognised the presence of ancient landscapes in Australia which complicated the simple climate–vegetation–soil relationships proposed by the Russians. Stephens (1962) modified Prescott's great soil groups and emphasised the need to group soils solely on observable soil features such as colour, texture, structure, consistence, the nature of horizon boundaries and various other features. The selection of soil features was based on their assumed relationship with soil-forming processes. Stephens also recognised the role of parent material. The latest development of the great soil groups is by Stace et al. (1968).

Other genetic systems of classification are in

Table 6.1 Examples of soil classification systems

Classification	Type	Basis	Examples
Australian Soil Classification Scheme	General purpose	Morphological and chemical properties of soil profile	Kurosols, Chromosols, Vertosols, Kandosols
Great Soil Group	General purpose	Morphological and chemical properties based on concepts of soil genesis	Red podzolic soils, krasnozems, podzols, yellow earths, black earths
Factual Key	General purpose	Morphological and chemical properties based on their ability to group soils into recognisable groups	Hard, pedal red duplex soils (Dr2), black cracking clays (Ug5.15), yellow massive earths (Gn2.23)
Soil Taxonomy	General purpose	Morphological, chemical and climatic features of soil based on concepts of soil genesis	Pelliustert, calcixeroll, natrixeralf, palexeralf, haplohumox, haplargid
Unified Soil Classification	General purpose (engineering)	Particle size grading, plasticity and organic matter content	Gravel, GC, GM, GP; sands, SC, SM; clays of low (CL) and high plasticity (CH)
Location	General purpose	Location, sometimes in combination with other classifications	Urrbrae loam, Moree cracking clays, Bathurst red duplex soils
Sodicity	Special purpose	Exchangeable sodium	Non-sodic, sodic, strongly sodic
Salinity	Special purpose	Salt content	Saline, non-saline
Expansive Soils	Special purpose	Shrink-swell potential	Expansive, non-expansive
Acid Sulfate Soils	Special purpose	Levels of oxidisable sulfur in the soil	Nil, low, moderate, high levels of oxidisable sulfur

Some commonly used, general classifications:*

Classification	Type	Basis	Examples
Parent Material	General purpose	Parent material of soil	Granite soils, basalt soils, shale soils
Texture	General purpose	Soil texture	Light, heavy; coarse, fine; sandy loam, loam, clay loam, clay
Topography	General purpose	Topography of soil formation	Mountain soils, flats, gilgai soils, plains, trap soils
Vegetation	General purpose	Species of vegetation growing on the soil	Ironbark soils, White box soils, Mulga soils, Mallee soils
Climate	General purpose	Climate of soil formation	Tropical, temperate, alpine

These soil classification systems are commonly used by agronomists, engineers, valuers and farmers

operation throughout the world, the most significant being Soil Taxonomy (Soil Survey Staff, 1975) and the FAO/UNESCO system (FAO, 1990).

In contrast to the great soil group systems, the Factual Key (Northcote, 1979) is an Australian-based hierarchical system or bifurcating key with clearly defined classes. Only soil features that can be observed and evaluated in the field are used to classify soils, and no assumptions are made about soil genesis. It was developed in the late 1950s and early 1960s, in order to produce the *Atlas of Australian Soils* (Northcote et al., 1960–68). It culminated in *A Description of Australian Soils* (Northcote et al., 1975).

Soil classification in Australia has been reviewed by Moore et al. (1983), Northcote (1983) and Isbell (1984, 1988). These reviews generally conclude that while the great soil group system and the Factual Key are useful, workable systems, both have deficiencies which need to be recognised when they are used for soil survey or in relation to land management.

More recently, Isbell (1996) has developed a new Australian Soil Classification System. The development of this classification involved not only field trips with a wide range of Australian soil scientists, but was based on a data base of over 14 000 soil profiles from throughout Australia. It is generally seen as a better approximation to the soils of Australia than the great soil groups or the Factual Key. However, it is relatively new and it will take time for people to build up the experience of using this classifica-tion for land management decisions and identifying soils. This classification has a strong focus on those soil properties considered most important for land management and it has been widely accepted by government land resource agencies around Aus-tralia. Over time it is expected to replace the Fac-tual Key (Northcote, 1979) and the great soil groups (Stace et al., 1968).

6.2 The Australian Soil Classification System (ASCS)

6.2.1 Origins and Foundations

The Australian Soil Classification System (Isbell, 1996) has been developed to overcome a number of perceived shortcomings in previous soil classifi-cation systems especially in regard to land use management. The ASCS is a multicategoric, hier-archical, general purpose scheme with mutually exclusive classes based on diagnostic horizons within a profile (Isbell, 1996). The system is based on a national database of over 14 000 profiles and has been tested by many soil scientists including various state government land assessment agencies prior to its publication. The classification is open-ended and classes can be included as knowledge is gained. One of the main criticisms of this classifica-tion is of its strong agricultural base and its limited application to urban and engineering applications. It is hoped this may be overcome as the classifica-tion is upgraded in the future. The Australian Soil Classification has been adopted by most soil survey agencies including the New South Wales Depart-ment of Land and Water Conservation.

6.2.2 Operation of the System

Like the Factual Key, the scheme is based generally on diagnostic field properties of the soils. However, laboratory properties (mainly chemical) are used at various levels in the classification to help distinguish classes with different land use properties (Isbell, 1996). For instance, at the highest level of the classifi-cation, the *order*, free iron content is used to distin-guish Ferrosols and exchangeable sodium percentage to distinguish Sodosols.

The classification uses a number of diagnostic terms, some of which will be new to the reader. A detailed glossary is provided in the glossary of the classification and needs to be closely followed. Fur-ther elaboration of the diagnostic features is provided in *Concepts and Rationale of the Australian Soil Classifica-tion* (Isbell et al., 1997).

A schematic diagram of the hierarchical classifica-tion (order, suborder, great group, subgroup, family) is outlined below. Note that the criteria will change depending on the soil order.

(i) Soil features used to classify soils at the order level:

Degree of pedological organisation, presence or absence of the following: strong texture contrast, sodicity in upper B horizon, free iron in upper B horizon, calcareous through profile, pH less than 5.5 in upper B horizon, cracks and slickensides, seasonal saturation, organic materials, podzol-type horizons, structure in B horizon, and degree of man-made disturbance.

(ii) Soil features to classify soils at the suborder level:

Colour, amount and type of calcium carbonate nodules/accumulations, presence of gypsum, degree of tidal influence, salinity, type of organic material, degree of saturation, stratifi-cation and depth of solum.

(iii) Soil features to classify soils at the great group level:

Character of underlying materials (rock, cal-crete, marl, siliceous pans), amount of exchangeable cations, presence and form of calcium carbonate and gypsum, presence and form of sulfidic materials, presence of humic/iron/aluminium compounds, presence of mottles and relative amounts of exchange-able sodium and magnesium.

(iv) Soil features to classify soils at the subgroup level:

Humic properties of the A horizon, presence of bleached A horizon, mottling in the B horizon, presence of free iron in lower B horizon, presence of sodium in the lower B horizon, presence and form of gypsum and calcium carbonate in the lower B horizon, presence of vertic and subplastic properties in the lower B horizon, presence of acidic lower B horizons, presence of manganese in B horizons, presence of bauxite-like materials in B horizon, and other criteria specifically for some great groups.

(v) Soil features to classify soils at the family level:

Depth of A horizon, texture or clay content of A horizon, texture or clay content of B horizon, amount of gravel in surface soil, depth of solum.

To classify a soil profile a number of keys have to be followed sequentially. The first selection is based on the *first order* of the key that includes the soil under examination. Thence a selection is made of the first appropriate class in *suborders, great groups* and *subgroups*. Finally, an appropriate selection is made of the *family* level. The order in which the classes are arranged is thought to be of decreasing order of importance and may be changed as new knowledge becomes available (Isbell, 1996).

Classification names where possible are based on Latin or Greek roots (see Isbell, 1996).

The order of listing of the nomenclature is generally: subgroup, great group, suborder, order, and family.

An example is:

sub-group	great group	sub-order	order	family
Bleached [AT]	Eutrophic [AH]	Red [AA]	Chromosol [CH]	thin [A], gravelly [G], sandy [K], clayey [O], shallow [U]

Bleached: soil has a conspicuously bleached A$_2$ horizon
Eutrophic: the soil has base status greater than 15 cmol(+)/kg clay in the major part of the upper 0.2 m of the B$_2$ horizon, and the B and BC horizons are not calcareous
Red is the dominant colour in the upper 0.2 m of the B$_2$ horizon
Chromosol: the soil has texture contrast between the A and B horizons and is neither strongly acid or sodic
thin: the soil has an A horizon 0.1 m thick

gravelly: 10–20% gravel occurs on the surface and A$_1$ horizon
sandy: has sandy A$_1$ horizon texture (S, LS, CS < 10% clay)
clayey: the B horizon maximum texture of clay (LC-MC-HC > 35% clay)
shallow: soil depth is 0.25–0.5 m

[ ] is the two-letter code (order, suborder, great group, subgroup) or one-letter code (family) for each class of the soil described.

The classification of a soil may be shortened if desired or if some levels of the hierarchy cannot be determined, for example Red Chromosol; Red Chromosol bleached; Red Chromosol, thin gravelly; and so on.

A two-letter code is provided for each class except for the family level which has a one-letter code. The codes are useful for recording on field sheets and will allow interrogation of databases to be carried out.

Provision is also made for situations where no suitable class can be identified (code ZZ), or where the class cannot be determined from the available information. A confidence level of the classification can also be used when class descriptions require analytical data.

Key to Soil Orders (based on Isbell, 1996; Isbell et al., 1997) 6.2.3

A. Soils resulting from human activities.

Anthroposols
Caused by a profound mixing, modification, truncation or burial of the original soil, or the creation of new parent materials. Common in urban and industrial areas. Limitations are varied but can include contaminated soils, soil erosion hazard, poor fertility, and so on.

B. Soils that are not regularly inundated by saline tidal waters and either:

1. have more than 0.4 m of organic materials within the upper 0.8 m. The required thickness may either extend down from the surface or be taken cumulatively within the upper 0.8 m; or

2. have organic materials extending from the surface to a minimum depth of 0.1 m; these either directly overlie rock or other hard layers, partially weathered or decomposed rock or saprolite, or overlie fragmental material such as gravel, cobbles or stones in which the inter-

stices are filled or partially filled with organic material. In some soils there may be layers of humose and/or melacic horizon material underlying the organic materials and overlying the substrate.

Organosols

Generally limited in extent to coastal swamps and alpine areas, these soils are often known as peats. Some less acidic organosols have been drained and sown to pastures for dairying or used for intensive vegetable growing. Waterlogging, peat shrinkage and decomposition are limitations for many land uses.

C. Other soils that have a Bs, Bhs or Bh horizon (see Podosol diagnostic horizons).

These horizons may occur either singly or in combination.

Podosols

Commonly known previously as podzols, these soils occur generally on Quaternary coastal sand bodies but can also occur on siliceous depositional material such as on quartz sandstones. Agriculture is extremely limited owing to the poor fertility, poor water retention and locally high water tables.

D. Other soils that:

1. have a clay field texture or 35% or more clay throughout the solum except for thin, surface crusty horizons 0.03 m or less thick, and
2. unless too moist, have open cracks at some time in most years that are at least 5 mm wide and extend upward to the surface or to the base of any plough layer, self-mulching horizon, or thin, surface crusty horizon, and
3. at some depth in the solum, have slickensides and/or lenticular peds.

Vertosols

Cracking clay soils in New South Wales occur on a variety of landforms from alluvial floodplains to rolling hills but are derived primarily from basaltic parent material. Mean average rainfall ranges from about 200 to 700 mm. These soils are mainly used for grazing of native and improved pastures and for dryland agriculture. Limitations to land use include high shrink-swell soils that pose problems for infrastructure, foundations and some crops. Irrigation problems due to restricted drainage may also occur.

E. Other soils that are saturated in the major part of the solum for at least two to three months in most years (that is, includes tidal waters).

Hydrosols

These wet soils occur principally in low-lying plains and basins especially along the coast of New South Wales. They require drainage to make them suitable for grazing. Along with drainage, furrowing has also been used to improve drainage for horticulture in some areas in the past.

F. Other soils with a clear or abrupt textural B horizon and in which the major part of the upper 0.2 m of the B$_2$ horizon (or the major part of the entire B$_2$ horizon if it is less than 0.2 m thick) is strongly acid.

Kurosols

Generally located in coastal and subcoastal areas (about 600 to 1200 mm average annual rainfall) on a variety of landforms. Parent materials are generally siliceous and include sediments, metasediments and granitic rocks. In high rainfall areas dairying on improved pastures and hardwood forests are common with grazing in lower rainfall areas.

G. Other soils with a clear or abrupt textural B horizon and in which the major part of the upper 0.2 m of the B$_2$ horizon (or the major part of the entire B$_2$ horizon if it is less than 0.2 m thick) is sodic and is not strongly subplastic.

Sodosols

Most extensive in areas with rainfall < 1200 mm. They occur commonly on plains and gently undulating to rolling landscapes often on colluvial or part-colluvial deposits. Used mainly for grazing and forestry, they have several major problems for use including extremely hardsetting topsoils, poor chemical and physical fertility, and are prone to sheet and tunnel erosion. Earthworks often require amelioration with gypsum to stabilise these sodic, often dispersible soils.

H. Other soils with a clear or abrupt textural B horizon and in which the major part of the upper 0.2 m of the B$_2$ horizon (or the major part of the entire B$_2$ horizon if it is less than 0.2 m thick) is not strongly acid.

Chromosols

These soils extend over a wide rainfall range from 300 to about 1200 mm and occur on a range of parent materials. These soils are prominent in the wheatbelt of southern New South Wales where they are used for cereal and seed growing. Soils are prone to structural degradation following long-established agricultural practices, particularly cultivation.

I. Other soils that are either calcareous throughout the solum or at least directly below the A₁ or Ap horizon or a depth of 0.2 m if the A₁ horizon is only weakly developed.

Carbonate accumulations must be judged to be pedogenic, that is, are a result of soil-forming processes in situ (either current or relict) in contrast to fragments of calcareous rock such as limestone or shell fragments. See also *calcrete*.

Calcarosols

These soils are generally confined to the arid and semi-arid areas of the state with 200–350 mm mean average rainfall. Most of the original mallee has been cleared for native pasture. Limitations can include poor chemical fertility such as alkalinity, boron toxicity, phosphorus deficiency and low trace elements.

J. Other soils with B₂ horizons in which the major part has a free iron oxide content greater than 5% Fe in the fine earth fraction (< 2 mm).

Soils with a B₂ horizon in which at least 0.3 m has vertic properties are excluded.

Ferrosols

These soils occur on basalt and to a lesser extent on dolerite and ultrabasic serpentine with 500–3000 mm mean average rainfall and on various landform patterns. Higher rainfall zones > 1000 mm originally supported rainforest and these have been extensively cleared for agriculture such as dairying on improved pasture. Some areas such as Lismore-Ballina support intensive horticulture. Although these soils are generally regarded as being among the most structurally stable soils, they can degrade under cropping from erosion and compaction. These soils are also highly aggregated and can pose sealing problems in earth dams.

K. Other soils with B₂ horizons that have structure more developed than weak throughout the major part of the horizon.

Dermosols

Occur dominantly in the wetter east coastal and sub-coastal zones on a variety of landforms and parent materials. They support a wide range of land uses, from hardwood forests and national parks in the coastal areas to extensive cropping for cereal production in central New South Wales.

L. Other soils that:

1. have well-developed B₂ horizons in which the major part is massive or has only a weak grade of structure (compare with tenic B horizon and cemented pans), and
2. have a maximum clay content in some part of the B₂ horizon which exceeds 15% (that is, heavy sandy loam, SL+).

Kandosols

Widespread in New South Wales occurring extensively on gently undulating plains related to lateritic profiles. Small areas are used for extensive agriculture such as the southern wheat-belt of New South Wales and on the Sommersby Plateau between Sydney and Newcastle.

M. Other soils with negligible (rudimentary) pedological organisation apart from the minimal development of an A₁ horizon, or the presence of less than 10% of B horizon material (including pedogenic carbonate) in fissures in the parent rock or saprolite.

The soils are apedal or only weakly structured in the A₁ horizon. There is little or no texture or colour change with depth unless stratified or buried soils are present. Cemented pans may be present as a substrate material.

Rudosols

Occur predominantly in arid areas but also on recent beach and foredune coastal deposits and recent nearstream alluvium. Land use is dependent on location and environment. Coastal sand ridges are generally used for recreation, while alluvium can be intensively used for crops and dairying, and dunes in the Riverine Plain are irrigated for citrus, vines and potato farming.

N. Other soils.

Tenosols

Tenosols have weak pedological organisation apart from the A horizon. They range widely in their physical environment from both high and low rainfall areas, from near desert to alpine areas. In New South Wales they occur mainly on hilly to mountainous erosional landscapes on all but the more easily weathered basic rocks. Owing to their general poor chemical and physical fertility, poor water retention and often stony nature, these soils are used for sparse sheep and cattle grazing, and some hardwood forests occur in higher rainfall areas.

6.3 Great Soil Groups

6.3.1 Origins and Foundations

In the great soil group system of classification, also referred to as GSG, the grouping depends on the presence of and types of morphological features observed in the field. The selection of these features and the weighting they receive are based on the concepts of soil genesis. The soil morphological features used are summarised in Table 6.2. Normally, laboratory data are not required to classify soils, although, more recently, data on exchangeable cations have been considered necessary by some, in order to classify a soil as a solodic soil. The notes in Appendix II will also assist in the identification of the great soil groups.

6.3.2 Operation of the System

The great soil group system operates using central concepts for each group, with ill-defined boundaries between groups, as shown in Figure 6.1.

The allocation of a soil profile to a great soil group is made using the knowledge and experience of the person examining the profile to ensure the soil profile has features consistent with the central concept of that group. No keys, defined boundaries or rules for allocating profiles exist. Intergrades between great soil groups are relatively common and probably should be recognised as such, rather than be allocated to one great soil group or another to which they do not fully conform.

An important point about the operation of GSG is that the allocation to a great group does not provide a very precise statement about a particular soil. Identification of a soil at the great group level is equivalent to identifying rocks as granite, basalt, and so on. There is scope for further refinement by qualifying the group name with the name of the locality where the particular member of the group is best represented, such as Cumberland red podzolic soil.

The types of soils recognised in GSG are presented in Table 6.3. The occurrence of these soils in New South Wales is discussed in Chapter 8. Equivalent groups of the Australian Classification Scheme, Soil Taxonomy and Factual Key are given in Appendixes I and II, respectively.

6.4 The Factual Key

6.4.1 Origins and Foundations

The Factual Key (Northcote, 1979) was developed in the 1960s to facilitate the production of a comprehensive map of Australian soils. Its development resulted from the lack of precise definition for the great soil groups, which saw similar soils being given different names, different soils being given the same name, and many soils which did not fit any great soil group. The Factual Key was formulated so that a specific set of soil features could be objectively observed in the field and then the soil classified on these. There were to be no inferences about soil genesis. The selection of soil features and the weighting they were given in the classification were based on their ability to group soils. The selection of these soil features was decided after examining a set of 500 profiles.

In practice, the Factual Key is very different from the great soil group system. Unlike the great soil group system which uses modal soil types with fuzzy

Table 6.2 Features used in the great soil group classification

Soil texture and texture profile

Soil colour

Soil structure — hardness and density of peds

        — degree of structure

        — type (blocky, polyhedral, columnar, lenticular)

        — presence of slickensides

Surface condition — crusting

           — self-mulching

Presence and type of A_2 horizons

Soil reaction (acid, neutral, alkaline)

Presence of calcium carbonate and distribution

Sequence of horizons

Soil depth

Organic matter levels

Presence of ironstone nodules

Parent material (terra rossa soils)

Seasonal cracking and gilgai

Exchangeable cations — Ca and Mg dominated

             — Na and Mg dominated (solodic soils)

Salt content

boundaries, both the Factual Key and the new Australian Soil Classification System (ASCS) are both hierarchical systems with mutually exclusive classes which rely on key diagnostic properties of the soil. Hence their strength lies in the emphasis on observable soil features and the way they define how these should be used to classify soils. This allows much more definitive statements to be made about soils, and it is because of this that the key has had wide usage and acceptance by organisations concerned with practical soil problems. One strength of the Factual Key is that it provides a convenient summary of the main features of a particular soil profile.

Operation of the Factual Key

General Aspects

The Factual Key depends on the recognition of a number of morphological features of the soil profile, and some simple chemical tests that can be easily car-

ried out in the field. Identification of a soil is by means of a key using the observed features of the soil profile. Each feature has been defined in a glossary. The method is essentially a field one, and does not depend on any laboratory determination.

Five main levels of classification have been included—those of division, subdivision, section, class, and principal profile form. The soil description is expressed as a letter/number combination. As each feature of the soil profile is determined, a letter or number is added to the coding to account for that feature, letters being used for divisions and subdivisions, and numbers for sections, classes and principal profile forms. Each letter and number (and their position) has, therefore, a defined meaning to anyone using the system. The soil profile features used in the Factual Key are outlined in Table 6.4.

Throughout the key, accent is placed on the characters of the A_2 and B horizons, as it was found during

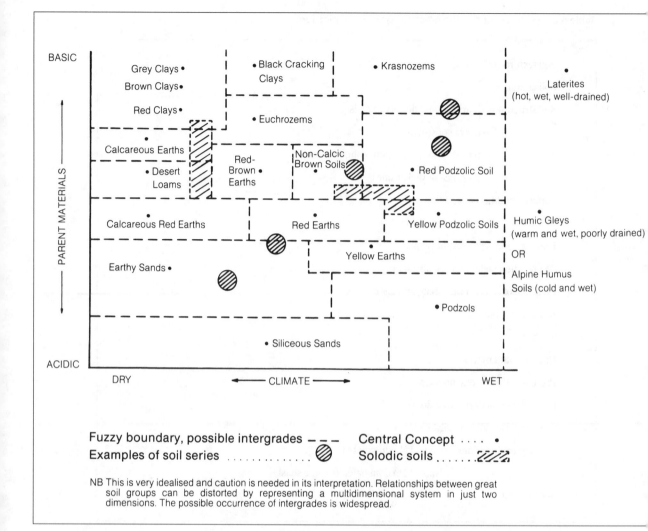

Figure 6.1 Suggested positions of the central concepts of the great soil groups

the study of profiles for this work that these were the most consistent and expressive features. However, other special features are used, such as the hardsetting nature of the surface and amount of calcium carbonate present in the profile.

Levels of Classification

The soil features used at each level of soil classification are described in the following section.

(i)   *Division:* Four soil divisions are recognised and, for precise definition, reference should be made to the glossary in the key itself.

Organic (O) soils are those dominated by organic matter.

Uniform (U) soils are soils of uniform texture throughout the profile. The change in soil texture does not exceed one texture group.

Gradational (G) soils show a gradual increase in texture (more clayey) down the profile, but without any sharp change. The total texture change in the soil profile exceeds one texture group.

Duplex (D) soils are those with a marked texture contrast between the A and B horizons. The texture contrast is at least 1.5 texture groups, within 10 cm or less.

(ii)   *Subdivision:* The subdivisions within each division are simply defined.

The U division is split into Uc—coarse textures; Um—medium textures; Uf—fine textures; and Ug—fine textures with seasonal cracking. The term 'seasonal cracking' applies to those clay soils that crack open in the dry season. Cracks should be at least 6

Table 6.3 Great soil groups in order of degree of profile development and degree of leaching (after Stace et al., 1968)

1. No profile differentiation	1. Solonchaks 2. Alluvial soils 3. Lithosols 4. Calcareous sands 5. Siliceous sands 6. Earthy sands
2. Minimal profile development	7. Grey-brown and red calcareous soils 8. Desert loams 9. Red and brown hardpan soils 10. Grey, brown and red clays
3. Dark soils	11. Black earths 12. Rendzinas 13. Chernozems 14. Prairie soils 15. Wiesenboden
4. Mildly leached soils	16. Solonetz 17. Solodised solonetz and solodic soils 18. Soloths (Solods) 19. Solonised brown soils 20. Red-brown earths 21. Non-calcic brown soils 22. Chocolate soils 23. Brown earths
5. Soils with predominantly sesquioxidic clay minerals	24. Calcareous red earths 25. Red earths 26. Yellow earths 27. Terra rossa soils 28. Euchrozems 29. Xanthozems 30. Krasnozems
6. Mildly to strongly acid and highly differentiated	31. Grey brown podzolic soils 32. Red podzolic soils 33. Yellow podzolic soils 34. Brown podzolic soils 35. Lateritic podzolic soils 36. Gleyed podzolic soils 37. Podzols 38. Humus podzols 39. Peaty podzols
7. Dominated by organic matter	40. Alpine humus soils 41. Humic gleys 42. Neutral to alkaline peats 43. Acid peats

Table 6.4 Soil properties used in the Northcote Factual Key

Division	Subdivision	Section	Class	Principal Profile form
O	Neutral or acid reaction			
U	Texture Uc (coarse) Um (medium) Uf (fine) Ug (fine and seasonally cracking)	Pedologic organisation of profile and character of A₂ horizon Pedality	Nature of material below A₁ and/or A₂ Colour Coherence CaCO₃ Pedality Hardpans	Colour of material below A₁ horizon Soil depth and underlying material
G	Presence of calcium carbonate Gc (calcareous throughout)	Structure (pedality) and fabric of B horizon	CaCO₃ concentration in A horizon	Colour of the A₁ horizon or Coincidence of maximum clay and CaCO₃ horizons
	Gn (not calcareous throughout)	Structure and fabric of B horizon	Colour of B horizon Character of A₂ horizon	Character of A₂ horizon Soil reaction trend
D	Colour of upper B horizon Dr (red) Db (brown) Dy (yellow) Dd (dark) Dg (gleyed)	Nature of A horizon and degree of mottling of upper B horizon	Pedality of B horizon Character of A₂ horizon	Presence of pans or laterite Soil reaction trend

mm wide and 30 cm deep. This phenomenon is typical of most black earths and other cracking clay soils. Gilgai microtopography is frequently associated with these soils.

The G division is split into Gc—those soils which are calcareous throughout the solum; and Gn—those which are not. Calcareous soils contain nodules of calcium carbonate (lime) which will effervesce clearly with two to three drops of hydrochloric acid (1N) applied from a dropper.

The D division is split into Dr—red; Db—brown; Dy—yellow; Dd – dark; and Dg—gleyed; these colours applying to the upper 15 cm (at least) of the clayey B horizon. Soil colour is of importance throughout the operation of the key, not only in the duplex division. It is, therefore, essential to have a clear understanding of how colour is determined, and not be colour blind.

(iii) *Section, Class and Principal Profile Form:* At these levels of classification, a number of key features of the soil are used for identification, some of which are common to all divisions and some of which are used in combination. The main ones are as follows:

— pedologic organisation
— character of the A₂ horizon
— structure (pedality)
— nature of surface soil
— mottling
— coherence
— soil reaction trend
— consistence
— pans.

Table 6.5   Major soils of the Factual Key (adapted from Northcote et al., 1975)

Sands — Uc	
Calcareous Sands	Uc1.1, Uc1.3
Siliceous Sands	Uc1.2, Uc1.4
Bleached Sands	Uc2, Uc3
Pale Sands	Uc4
Earthy Sands	Uc5
Weakly Structured Sands	Uc6
Loams — Um	
Calcareous Loams	Um1.1, Um1.3
Siliceous Loams	Um1.2, Um1.4
Bleached Loams	Um2, Um3
Pale Loams	Um4
Earthy Loams	Um5
Friable Loams	Um6
Organic Loams	Um7
Non-cracking Clays — Uf	
Non-cracking Clays of Minimal Development	Uf1.4
Non-cracking Strongly Structured Clays	Uf4.4, Uf5.1, Uf5.2, Uf5.3, Uf6.1, Uf6.2, Uf6.3
Non-cracking Poorly Drained Clays	Uf6.4, Uf6.5, Uf6.6
Cracking Clays — Ug	
Black Self-mulching Cracking Clays	Ug5.1
Grey Self-mulching Cracking Clays	Ug5.2
Brown and Red Self-mulching Cracking Clays	Ug5.3
Black Massive Cracking Clays	Ug5.4
Grey Massive Cracking Clays	Ug5.5
Calcareous Earths — Gc	
Massive Earths — Gn2	
Red Massive Earths	Gn2.1
Yellow Massive Earths	Gn2.2, Gn2.3
Brown or Mottled-Red Massive Earths	Gn2.4
Mottled-Yellow Massive Earths	Gn2.6, Gn2.7
Grey Massive Earths	Gn2.8, Gn2.9
Smooth-Ped Earths — Gn3	
Red Smooth-Ped Earths	Gn3.1
Brown Smooth-Ped Earths	Gn3.2
Black Smooth-Ped Earths	Gn3.4

Smooth-Ped Earths — Gn3 *(cont.)*	
Mottled-Brown and Red Smooth-Ped Earths	Gn3.5
Yellow Smooth-Ped Earths	Gn3.7, Gn3.8
Grey Smooth-Ped Earths	Gn3.9, Gn 3.0
Rough-Ped Earths — Gn4	
Red Rough-Ped Earths	Gn4.1
Brown Rough-Ped Earths	Gn4.3
Black Rough-Ped Earths	Gn4.4
Red Duplex Soils — Dr	
Crusty Red Duplex Soils	Dr1.1, Dr1.3, Dr1.4
Hard Pedal Red Duplex Soils	Dr2.1, Dr2.2, Dr2.3, Dr2.4
Hard Apedal Red Duplex Soils	Dr2.6, Dr2.8
Hard Pedal Mottled-Red Duplex Soils	Dr3.2, Dr3.3, Dr3.4
Friable Red Duplex Soils	Dr4.1, Dr4.2, Dr4.4
Brown Duplex Soils — Db	
Hard Pedal Brown Duplex Soils	Db1.1, Db1.2, Db1.3, Db1.4
Hard Pedal Mottled-Brown Duplex Soils	Db2.3, Db2.4
Friable Brown Duplex Soils	Db3.1
Yellow Duplex Soils — Dy	
Hard Pedal Yellow Duplex Soils	Dy2.1, Dy2.2, Dy2.3, Dy2.4
Hard Apedal Yellow Duplex Soils	Dy2.6
Hard Pedal Mottled-Yellow Duplex Soils	Dy3.1, Dy3.2, Dy3.3, Dy3.4
Hard Apedal Mottled-Yellow Duplex Soils	Dy3.6, Dy3.8
Friable Pedal Mottled-Yellow Duplex Soils	Dy5.1
Sandy Pedal Mottled-Yellow Duplex Soils	Dy5.4
Sandy Apedal Mottled-Yellow Duplex Soils	Dy5.6, Dy5.8
Black Duplex Soils — Dd	
Hard Pedal Black Duplex Soils	Dd1.1, Dd1.3, Dd1.4
Hard Pedal Mottled-Black Duplex Soils	Dd2.3
Gley Duplex Soils — Dg	
Hard Pedal Gley Duplex Soils	Dg2.4
Hard Apedal Gley Duplex Soils	Dg2.8
Friable Pedal Gley Duplex Soils	Dg4.1, Dg4.4
Friable Apedal Gley Duplex Soils	Dg4.8

4.3   Types of Soil Defined by the Factual Key

From the Factual Key, soils are given a code such as Dy3.43, Gn2.11, Um6.13 or Ug5.16. The codes are useful for specific information on soils, but people who do not use the key regularly find them difficult to interpret. Therefore, a system of descriptive names has been developed by Northcote and fully presented in *A Description of Australian Soils* (Northcote et al., 1975). A guide to soil codes related to each great soil group is to be found in Appendix II.

A summary of the names in *A Description of Australian Soils* is presented in Table 6.5. Most of these are at the section, subsection and class levels of the key. Some useful names can be gleaned from these including self-mulching cracking clays, hard pedal red duplex soils, friable loams, bleached sands and red massive earths. In many cases, they are more appropriate names for discussing soils than the principal profile form codes.

The distribution of the soil units of the Factual Key in New South Wales is described in Chapter 8 and Northcote et al. (1975).

6.5 Soil Taxonomy

6.5.1 Origins and Foundations

Soil Taxonomy (Soil Survey Staff, 1975) is a genetically based system of soil classification, published in the USA in the 1970s. It has a defined set of soil features, some of which are shown in Table 6.6, and rigidly defines how these are to be used to classify soils. Many of the soil features used can only be determined in the laboratory. The choice of soil features is strongly influenced by theories of soil formation developed on the great glacial plains of Europe and North America, and zonal concepts of soil formation are intimately incorporated.

Soil Taxonomy (ST) can claim some currency as a world classification system and much effort has been expended in extending the system outside the USA, particularly to tropical soils (Eswaran, 1983). As many international journals require soils to be classified in this system, some knowledge of it is useful to people with an interest in soils.

The application of Soil Taxonomy to Australian soils has been discussed in Moore et al. (1983) and Isbell (1984). Generally, the applicability of Soil Taxonomy in Australia is questioned because the data on which it was based did not include any Australian soils, which do show some unique features compared

to North American soils. The relevance of the zonal concept of soil formation is also questioned in Australia where soils and landscapes are considerably older than those of North America, and perhaps more influenced by geology and geomorphology. This brings into question the usefulness of moisture and temperature regimes defined in Soil Taxonomy to classify Australian soils.

6.5.2 Operation of the System

Soils are grouped according to the presence or absence of strictly defined soil horizons (diagnostic horizons, see Table 6.7) and soil features. These are chosen as indicators of the dominant processes of soil formation operating in the soils. The zonal concepts inherent in ST require the use of climatic data to classify soils. Both temperature and moisture regimes are defined (see Table 6.8). Watson (1980) has estimated soil temperature regimes for parts of New South Wales, and preliminary moisture regimes for the state have been estimated by Isbell and Williams (1980).

There are six categories in the system: order, suborder, great group, subgroup, family and series. There are 10 orders that are differentiated by the presence or absence of diagnostic horizons or pedological features. These features are indicators of

Table 6.6 Some soil features used for classification in Soil Taxonomy

1 Soil Morphology	**2 Climate**
Nature of soil boundaries	Moisture regime
Soil structure	Temperature regime
Depth to rock	Presence of permafrost
Presence of pans	**3 Parent Material**
Presence of clay skins	Volcanic ash
Seasonal cracking	**4 Chemical Properties**
Colour, bleached	Base saturation
mottled	Exchange capacity of clay fraction
black	Presence of primary aluminosilicates
bright red	Organic matter content
Presence of $CaCO_3$	Exchangeable sodium percentage
Presence of rock structure	Extractable iron and aluminium
Presence of gypsum	Salt content
Texture changes	**5 Physical Properties**
Gilgai	Particle size distribution
Slickensides	Bulk density
	Shrink-swell potential

Table 6.7 Diagnostic horizons used in Soil Taxonomy (after Soil Survey Staff, 1975)

A Horizons (epipedons)
Mollic (soft) — thick, dark-coloured, humus-rich, moderate to strong structure, low sodium
Umbric — similar to mollic, but acidic and base saturation less than 50%
Histic — very high organic matter
Anthropic and Plaggen — significantly changed by agricultural practices to be highly fertile and
    well structured
Ochric — other A horizons

B or Subsurface Horizons
Argillic — silicate clays accumulated to a significant extent by illuviation
Natric — sodic horizon (ESP > 15%)
Spodic — humus/sesquioxide deposits
Placic — iron/manganese pan
Cambic — weakly weathered
Oxic — strongly weathered, only iron and aluminium oxides with 1:1 minerals
Duripan — silcrete
Fragipan — weak, earthy pan
Albic — bleached
Calcic — calcium carbonate present

Table 6.8 Moisture and temperature regimes of Soil Taxonomy (after Soil Survey Staff, 1975)

Moisture Regime:
Aquic — a reducing regime where the soil is saturated by groundwater and by water of the capillary fringe
Aridic or Torric — soils which are largely dry when temperature is favourable for plant growth
Udic — humid regime where the soil is not dry in any part for as long as 90 days
Ustic — regime intermediate between Aridic and Udic
Xeric — regime typified by Mediterranean climate

Temperature Regime:
Pergelic — permanent frost, MAST* $< 0\,°C$
Cryic — very cold, MAST $0\,°C$ to $8\,°C$
Mesic — MAST $8\,°C$ to $15\,°C$
Thermic — MAST $15\,°C$ to $22\,°C$
Hyperthermic — MAST $> 22\,°C$ *MAST= mean annual soil temperature

differences in the degree and kind of soil-forming processes that have been dominant in the particular soil. The orders and approximately equivalent soils in the great soil group system are given in Appendix I.

6.5.3 Soil Units of Soil Taxonomy

Soil units are named according to a completely new terminology from that of the traditional great soil groups. Classical Greek and Latin words and phrases are coined to form soil names, some of which can appear quite foreign to Australian soil scientists. Names include cryofluvents, tropeptic haplorthox, aridic calcixeroll and typic ochraqualfs.

In terms of correlation with local classification (see Appendix I), it is evident that some suborders equate clearly with Australian great soil groups, while for others there is no clear relationship. Clear equivalence occurs between the alluvial soils and fluvents, the vertisols and the cracking clays, the chernozem/prairie soils and the mollisols, and the desert loams and the argids. There is reasonable equivalence between the calcareous, siliceous and earthy sands and the psamments; the podzols and the spodosols; and the krasnozems and oxisols. The red-brown earths are usually alfisols, but may be either xeralfs or ustalfs depending on the soil moisture regime. The red and yellow podzolic soils may be in the alfisols or ultisols, and the red and yellow earths can be in the alfisols, ultisols, oxisols or aridisols. Overall, many of the soil units of ST are difficult to interpret in Australia.

As can be seen, there are problems in applying Soil Taxonomy to Australian soils but, because of its growing international acceptance, they are problems that will have to be addressed to maintain lines of communication between Australia and the rest of the world on soils matters.

6.6 FAO/UNESCO Soil Classification System

The soil classification used in the FAO/UNESCO (FAO, 1990) system is designed to produce world soil maps at a scale of 1:5 000 000. Because the classification was designed to be an international one, many of the soil names selected were ones that did not change much on translation and had similar meanings in different countries. Where possible, traditional soil names are used, for example podzols, chernozems, solonetz, rendzinas and lithosols, but names which have become recently popular in some of the newer soil classifications are also included, for example vertisol from Soil Taxonomy. Where soil terms are particularly confusing in soils literature, such as 'podzolised' and 'brown forest soil', new terms have been selected. Generally, the classification has used many concepts from Soil Taxonomy. Some of the major orders of the classification and their Australian equivalents are presented in Table 6.9. Northcote et al. (1975) have listed equivalent or comparable soils between the Factual Key and the FAO/UNESCO classification.

The applicability of the soil classification system to Australia has been discussed in Moore et al. (1983) and Isbell (1984).

Table 6.9 Equivalent soils of the FAO/UNESCO and Australian soil classification systems

Great Soil Group	Examples from Factual Key	FAO/UNESCO
Siliceous Sands	Uc1.2, Uc4.2, Uc5.1	Arenosols
Earthy Sands	Uc5.2	Arenosols
Black Earths	Ug5.1, Gn3.4	Vertisols
Solodised Solonetz-Solodic	Dr, Dy, Db, Dd	Solonetz, Planosols, Luvisols
Red-Brown Earths	Dr2.13, Dr2.23	Luvisols, Solonetz
Red Earths	Gn2.11, Gn2.12	Yermosols, Ferralsols, Luvisols
Krasnozems	Gn3.11, Gn4.11	Cambisols, Ferralsols, Nitosols
Red Podzolic Soils	Dr2.21, Dr2.41	Nitosols, Ferralsols, Luvisols

6.7 Numerical Classification

6.7.1 Origins and Foundations

Systems of numerical classification use numerical or statistical techniques to calculate the most efficient or effective classification of a set of individuals. Such classifications are entirely objective once the attributes of the individuals have been chosen and rated. This has considerable appeal on a scientific basis. Usually the calculations are done by computer.

The techniques, such as classificatory programs which produce dendrograms and principal components analyses, were developed in the late 1960s, primarily for the classification of vegetation, but they have since been applied to soils. The primary aim of such systems is to increase the degree of objectivity in classification, and to look for any trends in the data that are not immediately apparent. Detailed accounts of the techniques and examples of their use are given in Williams (1976), Webster (1977), McBratney and Moore (1985), Joliffe (1986), and Austin and McKenzie (1988). This section is not intended to be a detailed review of the basis or interpretation of these techniques, but is intended to alert readers to the existence of these techniques.

6.7.2 Operation of the System

The major operation of numerical classification is to calculate mathematical measures of the relationship between soil individuals. Such mathematical measures include Euclidean distance or Canberra metric. The calculations are based on a set of data selected by the soil scientist. Although the actual calculation methods are objective, the selection of the data set and how it is presented, and use of the particular numerical technique, are all subjective decisions. These subjective decisions can have a large effect on the end result.

While there are many techniques for numerical classification, three are briefly described here as an introduction. Other techniques are discussed in the references at the end of the chapter.

Classification Techniques

Classification techniques calculate the similarity between individuals or groups of individuals using the available data. Similarities are calculated using non-dimensional mathematical measures such as Euclidean distance. In most techniques, the two most similar individuals are fused, then two most similar groups and so on, until the whole population is one group. In this way, a dendrogram as in Figure 6.2 is produced. In this figure, the affinity between the surface soils from similar geological origin is apparent, with the granite and the Ordovician sediments and volcanics clearly forming two groups. The surface soil properties included in the data set were clay, silt, fine sand, coarse sand, gravel, pH, and Emerson aggregate class. Other interpretations are put on the other groups. Other examples are given in Williams et al. (1983) and Williams (1976).

Such dendrograms are rarely considered ends in themselves; their main function is to provide insights into a set of data which were not previously apparent, hence they are hypothesis generating rather than hypothesis testing.

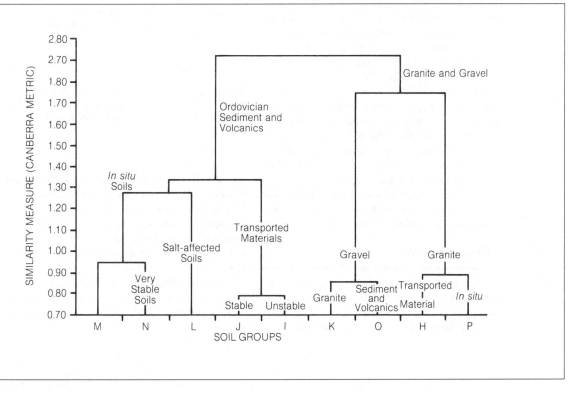

Figure 6.2   Dendrogram based on the numerical classification of surface horizons in the Bathurst–Orange region (after Murphy, 1984)

Ordination/Principal Components Technique

Principal components analysis has the major objective of simplifying large data sets to the level where only two or three principal components are necessary to describe the variation within the data set. Thus, the variation within a population of 100 profiles having data for 20 soil properties each, for example, may be represented by 20 axes. However if there is some relationship between the different soil properties, it may be possible to represent this data in only two or three principal axes or components. Those principal components are calculated from the data set. The first component (PCI) accounts for the largest proportion of the variation, the second (PCII), the second largest, and so on. Commonly, the first and second principal components can be represented on a two-dimensional graph as in Figure 6.3. This figure shows the principal components for the same data set as for Figure 6.2. These two components account for about 50% of the variation within the data set. The inclusion of the third principal component brought the total variation accounted for to about 60%.

The meaning of the principal components is not always apparent, although the first component in Figure 6.3 is clearly a geological factor, with granite being at one end of the y-axis (PCI) and the Ordovician sediments and volcanics at the other end. No clear explanation could be established for the variation in the x-axis (PCII). Other examples are presented in Webster (1977) and Williams (1976).

The ability of these techniques to represent large data sets with as few as three principal components, and to show relationships between soil profiles, makes this a potentially useful technique for developing general purpose soil classification schemes. Again, this technique is often not an end in itself but is a hypothesis-generating technique.

Other related techniques to analyse large data sets include Principal Co-ordinates Analysis, Biplot, and Correspondence Analysis (Joliffe, 1986).

Affinity Analysis

This technique has potential for allocating soil profiles to existing soil classification systems, as it calculates the affinity of a profile to various central concepts of soil taxonomic units. For example, if the central concepts of red podzolic soils and black

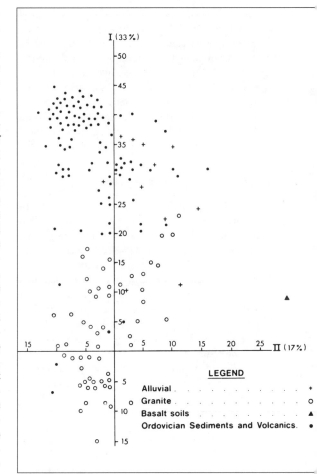

Figure 6.3 Principal component axes for the Bathurst–Orange region showing the relationship between surface horizon data and parent material

cracking clays are well defined, a profile can be said to have 80% affinity to the cracking clay and only 20% to the red podzolic soil. This technique has the potential to overcome the previous classification problems with rigid boundaries in soil classification, and is discussed more fully by McBratney and Moore (1985).

However, while numerical classification can be a useful tool, particularly for hypothesis generation using large data sets, it should generally be used with caution.

6.8 Soil Classification for Sustainable Soil Management

6.8.1 General Considerations

The wide range of soil properties that would need to be included in a soil classification scheme for sustainable soil management makes it difficult to develop. Effective land management is likely to require information on soils in relation to any of the following at a particular time:

— soil erodibility, both surface and subsoil
— soils in relation to earthworks
— soils in relation to vegetation management
— surface soils and crop management
— soils in relation to urban development
— soils in relation to degradation and loss of productivity
— soils in relation to catchment management, considering water quality and quantity, and salinisation.

To address these issues adequately, information is required on the following soil properties:

— erodibility of surface soils and subsoils; vulnerability of soil quality to soil erosion
— soil structural stability; influence of organic matter levels and exchangeable sodium levels
— specific limitations of cropping particular soils, workability, friability, surface sealing
— acidity and susceptibility to acidification
— salinity, natural salinity levels and soil hydraulic properties
— infiltration and permeability as they affect the flow of water and chemicals through the soil
— edaphic features of the soil that influence plant growth such as water-holding capacity, resistance to root growth, chemical fertility
— links of soils to particular vegetation types and the management of native vegetation
— engineering properties such as particle size grading, shrink-swell potential, plasticity and shear strength, dispersibility
— identification of acid sulfate soils.

The measurement of these soil properties and their interpretation for sustainable soil management are presented in Chapters 10 to 14.

No one soil classification system exists that incorporates all these properties. Arguably the scheme that should come closest is the new Australian Soil Classification System (Isbell, 1996), which takes into account many of the soil properties considered important for land management. However, it has been in operation only for a short time, and its effectiveness as a scheme for sustainable soil management has yet to be established. It is possible that an effective technical classification for sustainable soil management is best achieved by having several subsystems for particular aspects of soils, including erodibility, soil degradation, earthworks, hydrology, and so on.

None of the general soil classification schemes, such as the great soil group system (Stace et al., 1968) or the Factual Key (Northcote, 1979), can account for all the soil properties that are important for sustainable soil management, although some general inferences can be made (see Chapters 8 and 9). More information on these properties is available at the soil family, soil series or soil profile class level of classification. For example, localised varieties of the great soil groups (such as Lithgow yellow podzolic soils), the Factual Key units (such as Bathurst red duplex soils), or the units of the Australian Soil Classification System (ASCS) (such as Wellington Red Dermosol or Gunnedah Black Vertosol), may be reasonably uniform in many properties relevant to sustainable soil management. However, in many instances, specific site investigation, or special purpose surveys, will be the only means of obtaining sufficient information on many of the soil properties listed, to make effective decisions for land management.

As mentioned, none of the existing soil classification schemes adequately caters for all these requirements. The order of the systems as they relate to sustainable soil management is most likely:

ASCS > Factual Key > Australian great soil groups > Soil Taxonomy.

The reasons for this rating are as follows.

(a) Australian Soil Classification System (ASCS)

The ASCS has been based on both morphological and chemical properties. It includes a series of soil properties that relate to sustainable soil management. As well as considering general soil properties such as texture and colour which can often be generally related to soil management, the scheme takes into account a range of soil properties that specifically relate to sustainable soil management that have not been considered in other classifications. These soil properties include:

—soil sodicity and hence dispersibility (Sodosols and sodic subunits of other groups)
—highly acidic subsoils (Kurosols)
—soil salinity levels
—self-mulching surface soils (self-mulching Great Group in the Vertosols)
—base saturation and degree of leaching of soils
—the level of free iron which relates strongly to highly aggregated soils (Ferrosols)
—levels of organic matter in the soil (Organosols)
—soils with high shrink-swell potential that seasonally crack (Vertosols)
—identification of acid sulfate soils (Sulfuric and Sulfidic Hydrosols)
—horizon depth and stoniness at the family level.

The degree of usefulness of the classification is still too early to tell, as a history of experience has not been built up for the units within this classification as had been the case for the great soil groups and the Factual Key. However, the selection of soil properties and the way they are used in the classification would suggest that the ASCS will be a useful tool for sustainable soil management in the future.

Some of the limitations apparent are the lack of hydrological and engineering soil properties in the classification, as it is very much an agriculturally based classification. There is a lack of soil physical and engineering criteria to assess soil stability, suitability for earthworks, and rural and urban capability. The emphasis on environmental management in more recent years may mean that the ASCS has more limitations than expected for making decisions about sustainable soil management and that it will need to be developed to take these into account.

(b) Factual Key

The dominant use of texture and profile form in the Factual Key enables some account to be taken of all the properties important for sustainable soil management. It is one of the strong points of the Factual Key. An extension to the principal profile form (Northcote, 1979) can greatly increase the precision on soil texture.

The descriptions of the surface soil condition as self-mulching, massive, friable and, in particular, hardsetting are useful for soil erodibility, soils in relation to plant growth, and susceptibility to degradation. This is also a strong point of the Factual Key. Highly aggregated soils that may leak when used in earthworks are usually Gn3.1, Gn4.1, Uf5 or Uf6 soils.

Profile drainage is taken into account by the key in the use of colour, as is permeability to a lesser extent in the use of texture, colour and structure.

However, the Factual Key cannot be considered a particularly effective classification for sustainable soil management as it does not take specific account of sodic, dispersible soils, although some units such as Dy3.43 are usually dispersible. Also, many of the relationships between soil physical properties, such as soil erodibility and texture, are only of a general nature. No provision is made to identify acid sulfate soils.

Overall, the Factual Key has some useful features for sustainable soil management, but it is not a completely effective system in this regard.

(c) Great Soil Groups

These define broad soil conservation groups, but tend to lack definition for some soil features important for sustainable soil management. Deficiencies include:

—the level of exchangeable sodium varies widely within great soil groups; although sodicity is more common in some groups than others (solodics), it is not always possible to predict sodicity from a great soil group name alone (red-brown earths may or may not be sodic and some solodic soils are non-sodic)
—soil textures, surface soil textures in particular, can vary widely within one great soil group
—surface soil features of major importance for soil management such as self-mulching or sodicity are given little recognition in the great soil group system
—subsoil plasticity is not considered
—no provision is made for acid sulfate soils
—soil hydraulic properties can vary widely within each of the great soil groups.

While considerable experience has built up in using great soil groups, and the addition of local names or knowledge of local variations of great groups has improved its effectiveness, the scheme does not take enough account directly of those soil properties that are important to know for sustainable soil management

(d) Soil Taxonomy

Soil Taxonomy is probably not a good system for sustainable soil management in Australia. It places emphasis on pedogenesis and a narrow set of chemical aspects of soil. It places more emphasis on the nutrient fertility of soils than on the physical properties. Textures vary widely in most soil groups, probably as a consequence of the use of the argillic B horizon in classification.

While the mollic horizon is a diagnostic surface horizon, most other Australian surface horizons important for soil erodibility and susceptibility to degradation are not considered. The self-mulching massive, friable and hardsetting surface horizons all appear under ochric A horizons. The conceptual nature of many soil units and the non-use of much readily observable field data put it at a disadvantage to the new Australian Soil Classification System, the Factual Key and the great soil group system. A further problem is the use of a 15% ESP (exchangeable sodium percentage) to define sodic soils, which appears inappropriate in Australia (Northcote and Skene, 1972). Overall, Soil Taxonomy is the least suitable for sustainable soil management.

5.8.2 Soil Classification System for Western New South Wales

The Soil Conservation Service system for soil classification in the Western Division of New South Wales is used mainly in broad-scale mapping, for land system surveys (scale 1:250 000), and Western Lands Lease Management Planning (scale 1:50 000). The former surveys are used for basic descriptions of large areas of pastoral country, while the latter are used to map each pastoral holding in the Western Division into its land and vegetation types. These are then used to determine erosion susceptibility, stocking capacity and management requirements for maintenance or improvement of the resources available.

In the semi-arid to arid environment, drought is the rule rather than the exception, and vegetation has become delicately balanced with the environment. Overgrazing upsets this balance, sometimes irreversibly, causing degradation of pastures and then erosion of susceptible soils, especially when the overgrazing occurs in dry or drought conditions when vegetation has already become naturally depleted. Hence, it is the ability of the soil to produce vegetation and maintain it into dry periods and the susceptibility of the soil to erosion, that are of prime concern for sustainable soil management in the west.

Soil physical properties such as infiltration rate, water-holding capacity and presence of root-restricting layers, along with micro-topography (for example, gilgai where extra water accumulates), erodibility and inherent fertility are the main soil properties that need to be characterised. Accordingly, a soil classification system, suitable for use at the mapping scales required, has been developed to map the soils on each pastoral holding. This system takes into account the preceding features, using soil texture as an indicator of infiltration rate and water-holding capacity.

Sandy soils have a high infiltration rate, but low water-holding capacity and inherent fertility, and are highly susceptible to erosion because of their easily detachable surface. Heavy clays normally have relatively higher fertility and water-holding capacity and lower erodibility.

While vegetation growing on sandy soils responds rapidly to small falls of rain, the low water-holding capacity and low fertility of the soils do not favour the growth of perennial plants in dry periods. The soil thus becomes bare and susceptible to movement by wind. Vegetation on heavy clays responds to larger falls of rain, and once these soils are wet they retain moisture for long periods and are, therefore, more able to support perennial vegetation. However, these soils can become completely bare in droughts or after heavy stocking, but very little erosion occurs. Hence, generally speaking, as texture becomes finer, the soil is better able to produce perennial vegetation, has higher fertility and is less susceptible to erosion.

As most plant roots are in the surface soil layers, the depth of the surface soil is also taken into account in the classification system. Presence of layers which impede root penetration such as hardpans, stone layers, bedrock, or dense clay B horizons are also integrated into the system.

The presence of calcium carbonate was developed to meet the particular need for describing soils in the detail suitable for range assessment and management work. Previous broad-scale surveys of the area were not sufficiently detailed, while mapping at a more detailed scale was not warranted.

Further, the classification has a 'practical' basis. For example, soil studies have revealed that many soil profiles comprise two or more layers of different ages and, in the Factual Key, only the topmost layer is usually classified and noted as overlying other layers. The present system describes all profiles as single profiles, even if layers of different ages and origins occur within the profile. This is of most significance where a sandy layer overlies an older clay layer. The present system classifies such a profile as a 'texture contrast' soil rather than (for example) 'Uc5.11 over clay'.

The broad groups used are as follows:

SR	Soft Red Soils
HR	Hard Red Soils
SB	Solonised Brown Soils
TC	Texture Contrast Soils
SK	Skeletal Soils
HC	Heavy Clays
GR	Granite Soils
BG	Brown Gibber Soils
DL	Desert Loam Soils

(a) Soft Red Soils

These are red and red-brown massive sands to clay loams showing little or no change down the profile except a gradual increase in clay content. The soils are non-calcareous at the surface, but may be calcareous at depth. They are subject to windsheeting.

The main soils included are sands (Uc1, Uc5), loams (Um5.5) and red massive earths (Gn2.1).

(b) Hard Red Soils

These soils have red and reddish brown sandy loam to clay loam topsoils with little or no change down the profile except for a gradual increase in texture. They are acidic, compact, gravelly soils, frequently with an earthy hardpan at 20 to 100 cm depth. Structure is massive. They are subject to watersheeting frequently with minor to moderate rilling and gullying of the lower slopes.

The main soils included are red massive earths (Gn2.12, Gn2.11) and earthy loams with a red-brown hardpan (Um5.3).

(c) Solonised Brown Soils (Brown Calcareous Earths)

These soils have red, reddish brown to yellowish brown sand to clay loam topsoils with little or no change down the profile except a gradual increase in clay and calcium carbonate content. Light to heavy amounts of nodular limestone are found to 90 cm. They are massive or, occasionally, well structured. These soils are subject to windsheeting.

The main soils are calcareous earths (Gc1.12, Gc1.22 and Gc2.12). Some slightly heavier textured soils (Uf) have been included in this group.

(d) Texture Contrast Soils

These soils are typified by an abrupt change in structure and texture. The sandy to clay loam surface horizon is generally massive or weakly structured and extends to 30 cm in depth, overlying a clay with either a nutty or compact structure. Colour is generally reddish brown or grey over a red, reddish brown or grey subsoil. These soils are subject to scalding.

The main soils include the red duplex profiles (in western New South Wales mainly Dr2.33, Dr2.32, Dr2.23, Dy3.42 and Db1.33, Db1.43 and many others including Dr2.53, Dr4.53, Dy4.53, Dy5.13). These fit into the red-brown earth great soil group. Some actually consist of two profiles—a sandy profile over a clayey profile.

(e) Skeletal Soils

These are immature and poorly developed shallow, stony, gravelly and rocky soils associated with steep hills and ranges.

(f) Heavy Clays

The grey, brown and red-brown clays and silty clays form a very broad group of soils whose common properties are determined by their high clay content. Typically, they are moderately deep to very deep soils with little change in colour or texture as depth increases. Any marked change is in structure differentiation. Some lime and/or gypsum occurs in the subsoil.

These soils occur along the floodplains of the major rivers, the Riverine Plain and in gilgai or drainage sinks in 'red' country. They are subdivided according to the presence of gilgai, nature of the surface, or micro-topography.

The main soils include cracking clays (Ug5.24 and Ug5.25) on the Riverine and Macquarie-Upper Darling floodplains, and Ug5.28 and Ug5.29 along the middle Darling. Other types occurring include Ug5.38, Ug5.35 and Ug5.34 (brown cracking clays), and Ug5.4 and Ug5.5 which are cracking clays having coarsely structured to massive surface soils.

(g) Granite Soils

These are derived from granite and have surface soils that are reddish brown massive gritty sands to clay loams with a mottled red and brown clayey subsoil that often contains calcium carbonate nodules. The soil surface is non-crusting and usually has a light veneer of small quartz gravel, which is also commonly present throughout the profile.

The soils are moderately resistant to windsheeting, but slopes are susceptible to watersheeting and rilling, and drainage lines are highly susceptible to gullying.

These soils include red earths (Gn2.12, Gn2.13), red and yellow duplex soils (Dr2.43, Dy3.43, Dr3.33) and siliceous sands (Uc).

(h) Brown Gibber Soils

These soils have a surface crust with a heavy mantle (at least 20% cover) of silcrete, ferricrete, quartz or quartzite stone and gravel. Beneath the stone there is a thin red to reddish brown loam to clay abruptly overlying a stone-free red or brown, strong, fine-structured clay changing gradually to a coarse-structured, paler clay, often with gypsum and lime at depth. Severe gullying occurs in the highly dispersible subsoil when the surface stone is disturbed.

They are predominantly thin red duplex soils often with a hardpan layer (Dr1.1, Dr1.3, Dr1.4, Dy1.1, Dy1.3 and Dy1.4). Uniform red and brown cracking clays (Ug5.3) usually occur in gilgai depressions.

(i) Desert Loams

These soils have a thin, crusty, massive, light brown to reddish brown, loam surface, clearly separated from a moderately to well-structured subsoil with lime and gypsum at depth. They frequently have a variable surface pavement of rounded or subangular gravel and stone, which is less obvious than in brown gibber soils. They are susceptible to windsheeting and minor watersheeting erosion, and form soft scalded surfaces when the structured saline subsoil is exposed.

They are predominantly thin red and brown duplex soils, occasionally with a hardpan layer (Dr1.3, Dr1.4, Dy1.1, Dy1.3, Db0.2, Db0.3, Db0.4). Uniform brown and red cracking clays (Ug5.3) occur in gilgai depressions.

Desert loams can be distinguished from brown gibber soils by their lower position in the landscape and, in general, by the absence of a heavy mantle of stone.

BIBLIOGRAPHY

Austin, M.P. and McKenzie, N.J. (1988), 'Data Analysis' in R.H. Gunn, J.A. Beattie, R.E. Reid, and R.H.M. van de Graaff, *Australian Soil and Land Survey Handbook—Guidelines for Conducting Surveys*, Inkata Press, Melbourne.

Chittleborough, D.J., Masmecht, D.J., and Wright, M.J. (1976), *Soils of the Monarto Town Site*, Soil Survey SS16, Department of Agriculture and Fisheries, South Australia.

Eswaran, H. (1983), Recent efforts to refine soil taxonomy for the classification of the soils of the tropics. International Workshop on Soils Research in the Tropics, ACIAR, Townsville, Queensland.

FAO–UNESCO (1990), *Soil Map of the World: Revised Legend*, World Soil Resources Report 60, FAO, Rome.

Isbell, R.F. (1984), Soil classification in Australia, Review paper, National Soils Conference, Australian Society of Soil Science Inc., Brisbane, Australia.

Isbell, R.F. (1988) 'Soil classification', in Gunn R.H., Beattie, J.A., Reid, R.E. and van de Graaff, R.H.M., *Australian Soil and Land Survey Handbook: Guidelines for Conducting Surveys*, Inkata Press, Melbourne.

Isbell, R.F. (1996), *The Australian Soil Classification*, CSIRO, Melbourne.

Isbell, R.F., McDonald, W.S. and Ashton, L.J. (1997), *Concepts and Rationale of the Australian Soil Classification*, ACLEP, CSIRO Land and Water, Canberra.

Isbell, R.F. and Williams (1980), 'Dry soils of Australia: characteristics and classification', in *Proceedings of the Third International Soils Classification Workshop*, F.H. Berinroth and H Osman (eds), Damascus, Syria.

Joliffe, I.T. (1986), *Principal Component Analysis*, Springer-Verlag, New York, Berlin, Tokyo, Heidelberg.

McBratney, A. and Moore, A.W. (1985), 'Application of fuzzy sets to climate classification', *Agriculture for Meteorologists* 35, 165–85.

Moore, A.W., Isbell, R.F., and Northcote, K.H. (1983), 'Classification of Australian soils', in *Soils: An Australian Viewpoint*, CSIRO, Melbourne (Academic Press, London).

Northcote, K.H. (1979), *A Factual Key for the Recognition of Australian Soils*, Relim Technical Publications, Glenside, South Australia.

Northcote, K.H. (1983), *Soils, Soil Morphology, and Soil Classification: Training Course Lectures*, CSIRO Division of Soils, Relim Technical Publications, Glenside, South Australia.

Northcote, K.H., with Beckmann, G.G., Bettenay, E., Churchward, H.M., Van Dijk, D.C., Dimmock, G.M., Hubble, G.D., Isbell, R.F., McArthur, W.M., Murtha, G.G., Nicolls, K.D., Paton, T.R., Thompson, C.H., Webb, A.A., and Wright. M.J. (1960–68), *Atlas of Australian Soils*, sheets 1 to 10 (with explantory data), CSIRO and Melbourne University Press, Melbourne.

Northcote, K.H., Hubble, G.D., Isbell, R.F., Thompson, C.H., and Bettenay, E. (1975), *A Description of Australian Soils*, CSIRO, Melbourne; Wilke and Co., Clayton, Victoria.

Prescott, J.A. (1931), *The Soils of Australia in Relation to Vegetation and Climate*, Bulletin No. 52, Coun. Sci. Ind. Res. Aust., Melbourne.

Stace, H.C.T., Hubble, G.D., Brewer, R., Northcote, K.H., Sleeman, J.R., Mulcahy, M.J. and Hallsworth, E.G. (1968), *A Handbook of Australian Soils*, Rellim Technical Publications, Glenside, South Australia.

Stephens, C.G. (1962), *A Manual of Australian Soils*, Third Edition, CSIRO, Melbourne.

Williams, J., Prebble, R.E., Williams, W.T. and Hignett, C.T. (1983), 'The influence of texture, structure and clay mineralogy on the moisture characteristic', *Australian Journal of Soil Research* 21, 15–32.

Soil Survey and Mapping

G.A. Chapman and G. Atkinson

Soils are essentially a non-renewable and vital resource. Their precious nature means land use and land management decisions should not be taken without consideration of soil behaviour and land capability. Understanding the distribution and properties of soils is an essential part of making sustainable land use and management decisions (Sharley, 1987). This may be done through what has proven to be in the past an expensive process of trial and error. Soil survey is a deliberate process of building an understanding of the distribution and properties of soils. From a soil conservationist's viewpoint the hallmark of a good soil survey is based on not just the accuracy of the outputs but also the continuity of impact it has on land use and management.

There are numerous examples of a wide variety of purpose and applications in soil survey. They include hydrology assessments for abatement of dryland salinity (Abbs and Littleboy, 1998), intensive cultivation (Lawrie, 1988), irrigation (Hall et al., 1992), effluent disposal (Geary, 1994; Tulau and Wild, 1996), pollution control planning (Chartres et al., 1991), acid sulfate soil risk (Naylor et al., 1995), forest and plantation management (Turvey and Lieshout, 1983), and catchment management and rural land use planning (Harper, 1994; McAlpine and Keig, 1990).

7.1 General Principles

1.1 Literature Review

The principles and practice of soil survey and mapping have been discussed extensively in the published literature and can only be summarised here. Reference to this literature is, therefore, essential for anyone undertaking soil surveys in New South Wales. In the Australian context, Gunn et al. (1988) is a comprehensive manual on the techniques of soil survey, and Taylor (1970), Beckett and Bie (1978), Gibbons (1983) and McKenzie (1991) provide excellent insights into the history and context of soil survey activities in Australia. A useful set of references on the subject is the series of Monographs on Soil Survey published by Oxford University Press. This series includes Western (1978), on contracts and quality control; Butler (1980), on soil classification; White (1977), on aerial photography and remote sensing; and Webster (1979), on numerical methods. Other useful references include the general text on soil survey and land evaluation by Dent and Young (1981) and a paper by Webster (1985) on quantitative spatial analysis of soils in the field.

Comprehensive bibliographies of Australian surveys are contained in the National Land and Soil Survey Resources Data Directory (see Barson and Shelley, 1994) and the New South Wales Natural Resources Data Directory (http://www.nrims. nsw.gov.au/nrdd), both of which are updated regularly. Other lists of soil surveys include Northcote et al. (1960–68), Beckett and Bie (1978), and Blain et al. (1985). Hazelton and Atkinson (1985) have listed soil surveys available in the Sydney region. SCS (1980) has listed relevant studies undertaken by the former Soil Conservation Service in New South Wales.

Thorough researching of the literature to find all previous work on an area is a fundamental prerequisite for producing a good soil survey. Failure to do this could lead to the wrong approach being taken or, at worst, an embarrassing duplication of effort.

One of the principal reasons for the development of a Soil Data System for New South Wales was to avoid this unnecessary duplication of effort. This system provides access to soil morphology data for any specified area, and the laboratory soil test results. A prime advantage of the New South Wales Soil and Land Information System is that it is digital. Digital data can be accessed, reused and recombined for many different purposes (Martin, 1997). This is a prime and unexpected feature of digital storage of soil survey data that has greatly increased the application and user acceptance of soil survey outputs.

7.1.2 The Purpose of Soil Survey

It would generally be conceded that the purpose of soil survey is to:

study, classify, describe and map soils so that predictions can be made about their behaviour for various uses and their response to defined management systems. (USDA, 1950)

This is usually achieved via the medium of a soils map and an accompanying report, although computer technology is providing access to many other formats and methods of visualisation including specific theme and chlorpleth maps (for example, Bregt et al., 1991). These are generated by geographic information systems and by making use of digital elevation models, web pages and hypertext linked documents (Fisher, 1998).

Following a major review of soil survey in Australia (Beckett and Bie, 1978), it was concluded that: *many soil and land system surveys have been requested or performed with no clear idea of just what question the soils map and/or report was to answer.*

Fundamental to the success of any survey is a clearly defined purpose, and it is the soil surveyor's first responsibility to determine that purpose. Close consultation with the ultimate or major users of a soil map is invaluable, especially as there is a lack of understanding in the community of the potential utility of modern soil surveys for a variety of purposes.

Much has been written on the distinction between general purpose soil surveys and special purpose surveys (see Dent and Young, 1981). General purpose, or multipurpose, soil surveys are more common, particularly as published maps. They are based on the rationale that it is possible to delineate natural soil units in the landscape which can be described using common morphological properties and which can be combined, analysed and interpreted in different ways for practical purposes without the need for resurvey.

The useable life of quality soil surveys is in the order of 50 years. If a broad range of data collected is stored in a readily extractable format then it can be filtered, recombined and manipulated to provide interpreted answers to a large range of questions, many of which relate to issues which were not antici-

pated at the time of the soil survey. For government agencies measurement of a broad range of morphological, chemical and physical data is therefore often found to be of considerable benefit for the public good (ACIL Economics and Policy, 1996).

Special purpose surveys have a limited number of specific objectives and the requirements of a particular user are incorporated into the survey design. Often a limited range of specific attributes is described and the survey may not result in the production of a 'soils' map. Because special purpose soil surveys are often commissioned by groups with a specific agenda, there is often an economic reluctance to include soil attributes that do not directly relate to the soil/land management issues in question at the time.

General purpose surveys conducted by the New South Wales Department of Land and Water Conservation are discussed later in the chapter, as are special purpose surveys.

Regardless of which type of survey is conducted, the guiding principle is to make the classification and mapping units relevant to the stated purpose. A basic set of questions that should be addressed by a general purpose soil survey is:

(i)   What soils occur in the area, and how can they be readily recognised by end users?
(ii)   Where do these soils occur? What land attributes indicate different soil conditions?
(iii) What are the properties of these soils?
(iv)   What are the relative proportions of the various soils?
(v)   What type of soil can be expected at a particular site?
(vi)   What can the soils be used for?
(vii) What are their significant qualities that may affect a range of uses?
(viii) How reliable are the statements and predictions made?

For a more specific purpose survey, a user might ask more detailed questions, such as:

(i)   What foundation hazards will be encountered?
(ii)   Is the soil deep enough to lay a pipeline and what is the best route it should follow?
(iii) Is the soil suitable for septic effluent disposal?
(iv)   Will the soil be suitable for irrigation?
(v)   What amelioration is required to grow a certain crop?
(vi)   How erodible is the soil when cultivated?

The soil surveyor must identify the questions that have to be addressed and tailor the survey to suit these specific problems. This will involve measuring 'useful and relevant properties of the local soil' (Butler, 1980), as well as using survey techniques and data analysis methods that will best meet the survey objectives.

The measuring of 'useful and relevant properties' involves not only the identification of what attributes of the soil are to be measured and recorded, but also a decision on what class intervals are appropriate for each attribute. The choice of site attributes and soil morphological properties to be described in the field can be made from those listed in Abraham and Abraham (1992) and McDonald et al. (1990), and class intervals listed in these handbooks, or some combination of them, are recommended in the interests of standardisation. Other properties requiring measurement in the field or laboratory should also follow standard field (McDonald et al., 1990) and laboratory procedures such as listed in Rayment and Higginson (1992). The Australian Collaborative Land Evaluation Program (ACLEP) has responsibility for determination and setting of such standards for land resource assessment in Australia. In addition ACLEP is responsible for fostering research in land resource assessment. The ACLEP website is at http://www.cbr.clw.csiro.au/aclep/.

Some soil properties such as infiltration, water retention or Atterberg limits may often be too expensive or too laborious to determine for more than a few locations. However, it is a truism to suggest that the most expensive data to determine is also the most useful. This is particularly the case for soil hydrological properties such as saturated hydraulic conductivity and water retention curves (McKenzie et al., 1991).

To overcome the expense of making these measurements, the soil surveyor often looks for soil properties that may or may not be useful in themselves but are easily assessed in the field and can be used as surrogates for the more expensive measurements. These are the common morphological properties of texture, colour, structure, depth, stoniness, concretions and so on. It is important that the assumption of correlation between these surrogates and the relevant properties be tested in the initial stages of the survey. There has been much work in the translation and manipulation of such surrogates into 'pedotransfer functions' which reliably predict important but not easily measured soil characteristics.

Any soil survey is limited in the amount of data it can collect, as Riddler and Lawrie (1988) point out:

. . . no survey can hope to collect, analyse and present all the data which could affect all of the uses to which land may be put. Surveys are therefore selective in data collection and presentation.

A very large body of data is generated in a soil survey. Often only a small percentage of these data is used for immediate decision making.

While the choice of data to collect is an issue in soil survey, it is overshadowed by the need to actually observe and sample the soils in the study area. Except for very intense soil surveys for very high value or academic purposes, this is often the limiting factor in the design of soil survey and often determines the scale at which the map can be presented with confidence.

In situations where there is an opportunity to collect sufficient soil data to examine it on a statistical basis, means, standard errors, coefficients of variation and frequency distributions can be useful. Use can also be made of multivariate or numerical classification techniques that can assist with data reduction, classification and allocation. Further details on these methods can be found in Chapter 6 and in Webster (1979, 1985), Webster and Oliver (1990) McBratney (1992) and in the ecological literature. Major applications of multivariate statistics include determining surrogates and pedotransfer functions; manipulating remotely sensed raster data such as radiometrics (Bierwirth et al., 1996); Thematic Mapper; and Magnetics and Digital Elevation Models (Odeh et al., 1993). Given the essential but relatively expensive nature of field sampling, it is important to ensure sampling sites are as representative as possible. Objectively derived sampling points and explicit sampling schemes based on multiple digital variables are now possible (Gessler et al., 1995; Ryan et al., 1996; Cook et al., 1996).

To be useful, a soil survey should be able to predict at least some of the properties of soil at a site without having to visit it. Its ability to do this is a measure by which the quality of the survey is judged. East (1994) reviews the methods by which predictive accuracy of soil mapping can be judged. Measures of predictive accuracy should accompany soil maps but rarely do, primarily because of paucity of recorded soils data.

The most useful classification systems are those with the ability to predict, within the context of the survey objectives, the properties and behaviour of the soil at any site. There is a well-recognised need to go beyond presentation of soil classifications to provide map users with information that will directly answer their questions (Woode, 1981). It is less likely that this will be achieved using broad or national scale classifications than if classifications tailored for the purpose are used. As a rule, the more general the survey objectives, the more general the classification and the less the predictive power of the map. The deficiency of general purpose surveys is that they cannot serve all purposes equally well. The corollary to this is that a more closely tailored survey will be useful for its stated purpose, but is less likely to be of value for other purposes.

To make the best use of general purpose soil maps, a large number of special purpose soil classifications based on the data collected to produce these maps, is available. These are usually in table formats, to determine a map unit's capability or suitability for particular purposes. One measure of the success of a general soil survey is the number and types of capability tables that can be confidently applied. This in turn is based on the extent and breadth of data collected, the ease of accessing and recombining the data, and the sensitivity of the original soil map and other maps in showing the variability of soil properties within the map units.

Capability tables have been developed in parallel by the United Nations Food and Agriculture Organisation (FAO, 1976 and 1983) and United States Department of Agriculture (1993). Australian examples of soil information interpretation and FAO-style suitability tables include Hazelton and Murphy (1992), Wells and King (1989) and Moore (1998). Chapman et al. (1992), McKenzie (1991) and Smith and Thwaites (1998) suggest means to further improve such systems. Shaxson (1992) has provided a set of tables for identifying problems that are likely to arise in managing soil moisture.

As an example: a survey designed to predict subgrade suitability for a road would be of little use in predicting agricultural suitability, whereas a more general survey based on the distribution of soil materials within a landscape may be able to offer useful information on both.

7.1.3 Soil Entities and Soil Classification

Soil is a continuum varying in three dimensions. It is usually anisotropic, as the rate of change of properties down the profile is much greater than the rate of change of properties across the landscape. Although the soil exists as a continuous mantle, when it is sampled it is regarded as being made up of a discrete number of elements that are in some way representative of the total population. In a typical soil survey the amount of soil inspected is a fraction of one millionth of the soil volume to be mapped. Most soil survey mapping methodology and dilemmas spring from this fact.

What constitutes a soil entity is somewhat arbitrary because soils do not have readily identifiable individuals as do plants and animals. The entity is generally considered to be the soil profile, but some classifications are based on the horizon or soil layer as the soil entity (FitzPatrick, 1971). Soil classification attempts to bring these entities into acceptable conceptual groups as a way of being able to make general statements and predictions about the groups (or classes).

Soil individuals may also be grouped on a landscape basis such that parcels of land contain soils which are similar (soil series, soil family or soil profile class) or which have recognisable and repeatable patterns (soil landscape, soil association, soil complex). These parcels of land are known as mapping units. Mapping units and soil classes are not the same thing, a fact that Dent and Young (1981), Webster (1979) and, particularly, Butler (1980) go to great pains to point out. Butler calls this the 'taxonomic gap'. Soil classes are taxonomic units (see Table 7.1). They are conceptual groups of soil, whether defined by limiting attributes or central concepts, in which the membership criteria for the class can be precisely defined. They generally have a hierarchy leading from great soil group or Factual Key group to soil family, soil series and soil phase (see Table 7.2). Mapping units (see Table 7.3) delineate parcels of land that ideally will contain a single taxonomic unit, or their boundaries will be placed so as to delineate units which are as homogeneous as possible. However, inevitably they will contain impurities, that is, soil units that do not conform to the definition of the taxonomic unit by which they are represented. This happens because a gradual boundary must be represented on a map by a line, or because the soil pattern is too complex or contains inclusions too small to map (Dent and Young, 1981). The broader the scale of the map, the more complex the mapping units are, and several soil classes may be defined within a mapping unit. Such mapping units are compound units and include soil landscapes, soil associations or soil complexes. In discussing this distinction between soil mapping units and soil classification units, Butler (1980) observes:

The search for a merging or reconciliation between soil taxonomic entities (soil classes) with landscape area units (mapping units) has not been very successful . . . either we adopt a soil classification from the landscape itself, in which case it fits the landscape but not the rest of a soil taxonomic system designed to cover a wider area, or we impose the classes of a general taxonomic system on the landscape, and then it is unlikely to fit the local soil modes or the individual character of the local landscape scene.

Butler places particular emphasis on the local uniqueness of soils and he opts for the former course, that is, to 'develop and use the soil classification that arises from the landscape itself'. Butler contrasts this approach with national classifications, which may fortuitously coincide with the structure present in the local soil data but, more often than not, will be less than optimal and in some cases will be misleading. Groupings of soils in the field do not always correspond to groupings in a national scheme and may often cross class boundaries in the national scheme

Table 7.1 Soil taxonomic units used in soil mapping

i Soil Series
A unit of soil classification and soil mapping comprising or describing soils which are alike in all major profile characteristics. Each soil series is developed from a particular parent material, or group of parent materials, under similar environmental conditions. It approximates to the principal profile form on a particular parent material. The name given to a soil series is geographic in nature, and indicates a locality where the soil series is well developed. (USDA, 1950)

ii Soil Family
Taxonomic unit intermediate between soil series and great soil group. Frequently comprises a number of soil series.

iii Soil Profile Class
A soil profile class is a group or class of soil profiles, not necessarily contiguous, grouped on their similarity of morphological characteristics and possibly some laboratory-determined properties. The profile class is defined at various levels of generalisation (such as series, family, great soil group or phase of each) depending upon the information available, the purpose of the survey and the map scale. (Gunn et al., 1988)

iv Soil Phase
A subdivision of a soil taxonomic unit based on characteristics that affect the use and management of the soil, but do not change the classification of the soil. Such characteristics include slope, erosion, depth, stoniness and rockiness, drainage, depositional layers, gilgai or scalding. (USDA, 1950)

Table 7.2 Relationship of family, soil series and phase in the red duplex soils on the Bathurst Granite

Factual Key Group/Great Soil Group: Neutral red duplex soils/non-calcic brown soils
Soil Family: Red duplex soils (non-calcic brown soils) on the Bathurst Granite
Soil Series: Dr2.42 (bleached A_2)
  Dr2.22 (non-bleached A_2)
  Dr2.12 (no A_2)
  Dr2.62 (A_2 apedal B horizon)
Soil Phases: As for series, plus:
  shallow phase (< 50 cm to weathered granite)
  deep phase (> 50 cm to weathered granite)

Table 7.3 Units used in soil mapping

Compound Mapping Units

Mapping units that contain several soil taxonomic units (soil series, soil families or soil profile classes).

i Soil Association
A soil mapping unit in which two or more soil taxonomic units occur together in a characteristic pattern, such as a toposequence. The units are combined because the scale of the map, or the purpose for which it is being made, does not require delineation of individual soils. The soil association may be named according to the units present, the dominant unit or be given a geographic name based on a locality where the soil association is well developed.

ii Soil Complex
A soil mapping unit in which two or more soil taxonomic units occur together in an undefined or complex pattern. The soils are intimately mixed and it is undesirable or impractical to delineate them at the scale of the map. The soil complex may be named according to the units present, the dominant unit or be given a geographic name based on a locality where the soil complex is well developed.

iii Soil Landscape
A unit of soil mapping having a characteristic landform pattern with one or more soil taxonomic units occurring in a related sequence. It is often associated with the physiographic features of the landscape and is similar to a soil association but, in a soil landscape, the landform pattern is specifically described. The soil landscape may be named according to the soil taxonomic units present, the dominant unit or be given a geographic name based on a locality where the soil landscape is well developed.

Simple Mapping Units

Mapping units containing one soil taxonomic unit with minor impurities of one or more other soil taxonomic units.

(Butler, 1980; McKenzie and Austin, 1989). These are referred to as intergrades where soils such as krasnozem/chocolate soils, podzolic/siliceous sands or red-brown earth/non-calcic brown soils may form localised soil classes. It is better to interpret soil usage based on field and laboratory observed attributes rather than to rely on sometimes dangerous generalisations from national classification schemes. For some old soil surveys the original data has been lost or cannot be linked to map units. In these cases reliance must be placed on soil classification.

Conversely, national schemes are not built up from the soil population of the survey area. However, they are based on a considerable volume of data and incorporate soil properties which are often relevant to land use. They do suggest what differences to expect in soils and, to some extent, what properties to use to differentiate between them. Often the properties used to differentiate soils and the way those properties are weighted to establish local soil classes are influenced by existing national classification schemes. Therefore, unless a highly specific purpose is envisaged, soil classes will be based, to an extent by default, on national classification schemes. A useful compromise for defining soil classes is to incorporate their local uniqueness while at the same time recognising their affiliations with a national scheme. This is done by using a combination of a local or geographic name with the taxonomic name. Typical soil class names then become 'Hawkesbury yellow earth', 'Dorrigo Red Ferrosols' or 'Narrabri black cracking clay'. An important distinction between this scheme and that of the USDA soil series scheme is that no soil outside the locality or map unit would be given the same soil class name. It only applies to the local tract of land or map unit. These soil classes are not as tightly defined as soil series in the USDA scheme and probably coincide more with the soil family level of classification.

In the soil mapping undertaken by the New South Wales Department of Land and Water Conservation, a pragmatic approach to soil classification and the naming of map units is usually adopted. Some variation occurs in the way mapping units are named, but they are usually defined and described using soil features related to the stated objective of the study, and take into account the range of soil property variation and the hazards and opportunities present within the study area. In these surveys, the commonly occurring taxonomic units of the Australian Soil Classification System (Isbell, 1996) and great soil groups (Stace et al., 1968) are usually identified within each mapping unit as a means of communicating the more obvious morphological features of the unit.

Soil Mapping Procedures 7.1.4

Principles and Methods

Whether a soil survey is for a general or specific purpose, it retains the same basic objective of extrapolating from a limited number of soil profile observations to the landscape. As Dijkerman (1974) noted, usually less than 0.0001% of the total survey area is field checked. From this he concluded:

It is clear that the soil surveyor needs to be an excellent scientist who is able to develop and use reliable hypotheses in order to make a good soil map!

Soil surveyors make and test hypotheses about the soil distribution in order to extrapolate from relatively few soil profile observations. It is the relationship of these hypotheses to the real landscape that determines the usefulness or value of a soil map.

In general, the development of a soil map can be seen as an interactive process in which hypotheses regarding the soil distribution are generated from theories of soil genesis, existing soil profile observations and other supplementary information (geology, geomorphology, topography, climate), and these hypotheses are subsequently evaluated by field checking to confirm or reject them. Any number of cycles of hypothesis generation and evaluation can be followed depending upon resources, time and the reason for approximations. Indeed, a soil map should be seen as a model of the soil distribution within the landscape based on the best available information at the time. This cyclic process, represented in Figure 7.1, is an ideal example of the scientific method and therefore makes detailed soil survey much more a research activity than a purely routine technical activity.

McKenzie (1991) examines the major methods of soil survey used by state and territory agencies in Australia. These include integrated survey where reliance is placed on using remotely sensed environmental variables to stratify field sampling and delineate mapping boundaries, and free survey where much field investigation is undertaken prior to building a map legend and mapping boundaries. Department of Land and Water Conservation soil surveyors use a mixture of integrated and free survey methods.

Sampling Techniques

Two basic sampling techniques are used in soil surveying. The first and most common is stratified sampling (Schelling, 1970). Information sources such as geology maps, topographic maps and aerial photographs are used to make hypotheses about soil distribution, and potential mapping units are drafted before undertaking any field work. The soil profile

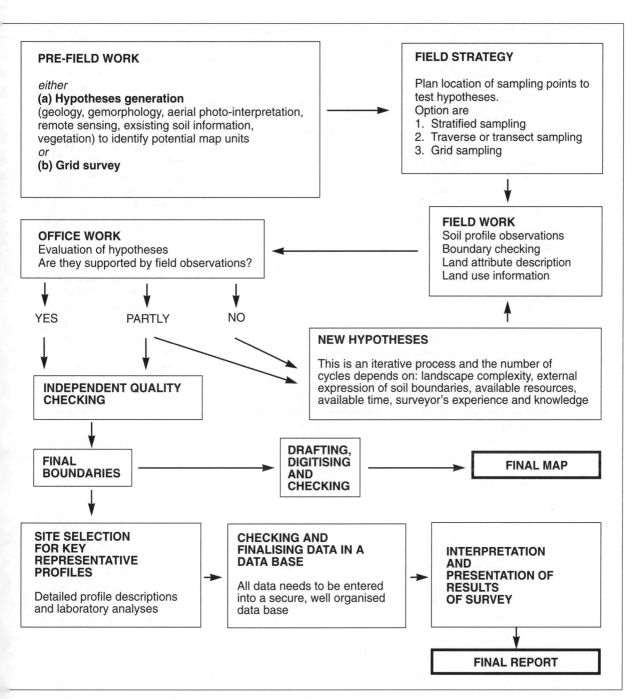

PRE-FIELD WORK

either
(a) Hypotheses generation
(geology, gemorphology, aerial photo-interpretation, remote sensing, exsisting soil information, vegetation) to identify potential map units
or
(b) Grid survey

FIELD STRATEGY

Plan location of sampling points to test hypotheses.
Option are
1. Stratified sampling
2. Traverse or transect sampling
3. Grid sampling

FIELD WORK
Soil profile observations
Boundary checking
Land attribute description
Land use information

OFFICE WORK
Evaluation of hypotheses
Are they supported by field observations?

YES   PARTLY   NO

NEW HYPOTHESES

This is an iterative process and the number of cycles depends on: landscape complexity, external expression of soil boundaries, available resources, available time, surveyor's experience and knowledge

INDEPENDENT QUALITY CHECKING

FINAL BOUNDARIES

DRAFTING, DIGITISING AND CHECKING

FINAL MAP

SITE SELECTION FOR KEY REPRESENTATIVE PROFILES

Detailed profile descriptions and laboratory analyses

CHECKING AND FINALISING DATA IN A DATA BASE

All data needs to be entered into a secure, well organised data base

INTERPRETATION AND PRESENTATION OF RESULTS OF SURVEY

FINAL REPORT

Figure 7.1 Flow chart for producing a soil map and report

observations are then located to sample these potential mapping units. The hypotheses predict that if soil differences occur, then the drafted boundaries are the most likely areas where differences will be found. This can be a very effective method of soil surveying, provided that those setting up the initial hypotheses are experienced and preferably have some local knowledge.

The second type of sampling technique is grid sampling, where soil profile observations are made independently of any external expression of the soil distribution or where there is no external expression. Grid types include regular, or random or stratified random (Webster, 1985). A third technique, which is a hybrid of stratified and grid sampling, is the traverse. In the traverse, soil profile observations are made on a regular or pre-planned basis along a traverse, but the location of the traverse can be strongly influenced by external expressions of the soil distribution, and is often planned to cross the most number of potential soil map units.

The appropriateness of these procedures depends on the landscape and scale of the mapping. Where there is a clear external expression of the soil distribution, as with basalt and granite-derived soils, grid techniques are probably not justified. However, in complex areas or where there is little external expression of the soil distribution, as on a level riverine plain, grid methods may be the most appropriate. Grid surveys or transects may also be useful for checking the effectiveness or accuracy of existing soil maps.

Despite their apparent scientific merit, soil maps based on grid sampling still need hypotheses to extrapolate between grid points. These hypotheses may include:
—assumed linear rate of change between grid points
—the influence on external factors of the soil distribution such as geomorphic processes
—the use of sophisticated statistical techniques, such as Kriging, to extrapolate between grid points.

A simple example of extrapolating from a limited number of soil profile observations using hypotheses is given in Figure 7.2. Sampling is based on transects that cover the two main geological units.

To extrapolate from soil profile observations, hypotheses are made about what happens between these points. The two soil profile observations on the alluvial plain, A and B, have very similar soils and there is no external expression of change in between. Thus, a reasonable hypothesis is that the soils do not change significantly between A and B. Depending on map scale and the intensity of soil information required, further sampling along the line AB may not be necessary. This will depend on time available, landscape complexity, and soil surveyor experience and confidence in the hypothesis. It is also necessary to consider the soil survey specifications that define the number and type of soil observations according to map unit area, map unit importance and effort required to meet survey aims and objectives.

On the other hand, between points B and C there is a topographic change and presumably a change in geological materials. A reasonable hypothesis is, therefore, that the soils change somewhere between these points. To find this boundary one can use a field traverse (BC) making a series of soil profile observations, use existing topographic or geological information, or use some external expression of the soil boundary that may be detected using remote sensing information such as aerial photography, radiometrics, magnetics, digital elevation models, radar or electromagnetics. In practice, one might use all of these. Unfortunately, many real soil mapping situations are not as straightforward, but the general principle applies. Soil surveyors need to balance the need to observe as many soils as possible against budget and time constraints. There are always areas where the soil surveyor would like to do further field checks.

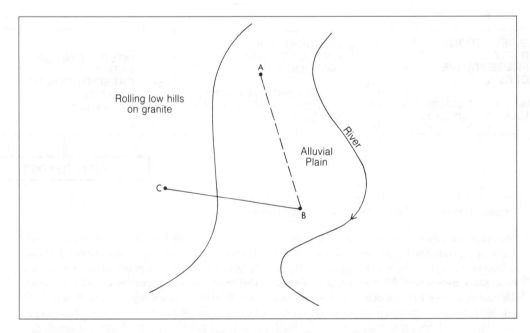

Figure 7.2 An example of soil mapping and extrapolation from soil profile observations

There are often doubts about the sufficiency and representativeness of sampling in complex mapping units, questions about how widespread 'anomalous' soils are which do not fit within soil landscape models, and fear of missing or misrepresenting the dominance of soil types. These dilemmas are usually rectified by further sampling. This highlights the need for efficient and well-planned field work, using appropriate soil sampling devices.

A theoretical discussion on the development of sampling strategies for predicting soil properties in small regions is presented by Webster and Burgess (1984).

Generating Hypotheses

There are a number of hypotheses about soil distribution that are commonly used in soil mapping. Those which allow observations to be made externally, without having to make soil profile observations, are the most useful but also the most dangerous. Such hypotheses should have a firm basis before being applied too widely. Constant observation and active constant questioning of soil and environmental relationships pays dividends. Sudden insights and subconscious thinking processes can also contribute to the soil survey process (Chapman, 1993). However, a number of obvious factors commonly influence the distribution of soils and, therefore, are useful for generating hypotheses. It is often useful to consider the interactive effect of these factors on any particular part of the landscape and its soils.

(i) Climate

While there are broad relationships between soils and climate, predictions at regional mapping scales are limited to areas where there are large changes in topography or elevation, or large changes in rainfall. The distribution of soils with particular climatic requirements such as alpine humus soils and humic gley soils is also strongly influenced by climate. More recently the application of climate prediction models based on location and digital elevation models in steep terrain has been used effectively to link microclimate variation with soil attributes (Ryan et al., 1995). Climatic differences due to aspect and exposure may influence soil attributes through indirect influences such as fire regime (Chapman et al., 1999). An example is the Hawkesbury Soil Landscape on the St Albans Sheet where steep northern slopes have appreciably more rock outcrop compared with south-facing slopes (McInnes, 1997).

(ii) Geology

This, through its influence on the distribution of soil parent material and the soil-forming processes, provides a wide range of valid, useful hypotheses about soil distribution. Geomorphological processes often interact with geology to develop the distribution of soil materials across the landscape. These processes lead to different landforms, and the effect on soil distribution is discussed in the next section.

Soil distribution control by rock type, geological formation and even broad structural units such as the Lapstone Monocline west of Sydney (McInnes, 1997), and the Molong Geanticline or Hill End Trough in central western New South Wales, are examples of useful geological hypotheses. Caution is required in using such hypotheses to map soils, as the relationship between geology and soils is not always clear. One geological unit may have several soil landscapes depending on landform and lithology. Conversely, relatively uniform soil landscapes may cross a number of geological formations. Therefore, a considerable amount of investigation is required before a geological boundary may be taken as a soil boundary.

Examples include the Yeoval Granite in central western New South Wales on which several contrasting soil landscapes occur (see Murphy and Lawrie, 1999), and the Cowra Trough which, despite having a large number of geological formations, has a relatively uniform soil landscape of undulating low hills of red-brown earths and non-calcic brown soils (see Kovac et al., 1989).

Further care should be taken with geology maps because unconsolidated soil parent materials such as alluvium or sand deposits are often ignored by geologists. Hard rock boundaries on hill slopes are often mapped higher up hills than where soil boundaries occur and are extrapolated from inferred geophysical phenomena deep below the soil profile.

There is a strong role for the use of soil maps in geological exploration (see Chapman, 1994). Roy and Matthei (1996) provide an excellent example of how soil surveyors and geologists can work together to gain a better understanding of earth resources.

(iii) Geomorphology— Topography—Landform

These also provide a number of useful hypotheses for mapping soils. First, the shape and pattern of the landform can provide information about soil distribution because these are often influenced by geology and geomorphological processes. Hypotheses based on the distribution of soil materials associated with geomorphological processes include such features as prior streams, ancient levees, back plains, playas, dunes and swales or alluvial fans. A list of geomorphological features or landform elements is given in McDonald et al. (1990). Each of these landform elements can be characterised by different kinds of soil parent materials.

Major Stability Problems: The Jordans Creek series (Dy3. 41, Dy3. 42) frequently shows severe gully erosion and the gullies are very deep (up to 10m). Gully erosion is active at the present time.

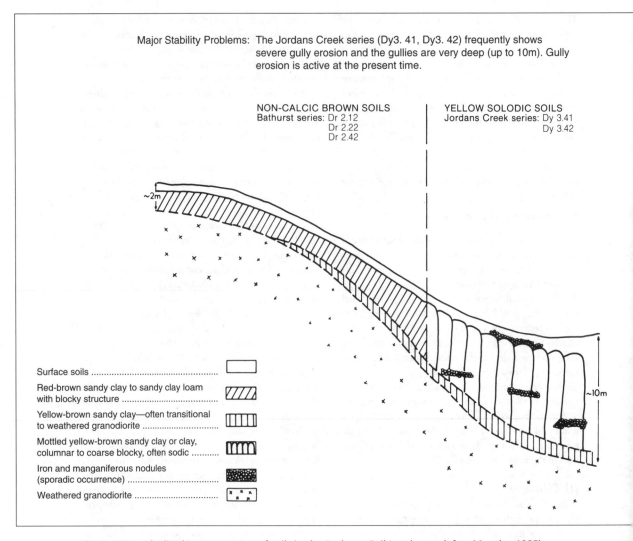

Figure 7.3 Idealised toposequence of soils in the Bathurst Soil Landscape (after Murphy, 1985)

For example, the regular dune-swale topography associated with aeolian landscapes characteristically has sandy soils. Similarly, narrow, level plains through rolling hills are characterised by alluvial soils. It is the close link between landscape-forming geomorphic processes and soil-forming processes that gives a soil landscape approach to soil mapping its strength as a mapping technique.

Second, topography can give rise to toposequences that are widely used in characterising variation within soil landscapes and soil associations. Toposequences reflect a regular pattern of soils down the slope as soil-forming processes, soil profile drainage and sometimes parent material change down the slope. Although it is often not possible to directly map soil variations within a toposequence, it is often possible to include diagrams within the report that show how the soils vary along the toposequence (see Figure 7.3). By using slope terrain maps or digital elevation models it is possible to display this soil variation at more intense mapping scales (Dewar et al., 1996).

(iv) Vegetation

Native vegetation, where present, can provide useful hypotheses for soil distribution, but this is often reduced by widespread clearing in agricultural lands. However, it is a very useful tool for mapping soils as vegetation, particularly certain species, often have a close association with soil properties. In certain conditions many species and communities have specific soil and water requirements. Observant soil surveyors often develop reliable models of what individual species indicate about soil conditions during the course of soil survey. See, for example, Tulau (1997).

(v) Land Use

Land managers may have many years of experience in learning to use land and soil characteristics to their advantage. Changes in land use often reflect variations in soil conditions.

Remote Sensing Techniques

Remote sensing techniques provide many useful hypotheses for soil mapping as well as being suitable for landscape boundary delineation. A major value for remote sensing is to provide a perceptive tool for patterns that cannot be determined from the ground. The patterns perceived can be used to provide indications of landscape-forming processes and landscape features which are useful for developing understanding of how the landscape evolved and for developing sampling strategies.

Aerial photography has traditionally been one of the most useful tools for soil surveying and should still be used routinely in soil surveys at scales of 1:250 000 or higher intensity. Stereoscopic viewing of air photos allows detailed, high resolution, three-dimensional perception of the landscape. This gives the capability to map many features related to the distribution of soils (see White, 1977).

Satellite imagery can be useful in broad-scale mapping (1:100 000 or broader) as it provides an excellent overview of a survey area and can be used for delineating landscape boundaries based on variations in texture, pattern, tone and shape in the image. On the older imagery some soil types can be recognised on MSS false colour imagery, such as red soils appearing green, black clays as dark blue and light sandy soils as pale tones, this instrument operating in the 0.4 to 0.9 micron wavelength region is not particularly sensitive to subtle variations in soil types. Of considerably more use is the Thematic Mapper, available on the new generation Landsat satellites, which has additional bands in the 1.65, 2.2 and 8 to 12 micron wavelength regions. These bands were designed for geological applications rather than the agricultural applications of the MSS instrument and, therefore, will yield information on soil texture, clay content, organic matter, iron oxide and mineralogy as well as having spatial resolution improved from 80 m to 30 m. Newer satellites are now available and the use of these has greatly improved the resolution and amount of data available.

Geologically sensitive satellite imagery bands are most useful where there is bare soil. Where soil is covered by vegetation, difference in vegetation growth response can indicate differences in soil type or fertility. The normalised differentiated vegetation index (NDVI) is useful for this purpose, relying on bands in the 0.6 to 0.7 and 0.7 to 0.9 micron wavelengths. The acquisition of imagery from drought periods is often best, as there is often more soil surface exposed and greater differences in the NDVI.

Increasing use is being made of geophysically derived imagery, which includes radiometrics, magnetics, and airborne and ground-measured penetrating radar, hyperspectral imagery and electromagnetics. While the use of radar and hyperspectral imagery is in its infancy, the use of electromagnetics is well established in salinity investigation work. Electromagnetics provides an indication of soil conductivity within various depth ranges. Soil conductivity is a function of moisture content, clay content and salt (Norman, 1989). Use of derivatives to discriminate surface magnetics features is being used for soil survey purposes in New South Wales for the first time.

Airborne gamma radiation (widely referred to as radiometrics) is proving to be an extremely useful tool for soil survey as the gamma radiation signals are derived from up to about half a metre depth, depending on moisture content and vegetation cover. The area covered by the signal can be split into counts for potassium, thorium and uranium isotopes as well as total count. Potassium and total count is particularly useful for soil mapping. Radiometrics cannot to date distinguish individual soil types, nor reliably be used to depict individual soil features such as pH or salinity without intensive ground truthing. It has proven useful to distinguish the location of different soil materials or materials of different origin. This is especially the case on flatter alluvial areas (Billings and Turner, 1998). Radiometrics has most use in tracing the fate of materials across the landscape and is best used as a perceptive tool for sampling possible soil differences in conjunction with other remote sensing tools. However, as with the use of all remote sensing tools, all hypotheses developed using them need to be checked with adequate field work.

Digital elevation models or computerised topographic information presented in a three-dimensional format are increasingly used for routine soil survey work. Their value depends on having an adequate topographic map base, and they are most effective when slopes average 10% or more. In these circumstances, digital elevation models provide an excellent means of visualising terrain features, and can be combined with other images, such as radiometrics, to obtain high resolution models of climatic variability, terrain features and soil distribution.

Evaluating Hypotheses

Usually hypotheses about soil distribution are checked by soil profile observations. Typically, standard statistical techniques are not used to evaluate hypotheses. This is done by observational evaluation in the field. Appropriate statistical techniques to evaluate hypotheses are described in Webster (1979, 1985). These statistical techniques are not commonly used in mapping soils in New South Wales as yet.

In collecting the base data for a soil mapping program, five levels of observation are usually employed.

(i)   Landscape transects: usually by vehicle and based largely on brief observations about soil classes, geology, geomorphology, landform and vegetation.

(ii) Brief soil profile descriptions to confirm allocation of a profile to a soil class: these observations are used to check map unit boundaries and confirm the soil classes present in the soil mapping unit.

(iii) Detailed soil profile descriptions identifying morphological properties: these observations are used to establish and describe the soil classes present in a map unit.

(iv) Detailed soil profile descriptions with physical determinations and laboratory analyses: these observations are used not only to establish the soil classes present in a map unit, but also to obtain more information about relevant physical and chemical properties of the soil classes. For example, infiltration, exchangeable cations and sodicity, free iron, nitrogen and organic carbon levels, aggregate stability, Atterberg limits and water-holding capacity may be determined. The assumption is made that the properties can, at least in part, be extrapolated to other soils of the same class which have similar morphological properties, parent materials and microclimate.

(v)   Reference sites—highly representative soil profile locations for important soil mapping units: reference sites include detailed soil hydraulic measurements, periodic sampling and laboratory testing to help determine changes and trends in soil condition and permanently maintained open pits for extension purposes. An example of a reference site is available for viewing at http://www.cbr.csiro.au/aclep.

The time taken and resources required for each observation increase from (i) to (v). As a consequence, the number of observations of each type for a soil map decreases in the same order. The number of detailed observations may be very few, and not all soil classes will necessarily have observations of type (iv). There are few reference sites in Australia at present.

For a soil map, the number of each type of observation should be recorded and its location be adequately specified so it can be relocated (for example, using grid references), and presented in the final report. In this way, users of the soil map have some idea of the map's reliability. For example, a soil landscape covering 100 ha and having 20 type (iii) observations and five type (iv) observations is very well known and reliable predictions should be possible. On the other hand, a soil landscape covering 10 000 ha and having 50 type (i) observations, the value of which depend very much on the experience and local knowledge of the surveyor, three type (ii) observations and one type (iii) observation is little known, and predictions about this soil landscape will be mainly preliminary or tentative.

Gray et al. (1997) outline the requirements for soil mapping associated with environmental impact studies. Similar mapping standards are outlined in Gunn (1988) and also by Moore et al. (1991).

7.2 General Purpose Soil Mapping

Most general purpose soil mapping is done as part of systematic mapping programs. This includes land systems mapping in the Western Division of New South Wales and Soil Landscape mapping at both 1:250 000 and 1:100 000 scales in the Eastern and Central Divisions. Some land resource studies have also been produced containing a general inventory of the soils of particular regions or areas, but most land resource studies have specific objectives and, therefore, the soil inventories are tailored to meet these objectives.

Land Systems Mapping 7.2.1

Land systems maps at a scale of 1:250 000 were completed by the Soil Conservation Service for the

Western Division of New South Wales. Land systems are 'recurring patterns of topography, soils and vegetation within an area of relatively uniform climate' (Dent and Young, 1981; Christian and Stewart, 1968). They can be subdivided into land units that are areas within which, for most practical purposes, environmental conditions are uniform.

Land systems maps are not strictly soil maps because boundaries are delineated on the basis of observable variations in terrain and vegetation. However, in the description of land systems, the soils within each land unit are classified and their relative area and patterns of occurrence recorded. With subsequent specialised field survey, a soils map can be generated from a land systems map. This was the case with Eldridge (1985) who produced maps of soil-landform associations from the land systems maps of the south-west corner of New South Wales, supplemented by extensive soil sampling. The land systems technique is particularly appropriate for mapping in the western zone where vast areas of native vegetation still exist.

The purpose of land systems mapping is to give a rapid overview of the physical environment of a region at reconnaissance level. With experience, this can be correlated with agronomic performance, and is now used as a basis for more detailed management plans for western land leases.

7.2.2 Soil Landscape Mapping

Soil landscape mapping programs are being undertaken by the New South Wales Department of Land and Water Conservation at scales of 1:250 000 across parts of the Central Division of New South Wales and at 1:100 000 in areas of greater land development pressure. The 100 000 program is also extended to areas where the soil distribution patterns are considered to be too complex for 1:250 000 soil mapping. This whole mapping program involves 15 soil scientists and is due to be completed in 2010 (see Colour plate 7.1).

In contrast to land systems maps, soil landscape maps place more emphasis on the soil component of the landscape. They describe associations or patterns of soils delineated by means of landscapes.

Northcote (1984) defined soil landscapes as:

natural areas of land of recognisable and specifiable topographies and soils that are capable of presentation on maps and of being described by concise statements.

Northcote showed the usefulness of the approach at broad scales in the *Atlas of Australian Soils* (Northcote et al., 1960–68), and quoted an example from the Murray River Basin to demonstrate the applicability of the method at varying scales (Northcote, 1984).

In this example he cites a unit consisting of dunes and swales—'dunes of brownish sands (Uc5.11)

with intervening swales of highly calcareous earths (Gc1.12)'—which would be a useful mapping unit at scales ranging from 1:25 000 to 1:250 000. If the same area were to be mapped at a scale of 1:10 000, the landscape units of dunes and swales and their component soils could be usefully divided into separate soil landscapes.

Butler (1980) discusses some of the underlying assumptions of soil landscape mapping. The principal assumption is that there is an underlying correlation between the distribution of soil classes and the landform factors of terrain, slope, relief and drainage. This is because these are genetically related through the action of geomorphic processes acting on similar parent materials under similar climatic conditions. Butler concedes that this is generally the case, but warns that the association is too important to be presumed and should be tested in any soil survey project.

Subsequent work carried out on the 1:250 000 and 1:100 000 soil mapping programs of the Department of Land and Water Conservation has supported Butler's comments and has shown the value of the soil landscape concept for mapping soils.

Soil landscapes can be used to distinguish mappable areas of soils because similar causal factors are involved in the formation of both landscapes and soils. Similarly qualities that are relevant to rural and urban development are related to both landscape and soil limitations. The soil landscape concept permits the integration of both soil and landform constraints into a single mapping unit.

For the two programs, soil landscapes have been defined as having:
(i)   specified landform patterns
(ii)   specified parent materials
(iii) specified soil classes at the series level, family level or soil materials level
(iv)   specified patterns of occurrence of soil classes
(v)   specified climatic zone.

A landscape includes all factors that distinguish the physical environment. It can be delineated by stereoscopic interpretation of aerial photographs and/or examination of the land surface. It is defined within a particular climate and by recurring patterns of underlying lithology, landform, slope and relief. The pattern of native vegetation, land use, erosion and drainage may also assist in its delineation. A soil landscape, as defined for these mapping programs, corresponds to a tract of land with relatively uniform landform pattern, climate, parent material and soil classes.

The Mapping Program

The Soil Conservation Service of New South Wales began a program to map the soil landscapes of the state in 1980. The principal objective of the New South Wales soil landscape series is to provide general, systematic coverage of the distribution and characterisation of the soils within the Eastern and Central Divisions of the state. This will complement the land systems series in the Western Division. The soil data collected during the course of this program are stored in the New South Wales Soil Data System.

These soil landscape maps will provide base data not only for the broad planning of land use, but also for agricultural, environmental and engineering practitioners including catchment managers, conservationists, district agronomists, environmentalists, valuers, educational institutions and engineering consultants. The maps will also provide useful information on the best locations for agricultural research work and the likely area over which these results may be extrapolated. They will also become a very useful educational resource, and have been used in many unexpected ways.

The program was developed to meet the increasing demand for soils information in developing areas following the inception of the Environmental Planning and Assessment Act 1979 in New South Wales. Requests for information have come from a broad spectrum of land managers, particularly from municipal and shire councils, government departments, private environmental consultants and engineers.

The soil landscapes used in this mapping program are developed using the methodology outlined in Figure 7.1. The Soil Landscape mapping program is constantly evolving and improving. There are differences between soil landscape reports that have been published at different times. Table 7.4 provides an indication of the contents of modern soil landscape descriptions. At the time of writing preparations are being made to publish Soil Landscape Products on the World Wide Web. Table 7.5 lists the contents of a typical soil landscape report.

Table 7.4 Contents of a soil landscape map unit data set

Section and Contents	Description and Purpose * follows items not included or of lesser detail in 1:250 000 mapping.
Map unit name **Summary**	Brief introduction to landscape*, soil-forming processes, soil classifications*, their distribution* and major land and soil qualities.
Landscape Photos	Provide an indication of the landscape.
Location and Significance	Major locations on the current and other published maps*. Type location: map reference for an accessible and representative location. For those who wish to visit a typical example of the soil landscape and also for future reference sites.
	Links the soil landscape to a physiographic region (for aggregating to broader scale mapping).
	Geographic significance, e.g. species or soil type range limits, unique properties, strategic land uses, areal extent*.
Geology and Regolith	Major lithologies and formations. Brief description of regolith types, depth and alteration*. Short notes on landscape evolution where applicable*.
Climate and Hydrology	Special climate considerations*, e.g. wind exposure, frost frequency. Seasonal soil moisture conditions and water tables*. Recharge and discharge area identification*.
Terrain	Land surface shape*, elevation range, slope angles, surface fragments, rock outcrop and microrelief description of major land form elements and their dimensions.* To provide a picture of the landscape and describe any soil sublandscapes (unmapped but known soil variations that can be delineated according to landform elements).

Table 7.4   Continued

Section and Contents	Description and Purpose
	* follows items not included or of lesser detail in 1:250 000 mapping.
Vegetation	Extent cleared, major vegetation communities and where they occur within the soil landscape (linked to soil sublandscape where possible)*. Major and indicator species within each plant community type*. Comments on weeds*.
Land Use/Land Degradation	Current and previous land use. Influences land use and previous land use or land history may have had on soil condition*. Types and severity of land degradation present. Comments on paired site comparisons including estimates of amounts of topsoil lost*.
Included Soil Landscapes and variants	Known mapping impurities of related soil landscapes and where they occur.
	Descriptions of soil landscapes that vary by one or two properties from the major soil landscape and the properties they differ by.

Soils

Soil Variation	Patterns of soil distribution and their association with landscape features*.
Soil Materials	Soil material names and abbreviations. Brief descriptions of dominant and associated soil materials, their origins and variable and constant properties*.
Soil Distribution Diagram	Toposequence diagram showing soil sublandscapes*, type profile locations, distribution and relative thicknesses of soil materials in comparison with parent materials and land surface shapes. Includes soil classifications.
Type Profile Descriptions	Detailed soil profile descriptions for typical or dominant soil profiles* in each soil sublandscape including soil materials allocated to various horizons*.

Land Use Qualities and Limitations

	Erosion Hazard, Rural Capability and Recommended Land Management, Urban Capability* and Engineering Hazards*.
Soil Qualities and Limitations (for each soil material)	Acidification Hazard, Acidity, Acid Sulfate Soil Potential, Alkalinity, Aluminium Toxicity, Erodibility (sheet, concentrated flow and wind), Fertility, Soil Fire Hazard, Hardsetting soils, Non-cohesive soils, Organic Soils, Periodically Frozen Soils, Permeability, Plant-available Water Capacity, Plasticity, Poor seed bed conditions, Salinity Hazard, Shrink-swell Potential, Sodicity/Dispersion, Stoniness, Structure Decline Hazard, Water Repellence, Wet Bearing Strength.
Significant Landscape Qualities	Listing of severity and extent of: complex terrain, complex soils, dieback, drainage, engineering hazard, fertility, flood hazard, gully erosion hazard, groundwater pollution hazard, inherent sheet erosion hazard, mass movement hazard, water tables, poor moisture availability, potential recharge, discharge areas, productive arable land, rock fall hazard, rock outcrop, runon, salinity hazard, seasonal waterlogging, scalds, shallow soils, steep slopes, wave erosion hazard, wind erosion hazard and woody weeds.

Table 7.4 Continued

Section and Contents	Description and Purpose
	* follows items not included or of lesser detail in 1:250 000 mapping.
Soil Test Results	
(for each soil material represented in each type profile)	Soil tests for Bray Phosphate, Phosphorus Sorption, Cation Exchange Capacity and Exchangeable Cations (Ca, Mg, K, Na, Al), Dispersion Percentage, Electrical Conductivity, Organic Carbon, Particle Size Analysis, pH in 1:5 Soil:Water, pH in 1:5 Soil:0.01 M $CaCl_2$, Unified Soil Classification, Volume Expansion, Non Wind-erodible Aggregates and Water Repellence. Lactate Phosphate, Linear Shrinkage, non-dispersed Particle Size Analysis and the Emerson Aggregate Test are performed on all suitable samples.

Table 7.5 Contents of soil landscape reports

Section Title	Description and Purpose (bold indicates chapter headings)
Title Page	Study title and reference details, Author address, Map showing location of study area.
Foreword	Introduction by a distinguished and knowledgeable individual.
Contributions/ Acknowledgments	Provides an indication of the extent of work and the extent of community and expert contributions.
Introduction	Reason why the area was mapped, important local issues, aims and objectives of the work. How to use the map and report.
Physical Attributes	Location. Size and boundaries of the mapped area and major settlements within it.
Physiography	Broad description of the area. Breakdown to physiographic regions. Map and brief description of physiographic regions.
Geology and Regolith	Reference to previous geology studies in the area. Broad description of geology and structural units. Geology of each physiographic region.
Climate	Reference to previous climate studies. Sources of climatic data. Information on rainfall, rainfall erosivity, rainfall reliability, and variation across the mapped area. Temperature variation and frost season. Wind patterns.
Vegetation	Previous work. Botanic conventions used. Notes on indicator species. Vegetation communities by physiographic regions and their major species. Degree of modification and modifying factors. Weeds.
Land Use	Distribution and types of land use. Brief history of land use and how this has impacted on soil and landscape.
Methods	Definitions of specialist terms, explanation of mapping symbols and mapping conventions. Previous soil surveys. Mapping methods, numbers of profiles described and sampled, sampling procedures, access to data collected. Data reliability, data quality, laboratory sampling, laboratory test methods, soil test interpretations.
Limitations and Capabilities	Explanations of soil and landscape qualities and limitations. Table of soil qualities and limitations for each soil material in each soil landscape. Table of landscape qualities and limitations for each soil landscape. Guide to urban capability and rural capability. Table outlining urban and rural capability for each soil landscape or soil sublandscape.

Table 7.5   Continued

Section Title	Description and Purpose (bold indicates chapter headings)
Soil Landscape Descriptions	See Table 7.4.
Bibliography and References	
Appendices	Common and Scientific Names of Plant Species. Guide to use of soil test results. Explanations of soil test results and their interpretations. Soil test results and their interpretations—linked to profile information and grid references. Soil Fertility Guide. Soil Erodibility Guide, Erosion Hazard Guide, Foundation Hazard Guide.
Glossary	

Designation of Soil Landscapes

Current national soil classification taxonomic units incorporating both those of the Australian Soil Classification System (Isbell,1996) and the great soils groups (Stace et al., 1968) are used. Allowance is made for the local uniqueness of soils, and soil landscape names are adopted from local geographical areas. Hence, soil landscapes may be given names such as the Bathurst soil landscape or the Byng soil landscape.

National Soil Classifications are extremely useful for communication between soil scientists and map unit correlation for aggregation purposes.

The 'Soil Materials' Concept

Users of soil maps require a knowledge of the types of soil material present, their distribution and their behaviour. National soil classifications, while intended to aid communication, were introduced to soil landscape mapping to simplify terminology and to make the relationships between the soil and the landscape, which are often stratigraphic, more easily understood by the non-technical user. The concept enables the behaviour of different materials to be discussed separately and permits non-solum material, such as stratigraphic layers, weathered bedrock and artificial fill, to be included in unit descriptions. This gives it a degree of flexibility not available with traditional soil classifications.

'Soil materials' are defined as 'three-dimensional soil entities which have both a degree of homogeneity and lateral continuity across the landscape' (Atkinson, 1993). At any given site, the soil will consist of one or more layers of 'soil material' in succession. These layers often correspond with soil horizons, but they may be of stratigraphic origin and they may cross traditional soil taxonomic boundaries. Each 'soil material' is described in terms of its morphological character-

istics. No account need be taken of its position in the profile and no assumptions are made regarding its genesis.

It has been found that, to present the information on the distribution and performance of soil materials that was being sought by many users, it was not necessary to introduce the added complexity of the more traditional soil classifications. This was the case even though national schemes are useful for communication between soil scientists and for correlation between soil map units.

In the Sydney region, it was recognised that there was a limited number of materials occurring within each landscape (Chapman and Murphy, 1989), and that it was easier to explain the occurrence and relationships of these materials than it was to explain the complex relationships of soil types in the landscape. Similarly, in many of the landscapes around Sydney, and particularly on the Hawkesbury Sandstone, many of the soils are made up of layers that are not genetically related to one another and, therefore, do not conform to classical ideas on soil genesis, and may not relate readily to traditional taxonomic systems (Paton, 1978).

The 'soil material' is uniquely defined for each landscape and each is identified by the landscape code letters followed by a sequential number. Thus, for the 'Hawkesbury Soil Landscape' (Chapman and Murphy, 1989), the dominant soil materials are labelled Ha1 to Ha3. The definitions are, therefore, specific to the particular landscape unit. However, that is not to say that similar materials do not occur in other similar landscapes.

As the 'soil material' is defined using observable and measurable morphological properties, it is possible to construct genetic theories to explain their distribution. Similarly, when engineering or agronomic performance data can be correlated with the morphological properties, reliable predictions can be made about the performance of the soil. Soil morphological data can be amalgamated into profiles and used to classify the soils into the units of the Australian Soil Classification, great soil groups or other classifications as required.

7.2.3 Derivative Map Products

Derivative maps are maps that combine and filter soil landscape information to show how soil and land varies in its behaviour or suitability for specific purposes.

One of the benefits of a good general purpose soil survey is its capacity for the data used to be recombined, filtered and reinterpreted, either independently or in combination with other data sets, to generate new specific purpose or derivative map products. In effect, production of soil landscape derivative maps involves changing of map unit colours and allocating a new legend, disclaimer and map title.

In order to produce derivative maps properly it is crucial that the information collected according to a recognised standard (for example, McDonald et al., 1990; Rayment and Higginson, 1992) during the course of soil survey is stored in a disciplined manner. The information used on the derivative map must be traceable back to the data sources where the data was originally collected. The types of derivative maps that can be produced from any soil survey are dependent on the type and extent of data collected. The rules that were used to derive digital data combinations should be well thought out and explicitly included on the map. As the rules for any particular map theme may change, it is useful for the date and edition number to be recorded on the map sheet and the rule set for the map to be recorded. To date over 80 different derivative map themes have been produced, using data and map boundaries from the New South Wales Soil Landscape mapping series.

Derivative maps are produced using a set of rules to allocate map units to legend classes. Rules may be based on simple class boundaries for readily observed attributes such as surface soil texture. More sophisticated interpretations may be based on consideration of many factors which could influence the sustainability of the land management system or land use.

The most common set of rules is based on a series of limiting factor tables and their associated definitions such as presented by the United Nations Food and Agriculture Organisation Land Suitability System of 1976 and 1983, although other organisations, such as the United States Department of Agriculture (1993), have independently developed very similar systems. An example of an FAO-style rating table used by DLWC soil survey is shown in Table 7.4. Table 7.6 outlines the features, advantages and disadvantages for the application of the FAO system and provides references on how such drawbacks may be overcome.

Table 7.6 Features, advantages and disadvantages for improvement of the FAO-style system of land capability/suitability rankings

Features	Advantages and Disadvantages
Relies on 'most limiting factor' rather than additive or multiplicative systems. Avoids compounding of errors.	Does not easily handle interactions between factors. Tends to rank conservatively (Burrough, 1989). Boundaries tend to be sharp, may be somewhat arbitrary (depending ultimately on the opinions and knowledge of those who develop and authorise the use of the tables). Ranking limits may be taken literally by local authorities (Van de Graaff, 1998). Limiting factors can tend towards pessimistic and defensive thinking at the expense of opportunity and user acceptance.
Easy to understand. Relies on detailed definitions of attributes and land uses.	Locations for land uses can be determined. Forces a disciplined, accountable approach to land use decisions. Easily arranged for computer-automated production of maps. There is, however, no clear solution for dealing with or presenting information where there is within-map-unit variability.
Ability to make precise statements about many particular land uses and to relate these to broad capability statements.	Ability to tailor rating systems to match with particular user requirements, e.g. types of machinery used. Can nest ratings within ratings (Chapman et al., 1992). Has an ability to relate risk factors (Johnson and Cramb, 1992) and land management and farming system operations (Smith and Thwaites, 1998).

.2.4 Numerical Methods in Soil Landscape Mapping

Although numerical methods have been used to define soil materials in special purpose studies (Atkinson, 1982), they have not yet been widely applied in soil landscape mapping in New South Wales. There is considerable opportunity to introduce appropriate numerical techniques to better define mapping units and to help design more efficient sampling strategies. The most obvious would be to use cluster analysis or, preferably, 'fuzzy sets' (McBratney and Moore, 1985) to define the composition of soil classes and then use multiple discriminant analysis to allocate new data to established classes. Principal component analysis may help explain the variation between different soil classes, and methods incorporating terrain data may be able to differentiate the more obscure soil landscapes. Quantitative spatial analysis techniques such as semi-variograms and Kriging could be useful in defining the variability within soil landscapes and in refining sampling procedures (Webster, 1985) (see also section 6.7).

7.3 Special Purpose Soil Mapping

As the name suggests, special purpose soil mapping is undertaken with clearly identified and narrowly defined objectives. They are usually undertaken for a specific user and accommodate the user's requirements in the survey design. It is in special purpose surveys, more than any other, that the soil surveyor has the opportunity to select from the vast array of techniques and strategies available.

It is not appropriate here to discuss the full range of techniques suitable for a special purpose survey. Rather, anyone designing such a survey is advised to consult the literature referred to in Literature Review at the beginning of this chapter.

The FAO system of land suitability assessment presented is clearly well suited to use in special purpose soil surveys. The aim is to reduce within-map unit variability for land attributes that are relevant to the issue or land use at hand. This cannot always be done for specific purpose surveys.

The Department of Land and Water Conservation has undertaken many special purpose soil surveys, some in association with other land resource investigations, some exclusively as soil surveys. A range of examples includes:

—acid sulfate soil risk (Naylor et al., 1995; Tulau, 1999)
—capability of rural land for agricultural use (Attwood et al., 1985)
—soil characteristics for effluent disposal (Tulau and Wild, 1996; Geary 1994)
—suitability for urban development (Hannam and Hicks, 1980; Murphy and Atwood, 1985; Hicks et al., 1984)
—suitability of topsoil for reuse after mining (Veness, 1980)
—input to forest growth and yield prediction (Murphy et al., 1998; Turvey et al., 1991)
—erosion hazards associated with quarrying (Atkinson, 1982)
—sedimentation hazards of proposed new dams (Hazelton et al., 1985)
—susceptibility of parkland soils to trampling (Wheen and Johnston, 1986).

For each of these uses, a different set of soil properties will be of critical importance. In addition, associated with each use will be some form of soil erosion. The extent and form of this erosion will also depend on soil properties.

The New South Wales Department of Land and Water Conservation is concerned not only with those properties that will allow a soil to support a certain form of land use, but also to sustain such use without long-term degradation. Two main areas of land use where soil erosion can be a major problem are agricultural production and urban development.

7.3.1 Soil Survey Interpretation for Rural Capability

The rural capability classification scheme (see Chapter 15) involves considerations of environmental factors that may limit the agricultural use of the land. The factors considered include those with general or widespread influence, such as climate, landform, slope gradient and soil erodibility, as well as those of a more local influence, such as soil depth, periodic inundation, rockiness, existing erosion status and possible salinity, sodicity or alkalinity.

Each of these factors is considered as part of a limiting factor analysis, in which one or a combination of factors may limit the capability of an area of land. Based on this type of analysis, land is classified into one of eight classes. The limitations to use, risks of damage and need for soil conservation measures become greater from Class I to Class VIII (Emery, 1985).

All agriculture depends, ultimately, on the capacity of the soil to sustain the growth of plants. Land capability for agriculture will, therefore, relate to the soil properties concerned with the capacity to grow plants, to accept regular disturbance during cultivation and to permanently sustain these activities without deterioration (see Chapter 15).

Generally speaking, agriculture is concerned with the use of the topsoil, since it is the surface horizons in which plant roots concentrate that are involved in the majority of cultivations and that are eroded most readily if not protected. Subsoil is of secondary importance, but it does have significance in relation to water-holding capacity, drainage, salinity and susceptibility to gully and tunnel erosion.

The soil attributes that need to be assessed in relation to rural land capability are:

(i) Physical fertility—texture, degree and size of structure, infiltration and drainage, surface condition, water-holding capacity, aeration, stoniness.

(ii) Chemical fertility—nutrient supply, organic matter, acidity, alkalinity, sodicity, salinity and toxicity.

(iii) Potential for soil erosion—soil erodibility, texture, degree and size of structure, aggregate stability, soil dispersibility, infiltration and permeability.

(iv) Ability to sustain repeated cultivation—soil texture, degree and size of structure, surface condition, aggregate stability, organic matter, consistence, wet and dry strength, depth of surface horizon, profile drainage, rockiness or stoniness.

The role of these soil attributes is discussed further in Chapters 12, 13, 15 and 16. With most land resource studies, the slope and soil attributes are mapped independently. These then can be integrated to give land capability by using a decision matrix to allocate the capability class (see Emery, 1981; Logan and Luscombe, 1984).

An example of a rural land capability decision matrix, taken from Emery (1981), is presented in Table 7.7. In this two-way table, soil associations are on the vertical axis and terrain on the horizontal axis. A separate table is produced for each slope class. Larger scale land resource studies can use a single matrix with soil and terrain on the vertical axis and slope on the horizontal axis because of the close relationship between soils and terrain at larger scale (see Atkinson, 1984).

Soil Survey Interpretation for Urban Capability

7.3.

In an urban context, the capacity of the soil to grow plants becomes less important although the long-

Table 7.7 Land capability decision matrix incorporating soils and terrain data (for slopes of 1–5%) (Emery, 1981)

| Terrain | Crests | | Sideslopes | | Footslopes | | Floodplains | | Drainage Plains | |
Soil Erosion	A^1	B^2	A^1	B^2	A^1	B^2	A^1	B^2	A^1	B^2
Soil Associations										
Havelock	IV	—	IV	VI	III	III	—	—	—	—
Murrumbateman	IV	—	IV	V	III	III	—	—	—	—
Hall	IV	—	III	III	III	III	—	—	—	—
Cavan	—	—	IV	VI	II	III	—	—	—	—
Yass	II	—	II	III	II	III	—	—	—	—
Kittys Creek	—	—	II	III	II	III	IIW	VIW	IIW	IIIW
Murrumbidgee	IV	—	IV	V	III	III	—	—	—	—
Mundoonen	IV	—	IV	VI	III	III	—	—	IVW	VW
Fairfield	IV	—	IV	V	III	III	—	—	—	—
Sutton	—	—	II	III	II	III	—	—	IIW	IIIW
Back Creek	—	—	III	III	II	III	IVW	VIW	IVW	VW
Lake George Terrace	—	—	—	—	IV	V	—	—	—	—
Yass River	—	—	—	—	II	III	IIW	VIW	IIW	IIIW

A^1 = soil classes with low soil erodibility
B^2 = soil erosion classes with high soil erodibility
W = a restriction to capability class due to periodic wet ground conditions

term stability of the soil will depend on the use of vegetation. The engineering properties of soil become dominant as there is far more intensive use of the land. Whereas agriculture is more concerned with topsoils, urban developments involve subsoils to a much greater degree. This is because topsoil is generally removed and replaced afterwards, and the majority of structures which typify the urban scene are built on, or involve the excavation of, subsoil.

Land capability for urban development requires soil information to determine:

(i) engineering suitability (or limitations) for building foundations, storage reservoirs, sewage disposal facilities, embankments and the location of roads, railways, pipelines and so on

(ii) suitability for sand, gravel and other mineral supplies

(iii) erosion hazard during the development phase

(iv) suitability of soil materials for revegetation.

Hannam and Hicks (1980) detail the urban capability classification used by the former Soil Conservation Service and describe the survey techniques used to apply it, and Chapman et al. (1992) and Morse et al. (1992) have suggested modifications and improvements to this approach. Methods of control of urban erosion and resulting sedimentation are described in Quilty et al. (1979) and more recently DLWC (1996). The limitations of soils for urban land use are discussed further in Chapter 21.

The soil properties that need to be understood in relation to urban development are:

(i) susceptibility to mass movement

(ii) depth and rockiness

(iii) suitability for foundation material (including the occurrence of expansive soils)

(iv) soil erodibility and dispersibility

(v) soil drainage and permeability

(vi) salinity

(vii) fertility

(viii) acid sulfate soils.

Table 7.8 Soils in the Coffs Harbour region ranked according to their degree and type of limitation to development (after Atkinson and Veness, 1981)

	Degree of Limitation Caused by:			
Soil Unit	Foundation Hazards	Drainage and Effluent Disposal Hazard	Erosion Hazard	Flooding and Inundation Hazard
A1	Bp	Dp	A	B
A2	Dyg	Dy	A	D
A3	A	B0–1% A1–5%	A	B
A3/1	Bp	Cpt	A	B
A4	A	Dt	Ct	D
A5	A	A	Bt	A
A6	Cpt	Dw	A	C
S1	Det	Dc	Dte	Dt
S2	Aeg	Dc	Ce < 15% De > 15%	0
S3	Ag	Dcw	Cte	0 < 5% B > 5%
S4	Dgt	Dy	Ct	D
S5	Aw–Bw	Cp	Be	B
S6	Dy	Dy	A	D

Degree of Limitation		Soil Limitation		Other Limitations	
0	none	**c**	very high permeability	**t**	topographic feature
A	minor	**g**	low wet strength	**s**	slope
B	moderate	**p**	low permeability	**y**	swamp
C	severe	**d**	shallow soil	**w**	seasonal waterlogging
D	limiting	**e**	erodibility		

Table 7.9   An example of data for a mapping unit for an urban capability study (after Hicks et al., 1984)

Soil Map Unit:	D3 HIGHLY PLASTIC YELLOW DUPLEX SOILS
Principal Profile Form:	Dy3.41, Db2.22
Great Soil Group:	Yellow Solonetzic Soil
USCS Code:	CH, SC

General Features

These are deep soils (more than 100 cm) with duplex texture profiles having highly plastic clay subsoils. The surface soil is a brown (10YR 4/3) fine loam with weak structure and a pH of 5.0 to 6.5. At 10 cm there is a clear boundary to a pale brown or bleached (10YR 6/3 wet) A_2 horizon with fine loam texture.

There is a clear boundary at 30 cm to a strongly structured yellow-brown (10YR 5/6) or brown (10YR 4/4) heavy clay. The clay is often mottled dark b[…]grey or yellow red. The pH is from 5.5 to 7.0, and the clay horizon extends to at least 100 cm. Variations include some dark soils (Dd1.42).

Distribution

The Highly Plastic Yellow Duplex Soils are probably closely associated with a highly basic component of the bedrock, and therefore have a limited distrib[…]They are confined to scattered occurrences on sideslopes and footslopes.

Drainage

These soils are imperfectly drained. The surface soils are moderately permeable, but the subsoils have very low permeability, resulting in interm[…]perched water tables in the bleached A_2 horizon during wet periods. Available water-holding capacity is moderate.

Shrink-Swell Potential and Plasticity

The shrink-swell potential and plasticity of the subsoils is high to very high (linear shrinkage: 17–21 per cent and liquid limits: 48–76 per cent). [...]attributes indicate that substantial foundation limitations exist which will require further investigation.

Erosion Hazard

These soils have a moderate soil erodibility and, on lower slopes where they are subject to significant amounts of runon, have a high erosion hazard, esp[…]if ground cover is removed.

Rockiness

There is no significant rock outcrop within this map unit. Weathered rock may occur within 150 cm of the surface on some sites.

SUMMARY OF LABORATORY DATA — SOIL MAP UNIT: D3 HIGHLY PLASTIC YELLOW DUPLEX SOILS

Horizon	A	B			
Average Sample Depth	0–20 cm	20–70 cm			
Number of Samples	1	6			
USCS Code	ML	CH			
		mean	**s.d.***	**c.v.***	**r**
Clay (%)	12	62	3.3	5	4
Silt (%)	31	16	2.0	12	1
Fine Sand (%)	38	16	2.4	15	
Coarse Sand (%)	19	5	1.5	29	
Gravel (%)	0	0			
Liquid Limit		64	2.4	4	5
Plasticity Index		34	2.3	7	2
Linear Shrinkage		19	1.3	7	1
Aggregate Test	3(1)	5			
Dispersion Percentage	14	13	2.2	18	
pH	6.0	6.0	0.8	13	5.

   *   **s.d.** = standard deviation
  ** **c.v.** = coefficient of variation

Variations in the occurrence of these criteria form the basis for identifying mapping units in a soil survey for urban capability assessment. Because of the detailed scales of urban capability investigations ranging from 1:25 000 to 1:4000, most soil mapping is at the soil series level. Mapping units are usually named descriptively using a combination of morphological and terrain descriptors such as 'siliceous dune sand' or 'alluvial loams and gravels'. In these cases, the common taxonomic units occurring within the mapping unit are listed prominently in the description of the unit. The engineering (USCS) classification of the subsoil may be added to the Australian Soil Classification unit (Isbell, 1996) and great soil group (Stace et al., 1968) classifications. When appropriate, mapping units may be named using principal profile form descriptors, such as 'hardsetting mottled yellow duplex' or using great soil group terminology such as 'yellow solodic soils'. Descriptions of the mapping unit highlight the significant hazards that may be encountered for land use. The normal subheadings in a unit description are:

—general features

—distribution

—drainage characteristics

—fertility

—foundation hazard

—erosion hazard.

In an example taken from the Coffs Harbour region (Atkinson and Veness, 1981), each soil unit is ranked according to the degree and type of limitation to development (see Table 7.8). An example of a soil map unit description is given in Table 7.9. Once the soil units are ranked in this way, it is possible to superimpose terrain and slope constraints to determine the urban capability of the unit.

In this type of detailed mapping, one should not expect the broad taxonomic units of national soil classification schemes to be adequate. The taxonomic units are usually localised variants of the great soil groups or Factual Key groups and correspond to soil series or soil family level. Examples of other urban capability studies demonstrating this are to be found listed in the bibliography at the end of Chapter 21.

The purpose of soil surveys for urban capability is to identify and locate areas with soil-related hazards for urban development. There are advantages in applying the 'soil materials' concept in certain situations. 'Soil materials' can be defined in terms of their hazard-related performance. Therefore, the use of mapping units based on the distribution of these 'soil materials' will allow more precise identification of the hazard and easier communication with users untrained in soil science. The method is at its best with large-scale mapping and with specifically defined soil problems. Both of these conditions are met in urban soil surveys.

Recent years have seen major developments in soil survey, both in methodology of collecting and producing soil survey information and soil maps, and in how the information is used. Improvements in remote sensing techniques, storage and analysis of data using computers, and expansion of the uses of soil survey information to environmental and engineering issues have also all moved forward and advanced the efficiency of soil survey operations.

BIBLIOGRAPHY

Abbs, K. and Littleboy, M. (1998), 'Recharge estimation for the Liverpool Plains', *Australian Journal of Soil Research* 36(2), 335–357.

Abraham, S.M. and Abraham, N.A. (eds) (1992), *Soil Data System—Site and Profile Information Handbook*, Department of Conservation Land Management, Sydney.

ACIL Economics and Policy (1996), 'An economic framework for assessing the benefits and costs of land resource assessment in Australia', *Australian Collaborative Land Resource Survey Newsletter* 5(3), 2–5.

Atkinson, G. (1982), *Soil Survey and Erosion Control Measures for the Penrith Lakes Scheme*, Soil Conservation Service of NSW, Sydney.

Atkinson, G. (1984), *Land Resources Study: Lord Howe Island*, Soil Conservation Service of NSW, Sydney.

Atkinson, G. (1993), 'Soil materials: a layer based approach to soil description and classification', *Catena* 20, 411–18.

Atkinson, G. and Veness, R.A. (1981), 'Soils of the Coffs Harbour region', *Journal of the Soil Conservation Service of NSW* 37, 97–115.

Attwood, R.D., Murphy, B.W. and Hicks, R.W. (1985), *Rural Land Capability: Dubbo City*, Soil Conservation Service of NSW, Sydney.

Bannerman, S.M. and Hazelton, P.A. (1990), Soil landscapes of the Penrith 1:100 000 sheet, *Soil Conservation Service of NSW*, Sydney.

Barson, M. and Shelley, P. (1994), 'On-line directory to Australian Soil and Land Resource Survey Data, *Australian Collaborative Land Evaluation Program Newsletter* 2(2), 3–5.

Beckett, P.H.T. and Bie, S.W. (1978), *Use of Soil and Land System Maps to Provide Soil Information in Australia*, Technical Paper No. 33, CSIRO, Division of Soils, Melbourne.

Bierwirth, P.N., Gessler, P. and McKane, D. (1996), 'Empirical investigation of airborne gamma-ray images as an indicator of soil properties—Wagga Wagga, NSW', *Eighth Australasian Remote Sensing Conference Proceedings*, Canberra.

Billings, S. and Turner, J. (1998), 'Validation of soil mapping within the Jemalong-Wylds Plains', (unpublished report).

Blain, H.D., Clark, J.L. and Basinski, J.J. (1985), *Bibliography of Australian Land Resources Surveys*, Divisional Report 85/4, CSIRO, Division of Water and Land Resources, Canberra.

Bregt, A.K., McBratney, A.B. and Wopereis, M.C.S. (1991), 'Construction of isolinear maps of soil attributes with empirical confidence limits', *Journal of the Soil Science Society of America* 55(1), 14–19.

Burrough, P.A. (1989), 'Fuzzy mathematical methods for soil survey and land evaluation', *Journal of Soil Science* 40, 477–92.

Butler, B.E. (1980), *Soil Classification for Soil Survey*, Clarendon, Oxford.

Chapman, G.A. (1993), 'Examples of the role of geomorphology in soil survey/land evaluation in New South Wales', *Australian Collaborative Land Evaluation Program Newsletter* 2(3), 15–16.

Chapman, G.A. (1994), 'The Australian Regolith Conference—general impressions from a land resource assessment perspective', *Australian Collaborative Land Evaluation Program Newsletter* 3(4), 15.

Chapman, G.A. (1994), 'Soil landscape maps and regolith information', poster paper abstract in *Proceedings of the Australian Regolith Conference, Broken Hill, November 1994*.

Chapman, G.A. and Murphy, C.L. (1989), *Soil Landscapes of the Sydney 1:100 000 Sheet*, Report, Soil Conservation Service of NSW, Sydney.

Chapman, G.A., Morse, R.J. and Hird, C. (1992), 'A framework for assessment of urban land suitability for New South Wales', *Proceedings of 7th International Soil Conservation Organisation Conference: People Protecting their Land,* Sydney, Australia, 27–30 September 1992.

Chapman, G.A., Murphy, C.L., Tille, P.J., Atkinson, G. and Morse, R.J. (1989), *Map of Soil Landscapes of the Sydney 1:100 000 Sheet,* Soil Conservation Service of NSW, Sydney.

Chapman, G.A, Zierholz, C. and Davies, B. (1999), 'Soil and land capability for sustainable fire regimes', *Proceedings of the Bush Fire Management Conference,* 27–28 February 1998, Nature Conservation Council, Sydney.

Chartres, C.J., Walker, P.H., Willett, I.R., Talsma, T., Bond, W.J., East, T.J. and Cull, R.F. (1991), *Soils and Hydrology of Ranger Uranium Mine Sites in Relation to Application of Retention Pond Water,* Technical Memorandum—Supervising Scientist for the Alligator Rivers Region.

Christian, C.S. and Stewart, G.A. (1968), 'Methodology of integrated surveys' in *Aerial Surveys and Integrated Studies,* Proc. Toulouse Conf., UNESCO, Paris, 233–80.

Cook, S.E., Corner, R.J., Grealish, G., Gessler, P.G. and Chartres, C.J. (1996), 'A rule based system to map soil properties', *Journal of the Soil Science Society of America* 60, 1893–1900.

Dent, D. and Young, A. (1981), *Soil Survey and Land Evaluation,* George Allen and Unwin, London.

Dewar, R.B., Anwar, S., Parker, J., Chapman, G.A. and Houghton, P.D. (1996), *Landscape Modelling for Natural Resource Assessment,* Department of Land and Water Conservation. Soil Survey Unit Miscellaneous Report No. 4.

Dijkerman, J.C. (1974), 'Pedology as a science: the role of data models and theories in the study of natural soil systems', *Geoderma* 11, 73–93.

DLWC (1996), *Urban Awareness Program—a strategy to implement sound land management policies and procedures for urban development,* NSW Department of Land and Water Conservation, Sydney.

East, T.J. (1994), *A Review of Reliability Measures for Australian Soil and Land Resource Maps,* Working Paper, Bureau of Resource Sciences, Department of Primary Industry and Energy.

Eldridge, D.J. (1985), *Aeolian Soils of South-western NSW,* Soil Conservation Service of NSW, Sydney.

Emery, K.A. (1985), *Rural Land Capability Mapping,* Soil Conservation Service of NSW, Sydney.

Emery, K.A. (ed.) (1981), *The Joint Shires Study,* Soil Conservation Service of NSW, Sydney.

FAO (1976), *A Framework for Land Evaluation,* Soils Bulletin 32, FAO Rome.

FAO (1983), *Guidelines: Land Evaluation for Rainfed Agriculture,* Soils Bulletin 52, FAO Rome.

Fisher, R. (1998), 'Dynamic/Interactive Land Resource Assessment in the Northern Territory', *Australian Collaborative Land Evaluation Program Newsletter* 7(1), 15.

FitzPatrick, E.A. (1971), *Pedology: A Systematic Approach to Soil Science,* Oliver and Boyd, Edinburgh.

Geary, P. (1994), 'Soil survey and the design of waste-water disposal systems', *Australian Journal of Soil and Water Conservation* 7(4), 16–23.

Gessler, P.E., Moore, I.D., McKenzie, N.J. and Ryan, P.J. (1995), 'Soil-landscape modelling and spatial prediction of soil attributes', *International Journal of Geographic Information Systems* 4, 421–32.

Gibbons, F.R. (1983), 'Soil Mapping in Australia' in *Soils: An Australian Viewpoint,* CSIRO, Division of Soils, Melbourne; Academic Press, London.

Gray, J., Chapman, G.A., Murphy, C.L. and Noble, R. (1997), *Soil and Landscape Issues in Environmental Impact Assessment,* Department of Land and Water Conservation Technical Report No. 34.

Gray, J.M. and Murphy, B.W. (1999), *A Guide to the Influence of Parent Material on Soil Distribution in Eastern Australia,* DLWC Technical Report No. 45.

Gunn, R.H., Beattie, J.A., Reid, R.E. and van de Graaff, R.H.M. (eds) (1988), *Australian Soil and Land Survey Handbook: Guidelines for Conducting Surveys,* Inkata Press, Melbourne.

Hall, D., Lubbers, P. and McBratney, A. (1992), 'Prevention of salinity in cotton growing areas', *Australian Cottongrower* 13(2), 8–11.

Hannam, I.D. and Hicks, R.W. (1980), 'Soil conservation and urban land use planning', *Journal of the Soil Conservation Service of NSW* 36, 135–45.

Harper, R. (1994), 'The nature and origin of the soils of the Cairlocup area WA, as related to contemporary land degradation', PhD Thesis, University of Western Australia, Nedlands, WA.

Hazelton, P.A. and Atkinson, G. (1985), *Bibliography of the Soils of the Sydney Region,* Aust. Soc. Soil Sci. Inc., NSW.

Hazelton, P.A. and Murphy, B.W. (1992), *What Do All the Numbers Mean? A Guide to the Interpretation of Soil Test Results,* Department of Conservation and Land Management (incorporating the Soil Conservation Service of NSW), Sydney.

Hazelton, P.A., Tille, P.J. and Warburton, B.G. (1985), *Land Resources Study: Lower Cox's River Dam—Wallerawang Pipeline,* Soil Conservation Service of NSW, Sydney.

Hicks, R.W., Murphy, B.W. and Houghton, P.D. (1984), *Urban Capability Study: North Orange,* Soil Conservation Service of NSW, Sydney.

Isbell, R.F. (1996), *The Australian Soil Classification,* CSIRO Melbourne.

Isbell, R.F., McDonald, W.S., Ashton, L.J. (1997), *Concepts and Rationale of the Australian Soil Classification,* ACLEP, CSIRO Land and Water, Canberra.

Johnson, A.K.L. and Cramb, R.A. (1992), *An Integrated Approach to Agricultural Land Evaluation—Data Paper 5: Risk Analysis,* Department of Agriculture and the University of Queensland.

Kovac, M., Lawrie, J.A. and Murphy, B.W. (1989), *Soil Landscapes of the Bathurst 1:250 000 Sheet,* Soil Conservation Service of NSW, Sydney.

Lawrie, R.A. (1988), *Soils and Land Use of the Narrandera District*, Soil Survey Bulletin, Department of Agriculture, NSW.

Logan, P. and Luscombe, G. (1984), *Urban and Rural Capability Study: North-west Sector, Sydney*, Soil Conservation Service of NSW, Sydney.

McAlpine, J.R. and Keig, G. (1990), 'Resource information systems for rural development planning', *International Proceedings of the Symposium on Integrated Land Use Management for Tropical Agriculture,* September–October 1990, Cairns, Brisbane, Canberra, Queensland Department of Primary Industries and Bond University, Brisbane.

McBratney, A.B. (1992), 'On variation, uncertainty and informatics in environmental soil management', *Australian Journal of Soil Research* 30, 913–35.

McBratney, A.B. and Moore, A.W. (1985), 'Application of "fuzzy sets" to climatic classification', *Agricultural and Forest Meteorology* 35, 165–85.

McDonald, R.C., Isbell, R.F., Speight, J.G., Walker, J. and Hopkins, M.S. (1990), *Australian Soil and Land Survey Field Handbook*, Inkata Press, Melbourne.

McInnes, S.K. (1997), *Soil Landscapes of the St Albans 1:100 000 Sheet Report*, NSW Department of Land and Water Conservation.

McKenzie, N.J. (1991), *A strategy for coordinating soil survey and land evaluation in Australia*, Division of Soils Divisional Report No. 114, CSIRO, Division of Soils, Glen Osmond, SA.

McKenzie, N.J. and Austin, M.P. (1989), 'Utility of the Factual Key and Soil Taxonomy in the Lower Macquarie Valley, NSW', *Australian Journal of Soil Research* 27(2), 289–311.

McKenzie, N.J., Ringrose Voase, A.J., Smettem, K.R.J. (1991), 'Evaluation of methods for inferring air and water properties of soils from field morphology', *Australian Journal of Soil Research* 29(5), 587–602.

Martin, C. (1997), *The Digital Estate: Strategies for Competing, Surviving and Thriving in an Internetworked World*, McGraw Hill, New York.

Matthei, L.E. (1995), *Soil Landscapes of the Newcastle 1:100 000 Sheet*, Soil Conservation Service of NSW, Sydney.

Moore, G. (1998), *Soil Guide—A Handbook for Understanding and Managing Agricultural Soils*, Agriculture Western Australia Bulletin No. 4343.

Morse, R.J., Chapman, G.A. and Hird, C. (1992), 'Specific Urban Land Capability Assessment', *Proceedings of 7th International Soil Conservation Organisation Conference: People Protecting their Land,* Sydney, Australia, 27–30 September 1992.

Morse, R.J., Mitchell, P.B., Hird, C., Chapman, G.A. and Lawrie, R. (1991), 'Assessment of soil constraints in environmental impact statements', *Australian Journal of Soil and Water Conservation* 4(2), 12–17.

Murphy, B.W. (1985), 'Regional distribution, soil formation and instability problems of soils in the Bathurst–Orange region', MSc (Ag.) Thesis, University of Sydney.

Murphy, B.W. and Atwood, R.D. (1985), *Reconnaissance of Soil Survey Forbes Township*, Soil Conservation Service of NSW, Sydney.

Murphy, B.W. and Lawrie, J.A. (1999), *Soils of the Dubbo 1:250 000 Sheet*, NSW Department of Land and Water Conservation, Sydney.

Murphy, C.L., Fogarty, P.J. and Ryan, P.J. (1998), *Soil Regolith Stability Classification for State Forests in Eastern New South Wales*, Technical Report 41, Department of Land and Water Conservation, Sydney.

Naylor, S.D., Chapman, G.A., Atkinson, G., Murphy, C.L., Tulau, M.J., Flewin, T.C., Milford, H.B. and Morand, D.T. (1995), *Guidelines for the Use of Acid Sulphate Soil Risk Maps*, Soil Conservation Service of NSW.

New South Wales Department of Housing (1998), *Managing Urban Storm Water: Soils and Construction*, NSW Department of Housing, Sydney.

Norman, C.P. (1989), *Kyvalley EM38 Salinity Survey*, Research Report Series, Victorian Department of Agriculture and Rural Affairs.

Northcote, K.H. (1978), 'Soils and land use' in *Atlas of Australian Resources*, vol. 1, Division of National Mapping, Canberra.

Northcote, K.H. (1984), 'Soil landscapes, taxonomic units and soil profiles: a personal perspective on some unresolved problems of soil survey', *Soil Survey and Land Evaluation* 4, 1–7.

Northcote, K.H. et al. (1960–68), *Atlas of Australian Soils*, CSIRO and Melbourne University Press, Melbourne.

Odeh, I., McBratney, A., and Chittleborough, D. (1993), 'Utilisation of geomorphic attributes from digital elevation models for predicting soil properties', *Australian Collaborative Land Evaluation Program Newsletter* 2(3), 8–10.

Paton, T.R. (1978), *The Formation of Soil Material*, George Allen and Unwin, London.

Quilty, J.A., Hunt, J.S. and Hicks, R.W. (1979), *Urban Erosion and Sediment Control*, Technical Handbook No. 2, Soil Conservation Service of NSW, Sydney.

Rayment, G.E. and Higginson, F.R. (1992), *Australian Laboratory Handbook of Soil and Water Chemical Methods*, Inkata Press, Melbourne.

Riddler, A.H.M. and Lawrie, R.A. (1988), 'Soil mapping' in Gunn et al. (eds), *Australian Soil and Land Survey Handbook: Guidelines for Conducting Surveys*, Inkata Press, Melbourne.

Roy, P.S. and Matthei, L. (1996), 'Late Cainozoic clay deposits in the Port Stephens area', *Australian Journal of Earth Sciences* 43, 395–400.

Ryan, P., McKenzie, N.J., Loughhead, A. and Ashton, L. (1995), 'New methods for forest soil surveys' in *Environmental management: the role of eucalypts and other fast growing species:* Joint Australian/Japanese Workshop, 23–27 October 1995 Proceedings.

Schelling, J. (1970), 'Soil genesis, soil classification and soil survey', *Geoderma* 4, 165–93.

SCS (1980), *Soil and Land Resource Surveys*, Soil Conservation Service of NSW, Sydney.

Sharley, T. (1987), 'Soil survey: the key to making the right decisions', *Australian Grapegrower and Winemaker* 282, 30–35.

Shaxson, T.F. (1992), 'A matrix to link land husbandry and soil moisture management', *Proceedings of 7th International Soil Conservation Conference,* Sydney, 1992, NSW Department of Land and Water Conservation, Sydney.

Smith, C. and Thwaites, R. (1998), 'TIM: Evaluating the sustainability of agricultural land management at the planning stage', *Conference Proceedings, National Soils Conference: Environmental Benefits of Soil Management,* Brisbane, 27–29 April 1998, Australian Soil Science Society Incorporated.

Stace, H.C.T., Hubble, G.D., Brewer, R., Northcote, K.H., Sleeman, J.R., Mulcahy, M.J. and Hallsworth, E.G. (1968), *A Handbook of Australian Soils,* Rellim Technical Publications, Glenside, South Australia.

Taylor, J.K. (1970), *The Development of Soil Survey and Field Pedology in Australia 1927–1967,* CSIRO, Melbourne.

Tulau, M.J. (1997), *Soil Landscapes of the Bega-Goalen Point 1:100 000 Sheet (Bemboka, Murrah, Rocky Hall, Pambula Beach),* Department of Land and Water Conservation of NSW, Sydney.

Tulau, M.J. (1999), *Acid Sulfate Soil Management Priority Areas in the Lower Richmond Floodplain,* Department of Land and Water Conservation, Sydney.

Tulau, M.J. and Wild, J.A. (1996), *Soil Landscapes of the Jindabyne Area,* Soil Survey Unit Miscellaneous Report No. 5, Department of Land and Water Conservation.

Turvey, N.D., Booth, T.H., and Ryan, P.J. (1991), 'A soil technical classification for Pinus radiata (D. Don) plantations', *Australian Journal of Soil Research* 28, 813–24.

Turvey, N.D. and Lieshout, H. (1983), *Some Uses of Soil Survey Information for Improved Management of Planted Forests in South-eastern Australia,* General Technical Report, Pacific Northwest Forest and Range Experiment Station, USDA Forest Service, No. 163, 46–52.

United States Department of Agriculture, Soil Conservation Service (1993), *National Soil Survey Handbook,* Part 620, *Soil Interpretations, Rating Guidelines,* Washington DC.

USDA (1950), *Handbook of Soil Survey Investigations: Field Procedures,* US Dept of Agriculture.

Van de Graaff, R.H.M. (1988), 'Land evaluation', in R.H. Gunn, J.A. Beattie, R.E. Reid, and R.H.M. van de Graaff (eds), *Australian Soil and Land Survey Handbook: Guidelines for Conducting Surveys,* Inkata Press, Melbourne.

Veness, R.A. (1980), *Soil Survey: Mt Arthur North Coal Lease, Authority No. 168, Muswellbrook,* Soil Conservation Service of NSW, Sydney.

Webster, R. (1979), *Quantitative and Numerical Methods in Soil Classification and Survey,* Clarendon, Oxford.

Webster, R. (1985), 'Quantitative spatial analysis of soil in the field', *Advances in Soil Science* 3, 1–70.

Webster, R. and Burgess, T.M. (1984), 'Sampling and bulking strategies for estimating soil properties in small regions', *Journal of Soil Science* 35, 641–54.

Webster, R. and Oliver, M.A. (1990), *Statistical Methods in Soil and Land Resource Survey,* Oxford University Press, Oxford.

Wells, M.R. and King, P.D. (1989), 'Land capability assessment methodology for rural-residential development and associated agricultural uses', *Land Resources Series* No. 1, Western Australia Department of Agriculture.

Western, S. (1978), *Soil Survey Contracts and Quality Control,* Clarendon, Oxford.

Wheen, M. and Johnston, D. (1986), *Physical Landscape Characteristics and Soils Within Some Vegetation Communities, North Head Precinct, Sydney Harbour National Park,* Soil Conservation Service of NSW, Sydney.

White, L.P. (1977), *Aerial Photography and Remote Sensing for Soil Survey,* Clarendon, Oxford.

Woode, P.R. (1981), 'We don't want soil maps, just give us land capability. The role of land capability surveys in Zambia', *Soil Survey and Land Evaluation* 1, 24–26.

Soils of New South Wales

B.W. Murphy, D.J. Eldridge,
G.A. Chapman and D.J. McKane

Within New South Wales there is a wide range of soils. Some types of soil cover large tracts of land and are extremely important productive soils. Such soils are the red-brown earths (Haplic Red Chromosols), the black and grey cracking clays (Self-mulching Black and Grey Vertosols), and the red earths (Red Kandosols). Other soils cover only small areas but still may be important agriculturally, such as the krasnozems (Red Ferrosols) and euchrozems (Red Der-

mosols and Mesotrophic Red Ferrosols), or they may be just unusual and perhaps interesting, such as the humic gley soils (Hydrosols).

For discussion, it is convenient to divide the state into two zones—the more humid eastern zone and the arid western zone. The latter zone corresponds approximately with the Western Division of New South Wales.

8.1 Soils of the Eastern Zone

The main soils are described in terms of the great soil groups (Stace et al., 1968), the Australian Soil Classification System (ASCS) (Isbell, 1996), and the Factual Key groups (Northcote et al., 1975) outlined in Chapter 6. The Factual Key groups are those used in the *Atlas of Australian Soils* (Northcote et al., 1966, 1968). The correlation between the great soil groups and the Factual Key groups is largely that of Northcote et al. (1975), and the relationship between great soil groups and the Australian Soil Classification is based on Isbell (1996).

While detailed soils information is available for some areas in New South Wales, there are still large areas of the state for which there is little information. Therefore, heavy reliance is placed on the *Atlas of Australian Soils* in describing their distribution. Information can also be drawn from Atkinson and Melville (1987). The major soils are presented in this section, which is followed by sections on factors controlling soil distribution and the agricultural and environmental significance of the soils.

8.1.1 Major Soils

(a) Alluvial Soils

Great Soil Group:	Alluvial Soils
ASCS	Rudosols (Stratic and Clastic) and Tenosols (Chernic)

Factual Key Group(s):	Siliceous Sands (Uc1)
	Siliceous Loams (Um1)
	Deep Loams (Um5)

Alluvial soils are confined to recent alluvial deposits associated with rivers and creeks. They typically occur on alluvial plains, alluvial fans, deltas and as lacustrine deposits. Although alluvial soils occur throughout New South Wales, their total area is not large.

Alluvial soils are minimally affected by soil-forming processes and often an A horizon with some organic matter accumulation is the only evidence of these processes. Frequently depositional layers are present in the profile and these can show marked differences in soil texture, colour, stoniness and calcium carbonate content. Textures include sands, loamy sands, sandy loams, loams, silty loams, clay loams and sometimes gravels, depending on the velocity of water flow from which the soil materials were deposited. Buried profiles of other soils may be evident beneath some alluvial soils. In forested areas charcoal and decayed organic materials may be present.

(b) Shallow Skeletal Soils

Great Soil Group:	Lithosols
ASCS	Tenosols (Orthic, Leptic,

	Bleached Orthic, Bleached Leptic)
Factual Key Group(s):	Variety of Shallow Siliceous Sands and Loams (Uc1, Uc2, Uc4, Um1, Um2, Um4, Um6.2)

Skeletal soils commonly occur on crests and steeper slopes of undulating to hilly and mountainous terrain. They occur wherever natural erosion is sufficiently rapid to ensure that only a thin cover of soil is maintained.

Skeletal soils are shallow, stony soils, usually with the A horizon directly overlying weathered rock. The development of B horizons is very limited and often restricted to rock fissures or a thin transitional layer between the surface soil and the bedrock. Soil texture and structure are very variable depending on parent material. Most skeletal soils have gravel and stones in the form of fragmented rock that displays varying degrees of weathering throughout the profile. Soil cover is often discontinuous and rock oucrops are a common feature. An A_2 horizon sometimes occurs.

(c) Alpine Humus Soils/Soils in Periglacial Environments

Great Soil Group:	Alpine Humus Soils
ASCS	Chernic Tenosols/Organosols
Factual Key Group:	Organic Loams (Um7.1)

Alpine humus soils have a limited occurrence and are confined to colder areas higher than about 1100 m above sea level. The largest area occurs in the Snowy Mountains, but scattered areas also occur in other tableland locations.

These soils are characterised by a marked accumulation of humified organic matter throughout the profile and a uniform loam texture. The A horizons are highly organic. Soil colours are mainly dark brown to black or dark reddish brown. They occur in association with very organic soils or peats, which have large accumulations of organic matter (Organosols).

(d) Podzols

Great Soil Group:	Podzols
ASCS	Podosols
Factual Key Group(s):	Bleached Sands with a coloured B horizon or pan (Uc2.2, Uc2.3)

Podzols occur sporadically in the higher rainfall areas (above 700 mm average annual rainfall) on highly siliceous, sandy parent materials. Such materials include old coastal sand dunes, siliceous sandstones and some siliceous granites.

Podzols are acidic sandy soils, usually with a uniform coarse texture profile, but with strongly colour-differentiated horizons. The horizons include a dark-coloured A_1 horizon containing a lot of organic matter, a bleached or white A_2 horizon, and a yellowish brown to reddish brown B horizon.

Associated soils: Siliceous Sands, Humus Podzols and Peaty Podzols.

(e) Humic gleys/Poorly Drained Soils on Coastal Flats

Great Soil Group(s):	Humic Gleys and Gleyed Podzolic Soils
ASCS	Hydrosols (includes some Sulfuric and Sulfidic Great Groups)
Factual Key Group(s):	Friable Gley Duplex Soils (Dg4.1, Dg4.4, Dg4.8) Non-cracking Plastic Clays (Uf6.6)

These are not common soils in New South Wales, being confined to the more humid areas (more than 750 mm average annual rainfall) along the coast, and associated with Pleistocene and Holocene deposits on the coastal lowlands. Although relatively small in area they can have a major environmental impact, as they include some areas of acid sulfate soils. They are principally found on very poorly drained coastal plains, particularly in association with the lower reaches of coastal rivers and estuarine flats. Humic gleys also occur in poorly drained areas between sand dunes. On the north coast they are used for sugar cane, grazing and some horticulture.

The humic gley soils are characterised by dark A horizons with moderate to high organic matter contents, and a pale grey to bluish grey B horizon, which is often clayey. Structure of the A horizon is usually moderately developed fine blocky to crumb, while that of the B horizon is moderate to strong prismatic or coarse blocky. O horizons are common. Most profiles are 1 to 1.2 m deep, overlying layered alluvium. Soil reaction trend is acid to neutral. Nutrient levels, particularly nitrogen and phosphorus, are generally low, and some trace element deficiencies (such as copper) are likely.

The Tidal Hydrosols (ASCS) are of Holocene age and were originally formed under estuarine conditions, often in association with mangrove communities. These soils are often high in sulfidic materials, and if oxidised can become very acidic and produce very acidic leachates that flow into estuaries and lagoons (see Acid Sulfate Soils). Other Hydrosols on coastal lowlands may also have sulfidic materials.

Associated soils: Humus Podzols, Peaty Podzols.

(f) Red and Yellow Podzolic Soils

(i)

Great Soil Group:	Red Podzolic Soils
ASCS	Red Chromosols (Dystrophic), Red Kurosols
Factual Key Group(s):	Red Duplex Soils with acid reaction trend (Dr2.21, Dr2.41, Dr3.21, Dy3.22, Dy3.41, Dy3.42, Dy3.6, Dy5.8)

(ii)

Great Soil Group:	Yellow Podzolic Soils
ASCS	Yellow Chromosols (Dystrophic), Yellow Kurosols
Factual Key Group(s):	Yellow Duplex Soils with acid or neutral reaction trend (Dy2, Dy3.21, Dy3.22, Dy3.42, Dy3.6, Dy5.8)

Red and yellow podzolic soils are common throughout New South Wales, but are principally located in the more humid or seasonally humid parts of the state, with average annual rainfall above about 650 mm. There are some occurrences down to about 500 mm on inland ranges, such as the Catombal Range near Dubbo and the Hervey Range near Parkes. Frequently red podzolic soils occur in better-drained areas, in association with yellow podzolic or yellow solodic soils, which occupy the more poorly drained lower slope positions. However, in wetter areas, yellow podzolic soils are dominant in some landscapes, even occupying upper slope positions. A wide range of parent materials is encountered, from siliceous to relatively basic rocks, including granites, shales, metamorphosed sediments, siliceous volcanics (dacite and rhyolite) and intermediate volcanics (trachytes). Red and yellow podzolic soils may form *in situ* or on transported parent material, including alluvium and colluvium.

Red and yellow podzolic soils are texture contrast soils with light-textured A horizons overlying heavier textured, structured B horizons. A distinct pale A_2 horizon is usually, though not always, present and the profile is acidic. The B horizons are characterised by moderate polyhedral or angular blocky structure, and tend to be somewhat friable when moist. Unlike the B horizons of the solodic soils, they are not tough, hard and dense. Some podzolic soils, especially in high rainfall areas near the coast, have unusual subsoil properties, and have strongly acidic subsoils. These are Kurosols (ASCS).

Associated soils: Brown Podzolic Soils, Lateritic Podzolic Soils and Grey-Brown Podzolic Soils.

(g) Krasnozems and Euchrozems

(i)

Great Soil Group:	Krasnozems
ASCS	Red Ferrosols
Factual Key Group(s):	Acid Red Structured Earths including the Acid Red Rough and Smooth-Ped Earths (Gn4.11, Gn3.11)

(ii)

Great Soil Group:	Euchrozems
ASCS	Red Ferrosols, Red Dermosols
Factual Key Group(s):	Neutral and Alkaline, Red and Brown, Rough and Smooth-Ped Earths (Gn4.12, Gn4.13, Gn3.12, Gn3.13)

Krasnozems are widely distributed in subhumid to humid areas towards the coast and along the central tablelands (750 to 1800 mm annual rainfall). The closely related euchrozems occur along the western slopes in arid to subhumid areas (600 to 750 mm annual rainfall). Parent materials are dominated by the more basic igneous rocks such as basalt and, to a lesser extent, andesites, but also include some metasediments and sediments that are intermediate in composition. Most krasnozems are associated with well-drained sites on hilly uplands and on undulating plateaux.

Krasnozems and euchrozems are strongly structured red soils, often with fine shiny red polyhedral peds. Soil texture gradually becomes more clayey with depth. They are either acid soils (krasnozems) or neutral to alkaline soils (euchrozems), with high levels of sesquioxides.

Associated soils: Chocolate Soils, Xanthozems (yellow) and possibly some Terra Rossa Soils.

(h) Prairie Soils

Great Soil Group(s):	Prairie Soils, Chernozems
ASCS	Black Dermosols, Brown Dermosols

Factual Key Group(s): Black Friable Loams (Um6.11, Um6.31)

Black Structured Earths (Gn3.43)

Prairie soils are not common soils in New South Wales, but where they do occur, they are important agricultural soils. They are found in both humid and subhumid climates, with a wide range of annual rainfalls (between 600 and 1400 mm). The most notable occurrences are on alluvium derived from intermediate to basic parent materials, such as basalt and andesite. A number of the major rivers in New South Wales have alluvial plains dominated by these soils.

Prairie soils are characterised by uniform and occasional gradational loam or clay loam texture profiles, moderate to strong crumb or fine blocky structure, and weak development of horizons. They are deep soils and can be alkaline at depth.

(i) Solodic Soils

Great Soil Group(s):	Solodic Soils, Solodised Solonetz, Solods
ASCS	Sodosols, some Kurosols
Factual Key Group(s)	Some Mottled Yellow, Brown and Red Duplex Soils with bleached A_2 horizons (Dy3.4, Dr3.4, Db2.4)

Solodic soils occur at a wide range of locations in the semi-arid to humid rainfall areas (375 to 1200 mm annual rainfall). They have formed on a wide range of parent materials, including siliceous to intermediate igneous rocks (especially granite and rhyolite), sedimentary and metamorphic rocks, and on alluvium and colluvium. These soils are rarely associated with more basic rocks, such as basalt. Solodic soils typically occupy the mid to lower slopes of hilly lands, especially in the tablelands and on coastal plains. On the western plains sodic soils appear to be very common. The sources of sodium are very variable, and include sodium from weathering rocks, mobile sodium brought in by water flows through the landscape, and aeolian sources.

Solodic soils are characterised by strong texture contrast profiles with light-textured surface soils overlying tough, hard and dense B horizons, which are usually very unstable to wetting. The boundary between the A and B horizons is very abrupt. There is also a characteristic bleached A_2 horizon. These soils have low fertility, poor trafficability when wet, and are very erodible. Severe gully erosion is often a feature of these soils in the tablelands.

(j) Red Earths

Great Soil Group(s):	Red Earths, Calcareous Red Earths
ASCS	Red Kandosols
Factual Key Group:	Red Massive Earths (Gn2.1)

Red earths are common soils in New South Wales. They occur under a wide range of conditions with annual average rainfall varying from 200 mm up to more than 1000 mm. Parent materials are siliceous to intermediate, including granites, quartzose sandstones, intermediate igneous rocks (andesite) and river alluvium. Most occurrences are associated with well-drained sites on undulating hills and plains. In higher rainfall areas red earths can be deep and overlie lateritic-type or petroferric materials. In steeper terrain, the red earths are often associated with skeletal soils.

Calcareous red earths occur in the arid areas of the state, where rainfalls are generally less than 350 mm.

Red earths are a diverse group of soils, but all have the distinctive features of a massive, porous, earthy soil material, reddish brown to red colour and a gradual increase in clay content with depth. They have a gradational texture profile. The earthy fabric is evident as a full matt or dusty appearance on the face of freshly broken soil in contrast with the shiny appearance of most clay soils. Textures are usually sandy loams, loams or clay loams.

(k) Yellow Earths

Great Soil Group:	Yellow Earths
ASCS	Yellow Kandosols
Factual Key Group:	Yellow Massive Earths (Gn2.2, Gn2.3, Gn2.6, Gn2.7)

Yellow earths are widespread in the moister subhumid and humid areas (550 to 1400 mm annual rainfall) of eastern New South Wales. Major occurrences are associated with siliceous sandstones, such as the Pilliga Sandstone of the central western slopes and the Hawkesbury and Narrabeen Sandstones in the ranges west and north of Sydney, where the earths occur in association with shallow or skeletal soils. The yellow earths also occur on some granites. A common occurrence of the yellow earths is as part of a toposequence with the red earths described above, the yellow earths occupying the less well-drained slope positions.

Yellow earths are similar to red earths and have the same massive, porous, earthy materials and gradual increase in clay with depth, but they are predominantly yellow in colour. The yellow earths have a gradational texture profile. Textures are usually sandy loams, loams or clay loams.

Associated soils: Grey Massive Earths.

(l) Red-Brown Earths

Great Soil Group:	Red-Brown Earths
ASCS	Red Chromosols (Eutrophic, Mesotrophic), Red Sodosols
Factual Key Group(s):	Hard Pedal Red Duplex Soils with alkaline or less frequently neutral reaction trends (Dr2.13, Dr2.23, Dr2.33, Dr3.23, Db1.23)

Red-brown earths are widespread on the slopes of New South Wales and are one of the major wheat-growing soils. They occur in semi-arid to subhumid areas (350 to 700 mm annual rainfall) on a wide range of parent materials, including alluvial and colluvial deposits, sedimentary, metamorphic and siliceous to intermediate igneous rocks. Generally, they occupy the slopes, plains and broad hillcrests, which are ideal for growing crops.

Red-brown earths are texture contrast soils with a clear to sharp texture boundary between the light-textured A and clayey B horizons. The weakly structured, light-textured surface soil is of major interest, as it is this soil material that is subjected to soil structure degradation, acidification and erosion under cropping and grazing in the wheat-belt. The clayey B horizon is usually reddish brown, and has moderate to strong blocky structure. Evidence suggests that a proportion of the red-brown earths have sodic subsoils particularly on the western plains, and so are sodosols in the ASCS. Calcium carbonate is typically present in the lower subsoil.

(m) Non-calcic Brown Soils

Great Soil Group:	Non-calcic Brown Soils
ASCS	Red and Brown Chromosols (Eutrophic and Mesotrophic)
Factual Key Group:	Hard Pedal Red Duplex Soils with neutral reaction trend (Dr2.12, Dr2.22, Db1.12, Db2.22)

Non-calcic brown soils are common on the transitional slopes between the tablelands and the western slopes and plains. They also occur on some of the

drier areas of the tablelands (less than 700 mm annual rainfall), as at Bathurst. They are essentially soils that are intermediate between the red podzolic soils and the red-brown earths and hence have many similar features.

Non-calcic brown soils are texture contrast soils with a clear to sharp boundary between a lighter-textured A horizon and a clayey B horizon. As for the red-brown earths, the weakly structured, lighter-textured surface soil is also of interest because of its importance in agriculture and land degradation. The B horizons are reddish brown and have a moderate to strong blocky structure. Soil reaction trend is neutral and there is no calcium carbonate in the profile.

(n) Black Cracking Clays

Great Soil Group:	Black Earths
ASCS	Self-mulching Black Vertosols
Factual Key Group:	Black Self-mulching Cracking Clays (Ug5.1)

Black cracking clays are common soils on the north-western slopes and plains of New South Wales. Annual average rainfall varies from 500 to 850 mm. Parent materials are basic igneous rocks, such as basalt, diorite and some andesites; basic sedimentary rocks, including calcareous mudstone, clayey sandstone, lithic and felspathic sandstones; and clayey alluvium derived from the preceding rocks. Basalts and associated clayey alluvium are the commonest parent materials. Black cracking clays are most commonly associated with low hills and gently sloping alluvial plains.

These soils have a uniform clay profile, and colours are dark grey, very dark brown or black. The surface soil is self-mulching, being very sticky and plastic when wet but upon drying it forms a loose layer of granular to fine polyhedral aggregates 2 to 5 cm thick. Montmorillonite clays dominate the profile and, when dry, the soil forms wide (at least 6 mm) cracks at the surface that extend deep into the profile where the soil peds become very hard. Most profiles are alkaline at depth and, in uncultivated areas, gilgai are common.

Associated soils: Rendzinas, Weisenboden in poorly drained sites.

(o) Grey, Brown and Red Clays

Great Soil Group(s):	Grey, Brown and Red Clays
ASCS	Grey, Brown and Red Vertosols (various Great Groups)

Factual Key Group(s):   Grey Self-mulching
Cracking Clays (Ug5.2,
Ug6.2),

Brown and Red Self-
mulching Cracking
Clays (Ug5.3, Ug6.3),

Grey Massive Cracking
Clays (Ug5.5, Ug6.5)

Grey, brown and red clays represent a very broad group of soils.

Grey clays are self-mulching cracking clays, and are common on the broad alluvial plains in the semi-arid areas of midwestern New South Wales. The grey clays are formed on unconsolidated clay alluvium and colluvium derived from a wide range of basic rocks, including basalts and basic sedimentary rocks. The brown clays are intermixed with the grey clays on the broad alluvial plains in midwestern New South Wales. Smaller areas of red clays occur in the moister areas of the state in association with basalt.

Grey, brown and red clays are characterised by uniform clay texture profiles, weak horizon development and soil materials with high shrink-swell characteristics. Most have a self-mulching surface soil which is very sticky when wet, but upon drying it forms a loose layer of granular to fine polyhedral aggregates 2 to 5 cm thick. Other forms have hard, compact surface soils. Colours tend to become paler down the profile, and gilgai formation is a relatively common feature in uncultivated areas.

Variants from this general description include those with acid subsoils, or surface soils with massive to weak blocky structure at the surface (Ug5.5).

(p) Sands

Great Soil Group(s):   Siliceous Sands, Earthy
Sands

ASCS   Tenosols (Orthic,
Chernic)

Factual Key Group(s):   Siliceous Sands
(Uc1.2)

Earthy Sands (Uc5.1)

Significant areas of sands occur in the eastern part of the state in association with very siliceous parent materials. Such parent materials include siliceous granites, quartzose sandstones and newly deposited or active sand dunes along the coast. Earthy sands (Uc5.11) have been recorded on the Pilliga Sandstone at Dubbo. In some cases on deep sand deposits, sands have been described where relatively shallow exposures have been made in texture contrast soils with very deep A horizons.

(i) Siliceous Sands

Great Soil Group: Siliceous Sands

ASCS   Tenosols (Orthic,
Chernic)

Factual Key Group(s):   Siliceous Sands (Uc1.2)

Brownish Sands (Uc5.1)

These soils have variable depth and are characterised by uniform sand textures and a massive, single-grain structure. Most are lightly coloured, but some yellowish brown to reddish brown colours occur particularly in the subhumid part of the state. There is little or no change down the profile except for a minimal accumulation of organic matter at the surface.

(ii) Earthy Sands

Great Soil Group: Earthy Sands

Factual Key Group: Earthy Sands (Uc5.2)

These soils are usually deep and are characterised by uniform sand texture and a massive, single-grain structure. They are distinguished from the siliceous sands by the coherent nature of the B horizon. This indicates that there has been some pedological development in the form of sesquioxide accumulation and segregation in the B horizon. They occur in the humid and subhumid parts of the state. In the subhumid part of the state the B horizons have a reddish brown colour. The B horizons have an earthy appearance and may initially be mistaken for red massive earths until texturing reveals the uniform coarse texture.

8.2 Soils of the Western Zone

The Soil Conservation Service mapped the entire Western Division of New South Wales by the land system method during the period 1978–90 (Walker, 1991). The land systems of western New South Wales contain attribute and polygonal data on landform, geology and vegetation in addition to soil information. The soil data is not georeferenced.

Walker's 1991 publication gave rise to a continuous coverage of the Western Division for the first time. The report is now available in database form as a result of the Murray-Darling Basin Soil Information Strategy project (McKane, 1997; McKane and Oo, in press).

The other broad-scale maps available for semi-arid western New South Wales are the *Atlas of Australian Soils* (Northcote et al., 1966–68), Prescott (1944), Condon (1961) and James (1960), and the site-specific study by Eldridge (1985).

More specific detailed studies have been undertaken by Northcote (1951), Northcote and Boehm (1949), and Marshall and Walkley (1937) in the Coomealla area, and a number of erosion surveys by Beadle (undated).

The main soil types occurring in western New South Wales are described in terms of order and suborders of the Australian Soil Classification (Isbell, 1996) and the great soil groups (Stace et al., 1968). The Factual Key groups were also used for the *Atlas of Australian Soils* (Northcote et al., 1966–68). Correlation between the classifications is based on Isbell (1996) and McKane and Oo (in press). Table 8.1 shows the relationship between the Australian Soil Classification and the other schemes. The classification according to the western New South Wales scheme outlined in Chapter 6 is also presented in the following text.

8.2.1 Major Soils

(a) Sands

The sand group of soils is scattered throughout the western part of the western zone on plains and dunes in association with Red Kandosols on the plains, Red Chromosols in the swales associated with the Strzelecki dunefields, and Grey and Brown Vertosols and assorted Kandosols on the plains between the Darling River and the Riverine Plain. These soils have

been divided into three main groups on the basis of weak textural, colour and pH changes through the profile.

(i) Calcareous Sands

ASCS	Hypocalcic Calcarosol
Great Soil Group:	Calcareous Sands
Western NSW Group:	Soft Red Soils

These are very deep, lightly coloured massive soils. There is little or no change down the profile except for a minimal accumulation of organic matter at the surface. The sands may be weakly cemented with carbonate where some leaching has occurred.

(ii) Siliceous Sands

ASCS	Arenic Rudosol
Great Soil Group:	Siliceous Sands
Western NSW Group:	Soft Red Soils

These are similar to the preceding group, but have no apparent carbonate in the profile. Sometimes clay content increases slightly deep in the profile, from a loamy sand on the surface to a clayey, slightly compacted sand at depth.

(iii) Brownish Sands

ASCS	Orthic Tenosol
Great Soil Group(s):	Siliceous Sands and Earthy Sands
Western NSW Group:	Soft Red Soils

These are deep, reddish brown massive sands with an increasing carbonate content at depth.

A core of cemented carbonate or 'kunkar' often occurs at depth. These soils show weak textural changes with depth, often from a loamy sand to sandy clay loam.

All the sands have a low water-holding capacity and are highly permeable. Organic matter is low and is concentrated in the A_1 horizon. This often results in these soils being non-wetting. Fertility of the sands group is low, salts and bases are generally low, and the pH is mildly acid to neutral at the surface.

Table 8.1 Relationship between Factual Key groupings, Australian Soil Classification, Great Soil Groups and the US Soil Taxonomy in western New South Wales

Factual Key Grouping	Northcote Coding	Australian Soil Classification	Great Soil Group	US Soil Taxonomy
Calcareous sands (a)(i)	Ucl.1	Hypocalcic Calcarosol	Calcareous sands	Entisols
Siliceous sands (a)(ii)	Ucl.2	Arenic Rudosol	Siliceous sands	Entisols/aridisols
Brownish sands (a)(iii)	Uc5.1	Orthic Tenosol	Earthy sands, siliceous sands	Entisols
Calcareous loams (b)	Um5.4	Leptic Rudosol	Skeletal soils or Lithosols	
Grey clays (c)	Ug5.2	Grey Vertosol	Grey clays	Vertisols
Brown clays (c)	Ug5.3	Brown Vertosol	Brown and red clays	Vertisols
Massive earths (d)	Gn2.1	Calcarosol	Calcareous red earths	Aridisols
Calcareous earths (e)	Gcl, Gc2	Hypo, Litho or Hypercalcic Calcarosol	Solonised brown soils	Aridisols/mollisols
Crusty red duplex soils (f)	Drl	Red Sodosol	Desert loams	Aridisols
Hard red duplex soils (g)	Dr2	Red Sodosol or Red Chromosol	Red-brown earths, Non-calcic brown soils	Alfisols/aridisols

(b) Shallow Loams

ASCS	Leptic Rudosol
Great Soil Group:	Lithosols or Skeletal Soils
Western NSW Group:	Skeletal Soils

The Leptic Rudosols occur in association with low ridges and plateaux throughout the region, especially on the Barrier Ranges around Broken Hill. They occur in association with Arenic Rudosols on the ranges, Red Chromosols or Red Sodosols on the footslopes, and Kandosols on the plains.

These soils are typically shallow, dark red loams to clay loams with seasonally hardsetting surfaces of massive to weak subangular blocky structure. The soil reaction trend of these soils is neutral to slightly acidic.

(c) Grey, Brown and Red Clays

ASCS	Grey and Brown Vertosols
Great Soil Group(s):	Grey, Brown and Red Clays
Western NSW Group:	Heavy Clays

Grey and Brown Vertosols occur over extensive areas of western New South Wales associated with the floodplains of the Darling, Murray, Murrumbidgee, Bogan, Macquarie, Castlereagh, Namoi, Barwon, Gwydir and MacIntyre Rivers. Other large areas of clays occur north-west of the state near the Bulloo Overflow and on the Riverine Plain in association with Chromosols and Sodosols.

Grey Vertosols occur regularly on flooded and poorly drained floodplains, while Brown Vertosols are more common along the major drainage systems such as the Paroo and Bulloo Rivers where they occur on higher, less frequently flooded country. They also occur on the alluvial plains in association with Chromosols and Sodosols.

These soils are typically deep, grey to yellow-grey clays with a self-mulching to cracking surface and little change with depth except a gradual increase in carbonate content. The brown clays are self-mulching with a weak to strong surface crust, often mantled with silcrete or ferricrete stones. These soils have a fine blocky B horizon overlying hardpans, sandstones and mudstones at depth. The subsoils are gypseous, calcareous and saline.

The red clays are moderately deep yellowish red self-mulching clays with surfaces similar to the brown clays. Surface textures range from light clays to heavy clays, the fine blocky B horizon overlying kaolin sandstone and gravel. They are moderately gypseous and saline, and may be calcareous at depth.

(d) Red Earths

ASCS	Calcarosols and Red Kandosols
Great Soil Group(s):	Calcareous Red Earths
Western NSW Group:	Soft Red Soils and Hard Red Soils

Extensive tracts of Calcarosols and Kandosols occur west of Bourke, on the Cobar Pediplain and south-east of Condobolin. These soils have gradational profiles up to 3 m deep and a loam to clay loam surface texture. Usually a hardpan is present (Lawrie, 1974).

The true soil profile is considered above this hardpan (Walker, 1978). Most of these soils have formed from wind-worked and water-deposited alluvial deposits.

The surface textures range from loams to clay loams and structure is usually massive to weakly subangular blocky or platy. The soil reaction trend is alkaline, and carbonate often occurs at depth.

(e) Brown Calcareous Earths

ASCS	Hypocalcic, Lithocalcic or Hypercalcic Calcarosol
Great Soil Group:	Solonised Brown Soils
Western NSW Group:	Solonised Brown Soils

Calcarosols occur over extensive areas of western New South Wales in an area bounded by the Riverine Plain, the Cobar Pediplain and the footslopes of the Barrier Ranges. They occur in association with sands on the dunefields, Red Kandosols and Grey Vertosols. These soils are characterised by large amounts of calcareous material in the profile, as hard or soft concretions or as fine earth. The carbonate layer is the most distinctive feature of these soils.

Surface textures range from loamy sand on the rises and dunes, to loam on the plains and clay loam in the depressions. There is usually a gradual increase in texture to a massive porous or weakly subangular blocky clayey subsoil.

(f) Desert Loams

ASCS	Red Sodosol
Great Soil Group:	Desert Loams
Western NSW Group(s):	Desert Loams and Brown Gibber Soils

These soils occur in the north-western corner of New South Wales, north of Broken Hill. They occur on the footslopes and plains in association with Grey Vertosols and a variety of Chromosols and Vertosols. North of Tibooburra they are found on gentle upslopes with stony mantles, in association with self-mulching clays in the gilgai formations.

These soils range in surface texture from loams to clay loams. There is a well-defined increase in clay content to a light or medium clay in a strong, fine blocky B horizon overlying a coarse blocky pedal subsoil. The surface soil has a strongly vesicular surface crust with an often well-developed gilgai microrelief. The soil reaction trend is alkaline or slightly neutral.

(g) Red-Brown Earths

ASCS	Red Sodosol or Red Chromosol
Great Soil Group(s):	Red-Brown Earths, Non-calcic Brown Soils
Western NSW Group:	Texture Contrast Soils

The Red Sodosols or Red Chromosols occur on the eastern edge of the western zone and on the Riverine Plain. They occur in association with Grey Vertosols west of Hay and, around Condobolin and Nyngan, in association with various other Sodosols and Chromosols.

These soils have brown sandy loam to loam surface textures that set hard when dry. The subsoils are red to reddish brown, highly pedal clays with a prismatic to blocky structure. Often small amounts of amorphous or nodular limestone are present in the profile.

Overall, infiltration rates of the Sodosols and Chromosols are moderate, but this depends on surface texture. They have a good water-holding capacity, especially in the B horizon.

8.3 Factors Affecting the Distribution of Soils

Several factors have influenced the distribution of soils in New South Wales. Parent material and current climate in combination with topography as it affects drainage can explain much of the distribution of soils (see Figure 6.1). However this simple three-factor model is complicated by the fact that New South Wales has been subjected to changes in climate since the end of the Tertiary period. This has resulted in the deposition of wind-blown dust in some parts of the landscape, particularly in the south-west of the state. Changes in sea level have resulted in the movement and deposition of coastal sediments, and parts of the south-west of the state were under salt water for extended periods. Changes in the base level of rivers has led to variation in erosion and deposition rates, with lower base levels accelerating erosion and increasing the amount of deposition in alluvial basins. Wet periods in the eastern part of the state have seen the build-up of swamps and lakes, which have subsequently disappeared with climatic changes although remnants of their existence are present in the geomorphology of the landscape. The important fact is that the soils of New South Wales have been subjected to a complex history.

However, as pointed out in Chapter 1, a cold phase in the late Quaternary (20 000 to 30 000 years ago) rejuvenated extensive erosion on the eastern highlands and slopes of New South Wales. Many of today's soils in this area date from surfaces exposed and materials deposited in this time. Therefore for a brief summary, the distribution of the soils is discussed using the simple three-factor model of soil distribution.

On basic parent materials, there is a general climatic sequence from krasnozems in humid areas to euchrozems and black cracking clays in subhumid areas, to red and brown cracking clays in arid areas. However, local topography, especially as it affects drainage, can lead to toposequences of these soils occurring in close proximity (see Beckmann et al., 1974). Sequences of microclimates occur in such toposequences, with generally wetter environments occurring downslope. The microclimate experienced by a soil at a site is also affected by aspect, with northerly aspects generally being warmer and drier than southerly aspects. The black structured loams

(prairie soils and chernozems) probably also fit into this parent material group.

On acidic or siliceous parent materials there is a climatic sequence from podzols or yellow earths in humid climates, to siliceous sands or earthy sands in subhumid climates, and siliceous sands, earthy sands or calcareous sands in arid areas. On intermediate parent materials there is a climatic sequence from red and yellow podzolic soils in humid areas to red-brown earths in subhumid areas and desert loams in arid areas.

The red earths occur mainly on intermediate to acidic parent materials, across a wide range of climates, becoming more calcareous in arid areas.

The solodic soils are associated with areas where sodium has accumulated in disproportionate amounts in soils from one or more sources including:

—sodium from the weathering of rocks (such as granites high in sodic plagioclase or marine sediments)

—seepage flows

—atmospheric accession in rainfall or dry deposition. This can be important, even in low amounts, in soils with low cation status, such as the sandy solodic soils of the Pilliga Sandstone on the eastern margins of the plains. Aerial additions of sodium are considerable in soils adjacent to the coast, although high rainfall can also wash the sodium out of the soil.

Other soils, including the alpine humus and humic gleys, occur in specific climatic or microclimatic niches (Corbett, 1969).

The alluvial and skeletal soils (lithosols) are characterised more by their lack of soil development and their geomorphological origin than any parent material or climatic factors. They occur throughout the state on a wide range of parent materials and in a wide range of climates.

8.4 Fertility of New South Wales Soils

The soils that are found in the agricultural parts of New South Wales have been grouped into five fertility groups in Table 8.2. The table is a general one and relates to those areas with potential for cropping and/or pasture improvement, indicating a ranking of soils according to their ability to support these uses from the aspect of their chemical fertility. The group-

ings are based on the physical and chemical features of great soil groups in their natural, undegraded condition.

The assessment of fertility is based on plant productivity in general. The natural focus is on the production of crops and pasture species, although tree growth may also be included. It is necessary to

Table 8.2 Relative fertility of New South Wales great soil groups

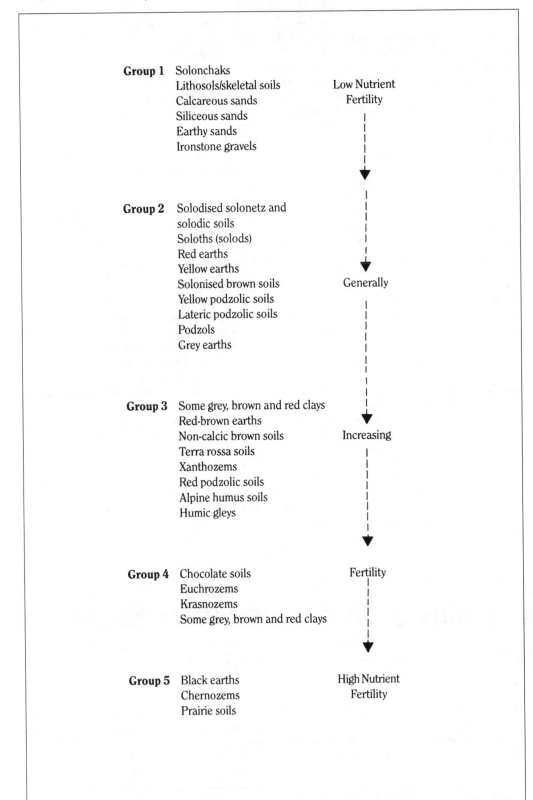

Group 1 Solonchaks
Lithosols/skeletal soils
Calcareous sands
Siliceous sands
Earthy sands
Ironstone gravels

Low Nutrient
Fertility

Group 2 Solodised solonetz and
solodic soils
Soloths (solods)
Red earths
Yellow earths
Solonised brown soils
Yellow podzolic soils
Lateric podzolic soils
Podzols
Grey earths

Generally

Group 3 Some grey, brown and red clays
Red-brown earths
Non-calcic brown soils
Terra rossa soils
Xanthozems
Red podzolic soils
Alpine humus soils
Humic gleys

Increasing

Group 4 Chocolate soils
Euchrozems
Krasnozems
Some grey, brown and red clays

Fertility

Group 5 Black earths
Chernozems
Prairie soils

High Nutrient
Fertility

Notes on Table 8.2

Group 1 includes soils which, due to their poor physical and/or chemical status, only support limited plant growth. The maximum agricultural use of these soils is low intensity grazing. This group includes shallow and sandy or stony soils which, by virtue of their poor water retention characteristics, can support only limited plant growth. The solonchaks, due to salinity, are chemically inhibitive to plant growth, while the calcareous sands are so deficient in cobalt and copper that animals grazing on them are affected (Stace *et al.*, 1968).

Group 2 includes soils with low fertilities, such that, generally, only plants suited to grazing can be supported. Large inputs of fertilisers are required to make the soil usable for arable purposes.

The podzols, due to their poor water retention characteristics, are borderline between this group and Group 1. The remaining soil types within this group are generally chemically deficient in nitrogen, phosphorus and many other elements. The solodised solonetz, solodic soils and soloths also have poor physical conditions and difficult moisture relationships (Stace *et al.*, 1968). The red earths are a highly variable group of soils and some of the finer textured types could be placed in Group 3.

Group 3 soils have low to moderate fertilities and usually require fertiliser and/or have some physical restrictions for arable use.

Except for the terra rossa and alpine humus soils, the soils within this group are moderately deficient in nitrogen and phosphorus and some other elements. The terra rossa soils and alpine humus soils have better chemical fertilities, but their shallowness limits their potential for plant growth. The grey, red and brown clays also have a somewhat better chemical status than the other soils within this group, but many of them are sufficiently hardsetting for seedling emergence to be restricted, and the high clay contents and strongly coherent nature of some subsoils restrict water and root penetration. The red podzolic soils are a highly variable group and some of the lighter textured types may be more appropriately placed in Group 2.

Group 4 soils have a high level of fertility in their virgin state, but this fertility is significantly reduced after only a few years of cultivation.

The chocolate soils and euchrozems have particularly good fertility in their virgin state, but phosphate fertilisers are generally required after only a few cultivations. Physically, the krasnozems are better than most other soils; their high clay content gives them a high water-holding capacity; their structural condition and comparatively low swelling potential enable water to penetrate easily.

Chemically, krasnozems have some undesirable features. For example, they have a moderately acid pH which, if too low, allows solution of iron and aluminium leading to serious phosphorus fixation.

Group 5 soils have high fertilities and these soils generally only require treatment with chemical fertilisers after several years of cultivation. These soils generally provide good chemical and physical conditions for plant growth. The black earths are well known for their high fertility and have produced high yields without added fertilisers. However, the use of nitrogenous fertilisers is now increasing and responses to phosphates are being obtained on some soils (Stace *et al.*, 1968).

consider that different species and even cultivars of the one species have different soil requirements. As an extreme example, a soil that is suitable to grow rice may not be suitable for orange trees, or be used to propagate banksias. There can be a wide variety of soil fertility within one great soil group. The fertility within podzols, red podzolic soils and red earths can vary widely depending on location and the parent materials of the soil.

8.5 Special Soil Groups

Several special soil groups occurring in New South Wales are of particular relevance to soil conservation. Such special soil groups are often used as a basis for problem-oriented works, investigations or research. They can be of critical importance and existing general soil classification systems do not always adequately account for them. A list of these special soil groupings that may be of importance for soil conservation is, therefore, presented here. Note that these groups are not mutually exclusive.

8.5.1 Sodic and Dispersible Soils

Sodic soils contain sufficient exchangeable sodium to adversely affect soil stability, plant growth or land use. Affected soils usually have very coarse, hard, dense structural units (coarse blocky or prismatic or columnar) with very low permeability. Such soils place severe restrictions on plant growth through waterlogging in wet times and lack of stored water in dry times. They are of sufficient importance that the new Australian Soil Classification recognises them as a specific order of soil, the Sodosols, and also recognises sodic subgroups within other orders (Isbell, 1996).

Sodic soils are dispersible and the consequent instability to wetting makes these soils susceptible to severe gully erosion and tunnelling of earthworks (see Chapter 20). Sodic soils are specifically identified by having exchangeable sodium percentages more than 6%. Whether they disperse or not is dependent on total salt content, organic matter, clay content and type, and cementing agents. An alternative approach is to assess the susceptibility of the soil to dispersion using dispersion tests (see Chapter 10). Tests to identify dispersible soils include the Emerson Aggregate Test and the percentage dispersion test. These tests and their interpretation are discussed in Chapter 10.

In the field, sodic and dispersible soils may be indicated by the joint occurrence of a number of the following features:

(i) soil morphological properties including:
— coarse columnar or prismatic structure
— an abrupt A/B texture boundary
— a bleached A_2 horizon

(ii) a very high strength when dry (often sufficiently high to offer substantial resistance to a pick or mattock)

(iii) very low permeability, often approaching zero. The very low permeability may be measured, but it can contribute to two other distinctive features of sodic soils:
— a 'wormy' appearance of the B horizon on vertical exposed cuttings
— a 'ball-bearing' texture because only the outsides of peds are wetted up and water only penetrates very slowly into the interior of peds. Thus, many dry sodic soils are difficult to texture without a mortar and pestle to break up the soil peds.

(iv) very high bulk density, often more than 1.8 g/cm^3.

It needs to be emphasised that the presence of these features does not replace the need for laboratory testing to confirm the occurrence of sodic and dispersible soils.

8.5.2 Hardsetting Surface Soils or Fragile, Light-textured Surface Soils

Hardsetting surface soils form a compact, hard and apparently apedal condition when dry (Northcote, 1983, p. 46). They may be quite soft when moist, but become hard as the soil dries out. Soil textures are usually loamy sands to clay loams and the soils have only a weak structure. The degree of hardsetting increases with increasing clay content, decreasing organic matter and increasing intensity of cultivation.

Hardsetting surface soils cover a large proportion of the cultivated land in New South Wales. Because these soils are often susceptible to wind erosion, and tend to give rise to high rates of runoff and subsequently erosion, they are of particular interest to conservationists. Many of these soils become degraded under intensive cropping and grazing leading to increased runoff and erosion and reduced soil productivity.

The amelioration of these soils requires a reduction in runoff by increasing infiltration, increasing organic matter, increasing moisture storage and developing a softer, more amenable environment for germination, emergence and root growth. This can be achieved by conservation farming practices such as crop/pasture rotations and conservation tillage. Relevant conservation tillage practices include direct drilling, reduced cultivation and no-tillage (see

Chapter 16). These practices have the potential to reduce erosion losses and to improve soil productivity. A review of hardsetting surface soils is given in Mullins et al. (1990).

8.5.3 Self-mulching Soils

Self-mulching soils have a high degree of pedality at the soil surface. Peds are usually in the 1 to 10 mm size range and form a loose surface mulch (McDonald et al., 1990). They are associated with the cracking clays (Vertosols). These soils are very favourable for crop growth, but because the surface is a loose mulch of incoherent soil particles, they can suffer severe erosion if the soil surface is not protected from heavy rainfall by actively growing plants or stubble. Stubble management and strip cropping are important in the conservation of these soils. The high shrink-swell characteristic of these soils also gives rise to problems with the establishment of pastures using small-seeded (grass) species (see Chapter 19 and Watt 1972, 1974).

8.5.4 Acid Surface Soils

Acid surface soils are sufficiently acid for plant growth to be restricted. Aluminium and manganese toxicity are usually considered the reasons for poor plant growth. When pHw in soil falls below 5.5, aluminium and manganese come into solution and may damage actively growing roots. These soils are most common in high rainfall areas on the tablelands where, it is thought, a long history of pasture improvement has induced this condition (see Chapter 13).

The problem of agriculturally induced acid soils is being addressed by the New South Wales government's Acid Soil Action Campaign. This project is a collaborative project between New South Wales Agriculture and the Department of Land and Water Conservation, and finances community research and extension programs.

8.5.5 Acid Sulfate Soils

Acid sulfate soils have formed in low-energy marine environments and isolated areas where pyrite has accumulated as a result of high sulfur contents in environments lacking oxygen. When exposed to oxygen, through excavation or drainage, the pyrite converts to sulfuric acid. Besides altering soil chemistry and causing extreme acidity, acid leachate with high levels of soluble aluminium and iron can adversely affect marine ecosystems. Acid sulfate soils are a major potential problem along the coast. Their extent has been identified by a soil mapping program (Naylor et al., 1995), and their management is being addressed by the Acid Sulfate Soil Management Advisory Committee. The Committee has provided funding for awareness, education, amelioration and research programs, and has also produced a comprehensive manual to guide the management of acid sulfate soils.

8.5.6 Aggregated or Subplastic Clays

Aggregated clays have strongly bound soil aggregates that, in terms of permeability, cause them to behave as sands in the field. Such aggregates are usually cemented by free iron. Affected soils are mainly the krasnozems and euchrozems (Gn3.1, Gn4.1, Uf5, Uf6). These soils cause problems when used for farm dams as they are too permeable to hold water (see Chapter 20). Identification is by the Emerson Aggregate Test (Class 6) or per cent dispersion which is generally zero. Subplasticity when field texturing is an indicator of aggregated clays. Field textures are substantially less clayey than would be expected from the clay content of the dispersed soil determined in the laboratory. For example, a clay loam field texture may have 70% clay when tested in the laboratory.

8.5.7 Expansive Soils

Expansive soils show substantial volume changes with changes in soil moisture content. Some of these have cracks extending to the surface when dry, if clayey surface soils are present (Ug soils), but many expansive soils do not show seasonal cracking. Expansive soils are common on, but not restricted to, basalt and basalt-derived parent materials, and the degree of shrinking and swelling is dependent on the amount and type of clay present.

Expansive soils are those having sufficient shrink-swell potential to be a problem as foundations for masonry walls, roads, buildings, underground services and soil conservation works unless appropriate precautions are taken. They are identified by a number of engineering tests, but for soil conservation purposes and general soil survey for urban development, linear shrinkage, volume expansion and Atterberg limit tests are used to identify them. Soils having linear shrinkage values of more than 17% are considered to have a high shrink-swell potential (see Chapter 11).

8.5.8 Saline Soils

Saline soils are those that contain sufficient soluble salts to adversely affect plant growth and/or land use. Such soils are commonly identified by electrical conductivity tests on either a 1:5 extract (greater than 1.5 dS/m) or a saturation extract (greater than 4.0 dS/m).

These soils occur in pockets in the tablelands, but are more common in the drier western region. Newly discovered areas where soils are affected by rising water tables are increasing in the tablelands and slopes of New South Wales. When saline water tables reach the surface, evaporation occurs, and salt becomes concentrated at the soil surface. Productive pasture species become replaced by salt-tolerant species and the ground cover becomes more patchy, surface protection is lost, and erosion is initiated. Irrigation in the Riverina has increased water table levels and saline soil problems are also developing there.

Soil salinity has come to be considered a major long-term hazard to farming in many parts of New South Wales. The New South Wales government has for several years financed the Salt Action program to help address, ameliorate and prevent soil salinity problems in New South Wales. The problem of soil salinity is discussed further in Chapter 13.

8.6 Major Soils and Land Use

The major wheat soils of the state are the red-brown earths and the black cracking clays. Summer cropping (sorghum, sunflowers and soya beans) is also common on the black cracking clays. Some grey cracking clays, euchrozems and areas of the red and brown cracking clays in the subhumid part of the state are also successfully used for wheat. Some areas of red earths are also used for wheat production. Large areas of red-brown earths have been pasture-improved and used for grazing cattle and sheep for fat lambs.

The valuable grazing lands of the tablelands are mainly on red earths in the central tablelands, and on red podzolic soils and krasnozems, especially in the northern tablelands. Significant areas of low-fertility solodic soils on the tablelands are considered excellent fine wool-growing country. Such areas include Yass, the area north-east of Orange and some areas in the vicinity of Armidale.

Intensive agriculture (vegetables and fodder crops) is largely confined to the black structured loams (prairie soils and chernozems) and the alluvial soils. Areas of red earths and krasnozems on the tablelands have been considered excellent potato-growing country. In the north-east of the state, krasnozems have been used for bananas and sugar cane, while elsewhere (for example, the Orange area) they have been used for cherries, apples and pears. The valuable cherry-growing area near Young is on a deep red friable soil that is very close to a krasnozem.

In the western semi-arid part of the state, extensive grazing is practised on the desert loams, calcareous earths (solonised brown soils) and sands.

Many irrigation developments require large areas of flat land. For this reason, irrigation is largely confined to the plains, primarily on the grey cracking clays, red-brown earths and, to a lesser extent, black cracking clays. Such irrigation is used to grow rice and cotton, and crops such as sunflowers, sorghum, soya beans and citrus fruits. In the southern part of the state, areas of the calcareous earths are irrigated. In the vicinity of Griffith, some red earths are irrigated as well as alluvial soils close to the river. New plantings for wine grape production, often using sophisticated drip irrigation systems, have been established in many undulating areas of the slopes of New South Wales.

The main soils used for the forestry industry of New South Wales are discussed in Murphy et al., (1998).

Further discussion of the effects of soils on land use and land capability will be found in Chapter 15.

BIBLIOGRAPHY

Atkinson, G. and Melville, M.D. (1987), 'Soils' in *Atlas of New South Wales—Portrait of a State*, Central Mapping Authority, Department of Lands, NSW.

Beadle, N.C.W. (undated), *Erosion Survey of Counties Peny and Wentworth, Manara and Livingstone, Taila and Kilfera,* unpublished reports, Soil Conservation Service of NSW, Sydney.

Beckmann. G.C., Thompson, C.H. and Hubble, G.D. (1974), 'Genesis of red and black soils on basalt on the Darling Downs, Queensland', *Journal of Soil Science* 25, 265–81.

Condon, R.W. (1961), 'Soils and landforms of the Western Division of New South Wales', *Journal of the Soil Conservation Service of NSW* 17, 31–46.

Corbett, J.R. (1969), *The Living Soil: The Processes of Soil Formation*, Martindale Press, West Como, New South Wales.

Dregne, H.E. (1976), *Soils of Arid Regions*, Elsevier Publishing Co., New York.

Eldridge, D.J. (1985), *Aeolian Soils of South-western New South Wales,* Soil Conservation Service of NSW, Sydney.

Isbell, R.F. (1996), *The Australian Soil Classification,* CSIRO Publishing, Melbourne.

James, J.W. (1960), 'Erosion survey of the Paroo-Upper Darling Region. II. Land form and soils', *Journal of the Soil Conservation Service of NSW* 16, 41–53.

Lawrie, J.W. (1974), 'Soils' in *Condobolin District Technical Manual,* Soil Conservation Service of NSW, Sydney.

McDonald, R.C., Isbell, R.F., Speight, J.G., Walker J., and Hopkins, M.S. (1990), *Australian Soil and Land Survey Field Handbook,* Inkata Press, Melbourne.

McKane D.J. (1997), *Storage and Manipulation of the Land Systems of Western New South Wales: A Relational Database for Polygonal Data,* Technical Report 18/97, CSIRO Land and Water, Canberra.

McKane D.J. and Oo, M.M. (1998), 'Program for determining the dominant principal profile form for a land system', in E. Bui (ed.), *Soil Information for the Murray Darling Basin,* CSIRO Land and Water, Canberra.

Marshall, T.J. and Walkley, A. (1937), *A Soil Survey of the Coomealla, Wentworth (Curlwaa) and Pomona Irrigation Settlements,* NSW CSIRO Aust. Bull. No. 107, CSIRO, Melbourne.

Mullins, C.E., MacLeod, D.E., Northcote, K.H., Tisdall, J.M. and Young, I.M. (1990), 'Hardsetting soils: behaviour, occurrence and management', *Advances in Soil Science,* 11, 37–108.

Murphy, C.L., Fogarty, P.J. and Ryan, P.J. (1998), *Soil Regolith Stability Classification for State Forests in Eastern New South Wales,* Department of Land and Water Conservation, Technical Report No. 41, Sydney.

Naylor, S.D., Chapman, G.A., Atkinson, G., Murphy, C.L., Tulau, M.J., Flewin, T.C., Milford, H.B. and Morand, D.T. (1995), *Guidelines for the Use of Acid Sulfate Soil Risk Maps,* Department of Land and Water Conservation, Sydney.

Northcote, K.H. (1951), *A Pedological Study of the Soils Occurring at Coomealla, New South Wales,* CSIRO Aust. Bull. No. 264, CSIRO, Melbourne.

Northcote, K.H. (1966), *Atlas of Australian Soils, Sheet 3, Sydney-Canberra-Bourke-Armidale Area,* with explanatory data, CSIRO Australia and Melbourne University Press.

Northcote, K.H. (1971), *A Factual Key for the Recognition of Australian Soils,* Third Edition, Rellim, Glenside, SA.

Northcote, K.H. (1983), *Soils, Soil Morphology and Soil Classification: Training Course Lectures,* CSIRO Division of Soils, Rellim Technical Publications, Glenside, SA.

Northcote, K.H. and Boehm, E.W. (1949), 'The Soils and Horticultural Potential of the Coomealla Irrigation Area', *Australian Soils and Land Use Series* No. 1, CSIRO, Melbourne.

Northcote, K.H., Hubble, G.D., Isbell, R.F., Thompson, C.H. and Bettenay, E. (1975), *A Description of Australian Soils,* CSIRO and Wilke and Co., Clayton, Victoria.

Northcote, K.H., Isbell, R.F., Webb, A.A., Murtha, G.G., Churchward, H.M. and Bettenay, E. (1968), *Atlas of Australian Soils, Sheet 10, Central Australia (Kalgoorlie-Cloncurry-Oakover River-Broken Hill),* with explanatory data, CSIRO Australia and Melbourne University Press.

Prescott, J.A. (1944), *A Soil Map of Australia,* CSIRO Aust. Bull. No. 177, CSIRO, Melbourne.

Stace, H.C.T., Hubble, G.D., Brewer, R., Northcote, K.H., Sleeman, J.R., Mulcahy, M.J. and Hallsworth, E.G. (1968), *A Handbook of Australian Soils,* Rellim Technical Publications, Glenside, SA.

Walker, P.J. (1978), 'Soils' in *Cobar District Technical Manual,* Soil Conservation Service of NSW, Sydney.

Walker, P.J. (1991), *Land Systems of Western New South Wales,* Technical Report No.25, Soil Conservation Service of NSW, Sydney.

Watt, L.A. (1972), 'Some observations on the rate of drying and crust formation on a black clay soil in north western New South Wales', *Journal of the Soil Conservation Service of NSW* 28, 41–50.

Watt, L.A. (1974), 'The effect of water potential on the germination behaviour of several warm season grass species with special reference to cracking black clay soils', *Journal of the Soil Conservation Service of NSW* 30, 28–41.

Soil Landscapes of New South Wales

CHAPTER 9

B.W. Murphy, D.J. Eldridge,
D.J. McKane and J.M. Gray

Soils are part of the landscape and it is not possible to map soils per se except at the most intensive scale, only the landscape on which they occur. Hence, most 'soil maps' are soil landform maps and the parcels of land mapped are soil landscapes (Northcote, 1966). This has been discussed previously in Chapter 7. Therefore, it is appropriate to describe the soil landscapes of New South Wales as well as the types of soil.

Landform plays a part in describing soil landscapes, which are generally considered to have particular topography and collections or associations of soils (Northcote, 1966). A standard method for describing landform is useful and a method is given in McDonald et al. (1984).

The major collections of soil landscapes in New South Wales are described in the following pages. It should be realised that a soil landscape is characterised by a particular landform and set of soils at the soil series or soil family level. As such, it would occupy one particular geological or geomorphological unit. Therefore, the soil landscapes discussed at this broad level are collections of soil landscapes and are best considered as soil landscape groups.

Again for convenience of presentation, the state is divided into eastern and western zones. The discussion is based on the *Atlas of Australian Soils* (Northcote et al., 1960–68), a soil map prepared for the *Atlas of New South Wales* (Atkinson and Melville, 1987), and detailed soil landscape mapping carried out since the early 1980s.

9.1 Soil Landscape Groups of the Eastern Zone

The level of information on soils in eastern New South Wales is steadily increasing owing to the extensive program of soil landscape mapping carried out by the Department of Land and Water Conservation (DLWC), the former Soil Conservation Service, and other bodies such as the CSIRO and the Murray-Darling Basin Commission (MDBC). Significant parts of the zone are covered by DLWC soil landscape maps at a scale of 1:100 000 or 1:250 000, and the MDBC is well advanced in a soil landscape mapping project over the entire Murray-Darling Basin. The areas covered by the DLWC's soil landscape mapping program as at March 1999 are shown in Colour plate 7.1.

In the past, before the 1980s, detailed information was available for areas that have been marked for development, particularly irrigation development, such as parts of the Riverina. A list of published surveys is given in Northcote et al. (1975) and Blain et al. (1985). However, there are also a number of unpublished surveys by the Department of Agriculture, Water Resources Commission and other government bodies and universities. The CSIRO and the

DLWC have programs to develop lists of these. A bibliography of soil studies in the Sydney region is to be found in Hazelton and Atkinson (1985).

Detailed soil mapping in the eastern zone of New South Wales has been carried out over most areas with a high potential for urban land use or agricultural use. Thus soil information is available for most coastal areas, the productive land of the New South Wales wheat-belt and the grazing lands of the tablelands, as well as for the Hunter Valley. There are, however, still significant areas yet to be covered by detailed soil maps.

Soil Landscape Distribution 9.1.1

The soil landscapes in the eastern zone are considered to fall into five main categories:

(a) Soil landscapes on basic/intermediate parent materials

(b) Soil landscapes on intermediate/siliceous parent materials

(c) Soil landscapes on highly siliceous parent materials

(d)   Soil landscapes having soils with minimal profile development

(e)   Soil landscapes where climate or microclimate dominates the soil-forming processes.

Each category has one or more different soil landscape groups which are identified in the maps of Northcote et al. (1960–68) (*Atlas of Australian Soils*) and Atkinson and Melville (1987) (*Atlas of New South Wales*—Soils). The soils are described using Great Soil Groups (Stace et al., 1968), the Australian Soil Classification System (Isbell, 1996), and the Factual Key (Northcote, 1979), using names from *A Description of Australian Soils* (Northcote et al., 1975) (see Chapter 6).

(a) Soil Landscapes on Basic/Intermediate Parent Materials

Basalt, dolerite, andesite, diorite, basic tuffs.

(i) Alluvial Plains of Prairie Soils

These relatively flat areas of black structured loams mainly comprise prairie soils (Black Dermosols; structured loams) and occur on a number of the state's major rivers including the Hawkesbury, Hunter, Macquarie, Castlereagh, Peel, Murrumbidgee and Clarence Rivers. They occur on alluvium that has a substantial component of basic to intermediate parent material. The area covered is not large, but where they do occur they are important because of their productivity.

Associated soils include alluvial soils (Rudosols and Tenosols; Uniform sands and loams) on very recent alluvial deposits, and Red and Yellow Chromosols (red podzolic and yellow podzolic soils, red-brown earths or solodic soils; red and yellow duplex soils) or red earths (Red Kandosols) on older, usually higher, terraces.

Existing Soil Landscape Units

Atlas of Australian Soils
Gb
Atlas of New South Wales
Deep alluvial loams

(ii) Rolling to Steep Hills with Black Cracking Clays

These hills of black cracking clays (Black Vertosols) occur in association with the Liverpool Ranges, the Warrumbungle Mountains and other basaltic terrain in the northern tablelands. This soil landscape group often occurs on dissected plateau remnants, sometimes with boulder-strewn slopes.

Associated soils include euchrozems (Red Ferrosols, Red Dermosols; red structured earths), chocolate soils (Brown Dermosols; friable brown duplex soils), red and brown cracking clays (Red and Brown Vertosols) and, sometimes, non-calcic brown soils (Red Chromosols) and red earths (Red Kandosols).

Existing Soil Landscape Units

Atlas of Australian Soils
Kb, some Ke
Atlas of New South Wales
Shallow black self-mulching clays

(iii) Undulating Low Hills and Level Plains of Black Cracking Clays

This soil landscape group of deep black cracking clays (Black Vertosols) occurs on the north-west slopes of the state. The soil landscapes are associated solely with Tertiary basalt or alluvium derived from Tertiary basalt. Their occurrence is linked with subhumid climates. There are also a number of small areas in the northern tablelands around Inverell and Glen Innes. The largest area includes the slopes and plains north and south of the Liverpool Range, incorporating such localities as Quirindi, Gunnedah (see Colour plates 9.5 and 9.6), Breeza and Coolah. Other extensive areas occur in association with the Warrumbungle volcanics, including the plains around Coonamble. Another occurs to the north in the vicinity of Moree, but these soils are not as strongly self-mulching as those further south.

Associated soils include euchrozems (Red Ferrosols and Red Dermosols; red structured earths), brown and red cracking clays (Brown and Red Vertosols), and red-brown earths (Red Chromosols; hard pedal red duplex soils).

Existing Soil Landscape Units

Atlas of Australian Soils
Kc, Kd, Ke, Kh
Atlas of New South Wales
Deep black cracking clays

(iv) Undulating Low Hills and Level Plains of Grey and Brown Cracking Clays

This soil landscape group, incorporating the grey and brown cracking clays (Grey and Brown Vertosols), is largely associated with the north-western plains of the state. In this area, soil landscapes frequently show gilgai formation. The soils occur in a subhumid to semi-arid environment, and grade towards the western zone soil landscape unit of drainage lines and floodplains of grey cracking clays (Grey Vertosols). Typical localities include Moree, Walgett and Quambone. An extensive area of this soil landscape group is to be found on the south-western plains in the vicinity of Hay in the western zone.

Associated soils include red-brown earths (Red Chromosols; hard pedal red duplex soils), particularly on old levees of prior streams, black cracking clays (Black Vertosols) and massive grey cracking clays (Grey Vertosols). Small areas of red earths (Red Kandosols) and alluvial soils (Rudosols; sands and loams) also occur.

Existing Soil Landscape Units

Atlas of Australian Soils
CC
Atlas of New South Wales
Coarsely cracking grey and brown clays

(v) Rolling to Steep Hills with Krasnozems

These landscapes may include smaller areas of undulating low hills within the landscape and these are interpreted as being remnants of once more-extensive plateaux. This soil landscape group is mainly associated with Tertiary basalt parent material, but some other basic rocks such as dolerite and pre-Tertiary basalts or andesites are also included, as on the south coast at Moruya. All the soil landscapes occur in humid environments on the coast or the wetter tableland areas.

The soil landscapes with krasnozems (Red Ferrosols) are most common on the north coast and typical locations are in the vicinity of Murwillumbah, north of Dorrigo and at Yarrowitch east of Walcha. Other small areas occur west of Wollongong at Robertson, Kiama, in association with Mount Canobolas at Orange, at Bilpin on the Blue Mountains and at numerous other locations in humid climates where remnant basalt outcrops or basic rocks occur.

Significant areas of krasnozem-like soils, the brown structured earths (Brown Dermosols), occur on the south coast, north and west of Moruya.

Associated soils include brown structured earths (Brown Dermosols) and black structured earths (Black Dermosols), for which there are no corresponding great soil groups, and yellow structured earths and friable yellow duplex soils (Yellow Dermosols), which generally fall into the xanthozem great soil group. Other associated soils include red podzolic soils (Red Chromosols; acid red duplex soils) and skeletal soils (Rudosols, tenosols; shallow loams).

Existing Soil Landscape Units

Atlas of Australian Soils
Some Mj, some Me, some Mg, some Mf, some Mh, some Ml, Mp
Atlas of New South Wales
Deep structured red clay loams (humid climate)

(vi) Undulating Low Hills with Euchrozems

The undulating low hills of euchrozems (Red Ferrosols, Red Dermosols; red structured earths) are mainly associated with a range of basic parent materials including Tertiary basalts, Jurassic basalts and Palaeozoic andesites, basalts and dolerite. They are possibly more common than indicated on the *Atlas of Australian Soils*, a number of unmapped areas having been observed. Even so, they still only occupy relatively small areas. A minority of the landscapes have steeper terrain than undulating low hills. These soil landscapes occupy the subhumid parts of the state, including the drier parts of the tablelands and the western slopes.

An extensive area of this soil landscape group occurs on the northern tablelands in the vicinity of Inverell (Tertiary basalt). Other areas occur west of Gunnedah (Jurassic basalt) and in the vicinity of Wellington on the central western slopes (Palaeozoic basic rocks). Small but significant unmapped areas have been observed in the vicinity of Dubbo and Young, and on the south-western slopes, and appear to be associated with Palaeozoic basic rocks.

Associated soils include the closely related friable red duplex soils (Red Dermosols) and some black cracking clays (Black Vertosols). Other soils include non-calcic brown soils and red-brown earths (Red Chromosols), and non-sodic yellow duplex soils (Yellow Chromosols), for which there is no corresponding great soil group. In some soil landscapes, red earths (Red Kandosols) and yellow earths (Yellow Kandosols) have been recorded.

Existing Soil Landscape Units

Atlas of Australian Soils
Mo
Atlas of New South Wales
Deep structured red clay loams (subhumid climate)

(b) Soil Landscapes on Intermediate/Siliceous Parent Materials

Syenite, trachyte, granodiorite, dacite, greywacke, lithic sandstone, siltstone, mudstone, shales, schists.

(i) Undulating and Rolling Low Hills with Red Podzolic Soils or Non-calcic Brown Soils

Red podzolic soils (Red Chromosols and Red Kurosols; acid red duplex soils) and non-calcic brown soils (Red Chromosols; neutral red duplex soils) are generally associated with intermediate/siliceous parent materials such as granodiorites, shales and metamorphics with moderate levels of bases and intermediate volcanics. The two soil types are combined as they have so many features in common and the separation of them does not seem warranted at this broad level of description. There is a general tendency for the red podzolic soils (Red Kurosols, Red Chromosols) to occur near the coast and in the wetter tableland areas. The non-calcic brown soils (Red Chromosols) become more common in the drier tableland areas and on the western slopes. Typical areas of the red podzolic soils include the Cumberland Plain west of Sydney and areas at Bathurst, and in parts of the New England plateau and the southern highlands. The non-calcic brown soils are relatively common on the western slopes and grade into the red-brown earths (Red Chromosols, Red Sodosols; alkaline red duplex soils). Typical locations include areas north and east of Tamworth, east of Dubbo and significant areas east of Wagga Wagga.

Associated soils include, on lower slopes, yellow podzolic soils (Yellow Chromosols, Yellow Kurosols; acid yellow duplex soils), yellow solodic soils (Yellow Sodosols; sodic yellow duplex soils, Dy3.4), and non-sodic neutral and alkaline yellow duplex soils (Yellow Chromosols) for which there is no suitable great soil group.

Other soils include skeletal soils (Lithosols, Rudosols, Tenosols; shallow loams and sands), red earths (Red Kandosols) and yellow earths (Yellow Kandosols).

Existing Soil Landscape Units

Atlas of Australian Soils
Some Pb, some Pd, Qb, Qc, Qd
Atlas of New South Wales
Yellow and red texture contrast soils

(ii) Undulating Low Hills and Level Plains of Red-Brown Earths

This soil landscape group is extensive on the western slopes and includes a large proportion of the state's major cropping areas. It is associated mainly with intermediate siliceous parent materials, including shales, metamorphics, granodiorites, some sandstones and some intermediate volcanics, and a wide range of alluvium derived from these parent rocks. The soil landscape group is associated with subhumid to semi-arid climates of the western zone, and grades into the western zone soil landscape group of plains with hard pedal red duplex soils (Red Chromosols; red-brown earths).

The most extensive areas of this soil landscape group are in the central west and south-west slopes and plains including such locations as Gilgandra, Trangie, Parkes, Forbes, Condobolin, West Wyalong, Narrandera and Deniliquin. Other significant areas occur on the north-west slopes and plains at Tamworth, Manilla and Gunnedah, but there are also extensive areas of the black cracking clays in this part of the state. Red-brown earth soil landscapes have been recorded towards the east at Scone in the Hunter Valley.

Associated soils are often in complex patterns, depending on the geomorphology of the soil landscapes. Pockets of black, grey and brown cracking clays (Vertosols) occur in many of these landscapes, particularly those based on alluvial materials and showing prior stream development. Other soils include yellow solodic soils (Sodosols; sodic yellow duplex soils) and other non-sodic yellow duplex soils (Yellow Chromosols), euchrozem-like soils (Red and Brown Dermosols and Ferrosols; red and brown structured earths) and skeletal soils (Lithosols, Rudosols; shallow loams and sands) on ridges.

Existing Soil Landscape Units

Atlas of Australian Soils
Oc, Ob
Atlas of New South Wales
Yellow and red texture contrast soils

(iii) Undulating, Rolling and Steep Hills with Red Earths

This soil landscape group of red earths (Red Kandosols) tends to occur on intermediate/siliceous parent materials including granodiorites and intermediate volcanics such as trachyte and some andesites, and shales or metamorphics with moderate levels of bases. Geologically, they are most common on ancient rises or anticlines, as distinct from the solodic soils that are largely associated with ancient troughs or synclines (see Colour plates 9.3 and 9.4).

These soil landscapes are widespread around Orange and to the west of Bathurst in the vicinity of Sunny Corner. They are not common in the northern part of the eastern zone, but cover large areas of undulating to rolling low hills on the south-west slopes in the vicinity of Young and Cootamundra. Towards the west, these soil landscapes grade towards the calcareous red earths (Calcarosols).

Associated soils include yellow earths (Yellow Kandosols) and, in depressions or on lower slopes, yellow podzolic soils (Yellow Chromosols and Yellow Kurosols; acid yellow duplex soils), non-sodic neutral yellow duplex soils (Yellow Chromosols), and yellow solodic soils (Yellow Sodosols; sodic yellow duplex soils). Various red duplex soils including red podzolic soils (Red Chromosols and Red Kurosols) and non-calcic brown soils (Red Chromosols) also occur. Less commonly, euchrozems (Red Ferrosols and Red Dermosols) and krasnozem-like soils (Red Ferrosols) may occur.

Existing Soil Landscape Units

Atlas of Australian Soils
Mu, some Mw, Mz, My, Mx
Atlas of New South Wales
Massive red and yellow earths

(iv) Undulating and Rolling Low Hills with Yellow Podzolic Soils

The undulating and rolling low hills of yellow pod-zolic soils (Yellow Chromosols and Kurosols; acid yellow duplex soils) are associated with intermediate siliceous parent materials in the tablelands and on the coast. Parent materials include felspathic and lithic sandstones, often Permian, and shales and metamor-phic rocks of low base content. Typical localities include Lithgow (see Colour plates 9.1 and 9.2), Cessnock, Ulan, Casino, Jervis Bay, Goulburn, Yass and Armidale. The association of this soil landscape group with Permian sandstones has seen the develop-ment of many open-cut and shallow underground coal mines, as at Ulan and in parts of the Hunter Valley.

Associated soils include red podzolic soils (Red Chromosols and Kurosols; acid red duplex soils), yel-low earths (Yellow Kandosols) and skeletal soils (Lithosols, Rudosols; shallow stony loams and sands). There is also a significant proportion of yellow solodic soils (Yellow Sodosols; sodic yellow duplex soils) within some of these soil landscapes. Some of the yellow podzolic soils may have ironstone gravel (for example, *Atlas* unit Tb38).

Existing Soil Landscape Units

Atlas of Australian Soils
Ta, some Tb, some Wa
Atlas of New South Wales
Yellow and red texture contrast soils

(v) Undulating and Rolling Low Hills with Yellow Solodic Soils

The undulating and rolling low hills of yellow solodic soils (Yellow Sodosols) are also associated with siliceous and siliceous/intermediate parent materi-als, but tend to be in the drier parts of the tablelands and on the western slopes. However, the distinguish-ing feature of these soils, their high sodium contents, seems to be at least partly related to parent material. Many of these soil landscapes appear to be associated with granites, marine sediments of ancient trough zones or acid volcanics. Most of these materials have the potential, at least, to contribute large amounts of sodium to the soils. Atmospheric accession of sodium is also a source of sodium being particularly signifi-cant for the highly siliceous parent materials of the solodic soils on the Pilliga Sandstone (*Atlas* units X11

and Ya25). Typical localities for these soils include Armidale-Uralla, the Pilliga Sandstone north of Bara-dine, north of Orange, the Yeoval Granite near Dubbo, Goulburn and Tenterfield.

Associated soils include red podzolic soils (Red Chromosols and Kurosols; acid red duplex soils), non-sodic yellow duplex soils (Yellow Chromosols), which have no equivalent great soil group, yellow earths (Yellow Kandosols) and skeletal soils (Rudosols; shallow loams and sands). Other soils may be associated in specific soil landscapes.

Existing Soil Landscape Units

Atlas of Australian Soils
Some Tb, Ub, Va, some Wa, X
Atlas of New South Wales
Yellow and red texture contrast soils

(c) Soil Landscapes on Highly Siliceous Parent Materials

Quartz sands, quartzite, quartz sandstones, granite, rhyolite, adamellite, siliceous tuff.

(i) Undulating and Rolling Low Hills with Yellow Earths

This soil landscape group comprises undulating and rolling low hills of yellow earths (Yellow Kandosols). The majority of these are sandy and the largest area is associated with the Pilliga Sandstone, a quartzose sandstone. Some areas also occur on the Hawkesbury Sandstone such as the Somersby Plateau west of Gos-ford, which is used for citrus production. Other small areas occur north of Goulburn and in the northern tablelands.

Other soils include red earths (Red Kandosols) and earthy sands (Tenosols) on ridges and upper slopes and, sometimes, yellow solodic soils (Yellow Sodosols; yellow duplex soils) in depressions, partic-ularly in the Pilliga Sandstone. Yellow podzolic soils (Yellow Kurosols and Chromosols; acid yellow duplex soils) are commonly associated with these soils in the coastal and tableland regions.

Existing Soil Landscape Units

Atlas of Australian Soils
Ms, Mr, Mb, Mc, Mq
Atlas of New South Wales
Massive red and yellow earths, stony sandy loams, shallow loams

(ii) Rolling to Steep Hills with Podzols

The rolling to steep hills of podzols (Podosols; bleached sands) are almost solely related to land-scapes on siliceous granites. Tors and rock pavements are usually common. This soil landscape group is largely confined to the more siliceous granites of the tablelands in areas such as Tenterfield and Moonbi, north of Tamworth. Smaller areas occur in the central and southern tablelands near Bathurst and, possibly, in some parts of the Bega Batholith.

Associated soils include yellow podzolic and yellow solodic soils (Yellow Sodosols; yellow duplex soils), on lower slopes and in depressions, and skeletal soils (Lithosols, Rudosols; shallow sands) in some more exposed parts.

Existing Soil Landscape Units

Atlas of Australian Soils
Cd, some Cb
Atlas of New South Wales
Stony sandy loams, shallow loams

(iii) Undulating Low Hills and Level Plains of Podzols

These landscapes of podzols (Podosols; bleached sands) are largely confined to the older coastal sand dunes, although small pockets occur in association with the siliceous granites in the tablelands (for example, Mudgee). They extend all the way down the coast, although the largest areas are generally to the north of Sydney. Significant areas occur at Sydney, Mascot, Kurnell Peninsula, south of Jervis Bay, Port Stephens, near Newcastle and Myall Lakes.

Associated soils include unstable dunes of calcareous sands (Rudosols) and siliceous sands (Rudosols), swamp areas of humic gley soils (Hydrosols; gleyed duplex soils) and peats (Organosols). Some sandy yellow podzolic soils (Yellow Chromosols and Kurosols; yellow duplex soils) and sandy yellow earths (Yellow Kandosols) may also occur in older sand dunes. Some clay soils (Vertosols) have also been reported.

Existing Soil Landscape Units

Atlas of Australian Soils
A, B, Ca, some Cb
Atlas of New South Wales
Siliceous dune sands, calcareous sands

(d) Soil Landscapes Having Soils with Minimal Profile Development

Steep Terrain with Skeletal Soils

These landscapes involve steep hills and mountainous terrain of skeletal soils (Lithosols, Rudosols, Tenosols; shallow, stony loams and sands). Rock material is variable with rock outcrops in the form of granite boulders, sandstone shelves and cliffs, basalt boulders and vertically bedded shales, and metamorphics or other rock formations are common. Also included within the areas mapped as this soil landscape group are areas of more level terrain and narrow valleys. In these areas where soil development can occur, there is a wide range of soils including red and yellow earths (Red and Yellow Kandosols), red podzolic soils (Red Chromosols and Red Kurosols; acid red duplex soils), podzols (Podosols; bleached sands), and even krasnozems (Red Ferrosols; red structured earths) and black cracking clays (Black Vertosols).

The major areas of this soil landscape group are associated with the Great Dividing Range, and large areas occur west of Sydney (Blue Mountains), west of Port Macquarie, south-west of Sydney (the Budawang Ranges) and in association with the Snowy Mountains. Other localised areas occur west of the Dividing Range in Tertiary volcanic areas such as Mount Kaputar and the Warrumbungle Mountains. Other less prominent areas, but frequently as rugged, are the many long narrow ranges of the western slopes. Such ranges include Black Mountain at Canberra, the Hervey Ranges near Parkes, the Baldwin Range at Manilla, the Moonbi Range at Tamworth, the Catombal Range at Wellington, and so on. Many of these ranges have red pozolic soils (Red Chromosols and Red Kurosols; acid red duplex soils) in association with the skeletal soils.

Existing Soil Landscape Units

Atlas of Australian Soils
JJ, Fz, LK, F, Fa and in association with many others like Mb2, Mw16, Mw9, Pd4
Atlas of New South Wales
Shallow loams and stony sandy loams

(e) Soil Landscapes Where Climate or Microclimate Dominates the Soil-forming Processes

(i) Steep Hills with Alpine Humus Soils

The steep hills of alpine humus soils (Tenosols, Organosols; organic loams) are confined largely to elevations above 1100 m. Their main occurrence is in the Snowy Mountains. Rock outcrop is often common. This soil landscape group also occurs in other high altitude areas such as the summits of Mount Canobolas and Mount Kapular, the Barrington Tops and parts of the New England plateau near Ebor.

Associated soils include peats (Organosols) and skeletal soils (Lithosols, Rudosols, Tenosols; shallow stony loams and sands).

Existing Soil Landscape Units

Atlas of Australian Soils
KK
Atlas of New South Wales
(not included)

(ii) Level Plains of Humic Gley Soils

These are mainly coastal plains of humic gleys (Hydrosols; gley duplex soils). Usually the plains are low lying, poorly drained and are subject to flooding. Some wetlands and estuarine flats are included. An appreciable area occurs in the vicinity of Casino on the Richmond River. This soil landscape group occurs mainly on the north coast, but areas of this soil landscape group can be found all along the coast. Acid sulfate soils are common in this soil landscape group. When these soils are drained they can develop extreme acidity causing major environmental problems (see Chapter 13).

Associated soils include podzols (bleached sands, Uc2), peats (O) and yellow podzolic soils (yellow duplex soils, Dy5.6, Dy5.8).

Existing Soil Landscape Units

Atlas of Australian Soils
NZ, NY
Atlas of New South Wales
Massive grey and black coastal clays

Agricultural Productivity of Soil Landscapes 9.1.2

The agricultural productivity of the soil landscapes of New South Wales is dependent on a number of factors, listed in Table 9.1. As well as soil factors, landform and climate also play a part in determining land use and productivity. For example, no matter how productive the soil, a soil landscape with steep slopes and much rock outcrop cannot be used for cropping. Low rainfall may also prevent cropping.

Table 9.1 Factors affecting the susceptibility to degradation and the agricultural productivity of the soil landscapes of New South Wales

Soil Physical Property	Constraint
Rockiness	Excessive rock hinders cultivation
Soil depth	Depths < 60 cm restrict rooting capability of agricultural plants and water-holding capacity
Profile drainage	Poorly or imperfectly drained soils restrict plant growth and are difficult to manage
Existing erosion	Less than 10 cm of topsoil severely restricts crop growth. Gullies reduce field areas and hinder cultivation and access
Moisture regime	Poor infiltration and water-holding capacity restrict moisture available to plants and increase erosion
Soil structure	Texture < loams and/or organic matter < 2% make soils vulnerable to soil structure degradation and erosion
Chemical fertility	Insufficient nutrients restrict plant growth
Soil acidity	Soil pH < 5.5 may lead to restricted plant growth (soil:water 1:5)
Surface soil salinity	Soil EC_{sat} > 4dS/m restricts plant growth and increases erosion hazard
Subsoil salinity	Potential problem in irrigation soils
Toxicity	Excess of some elements reduces plant growth (minespoil, acid soils)
Subsoil sodicity	Excess Na^+ gives rise to soil dispersibility, low permeability, gully/tunnel erosion hazard
Organic matter content	Low levels restrict nutrient supply, reduce cation exchange and soil aggregation
Landform Property	
Slope	Steepness prevents cultivation
Dissection	Restricts intensity of agricultural activities
Soil Erosion Factor	
Slope steepness	Increases erosion hazard
Slope length	Increases erosion hazard
Rainfall intensity	High intensities increase erosion hazard
Surface soil erodibility	High silt contents, poor aggregation, low organic matter give rise to increased sheet/rill erosion hazard
Subsoil sodicity	Sodic subsoils are dispersible giving rise to high gully/tunnel erosion hazard; reduced profile permeability increases surface runoff

A legend for rating the productivity of soil land-scapes is presented in Table 9.2. This table relates mainly to dryland cropping and grazing. Any consideration of irrigation or horticultural cropping may change the ratings. A brief outline of the general productivity of the soil landscapes is presented in Table 9.3.

The most widely used and suitable soil landscapes for mixed farming (grazing and crops) are the undulating low hills and level plains of red-brown earths (Red Chromosols, alkaline). The most productive for cropping are the alluvial plains of prairie soils (Dermosols), followed closely by the level plains and

Table 9.2 Legend for rating the productivity of soils

Division	Dominant or Major Limitation
Division A	Soils with no limitations to agricultural productivity, except perhaps a need for N and P fertiliser under continuous use.
Division B	Soils with moderate limitations to agricultural productivity, overcome by readily available management practices (such as crop and pasture rotation, conservation tillage, fertiliser application and deep tillage).
Division C	Soils with substantial limitations to agricultural productivity, overcome partially by readily available management practices (such as crop and pasture rotation, conservation tillage, fertiliser application, deep tillage, gypsum or lime application).
Division D	Soils with substantial limitations to agricultural productivity that either cannot be overcome, or can only be overcome by major cost inputs or specialised management practices.
Division E	Soils with severe limitations to agricultural productivity that either cannot be overcome, or can only be overcome by major cost inputs or specialised management practices.
Division X	Soils with special limitations to agricultural productivity, making them unusable for general agriculture.

Table 9.3 Productivity ratings of the soil landscapes of New South Wales

Division	Soil Landscape
Division A	Alluvial plains of prairie soils
Division B	Level plains and undulating low hills with self-mulching black, grey and brown cracking clays
	Undulating low hills with euchrozems and krasnozems
Division C	Level plains and undulating low hills with red-brown earths
	Level plains and undulating low hills with red earths
	Undulating low hills with red and yellow podzolic soils
Division D	Undulating to rolling hills with solodic soils
	Level plains and undulating low hills with yellow earths (sandstone-derived)
	Steep and rolling hills, with significant rock outcrop, or shallow soils of black cracking clays, krasnozems and euchrozems
	Steep and rolling hills with red podzolic soils and significant rock outcrop or shallow soils
	Rolling and steep hills with podzols (granite-derived), often with significant rock outcrop and shallow soils
Division E	Undulating and rolling low hills with very sandy soils (coastal podzols)
	Steep hills and mountainous terrain with skeletal soils
Division X	Extremely rugged terrain with rock walls, large tors, ravines, gorges and rock outcrop

undulating low hills with black, grey and brown cracking clays (Black, Grey and Brown Vertosols).

The productive grazing soil landscapes include the soil landscape groups:
—undulating to rolling hills with krasnozems and euchrozems (Ferrosols and Dermosols) on the tablelands
—undulating to rolling and steep hills with red earths (Red Kandosols) on the tablelands
—undulating rolling and steep hills with red and yellow podzolic soils (Red and Yellow Chromosols) on the tablelands.

For fine wool production, soils of lesser fertility are sometimes favoured. This makes the soil landscape group of the undulating, rolling and steep hills of solodic soils (Sodosols) with associated red podzolic soils (Red Chromosols) valuable for this agricultural activity. Areas such as Yass, north of Orange and parts of the New England tablelands are examples of this.

Irrigation is practised largely in areas of level plains on a range of soil types including black, grey and brown cracking clays (Vertosols) and red-brown earths (Red Chromosols) in the central and northern parts of the state, and on the grey cracking clays (Vertosols), red-brown earths (Red Chromosols), calcareous earths (Calcarosols) and red earths (Red Kandosols) in the southern part of the state.

Horticultural crops can often be grown in small landscapes that make suitable niches with good growing conditions. For example, the undulating low hills of sandy yellow earths (Yellow Kandosols) (Somersby Plateau) are very successfully used for citrus production. The krasnozems (Red Ferrosols) on the footslopes of Mount Canobolas are successfully used for cherries, as are the euchrozem-like soils (Red Ferrosols and Dermosols) at Young.

Although not widespread, the level plains of black structured loams (chernozems and prairie soils) (Dermosols) are highly productive and are intensely used for vegetables and fodder crops.

Soil Landscapes and Soil Conservation

Major soil conservation and erosion problems occur on the following soil landscape groups. While serious erosion problems can occur on other soil landscapes, they are not as widespread or economically significant as these.

(a) Undulating Low Hills and Level Plains of Black Cracking Clays (Vertosols)

The major agricultural activity on these soils is cropping, and soil erosion leading to loss of soil productivity and off-site erosion problems, including silting of roads and water storages, is a major problem. Conservation farming techniques, particularly stubble retention, strip cropping and earthworks, are major methods for controlling this erosion.

(b) Undulating Low Hills and Level Plains of Red-Brown Earths (Red Chromosols)

Soil erosion and soil structural degradation are serious problems leading to loss of soil productivity. Conservation farming techniques, particularly crop-pasture rotation, reduced tillage, direct drilling and stubble retention, together with earthworks, are major methods for controlling this erosion. Increasing attention needs to be paid to the use of pastures on these soils as a means of improving soil structure.

(c) Undulating and Rolling Low Hills of Yellow Solodic Soils (Sodosols)

Gully erosion is a major problem on these soils. Some areas are severely gullied, reducing soil productivity and producing large volumes of sediment to cause off-site problems. Control is by pasture improvement, gully filling and stabilisation, and control structures in unstable drainage lines.

9.2 Soil Landscape Groups of the Western Zone

Detailed soils information is unavailable for much of western New South Wales because of the relatively low value of the land and the extensive grazing enterprises constituting a large proportion of the area. Mapping of the soils in terms of great soil groups (Stace et al., 1968) presents some difficulties in the western zone as there are a number of soils that do not conform to any of the existing groups. Therefore, emphasis is placed on the Australian Soil Classification (Isbell, 1996) and Factual Key groups. Correlations between these classifications are outlined in Table 8.1.

However, the use of soil landscapes provides a relatively easy method of characterising the soils of the area as this takes into account their complexity and the high correlation between soil type and landscapes (landforms). This method has been used successfully by Northcote (1966) in compiling the *Atlas of Australian Soils* and by Eldridge (1985) in delineating soils susceptible to wind erosion in south-western New South Wales. Walker (1991) used the land systems concept in compiling the Land Systems of Western New South Wales. Other sources of information are Lawrie (1974) and Walker (1978).

9.2.1 Soil Landscape Distribution

In western New South Wales, eight major soil landscape groups have been delineated and these are described in Table 9.4 (see *a–h* below). The dominant and associated principal profile forms in each soil landscape are also given in Table 9.4. The Australian Soil Classification correlation comes from Isbell (1996) and McKane and Oo (in press).

(a) Dunefields of Calcarosols, Rudosols and Tenosols (Calcareous, Siliceous and Brownish Sands)

Subparabolic, parabolic and aligned dunes of Hypocalcic Calcarosols, Arenic Rudosols and Orthic Tenosols (calcareous, siliceous and brownish sands) occur over large areas of western New South Wales, in the extensive south-western corner of the state and in an area bounded by the Darling River and the Riverine Plain. Associated with these dunes are plains and swales of Calcarosols (brown calcareous earths) and assorted Chromosols and Sodosols. Swamps and depressions of Grey Vertosols (cracking grey clays) occur between the dunes, often in association with scalded Chromosols and Sodosols.

Much of this area is dominated by mallee (*Eucalyptus* species) and inedible shrubs.

Soil Landscapes in *Atlas*: DD1, DD3, BA51–53

(b) Plains with Dunes of Tenosols (Brownish Sands) and Swales of Red Earths

Extensive plains with linear south-west to north-east trending dunes are found in the north-western corner of New South Wales. These dunes, which extend as far as Lake Frome in South Australia, are the eastern-most extent of the Strzelecki Dunefield. Associated are broad swales of Kandosols and Calcarosols (neutral and calcareous red earths), with scattered pans and depressions of Grey Vertosols (self-mulching grey cracking clays).

The soils of the dunes are principally deep Orthic Tenosols (brownish sands) which differ from those of the subparabolic dunes in having higher carbonate levels, slight horizon development and weak textural changes through the profile. Areas of crusty Red Sodosols (desert loams) occur on the margins between the dunes and swales.

The dunes are dominated by sparse mulga (*Acacia aneura*) and other trees and dense inedible shrubs such as hopbush (*Dodonaea attenuata*).

Soil Landscapes in *Atlas*: Mx 34–36

(c) Hills and Valley Plains of Shallow, Leptic Rudosols (Skeletal Soils or Lithosols)

These soils are associated with the Barrier Ranges centred on Broken Hill. Shallow, Leptic Rudosols and Arenic Rudosols (siliceous sands) occur on the ridges and hills in association with Calcarosols (calcareous earths) and assorted Chromosols and Sodosols on the plains. The soils along the major drainage lines are dominated by crusty Red Sodosols (desert loams).

The vegetation on the hills is sparse to scattered mulga and dead finish (*Acacia tetragonophylla*).

The valleys and slopes are dominated by assorted bluebushes (*Maireana* species).

Soil Landscape in *Atlas*: F5

(d) Plains of Grey Vertosols (Grey Cracking Clays)

(i) Drainage lines and Floodplains of Grey Vertosols (Grey Cracking Clays): Grey Vertosols (self-mulching, grey and yellow-grey cracking clays) are associated with the broad floodplains of the major river systems in western New South Wales. Associated with this functional and non-functional drainage are low dunes and levees of Brown and Red Sodosols (desert loams) and sandy rises of deep Arenic Rudosols (siliceous sands). The plains are dominated by Red and Brown Kandosols or Sodosols (solonised brown soils).

These soils are dominated by river red gum (*Eucalyptus camaldulensis*), black box (*E. largiflorens*) and coolibah (*E. microtheca*).

Scattered shrubs including lignum (*Muehlenbeckia cunninghamii*) and nitre goosefoot (*Chenopodium nitrariaceum*) occur adjacent to the floodplain.

Soil Landscapes in *Atlas*: I11, CC1, CC16–20, II4, MM71–74

(ii) Plains with Localised Depressions and Prior Streams: A mosaic of Grey Vertosols (grey cracking clays) and hard Red Chromosols occurs on the Riverine Plain between Hay and

Table 9.4 Description of the major soil landscapes of western New South Wales

Dominant Soils of the Soil Landscape	Soil Landscape Description
Sands of Minimal to Weak Development	
(a) Hypercalcic Calcarosols and Arenic Rudosols (calcareous and siliceous sands) and Orthic Tenosol (brownish sands)	Dunefield: parabolic, subparabolic and aligned dunes of deep Hypercalcic Calcarosol (calcareous sand), Arenic Rudosols (siliceous sands) and Orthic Tenosol (brownish sands); swales and 'plains' of Hypocalcic, Lithocalcic or Hypercalcic Calcarosols (brown calcareous earths), Red Kandosols (red sandy earths) and assorted Brown Chromosols and Red Sodosols; swamps and depressions of Grey Vertosols (grey cracking clays) and assorted Chromosols and Sodosols; smaller areas of Hypocalcic, Lithocalcic or Hypercalcic Calcarosols (calcareous earths) on the sandy rises.
(b) Brownish sands (Orthic Tenosol)	Plains with longitudinal dunes and broad swales: swales and plains of Red Kandosols (calcareous and neutral red earths) with areas of crusty Red Sodosols; dunes of deep Orthic Tenosols (brownish sands) and Arenic Rudosols (siliceous sands); some areas of Semiaquic Podosols (bleached sands), scattered swamps and plains of Grey Vertosols (grey cracking clays).
Loamy Soils of Weak Horizon Development	
Shallow dense loamy soils (Um5.41) (Lithosols or Skeletal Soils)	Hills and small valley plains: shallow Kandosols (dense loamy soils) with areas of Tenosols or Leptic Rudosols (shallow loams) and Arenic Rudosols (siliceous sands) on the hills and ridges; areas of Hypocalcic, Lithocalcic or Hypercalcic Calcarosols (calcareous earths) occur in the valleys with Red Sodosols along the margins of the drainage lines.
Vertosols (Cracking Clay Soils) (Ug5)	
(a) Grey and Yellow Vertosols (grey and yellow grey cracking clays)	Plains associated with functional and non-functional drainage; chief soils are Grey Vertosols (grey cracking clays) and Brown Vertosols (brown clays): associated are low domes and levees of Red and Brown Sodosols and some levees and dunes of deep Arenic Rudosols (siliceous sands); sandy elevations and plains of Calcarosols (calcareous red earths).
(b) Grey Vertosols (grey cracking clays) and Red Sodosols or Red Chromosols (red duplex soils)	Riverine plains with gilgai, localised depressions and prior streams: plains dominated by Grey Vertosols (grey cracking clays) with a mosaic of Red Chromosols; gilgai and swamps of deep Grey Vertosols (yellow-grey cracking clays) and some Brown Vertosols (brown self-mulching

Table 9.4 Continued

Dominant Soils of the Soil Landscape	Soil Landscape Description
	clays); prior stream beds of shallow Rudosols (calcareous sands) and levees of Red Chromosols; lunettes and rises of Yellow Chromosols and Ferrosols (granulated cemented clays).
Red Kandosols (Red Earths) (Gn2.1) Red Kandosols and Calcarosols (calcareous and neutral red earths)	Undulating plains with areas of dunes and low ridges: gently to slightly undulating plains of Red Kandosols and Calcarosols (calcareous and neutral red earths); northern areas often mantled with a layer of Arenic Rudosols (siliceous sands), small areas of deep Tenosols (friable loams) and Hypocalcic, Lithocalcic or Hypercalcic Calcarosols (brown calcareous earths); flat and undulating plains of Red Sodosols and assorted Chromosols; scattered ridges and hills of shallow Tenosols or Leptic Rudosols (siliceous loams) and Red Sodosols; dunes and low rises of Calcarosols (calcareous red earths), deep Orthic Tenosols (brownish sands) and Arenic Rudosols (siliceous sands); depressions and pans with Grey and Brown Vertosols (cracking clays) and Red Sodosols.

Balranald. The plain, which is almost level, is traversed by a number of prior streams having typically scalded margins and levees, and beds of shallow Hypocalcic Calcarosols. Associated with these prior streams are a number of source-bordering dunes of deep Hypocalcic Calcarosols and Arenic Rudosols, and lunettes and rises of Yellow Chromosols and granulated clays.

Yarran (*Acacia homalophylla*) and black box occur in small clumps throughout the area with the understorey dominated by bladder saltbush (*Atriplex vesicaria*) and bluebushes.

Soil Landscapes in *Atlas*: CC1–3, Ill, DD24

(e) Undulating Low Hills and Rises of Calcarosols (Calcareous Red Earths)

This extensive soil landscape is found in a broad area from Wanaaring, Bourke, Cobar and Condobolin to Hillston. The landscapes, which are slightly to gently undulating, are characterised by Red Kandosols (calcareous red earths), with small areas of Hypocalcic, Lithocalcic or Hypercalcic Calcarosols (solonised brown soils). In some areas, a deep mantle of Arenic Rudosols (siliceous sands) has been deposited on the

windward side of the ridges, reworked by wind activity to form a number of linear dunes. The scattered ridges and hills are dominated by Leptic Rudosols (skeletal soils) and Arenic Rudosols (siliceous sands). Areas of hard Red Chromosols and Red Sodosols (red-brown earths) and Grey Vertosols (grey cracking clays) occur on the plains and depressions throughout the area.

This soil landscape is dominated by mulga in the far north; white cypress pine (*Callitris glaucophylla*), bimble box (*Eucalyptus populnea*) in the central west; and mallee-belah (*Casuarina cristata*) in the south.

Soil Landscapes in *Atlas*: Mxl–5, Mx7–8, Mx35, Myl–2, My9, Fzl

(f) Plains of Calcarosols (Solonised Brown Soils)

This soil landscape occurs throughout extensive areas of western New South Wales in an area bounded by the soil landscape just described and the footslopes of the Barrier Ranges.

The plains are dominated by Calcarosols, Red Kandosols and Orthic Tenosols (brownish sands). Tracts of linear dunes and sandy rises of Arenic Rudosols (siliceous sands) and Orthic Tenosols (brownish sands) occur on the plains, with the swales and flats dominated by Calcarosols (solonised brown soils) and Red Chromosols and Sodosols. Grey Vertosols (grey cracking clays) occur in isolated depressions.

Soil Landscape in *Atlas*: DD3

(g) Stony Downs and Plateaux

(i) Stony Downs and Plateaux of Red Sodosols (Desert Loams): Rolling stony downs of crusty Red Sodosols (desert loams) occur on the footslopes of the Barrier Ranges and to the north-east of Broken Hill. These soils have a variable cover of quartz and silcrete gravel and are interspersed with gilgai and depressions of Brown Vertosols (brown clays). The scarp-slopes and crests of the ridges are dominated by deep Kandosols.

The vegetation on the downs and plateaux is dominated by scattered mulga and belah, with a shrub community of bladder saltbush and perennial bluebushes.

Soil Landscapes in *Atlas*: Nb4–5, Nb30–31, Na2–3

(ii) Stony Downs of Red Sodosols (desert loams) and Brown Vertosols (brown cracking clays): This soil landscape is similar to the preceding one except that the whole area exhibits a clearly defined gilgai pattern, Red Sodosols on the plains and the Brown Vertosols (brown clays) in the gilgai formations.

The ridges and crests are dominated by Kandosols and Calcarosols. The vegetation is also similar to the preceding soil landscape.

Soil Landscapes in *Atlas*: MM69–71

(h) Plains of Red Chromosols

(i) Plains Traversed by Present and Former Streams: This soil landscape occurs on the eastern margin of the Riverine Plain and in an area south of Nyngan. The chief soils are Red and Brown Sodosols (red-brown earths) in association with Grey and Brown Vertosols (grey and brown cracking clays), and Yellow and Brown Chromosols and Sodosols. Depressions of Brown Chromosols and Grey Vertosols (grey cracking clays) occur in association with lunettes, small rises and prior streams.

The vegetation is dominated by clumped to scattered yarran, myall (*Acacia pendula*) and black box, with an understorey of bladder saltbush, perennial grasses and herbage.

Soil Landscapes in *Atlas*: Ncl, Ocl, Oc3–5, Oc7, Oc9, Oc12–14, Ob13

(ii) Plains with Hard Pedal, Red Chromosols (Red-Brown Earths): This soil landscape is restricted to an area centred around Con-

dobolin. Chromosols and Sodosols occur on the plains, with Red Kandosols on the ridges and hills.

Hard, red alkaline and neutral duplex soils scattered throughout the area. Isolated flats and pans occur throughout this soil landscape.

The natural vegetation is similar to that of undulating low hills of Calcarosols (calcareous red earths).

Soil Landscapes in *Atlas*: Oc9, Ob13, Oc12–13

Soil Management

9.2.2

The management of ecosystems in western New South Wales is achieved through erosion prevention, by the careful manipulation of grazing rates, and reclamation, by mechanical intervention at the earliest possible indication of degradation. The management of each soil landscape group is discussed on the following pages.

(a) and (b) Sands

The siliceous, calcareous and earthy sands are highly erodible because of their massive (poorly aggregated) structure and topographic position in the landscape. Fortunately, much of the area dominated by these soils is protected from erosion by a dense vegetative cover. Dense mallee occurs on the parabolic dune-fields, and mulga and inedible shrubs occur over much of the plains. Removal of this vegetative cover by overgrazing and/or clearing results in a high erosion hazard. Under the action of the prevailing westerly winds, sand is moved by surface creep or saltation to become entrapped by fences or vegetation.

Water erosion, in the form of scalding and water-sheeting, occurs on duplex soils in the swales. Often large areas of the swales are scalded by a combination of wind and water erosion.

Safe carrying capacity has been assessed for these soil landscape units at 12 to 18 ha per sheep. Overgrazing on these sandy soils leads to a reduction in the more desirable perennial grasses and an increase in annuals. This is normally followed by an increase in undesirable woody weeds, leading to a situation of reduced infiltration, increased runoff and, ultimately, erosion.

(c) Loams

The shallow, dense loamy soils of the hills are not generally susceptible to water erosion, although runoff yields may be high. The soils around the base of the hills are often affected by gully erosion and watersheeting. This often only occurs where sheep tracks or fence lines concentrate the flow of water. Erosion on the crest of the ranges is prevented by a substantial outcropping of rocks and by the small size of catchments.

(d) Cracking Clays

The grey, brown and yellow-grey cracking clays that occur extensively in western New South Wales are highly resistant to wind and water erosion owing to the blocky structured nature of the soil. These soils may form pseudoscalded surfaces on drying out.

(e) Calcareous and Neutral Red Earths

Where these soils occur on the plains, they have higher infiltration, are more fertile and have better moisture relations than those occurring on the ridges. Around Cobar and Condobolin, these soils have been used for cultivation under dryland and waterspreadng conditions. Here they are usually uneroded because of a good ground cover. Soil Conservation Service research has shown that cultivation of these soils rapidly reduces their aggregate stability, but that this can be improved by the introduction of a pasture phase (Hunt, 1980).

The surface of these soils often sets hard after cultivation. They are also seasonally hardsetting and this characteristic gives them some resistance to erosion. The gradual increase in clayiness with depth prevents serious loss in fertility with small losses of topsoil.

Although infiltration rates are moderate, percolation through the soil profile is often reduced by an impermeable hardpan layer. This has little significance for pasture species, but may restrict the root penetration of some tree species.

These soils are not highly dispersible, so should not present major problems with cultivation where rainfall is adequate and conservative practices are followed. Normally an increasing clay content with depth ensures an adequate water-holding ability, except where a gravel layer is present.

(f) Calcareous Earths (Solonised Brown Soils)

The calcareous earths or solonised brown soils occur adjacent to the red earths and extend to the west. These soils occur on plains and dunes, being normally more erodible on the dunes. Erodibility of these soils to wind depends on their aggregate stability, surface texture and amount of carbonate. Erosion hazard also depends on management considerations such as the degree of protection by crop or pasture, wind velocity and so on. Generally, textures are coarser on the dune crest and flanks than in the swales and plains, and the soil is more susceptible to movement.

However, infiltration rate is usually higher on the dune crests compared with the swales, so that these areas respond much more rapidly to lighter falls of rain than do the heavier flats. The finer-textured soils of the flats often have a seasonally hardsetting surface that may resist water infiltration.

The soils on the dunes have a negligible amount of natural aggregation and any structure that does build up in the soil, through incorporation of residues, is quickly broken down under cultivation. Swale and plain soils have very low to moderate levels of aggregation but, like the red earths, this is destroyed by repeated cultivation. Some of the soils are highly calcareous and the profile often contains nodular limestone at the surface. In some situations, this nodular material may act to reduce wind velocity with a resulting decrease in wind erosion.

Windsheeting of these soils is common with the finer material being sifted out of the soil. If erosion is severe, these soils may be eroded down to the limestone layer to form a pseudoscald (limestone scald). These scalds can be readily reclaimed by standard techniques, but they naturally revegetate after good seasons (see Chapter 17).

(g) Desert Loams (Crusty Red Duplex Soils)

These soils (desert loams) form soft scalds when the B horizon is eroded. They can readily be reclaimed by pitting and furrowing. Usually the most appropriate method is a reduction in grazing intensity through the control of stock numbers (see Chapter 17).

Where the perennial vegetation on slightly undulating slopes has been removed by overgrazing, watersheeting and gully erosion may result.

These soils generally make good construction material because of the high gypsum content in the subsoil, which counteracts the dispersibility caused by a high sodium content (Lawrie, 1974).

(h) Red-Brown Earths (Hard Red Pedal Duplex Soils)

These soils are highly erodible to wind and water when protective cover has been removed. Both hard and soft scalds that form are very difficult to revegetate because of the unfavourable environment presented to seeds. Wind speed increases on these scalds and blows away seed, higher temperatures occur and infiltration is negligible. A hard scald is formed when only part of the A horizon has been removed, leaving an A_2 horizon. This will easily blow when cultivated. When the entire A horizon is lost, the exposed B horizon often has a nutty, mulched surface forming a soft scald. The surface is often covered by a thin surface crust of dispersible clays which is easily broken down by water. Ponding of water on these scalds will increase infiltration, leach out any salts in the soils and promote seedling establishment (Lawrie, 1974).

Infiltration rates on these soils are generally moderate, but depend on the fineness of the surface texture. They have a good water-holding capacity in the B horizon and, in some soil types, the B horizon may restrict percolation of water and lead to a perched water table. Often when these soils are cultivated there are problems with compaction on the surface.

The hard red duplex soils (red-brown earths) vary in their suitability for construction purposes. Because of their high clay content, and the fact that this clay does not expand greatly on wetting, these soils will generally hold water quite well. A gravel layer, sometimes present, may prevent these soils from holding water. Some soils have a high salt concentration that can be associated with tunnel and gully erosion.

BIBLIOGRAPHY

Atkinson, G. and Meville, M.D. (1987), 'Soils', in *Atlas of New South Wales—Portrait of a State*, Central Mapping Authority, Department of Lands, New South Wales.

Blain, H.D., Clark, J.L. and Basinski, J.J. (1985), *Bibliography of Australian Land Resources Surveys*, Divsional Report 85/14, CSIRO, Division of Water and Land Resources, Canberra.

Eldridge, D.J. (1985), *Aeolian Soils of South-western New South Wales*, Soil Conservation Service of NSW, Sydney.

Hazelton, P.A. and Atkinson, G. (1985), *Bibliography of the Soils of the Sydney Region*, Aust. Soc. Soil Science Inc., NSW Branch, Sydney.

Hunt, J.S. (1980), 'Structural stability of mallee soil under cultivation', *Journal of the Soil Conservation Service of NSW* 36, 16–22.

Isbell, R.F. (1996), *The Australian Soil Classification*, CSIRO Publishing, Melbourne.

Lawrie, J.W. (1974), 'Soils' in *Condobolin District Technical Manual*, Soil Conservation Service of NSW, Sydney.

McDonald, R.C., Isbell, R.F., Speight, J.C., Walker, J. and Hopkins, M.S. (1984), *Australian Soil and Land Survey Field Handbook*, Inkata Press, Melbourne.

McKane, D.J. and Oo, M.M. (1999), *Construction, Storage and Manipulation of the Attribute Database Established for the MDBSIS: A Relational Database for Attribute Data*, CSIRO Technical Report.

Northcote, K.H. (1966), *Atlas of Australian Soils, Sheet 3, Sydney-Canberra-Bourke-Armidale Area, with explanatory data*, CSIRO and Melbourne University Press, Melbourne.

Northcote, K.H. (1971), *A Factual Key for the Recognition of Australian Soils*, Third Edition, Rellim Technical Publications, Glenside, SA.

Northcote, K.H., Hubble, G.D., Isbell, R.F., Thompson, C.H. and Bettenay, E. (1975), *A Description of Australian Soils*, CSIRO, Wilke and Co., Clayton, Victoria.

Northcote, K.H. with Beckmann, G.G., Bettenay, E., Churchward, H.M., Van Dijk, D.C., Dimmock, G.M., Hubble, G.D., Isbell, R.F., McArthur, W.M., Murtha, G.G., Nicolls, K.D., Paton, T.R., Thompson, C.H., Webb, A.A. and Wright, M.J. (1960–68), *Atlas of Australian Soils, Sheets 1 to 10 (with explanatory data)*, CSIRO and Melbourne University Press, Melbourne.

Stace, H.C.T., Hubble, G.D., Brewer, R., Northcote, K.H., Sleeman, J.R., Mulcahy, M.J. and Hallsworth, E.G. (1968), *A Handbook of Australian Soils*, Rellim Technical Publications, Glenside, SA.

Walker, P.J. (1978), 'Soils' in *Cobar District Technical Manual*, Soil Conservation Service of NSW, Sydney.

Walker, P.J. (1991), *Land Systems of Western New South Wales*, Technical Report 25, Soil Conservation Service of NSW, Sydney.

Soil
Physical
Properties

CHAPTER 10

G.W. Geeves, B. Craze
and G.J. Hamilton

In simple terms, soil physics studies the movement of water, air, energy and physical objects within soil. Soil physical properties are the properties that control this movement. The movement of water, air, energy and other objects interacts with soil hydrology, soil chemistry, soil biology, agronomy and soil mechanics to create both opportunities and challenges for land managers. Other chapters in this book describe challenges posed by soil erosion, soil salinisation, soil acidification, soil structural degradation and soil engineering failures. The modern land manager needs a basic understanding of soil physical properties and how, when combined with appropriate theory, they can be used to predict the risk of these land management problems and evaluate potential solutions (see Figure 10.1).

Soil physical properties that are commonly required in land management investigations include soil permeability, soil water retention, soil strength, friability and penetration resistance. These properties are strongly influenced by three soil properties, namely soil texture, soil structure and clay mineralogy. This chapter explains these three soil properties and the important soil physical properties that they influence. For each physical property, suitable methods are listed or described, and spatial and temporal variability are discussed. Approaches for using the measured data to make decisions or help resolve the practical problems of soil management are also discussed. For detailed descriptions of sampling procedures and methodology, readers should consult *Methods of Soil Analysis, Part 1* (Klute, 1986) and *Soil Physical Measurement and Interpretation for Land Evaluation*, Volume 5 in the Australian Soil and Land Survey Handbook Series (McKenzie et al., 1999). The chapter concludes with a brief comment on choosing sampling techniques and methods appropriate to the purpose at hand.

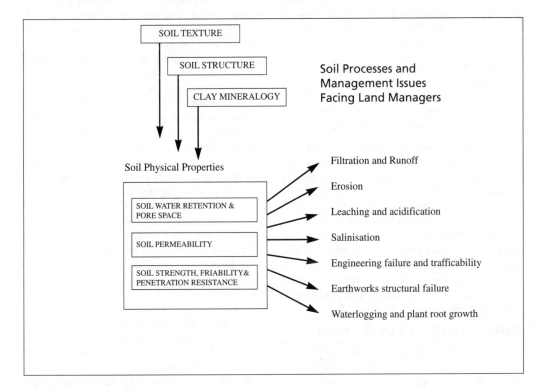

Figure 10.1 The relationship between soil physical properties and soil behaviour (boxes denote soil properties discussed in this chapter)

10.1 | Soil Texture

Soil texture refers to the fineness or coarseness of soil particles. It is determined by the proportions of sand, silt and clay that make up the 'fine earth' (< 2 mm diameter) fraction of a soil. It is an important property that can strongly influence soil physical properties such as permeability, water retention, friability and plasticity as well as chemical fertility and a host of other soil properties. It is a key property to consider in making land management decisions.

Soils are composed of aggregates of particles which contain both solid material and spaces called pores or voids. These aggregates may be subdivided into their constituent particles by chemical and mechanical means. When the constituents retain their physical and chemical identity, and cannot be subdivided further, they are called ultimate or primary particles. Ultimate particles differ in shape and size as well as in composition. For most soils, the majority of ultimate particles are less than 2 mm in size. It is conventional to call all material larger than 2 mm 'coarse fragments' and to subdivide the range of smaller particles into four principal groups as defined in the International System:

coarse fragments	greater than 2 mm	(> 2000 μm)
coarse sand	0.2–2.0 mm	(200–2000 μm)
fine sand	0.02–0.2 mm	(20–200 μm)
silt	0.002–0.02 mm	(2–20 μm)
clay	less than 0.002 mm	(< 2 μm)

Soil texture is generally measured in a laboratory using particle size analysis techniques. Alternatively, field texture can be determined in either the field or laboratory from the feel and behaviour of a moist bolus of soil according to methods described below.

Soil texture can vary markedly with depth, changing from sand to clay over a few centimetres in duplex soils. Considerable variation across an area can also be found in cases where prior streams cut across clay plains or where old erosion features have been infilled. Texture is generally regarded as a relatively stable soil property over time.

10.1.1 | Particle Size Analysis Methods

In most soil, ultimate particles are linked by physical, chemical and biological bonds. Bonding results from the great surface charge per unit volume of the smaller ultimate particles and organic matter, as well as from polymerised hydroxides of iron and alu-

minium (sesquioxides) in various stages of hydration, and other cementing agents. To disperse the particles in order that their amounts can be measured, the effects of these bonding agents have to be eliminated. All methods of dispersion involve violent agitation of the soil in a soil-water suspension and treatment with reagents such as sodium tripolyphosphate (Bowman and Hutka, 1999).

Soils high in organic matter are sometimes pretreated (prior to adding dispersing agent) with hydrogen peroxide to decrease the effects of excesses of organic matter. Likewise soils high in free carbonate, soluble salts and gypsum may be prewashed with water to prevent interference in the total dispersion of primary particles. Subplastic soils that are strongly aggregated and resist dispersion often require ultrasonic treatment to ensure complete physical dispersion. However, it is common for there to be no pretreatment, as aggregate sizes effectively operating in the field may be better approximated by not subjecting the soil to so many treatments. In all such cases it is necessary to make clear what has been done, or not done, to the soil.

After pretreatment, dispersing agent is added and the soil-water suspension shaken vigorously overnight. A mixture of dilute solution of sodium tripolyphosphate and sodium hydroxide or sodium carbonate is a good all-round dispersant. The latter two chemicals are used to bring the pH of the dispersing solution to about 8.5. Sodium hexametaphosphate can be substituted for sodium tripolyphosphate in this mixture, as they are equally effective dispersing agents. Effective sample dispersion in an aqueous solution implies that individual particles are detached from one another and are suspended in the liquid. In addition to the action of dilute concentrations of chemical reagents, aggregated particles are dispersed by various mechanical techniques using shakers or mixers that induce a shearing action or turbulent mixing of the suspension. Excessively vigorous mixing should be avoided since it may rupture individual grains. The amounts of clay and silt measured in a soil are an interaction of the method used for dispersion and the type of soil. The chemical treatment and the mechanical work done on the soil to produce clay are arbitrary. There is no absolutely 'correct' particle size distribution for a soil. The method that gives the most meaningful result will depend on the use to which the results are to be put.

There are three sedimentation techniques for measuring silt and clay fractions: pipette, hydrometer and plummet balance (a modified hydrometer).

The pipette method is the international standard. Specially constructed 20 mL pipettes are used to draw off suspension from a specified depth at various times after stirring. The appropriate time depends on the temperature of the suspension and the size fraction being determined. The samples are dried and weighed to determine the mass of the sediment. Correction must be made for the mass of dispersing agent contained in solution. Because this method measures directly the amount of suspended material, it is used as the standard of comparison for more rapid methods, for example the plummet balance and hydrometer methods.

The plummet balance and hydrometer methods measure the specific gravity of the suspension of soil in water, which depends on the amount and specific gravity of the suspended material and on the salt content. Both the plummet balance and the hydrometer can be calibrated to read directly in percentage of material in suspension.

Once determined, the clay and silt fractions are removed by decantation, leaving behind the sand fractions, which can then be sieved into fine and coarse grades and the amount of each determined by weighing. The gravel fraction is separated during the initial grinding of the air-dried field sample, and its percentage of the total field sample determined.

In addition to these sedimentation techniques there are automated methods that measure the density of the dispersed soil suspensions using laser light and X-rays. The devices used have generally been designed for quality control in industrial and manufacturing processes where particle size varies within a narrower range and the nature of the material is relatively constant. When these devices are used for

Plate 10.1   Particle size analysis using the hydrometer method—the different heights of the hydrometers indicate different clay contents

analysing natural soils there is potential for error resulting from variation in particle density, particle shape and clay mineralogy. When used correctly, with appropriate caution and calibration against the pipette method, these devices offer potential for rapid and convenient particle sizing.

10.1.2 Expressing Particle Size Data

The particle size distribution can be presented as a cumulative particle size distribution curve or grading curve (see Figure 10.2). Alternatively, the amount of clay, silt and sand can be interpolated from the curve and presented in simpler form as three or four percentages. A further simplification can be made by plotting the result on a texture triangle with vertices corresponding to sand, silt and clay as in Figure 10.3, thus identifying the texture of a sample with a single texture class. This figure is derived from Marshall's diagram, which showed the result of his analysis of Australian soils (Marshall, 1947). It differs slightly from similar overseas diagrams because it gives less prominence to the silt fraction. This diagram can be used to find rough percentages for field textures. It omits some texture classes, and further subdivisions into light, medium and heavy clays are often used.

Field Approximation of Soil Texture/Field Texture Grades

10.1

When particle size data are not available, soil texture can be approximated by determining field texture grade (McDonald et al., 1990). This is determined by first taking a sample of soil (excluding particles greater than 2 mm in diameter) sufficient to fit comfortably into the palm of the hand. The sample is moistened with water, a little at a time, and kneaded until the ball of soil just fails to stick to the fingers. More soil or water may be added to maintain this condition which is known as the sticky point. It approximates field capacity for the soil. Kneading and moistening are continued until there is no apparent change in the soil ball. This usually takes one to three minutes. The soil ball, or bolus, is now ready for shearing manipulation. The behaviour of the soil during bolus formation is indicative of its plasticity type. The behaviour of the bolus and the ribbon, produced by shearing (pressing out) between thumb and forefinger, characterises field texture with 15 grades recognised (see Table 10.1). While McDonald et al. give approximate clay content for each field texture grade, the relationship between field texture and particle size distribution is only an approximate one, as

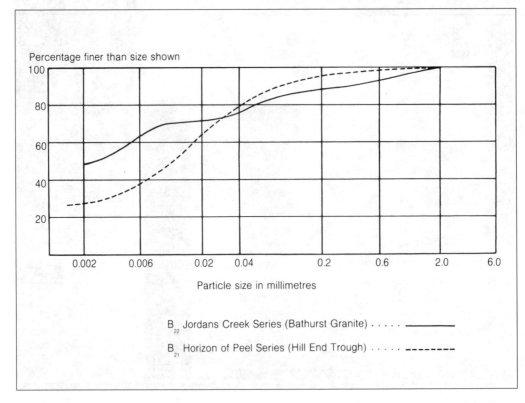

Figure 10.2 Grading curves of subsoils of two yellow duplex soils, the Peel and Jordans Creek series from Bathurst

field texture is influenced by organic matter, clay type, presence of carbonate, cations on the clay, and iron and aluminium oxides.

Inferring Soil Properties from Soil Texture

Soil texture is a relatively simple and widely understood basis for describing soils. It can provide useful indications about soil properties, expected soil behaviour and soil profile development, so it features strongly in most classification systems for soil profiles and soil materials. Because of the potential for correlation between soil texture and other soil properties, it is often used as the basis for the discussion of soils by farmers, agronomists, soil conservationists and research workers, and generalisations are often made.

In relation to soil erosion and soil conservation, it is often inferred that coarse materials such as coarse sand, gravel and stones result in high soil permeability, low potential for runoff and greater soil surface stability against erosion by low-velocity water flows. This coarse material is generally too permeable for constructing water storage structures unless a clay or plastic lining is used. The behaviour of fine material (silts and clays) depends strongly on the degree of aggregation. Fine material that is strongly aggregated may also be too permeable to hold water. Silts are materials of low plasticity; lumps of silty material will break down readily in the presence of water. Soils made up primarily of this component are often quite susceptible to surface sealing and to the erosive action of water, owing to the combination of fine particle size and the relatively weak attraction between particles. Clays have low resistance to deformation when wet, but when dry they offer much higher resistance than do silt particles. In contrast to silts, which may or may not be impervious to water, clay soils can be expected to be rather impervious to water, except where aggregated.

While such generalisations can be useful, it is necessary to remember that soil structure and clay mineralogy also affect soil physical properties, and their influence can override effects of soil texture. For example, a strongly aggregated and stable clay soil may have high permeability whereas a massive sodic clay soil may be effectively impermeable. Likewise, the presence of macropores created by biological activity can also drastically change hydraulic properties.

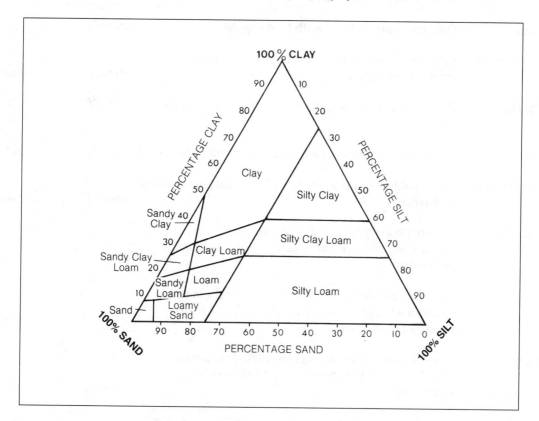

Figure 10.3 Triangular texture diagram based on international fractions (after Marshall, 1947)

Table 10.1 Field soil texture classes (after McDonald et al., 1990)

Texture Class	Bolus Behaviour	Ribbon Length (mm)	Approx. Clay %
Sand	Coherence nil to very slight; sand grains of medium size; single sand grains adhere to fingers	Can't be moulded	< 5
Loamy Sand	Slight coherence; sand grains of medium size	About 5	About 5
Clayey Sand	Slight coherence; sand grains of medium size; sticky when wet; many sand grains stick to fingers; discolours fingers with clay stain	5 to 15	5–10
Sandy Loam	Coherent but very sandy to touch; dominant sand grains are medium size and readily visible	15 to 25	10–20
Loam	Coherent and rather spongy, smooth feel with no obvious sandiness or silkiness; may be somewhat greasy to the touch if much organic matter present	About 25	About 1
Silty Loam	Coherent; very smooth to often silky	About 25	About 2
Sandy Clay Loam	Strongly coherent; sandy to touch; medium-sized sand grains visible in finer matrix	25 to 40	20–30
Clay Loam	Coherent plastic bolus smooth to manipulate	40 to 50	30–35
Clay Loam, Sandy	Coherent plastic bolus; medium-sized sand grains visible in finer matrix	40 to 50	30–35
Silty Clay Loam	Coherent smooth bolus; plastic and often silky to touch	40 to 50	30–35 silt > 2
Light Clay	Plastic bolus; smooth to touch; slight resistance to shearing	50 to 75	35–40
Light Medium Clay	Plastic bolus; smooth to touch; slight to moderate resistance to shearing	About 75	40–45
Medium Clay	Smooth plastic bolus; handles like plasticine, can be moulded into rods without fracture; moderate resistance to shearing	> 75	45–55
Medium Heavy Clay	Smooth plastic bolus; handles like plasticine, can be moulded into rods without fracture; moderate to firm resistance to shearing	> 75	50+
Heavy Clay	Smooth plastic bolus; handles like plasticine, can be moulded into rods without fracture; firm resistance to shearing	> 75	50+

10.2 Clay Mineralogy

Soil texture describes only the size distribution of primary mineral particles, not what they consist of. For larger primary particles in the coarse fragment sand and silt range, the mineralogy of particles has relatively little influence on soil physical behaviour. However, the physical properties of clays are strongly controlled by the effective surface area and charge density, and this can vary widely with mineralogy. A highly simplified summary of physical properties of some common clay types is presented in Table 10.2. The three common layer aluminosilicates—kaolin, illite and smectite—all consist of similar tetrahedral silica layers and octohedral aluminium hydroxide layers. The differences in physical behaviour result largely from expansion of the 2:1 layer clays, allowing greater adsorption of cations and water onto the large area of interlayer clay surfaces relative to the 1:1 layer clays. This expansion significantly increases soil water retention, promotes shrink-swell behaviour that creates macroporosity, and influences potential for clay dispersion by increasing cation exchange capacity.

While clay mineralogy is a basic soil property that can have a strong influence over some soil physical properties, it is expensive and not routinely measured. Clay mineralogy can be determined semi-quantitatively by X-ray diffraction (XRD) analysis in combination with X-ray fluorescence (XRF) analysis of a sample of clay obtained during particle size analysis, but these are costly specialist techniques that are often reserved for research purposes. Useful inferences about clay mineralogy can be made from measurements of cation exchange capacity but this can be misleading if organic matter is contributing significantly to the measurement. It may often be easier to directly measure the soil physical property of interest rather than infer it from measured clay mineralogy (for example, Chapter 11 explains how soil plasticity can be assessed by measuring Atterberg limits and how shrink-swell behaviour can be assessed by measuring linear shrinkage. Both of these properties are related to clay mineralogy).

Table 10.2 Generalised properties of some major clays

Clay Minerals	Structure	Cation Exchange Capacity (cmol/kg)	Shrink-Swell Potential	Other Comments
Layer aluminosilicates				
Kaolin	1:1 layer	1–10, partly pH dependent	low	low plasticity
Illite	2:1 layer	10–30	intermediate	moderate plasticity
Smectite	2:1 layer	80–150	high	high plasticity
Interstratified minerals	2:1 layer, 1:1 layer	variable with composition	variable	variable
Oxide minerals				
Goethite	crystalline	pH dependent	low	low plasticity
Hematite	crystalline	pH dependent	low	low plasticity
Amorphous aluminosilicates				
Allophane	amorphous	20–50, pH dependent	low	

Soil Structure

Soil structure refers to the arrangement of clay, silt, sand and gravel (the primary particles) and organic matter in a soil. More importantly, it also refers to the arrangement of pores lying between these particles. It is these pores that provide for storage and movement of water and air within soil. Kay (1990) clarified discussion of soil structure by defining three separate aspects. Soil structural *form* refers to the current arrangement of soil aggregates and pores. Soil structural *stability* measures the change in structural form occurring under an external stress such as tillage. Soil structural *resilience* refers to the tendency for a soil to regain its previous structural form by natural processes after its structural form has been disrupted.

10.3.1 Soil Structural Form

In most soils there is some grouping of primary particles into units referred to as aggregates or peds (see Figure 10.4). Aggregates can be bound together by electrostatic forces such as charges on clay minerals, cementing substances from organic matter or hydroxides of iron, or binding strands of fungal hyphae and roots. Soils with a high degree of aggregation will generally have a high degree of porosity formed by voids between packed aggregates. Where the primary particles are not bound together into aggregates and there is a reasonable range of primary particle sizes, then soils can pack to form a dense soil structural form with low porosity. Desirable soil structural form is dependent on the use to which the soil material is being put. For agricultural enterprises (other than rice growing), an aggregated, low density/high porosity condition with reasonable permeability is generally preferred. For the lining of a water supply dam, a massive, high density/low porosity condition with very low permeability is preferred. Soil management can often be altered or special treatments applied to promote the appropriate soil structural form. There are both direct and indirect techniques for measuring and describing soil structural form.

Soil structural form can be directly described by visual observations of the size and shape of peds and visible pores in a shallow soil pit or undisturbed soil core. This description is part of standard soil profile descriptions and is done using techniques described in Chapter 5 and more fully in McDonald et al. (1990). There is also a range of laboratory techniques, including wet and dry sieving, elutriation and sedimentation, that nominally determine the size distribution of aggregates, but these generally involve disruptive forces that significantly change the property they are measuring. A back-correction for aggregate abrasion caused by sieving can be incorporated into the dry sieving method (Kemper and Rosenau, 1986), but the remaining techniques are best regarded as measures of aggregate stability rather than aggregate size.

Direct measurements of the size and shape of peds and pores can also be made in the laboratory on undisturbed soil samples that have been impregnated

(a) Structureless condition
(single grain or massive)

Particles are packed as close together as possible. High density, low porosity

(b) Structureless or aggregated condition

Aggregate

Transmission pore

Water storage pore

Particles are formed into aggregates and are loosely packed. Low density, high porosity

Figure 10.4 Soil structure — the arrangement of primary particles

with resin and subsequently cut into thin sections. While these thin section techniques are not used routinely they have several advantages. First, they allow direct measurement of pores as well as aggregates. Second, examination of adjacent thin sections can provide information in three dimensions that allows an operator to assess the degree of connection between pores, which is important in determining their effectiveness at transmitting water and air. Third, they allow the use of microscopes capable of measuring structural features that are not discernible with the eye but are very important to soil behaviour. Fourth, they are suited to photographic recording and automated image analysis techniques.

Soil structural form can also be described through indirect measurements of total porosity and porosity in different size classes. Total porosity, which is the total volume of pore space in dry soils, generally varies between 30% and 60%. The total porosity of soil can be calculated directly from bulk density (BD) and specific gravity (SG) of the soil particles (usually assumed to be 2.65 g/cm³), using Equation 10.1. Bulk density is commonly determined from the oven dry weight of a soil sample taken using a thin-walled sampler of known volume driven into the soil and trimmed. In stony soils it may be difficult to drive a thin-walled sampler into the soil without damage to both soil and sampler. In these soils a sample can be obtained from a shallow excavation and the volume of the excavation determined by filling it with a measured volume of water contained in a plastic liner (Cresswell, 1990a).

$$\text{Porosity \%} = (1 - BD/SG) \times 100 \quad \ldots\ldots\ldots 10.1$$

Plate 10.2 Subangular blocky soil structure

Total pore space is made up of space in pores ranging over many orders of magnitude in size. Because capillarity depends on pore size, pores of different sizes tend to serve different functions (see Table 10.3). Large pores, or macropores (that is, pores of diameter greater than 0.05 mm), contribute largely to infiltration and drainage under saturated surface conditions and, because they drain freely, aeration. Aeration porosities of less than 10% are restrictive to root proliferation. Smaller pores are required for water storage. The size distribution of pores is clearly important to the land manager, and it can be estimated from the soil water characteristic,

Table 10.3 A generalised view of pore size groups and functions (columns one and two adapted from Greenland, 1981)

Pore Size (μm)	Function	Equivalent Particle or Aggregate Size[a] (μm)	Equivalent Soil Water Potential[b] (m)
> 500	Aeration and water transmission	> 1600 (mostly gravel size, some coarse sand size)	< 0.06
50–500	Water transmission (infiltration, permeability)	160–1600 (mostly coarse sand size, some fine sand size)	0.06–0.6
0.5–50	Storage (water-holding capacity, plant-available water)	1.6–160 (mostly silt and fine sand size, some clay size)	0.6–60
< 0.5	Residual	< 1.6 (mostly clay size)	> 60

[a] Equivalent particle size = 3.2 × pore size (assuming spherical, uniform-size particles)
[b] Pore diameter (mm)=3.0/Soil water tension (cm)

or water content–water potential curve discussed later in this chapter.

10.3.2 Soil Structural Stability

The arrangement of peds and pores (structural form) within a soil can change drastically under stress such as heavy rainfall or vigorous tillage. Current soil structural form may be ideal for current use but if it has low stability it will easily lose its ideal structural form. Two contrasting situations illustrate the importance of high structural stability. For dryland cropping it is generally agronomically desirable to maintain a lower density soil that contains stable aggregates and a good network of pores. These aggregates should be resistant to the disruptive forces of raindrop impact on dry soil or tillage and stock compaction on wet soil. They will maintain higher infiltration rates and will be less prone to surface waterlogging, cloddiness, root growth restriction, seedling emergence problems and hardsetting. In contrast, banks of some soil conservation earthworks are purposely constructed with high density and very low porosity. While aggregation may be undesirable, the soil should be resistant to dispersion that can lead to tunnelling where seepage flows occur.

When considering soil structural stability as a limiting factor for plant growth or soil preparation, there is a range of aggregate stability tests available. Most of these tests measure susceptibility to one or other or both of the two important processes involved in aggregate breakdown under wetting, namely slaking and dispersion. When a dry aggregate is wetted, water enters from all sides as a result of the moisture potential gradient, and encloses air within the aggregate. If the mechanical strength of the aggregate is sufficiently low, the compressed air shatters it, generally into smaller aggregates. This process is called slaking. A second process whereby water reacts with sodium and other cations on the clay exchange causing it to collapse completely and move into suspension as a cloud of 'muddy water' is called dispersion. Soil dispersion represents the spontaneous deflocculation of the clay fraction of a soil in water. Slaking and dispersion may occur independently or together, providing there is sufficient water present. Dispersion in topsoils is generally undesirable as it leads to poor physical conditions for plant growth, such as surface crusting, cloddiness, low water intake, poor drainage, waterlogging, low aeration and poor emergence of plants.

Plate 10.3 Apparatus for determination of water-stable aggregation of soils. The photograph shows the water bath in which the nests of sieves containing the soil aggregates are oscillated vertically, under standard conditions.

A commonly used method that combines both mechanical stress and stress from soil wetting is aggregate stability determined by wet sieving (Kemper and Rosenau, 1986). Plate 10.3 shows the apparatus used to oscillate initially dry aggregates under water on a 0.25 mm sieve. When sieving is complete the percentage by dry weight of original aggregates remaining on the sieve (after adjusting for the amount of primary particles greater than 0.25 mm) is reported. For most soils, stress resulting from rapid wetting of the initially dry aggregates far exceeds the additional mechanical stress of wet sieving and most breakdown occurs in the first minutes. Geeves et al. (1995) found values for lighter-textured soils in the wheat-belt ranging from 5% to 89% with higher values under woodland and pasture.

Aggregate breakdown under rainfall is a key process affecting surface soil structure under cropping. Numerous tests have been proposed measuring breakdown of aggregates and/or change in surface hydraulic conductivity under simulated rainfall (for example, Young, 1984; Loch, 1994). While these tests require specialised rainfall simulators, they test structural stability under a more realistic set of external forces than wet sieving.

Other laboratory techniques measure the stability of pores to wetting (for example, the improved High Energy Moisture Characteristic method of Pierson and Mulla, 1989). However, these methods are not generally convenient and so the commonly used routine methods generally measure the stability of individual aggregates and require the user to infer the stability of pores.

When considering the stability of soil in the banks of conservation earthworks, soil dispersion is often the major concern. Dispersion in the presence of seepage flow can create a serious hazard of tunnel erosion of soil conservation earthworks. A number of measures can be used to assess dispersibility and the likely development of tunnel erosion in farm dams. The most widely used tests are exchangeable sodium percentage (ESP), dispersion percentage (DP) and Emerson Aggregate Test (EAT). Although the tests are interrelated, they each reflect specific aspects of soil structural stability in relation to predicting soil behaviour.

Exchangeable Sodium Percentage (ESP)

Exchangeable sodium percentage is defined as the percentage of sodium on the exchange complex divided by the cation exchange capacity, both in units of centimoles of positive charge per kilogram of dry soil. It is routinely measured in soil chemical testing but can also be estimated in the field using simple devices such as the Soil Sodicity Meter (Cooperative

Research Centre for Soil and Land Management, Glen Osmond, SA). It is a useful measure because the main characteristic of a clay governing the susceptibility to dispersion is the quantity of adsorbed sodium cations on the clay surface (exchange complex) relative to the quantities of other main cations (calcium, magnesium and potassium). The relatively large size and single charge of sodium ions tends to decrease the attractive forces between clay particles, thus encouraging dispersion. A second factor governing susceptibility to dispersion is the total content of dissolved salts in the water. Natural catchment waters are usually well below the level necessary for flocculation in many soils.

Referring to Figure 10.5, clays with ESP of 6–14% have been found to be moderately dispersible and have been associated with the tunnelling failures of dams when the dam water is relatively pure. Clays with ESP of 15% or more may be highly susceptible to tunnelling and soil surface sealing. The figure also shows that a very high electrolyte concentration in the percolating water is necessary to keep a high ESP

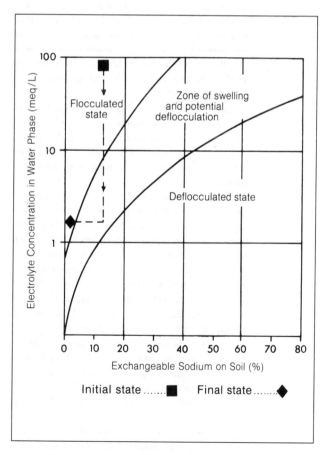

Figure 10.5 Factors affecting deflocculation dispersion (after Moore et al., 1985)

soil in a flocculated condition. The dashed line in this figure illustrates a common situation where tunnel erosion failures occur (Ingles et al., 1969). The soil starts out at the end of earthwork construction in a stable flocculated state, and final equilibrium should also be in a stable flocculated state. However, following initial dam filling, the soil may pass through a state of deflocculation and tunnelling could develop.

Soils with ESP of 6% or more within the profile are termed sodic and are potential structurally unstable soils.

Dispersion Percentage (DP)

This test estimates the amount of soil material that is easily dispersible in water. The dispersion percentage is the ratio of soil material less than 5 μm after limited mechanical dispersion without dispersants, compared with the total soil material less than 5 μm determined in the particle size analysis. The method is a useful, quick way of determining the degree of dispersion and tunnelling susceptibility of soils.

As it is necessary to know the total soil material less than 5 μm, this is estimated in conjunction with the particle size analysis. Measurements of the less than 5 μm fraction are made at the appropriate time of settling either by plummet balance, pipette or hydrometer. The fraction size measured is the same as the US Soil Conservation Service dispersion test (Decker and Dunnigan, 1977) and Australian Standard AS 1289 (1997).

The test was developed in order to identify soils that, when used in earthworks, are susceptible to failure by tunnelling. Such a failure is one that results from post-construction deflocculation (dispersion) and subsequent accelerated removal of the dispersed material.

There are a number of aspects that must be considered in respect of the interpretation of the dispersion percentage test. Failure of earthworks by tunnelling is a complex process depending on variables such as the density and moisture content of the soil, soluble salt content of the stored water and rate of filling, as well as on the properties of the soil (Rosewell, 1970). When limits are set, based on a dispersion percentage test, there must be a range of uncertainty in the prediction of earthwork performance as such limits are based on average conditions of construction and filling of the earthwork. In addition, the majority of soils that have been tested, and on which experience has been based, have been clays of low plasticity with clay contents in the range 25–50%.

There is no one value for dispersion percentage to indicate tunnelling susceptibility (see Table 10.4). Critical values depend on the amount of clay in the soil and its plasticity.

Emerson Aggregate Test (EAT) and its Derivatives

This simple test, originally devised by Emerson (1967), involves observation of the behaviour of an air-dried crumb of soil when it is placed in distilled water. The interaction in water of clay-sized particles in aggregates may largely determine the structural stability of the soil. The test procedure has been standardised by the Standards Association of Australia (AS 1289 (AS 1289, 1997), and a refined version is presented by Emerson (1999). From the results of this test, the soil is classified on a scale from 1 to 8 (see Figure 10.6).

The first separation of aggregates is made according to whether the dry aggregates break up (slake) when immersed in distilled water in a beaker. Most aggregates will probably slake owing to the stresses induced by entrapped air and by swelling. These aggregates are placed in classes 1 to 6. Those that do not slake are divided into two classes: Class 7 aggregates swell but remain coherent; Class 8 aggregates remain unchanged.

Once aggregates are immersed in water, an osmotic stress arises between the negatively charged clay particles. The stress increases as the soluble salts initially present in the aggregates diffuse out. The increase may be sufficient to cause dispersion of the clay.

Class 1 aggregates disperse completely, leaving only sand grains in a cloud of clay. Those of Class 2 show only partial dispersion. The dispersion of aggregates can be checked rapidly if the finest slaked fragments are observed.

Clay aggregates that are saturated with calcium and magnesium do not disperse when initially placed in water. However, clays that are saturated with these two ions, and are also capable of swelling significantly, will 'disperse' when shaken in a suspension of water. Under similar treatment, calcium-saturated kaolinite (a non-swelling clay) will remain aggregated or flocculated in a suspension of water. When a clay has been shaken in a suspension, energy has been applied to the system to aid in dispersion. There is also a certain water content at which dispersion will take place. This water content will be intermediate between the maximum water uptake of an initial dry clay and that corresponding to a suspension.

These principles are the basis for separating classes 3, 4, 5 and 6. Class 3 aggregates are defined as aggregates that, after remoulding at a water content equivalent to field capacity, show dispersion when immersed in water.

The clay present in the aggregates may still not disperse if there are minerals in the aggregates that dissolve rapidly enough to maintain the divalent ion

Table 10.4 Critical values for interpreting dispersion percentage (after Perry, 1975, and Decker and Dunnigan, 1977)

Total Fraction < 5 μm	USCS*	Critical Dispersion Percentage
25		20
30		33
35		42
40		50
50		60
	ML & SC	> 25
	CL	> 35
	CH	> 40

* Unified Soil Classification System (see Chapter 11)

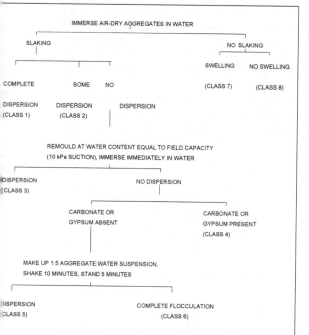

Figure 10.6 Emerson Aggregate Test classes

concentration above the flocculating or aggregating threshold. Thus, aggregates that do not disperse after remoulding at field capacity, but contain such minerals as calcite or gypsum, are placed in Class 4. Aggregates that disperse at a water content intermediate between field capacity and that of a suspension (1:5 aggregate:water ratio, ten-minute shaking) form Class 5. If a suspension of aggregates flocculates completely after five minutes' standing, the aggregates are placed in Class 6.

The test should be carried out on an air-dry aggregate 5 to 10 mm in size. The aggregate should be placed in 75 mL of distilled water contained in a 250 mL beaker. Subsamples should be tested in triplicate for Classes 1 and 2, and singly from each

reworked sample for Class 3. Dispersion should normally be rated after two hours and 20 hours, with the latter being used for interpretative purposes.

Reworking (Class 3) of a sample is carried out by mixing about 40 g of soil in an evaporating dish, using a spatula and working for 30 seconds at a moisture content in the plastic range. The wet sample should not be touched by hand. A piece of 5 mm glass tube and perspex plunger can be used to extrude a suitable sample.

Subdivision of Classes 2 and 3 is as follows, giving three subclasses for Class 2 and four subclasses for Class 3 (after Loveday and Pyle, 1973):

(1) slight milkiness, immediately adjacent to the aggregate
(2) obvious milkiness, less than 50% of the aggregate affected
(3) obvious milkiness, more than 50% of the aggregate affected
(4) total dispersion, leaving only sand grains.

The test has also been used to assess the structural stability of agricultural soil. A modified test known as ASWAT with a more convenient scoring time of 10 minutes has been proposed by Field et al. (1997) for Vertosols.

Application of the Emerson Aggregate Test and its Derivatives

10.3.3

Although the test is subjective, the Emerson test gives a useful first sorting of soils according to the observed coherence of their aggregates in water. The results are particularly useful in relation to the behaviour of tilled surface layers. Dispersion of surface soils can be prevented by applying gypsum, and Loveday (1974) found that the response to gypsum, including increased hydraulic conductivity, can be predicted from the modified test. In the context of irrigation, it would be desirable to make the observation both in distilled water (equivalent to rainwater) and in the relevant irrigation water, or its synthetic equivalent, because the electrolyte content of the water influences the degree of dispersion.

In the field, surface soil aggregates are mechanically stressed during cultivation and also under intense rainfall. If surface soils containing Class 3 aggregates are cultivated wet, then the clay present is liable to disperse when water is applied to the surface, resulting in crusting and possibly reduced germination of crops.

By applying gypsum to these soils, so as to maintain an appropriate calcium ion concentration in the soil solution, crusting can be reduced. The amount of gypsum required would be the amount sufficient to make the aggregates behave as Class 4 aggregates.

In neutral to acid soils, the application of excess calcium carbonate could also be effective as a long-term treatment, if the applied carbonate is subsequently reprecipitated throughout the aggregates.

For many years the original Emerson test has been used in combination with dispersion percentage and other tests to evaluate suitability of soils for conservation earthworks and water storages. In particular, it identifies high susceptibility to tunnelling (Class 1) and susceptibility to leakage due to strong aggregation (Class 6). The full criteria used and the prescribed modifications to standard construction methods are defined in Table 20.8.

10.3.4 Soil Structural Resilience

Soil structural resilience refers to the extent to which natural processes that occur within soil can re-establish the previous soil structural form after it has been disrupted. A major factor influencing structural resilience is the potential of the clay fraction to reform aggregates through repeated shrinkage and swelling owing to wetting and drying cycles. Chapter 11 describes the linear shrinkage test and other methods for measuring shrink-swell potential for engineering purposes. These methods can be used to identify structurally resilient soils but it is more common to rely on inferences from field observations of soil cracking, self-mulching, gilgai features and presence of significant depths of medium to heavy clay.

Cultural Practices Affecting Soil Structure　　10.3.5

Soils of New South Wales vary widely in their structural form, stability and resilience and in their response to cultural practices. High clay content Black Vertosols, for instance, generally have well-developed and relatively stable structure, and their strong tendency to shrink and swell with changing water content can restore structural form. Many of the soils with coarser-textured surfaces, including Chromosols, Kurosols, Sodosols, Dermosols, Kandosols and Podosols (Isbell, 1996), have weakly developed, and relatively unstable, structure that can be difficult to restore. In addition, these soils are subject to widely varying climatic conditions and are used for a wide variety of purposes. These sources of variation make generalisations about cultural practices and soil structure difficult but the following is clear.

For coarser-textured surface soils it is important to avoid excessive tillage which reduces organic matter content and reduces structural stability.

For clay soils it is important to:
—use gypsum and lime applications to maintain an appropriate cation balance to reduce dispersion
—avoid compaction caused by tillage and uncontrolled traffic on soil wetter than the plastic limit.

A detailed summary of effects of farming practices on soil structure in the New South Wales wheatbelt is presented in Chapter 16.

10.4 Soil Water Retention and Pore Space

As indicated in Table 10.3 soil pores range widely in size, and the strength with which they can retain water varies correspondingly according to their size. A graph of water retained versus water potential (for example, the graphs in Figure 10.7) is called the soil water characteristic. The soil water characteristic is an important soil property because it controls the availability of water for plants and influences the movement of water through soil.

Defining the soil water characteristic requires simultaneous measurements of volumetric soil water content and soil water potential. Volumetric soil water content is the volume of water held at a particular time by a certain volume of soil. It can be

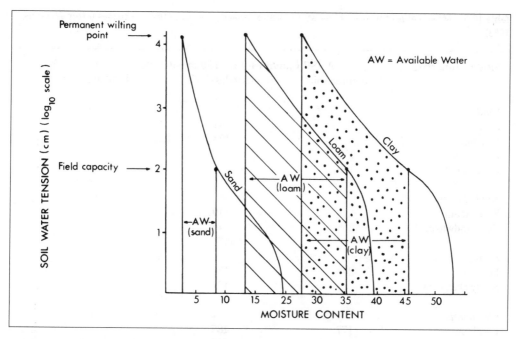

Figure 10.7 Soil water tension and moisture content curves showing field capacity, permanent wilting point and available water (after Salter and Williams, 1967, 1969)

expressed as a fraction or percentage of the volume or as a depth of water per depth of soil (that is, mm/m). It can be determined for field soil directly by sampling a known volume of soil (usually in a thin-walled metal core), weighing it before any water is lost from the sample, drying the soil sample overnight at 105°C and reweighing the sample. The loss in weight measures the original weight of water in the soil, and the gravimetric soil water content is calculated as the weight of water lost over the weight of dry soil. Volumetric water content is then calculated as gravimetric soil water content multiplied by bulk density.

Soil water potential can be measured in the field using tensiometers, gypsum blocks (Aitchison et al., 1951), and filter papers (Greacen et al., 1989). However, soil water characteristics are commonly determined in the laboratory by placing wet core samples on a tension plate or in a pressure chamber and measuring the volumetric water content at various stages as the water potential is reduced. Cresswell and Paydar (1996) have shown that soil water characteristics from a range of cropping soils follow log-linear relationships and therefore can be reasonably defined by measurements at only two water potentials, thus significantly reducing the cost of measurement.

The soil water characteristic is also an important indirect way of describing soil structural form because it can be used to estimate the number of soil pores in each size range using Equation 10.2.

$$\text{Pore diameter (mm)} = \frac{0.03}{\text{Soil water potential (m)}} \qquad 10.2$$

Hysteresis is an important property of soil water characteristics. The angle of contact between soil water and the surface of the surrounding soil pore will generally depend on whether the soil has reached its current water potential by getting wetter or by getting drier. So a soil water characteristic measured on a soil that is wetting up generally shows a lesser water content at a given water potential than a characteristic measured on the same soil drying out. While this effect can significantly affect water movement in soil, hysteresis is often ignored and soil water characteristic is generally determined only for drying soil. This saves the effort of additional measurements and avoids complexities in modelling the effect mathematically.

Available-Water Storage Capacity (AWSC)

The available-water storage capacity of soil is the volume of water that can be stored by soil that is available to plants. It is defined as the difference between the soil water content after drainage (field capacity) and the soil water content when plants have extracted all available water and have therefore permanently wilted (permanent wilting point).

Field capacity was originally defined as the water content measured in the field after wetting a soil and

Table 10.5 Mean values for available-water storage capacity for some texture classes (after Williams, 1983)

Field texture class	AWSC (mm/m) (-1 to -150 m potential)		
	Structured soils	Structureless soils	All soil in group
Sands			136
Sand		141	
Fine sand		153	
Loamy sand	155	123	
Clayey sand	140		
Sandy loams			155
Sandy loam		119	
Fine sandy loam	265	146	
Loams			158
Loam	243	136	
Sandy clay loam		127	
Clay loams			169
Clay loam	177	185	
Silty clay loam	127	127	
Light clays			138
Sandy clay		167	
Silty clay		92	
Light clay	115	140	
Medium to heavy clays			115
Medium clay	122		
Heavy clay	115		
Self-mulching clays			214
Self-mulching clays	214		

allowing it to drain for a fixed period; however soils drain at varying rates and drainage can continue for much longer than this time period. It is therefore common practice to determine the upper limit of available-water storage capacity in the laboratory by applying a fixed soil water potential, usually −1 m, to small core samples. Conventionally, permanent wilting point is generally determined by applying a soil water tension of −150 m to small soil cores. Direct observation of plant response can also be used. The soil water characteristic can be used to determine plant-available water, as in Figure 10.7.

Soil texture and structure both influence soil water retention and available-water storage capacity but the relationships are complicated by variation in organic matter levels, soil management history, clay type and soil structural form. Table 10.5 presents mean values for a wide range of Australian soils. Caution should be used when applying these values, and laboratory measurements should be made for any critical applications.

It is interesting to note that while clays often have high field capacities, much of this water is unavailable to plants, and available soil water capacity can actually be less than that for medium-textured soils. Depth of soil directly affects the amount of soil water available to plants as does rooting depth of the plants concerned.

Soil Permeability

Soil permeability is the capacity of soil to conduct gases and fluids, especially water. Many soil management issues such as dryland salinisation, soil acidification and water erosion are closely linked with movement of water through soils, and therefore permeability is a key soil property to know and understand.

Hydraulic Conductivity

The property generally used to measure soil permeability to water is hydraulic conductivity (labelled K). It is effectively the rate at which gravity moves water vertically through a soil layer with no flow restrictions above or below. Hydraulic conductivity depends on soil structure and, in particular, on the number of connected pores, the size of their narrowest sections and the tortuosity or windiness of the connected pathways. Pore connections are important because only pores that are connected in some way to both ends of a soil core can conduct water. The conductivity of a pore is assumed to be related exponentially to the pore radius, and hence hydraulic conductivity is very dependent on the frequency and distribution of larger pores.

Hydraulic conductivity also depends on soil water potential. When soil is saturated all of its pores are completely full and able to conduct the maximum flux of water (known as K_{sat}). As water potential decreases, and soil becomes drier, larger pores begin to partially empty. As they fill with air there is less cross-sectional area able to conduct water and the path the water must follow becomes a longer, more tortuous path, so hydraulic conductivity declines. The graph of hydraulic conductivity K versus soil water potential ψ is called the unsaturated hydraulic conductivity function $K(\psi)$. While hydraulic conductivity of relatively dry soil can sometimes be important for predicting drainage it can be difficult to measure, and most effort is generally focused on measuring K near saturation.

A wide range of methods has been used to measure hydraulic conductivity and each can offer advantages under certain conditions. In the field, saturated hydraulic conductivity (K_{sat}) can be measured using positive head infiltrometers, well permeameters, sprinkling infiltrometers, or inferred from rainfall simulations. Unsaturated hydraulic conductivity may be measured using tension infiltrometers or by cali-

brating soil water movement models to match field measurements of soil water content and soil water potential over the duration of infiltration and drainage events. In the laboratory hydraulic conductivity is generally measured on intact soil cores sampled in the field. Saturated hydraulic conductivity is determined by supplying water to one side of the saturated core with a known hydraulic gradient across the core and measuring the water flux. Unsaturated hydraulic conductivity can be measured in a similar way but with the water supply restricted by placing it under negative water potential or by reducing the metered flow.

Like many soil properties, hydraulic conductivity can be highly spatially variable and may range widely within a paddock under uniform management. Compaction under tractor wheels or stock tracks can lead to very low conductivity ($K_{sat} < 0.1$ mm/h), while growth of tap-rooted crops or macro-faunal burrowing activity can lead to very high values close by (> 300 mm/h). As well as varying over short lateral distances, hydraulic conductivity often varies with depth. Hydraulic conductivity generally declines with depth in most soils, but the impact of raindrops or spray drops on a bare soil surface can lead to a seal of very low conductivity in the surface few millimetres of many soils. Hydraulic conductivity can vary over time as well as space. For example, surface hydraulic conductivity under dryland cropping may be relatively high at sowing time when the soil is loose and porous after cultivation, but then may decline rapidly as soil slumps or seals under rainfall and increase again several-fold as plant growth and soil faunal activity create macropores (see for example, Murphy et al., 1993). As with other physical properties this temporal variation must be taken into account when measurements are being used to characterise a soil.

This high degree of spatial variability in K_{sat} has led to attempts to predict it from soil texture, structure and other visible features, such as the scheme in Table 10.6. More accurate relationships have been found for some soil data sets when important functional soil properties are measured. For example, McKenzie and Jaquier (1997) found areal porosity, bulk density and dispersion combined with field texture and grade of structure all significantly improved prediction of K_{sat}. However, these schemes will always have difficulty in

Table 10.6 Scheme for inferring the hydraulic conductivity range of soil horizons (Note that much greater values of K_{sat} can occur when large pores are created by biological activity, especially in clay soils.)

Texture	Structure Grade*	ESP	$EC_{1:5}$ (dS/m)	Low K_{sat} (mm/h)	High K_{sat} (mm/h)
Sands	1, 2, 3, 4, 5	–	–	120	> 700
Loamy Sands	1, 2, 3, 4, 5	–	–	60	700
Clayey Sands	1, 2, 3, 4, 5	–	–	2.5	60
Sandy Loams	4, 5	–	–	60	700
	1, 2, 3	–	–	5	60
Loams	4, 5	–	–	60	300
"	1, 2, 3	–	–	5	60
Clay Loams	4, 5	< 6	–	20	300
"	3	< 6	–	5	20
"	1, 2	< 6	–	2.5	5
Clay Loams	4, 5	> 6	< 0.7	0.1	2.5
"	3	> 6	< 0.7	0.1	2.5
"	1, 2	> 6	< 0.7	0.1	2.5
Clay Loams	4, 5	> 6	> 0.7	5	10
"	3	> 6	> 0.7	5	10
"	1, 2	> 6	> 0.7	5	10
Clay Loams	4, 5	> 15	< 1.9	0.1	1
"	3	> 15	< 1.9	0.1	1
"	1, 2	> 15	< 1.9	0.1	1
Clay Loams	4, 5	> 15	> 1.9	5	10
"	3	> 15	> 1.9	5	10
"	1, 2	> 15	> 1.9	5	10
Light Clays	5	< 6	–	5	40
"	4	< 6	–	2.5	5
"	1, 2, 3	< 6	–	0.5	2.5
Light Clays	5	> 6	< 0.7	< 0.1	2.5
"	4	> 6	< 0.7	< 0.1	2.5
"	1, 2, 3	> 6	< 0.7	< 0.1	2.5
Light Clays	5	> 6	> 0.7	5	10
"	4	> 6	> 0.7	5	10
"	1, 2, 3	> 6	> 0.7	5	10
Light Clays	5	> 15	< 1.9	< 0.1	1
"	4	> 15	< 1.9	< 0.1	1
"	1, 2, 3	> 15	< 1.9	< 0.1	1
Light Clays	5	> 15	> 1.9	5	10
"	4	> 15	> 1.9	5	10
"	1, 2, 3	> 15	> 1.9	5	10
Medium & Heavy Clays	5	< 6	–	2.5	40
"	4	< 6	–	0.5	2.5
"	1, 2, 3	< 6	–	0.5	2.5
Medium & Heavy Clays	5	> 6	< 0.7	< 0.1	2.5
"	4	> 6	< 0.7	< 0.1	2.5
"	1, 2, 3	> 6	< 0.7	< 0.1	2.5
Medium & Heavy Clays	5	> 6	> 0.7	5	10
"	4	> 6	> 0.7	5	10
"	1, 2, 3	> 6	> 0.7	5	10
Medium & Heavy Clays	5	> 15	< 1.9	< 0.1	1
"	4	> 15	< 1.9	< 0.1	1
"	1, 2, 3	> 15	< 1.9	< 0.1	1
Medium & Heavy Clays	5	> 15	> 1.9	5	10
"	4	> 15	> 1.9	5	10
"	1, 2, 3	> 15	> 1.9	5	10

* Structure Grade: 1 = single grain, 2 = massive, 3 = weak, 4 = moderate, 5 = strong

encompassing the wide variability of K_{sat} in the field and it is recommended that direct measurements be made for any critical applications such as earthwork design.

Hydraulic conductivity measurements alone may be useful in some cases as simple direct indicators of permeability or soil structure but they are generally most useful to land managers when used with other information to predict soil water movement.

10.6 Soil Water Movement

Water moving in soil can pass through soil aggregates and through packing pores between them (which together make up the soil matrix). It can also bypass this soil matrix and flow through macropores such as shrinkage cracks or channels and burrows made by roots and soil fauna.

In most cases water moving through the soil matrix is moved primarily by a combination of the force of gravity pulling down and the suction of dry soil pulling water from wetter soil. The rate of water movement is proportional to the combined force moving that water. It is also proportional to the soil hydraulic conductivity of the soil layer at that time. This complicates matters since hydraulic conductivity is itself dependent on water content, which may be changing as soil water moves through soil. These basic principles of soil water flow through the matrix are incorporated into the flow equation known as Richards' equation (see Equation 10.3). Prior to the development of modern computers, the difficulty of solving Richards' equation under varying conditions led most people to use simplified infiltration equations such as the Philip equation (1957), the modified Green and Ampt equation (1911), or other more empirical equations. Recently, computer programs have become available that predict soil water movement by solving Richards' equation numerically (for example, SWIM, Verburg et al., 1996). Land managers can now use these programs to predict soil water movement providing they have measures of hydraulic conductivity and the soil water characteristic, and providing they have some knowledge of the boundary conditions.

$$\partial \frac{\partial \theta}{\partial t} = \frac{\partial}{\partial x} K \left(\frac{\partial \Psi}{\partial x} + \frac{\partial z}{\partial x} \right) + s \qquad 10.3$$

In contrast to soil matrix, flow through macropores generally only occurs when soil is saturated. Macropore flow is generally more rapid than matric flow and much less uniform across an area. In addi-

tion, water flowing in macropores only interacts with the small proportion of soil that surrounds the macropore. In predicting macropore flow, the rate of flow is generally assumed to be an exponential function of macropore radius. This means that larger macropores can carry large flow rates providing they are not blocked at the surface end or elsewhere.

Infiltration and Runoff under Rainfall   10.6.1

Infiltration under rainfall, or the downward entry of rainwater into the soil surface, is a special case of soil water movement for two reasons. First, it is the critical process that divides precipitation between runoff (which may cause flooding or erosion) and infiltration (which may add to stored soil water for plant growth or may leach through the soil and drain to groundwater). Second, it is a more complex case of soil water movement. While infiltration follows the principles outlined above, it can be strongly influenced by soil surface processes such as surface sealing, filling of depressional storage, runoff routing due to surface roughness, water repellence, flow through 'biopores' (macropores formed by plants or animals), and deep crack-filling. Figure 10.8 illustrates how infiltration can be affected by these various surface processes.

In spite of the complexity, it is possible to say that when rain first begins falling the potential infiltration rate will generally be higher than the rainfall rate mainly due to the suction of a dry soil profile. Surface detention and depressional storage are also fully available to store any rainfall excess, so runoff is unlikely. As rain continues to fall infiltration rate declines owing to surface sealing and reduction in the sorptive effect of the subsoil. The various storages on the surface gradually fill and eventually runoff occurs. Accurately predicting the amount of runoff and when it will occur requires knowledge of all the surface processes in Figure 10.8. Rainfall simulators provide the opportunity to measure each of these separate processes, but rainfall simulations are relatively time-consuming and the results are generally specific to the antecedent conditions.

Drainage is less complex because many of the surface processes are irrelevant. As a wet soil begins to drain it loses water at a relatively rapid rate controlled by the saturated hydraulic conductivity of the soil layers below. The rate of drainage decreases as the soil dries because conductivity declines as larger pores partially empty and because capillary forces retain water in the smaller pores. Such water is gen-

erally described as being part of the soil's water-holding capacity. The idealised situation is shown in Figure 10.9. However, it is generally recognised that soil water is not likely to be redistributed completely in 48 hours, but the figure does present a useful model to describe the field behaviour of soils.

Because of the complexity involved in infiltration and runoff processes and the difficulty in predicting them from measured hydraulic properties, some workers use rainfall simulators to characterise soil behaviour under rainfall. Interpretation of soil parameters from measured runoff involves the use and calibration of simulation models.

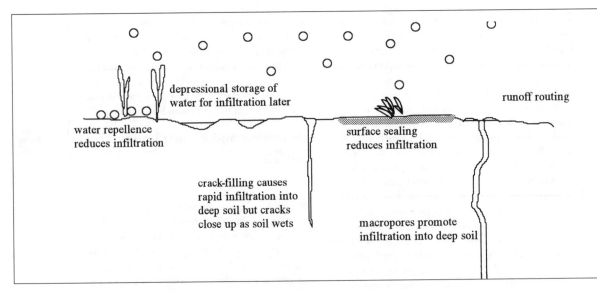

Figure 10.8 Surface processes influencing infiltration under rainfall

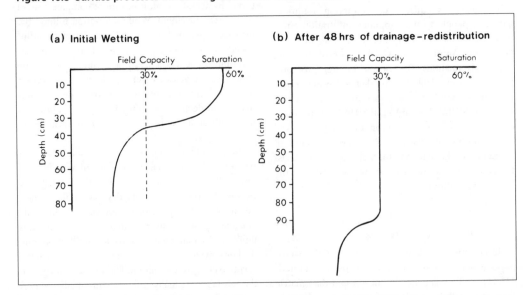

Figure 10.9   Idealised description of drainage and redistribution of water in a soil profile

10.7 Soil Strength, Friability and Penetration Resistance

Soil physical properties relating to penetration resistance, strength and friability are generally of interest to land managers because of the potential for high strength soil to limit plant growth or restrict the range of tillage options for soil preparation. A range of soil mechanics tests is available to test soil strength under various conditions but the three tests described below are commonly used for assessing soil suitability for plant growth. Soil strength measured in each of these tests is strongly dependent on water content, with dry soil generally being harder and stronger than moist or wet soil.

10.7.1 Penetration Resistance

Penetration resistance is the force required to push a penetrometer or micro-penetrometer tip into soil. It can be an important soil physical property for land managers to know because it can indicate the resistance experienced by elongating plant roots and shoots and hence indicate soil suitability for plant growth. Penetration resistance can be measured on soil profiles under field conditions but variations in soil water content with depth can strongly influence penetration resistance profiles and lead to false interpretations of compacted layers. It is generally desirable to measure penetration resistance under laboratory conditions on soil cores at a range of known soil water contents. This allows the land manager to predict soil and plant behaviour under the full range of moisture conditions that may occur in the field.

Precise interpretation of penetration resistance measurements can be difficult because plant roots do not behave in exactly the same way as metal rods and they may find less direct but easier pathways through soil. Despite this problem and the wide range in published critical penetration resistance values for plant growth (for example, from 800 to 5000 kPa, Erickson et al., 1974), useful correlations with plant root growth have been found and the technique is commonly recommended to identify potential problems.

0.7.2 Soil Strength and Modulus of Rupture

A range of soil mechanics tests exists for measuring soil strength under shear and compression forces.

Another approach useful in evaluating the potential of soil to be hardsetting is the modulus of rupture test. This test originally measured the force required to break soil briquettes made using rapid wetting followed by drying. The modulus of rupture test has been further modified by Cochrane and Aylmore (1999) to use rapid wetting and slow wetting with distilled water and dilute salt solutions to apportion the strength of the soil briquette separately to slaking and dispersion processes. It is in one sense a test of soil structural stability but in another sense a test of soil strength under a given set of conditions.

Friability 10.7.3

Aggregate strength can be measured using a modified penetrometer to determine the force required to crush each aggregate. Repeated measurements of aggregate strength over a range of aggregate sizes determine a soil's friability. Friability is an important property determining the likely effect of agricultural tillage operations. When a soil that is very friable is cultivated it will break into a wide range of aggregate sizes, whereas a non-friable soil may break only into large clods or fine powder and be unsuitable for seed germination.

Friability (k) can be determined from the negative slope of a linear relationship between the volume and the tensile strength of a large number of aggregates generally measured with a modified penetrometer (Utomo and Dexter, 1981). In very friable soils (k > 0.4), larger aggregates have a lower tensile strength on a volume basis and they tend to fracture more readily than smaller aggregates. Friability is not routinely measured but is often assessed qualitatively in the field by breaking large clods or by observing the response to tillage. Land managers should be aware that non-friable soils can be less suitable for direct drilling because a single sowing cultivation can often result in a cloddy seedbed. However, further cultivations aimed at breaking large clods will also tend to break any smaller clods into an excessively fine seedbed that may be less stable to wetting. Friability is dependent on soil water content and tends be greatest around the lower plastic limit (Mullins et al., 1990).

10.8 Appropriate Methods, Sampling and Interpretation

It will always be wise for land managers to choose carefully sampling and testing methods that are appropriate to the situation at hand. Important issues to consider include the following:

(a) the nature of the soil—that is, some methods listed in this chapter are less appropriate for swelling soils and highly permeable soils

(b) the spatial and temporal variability of the soil physical properties

(c) the spatial and temporal scales over which the underlying physical processes operate

(d) the nature of the issues or decisions facing the land manager and the methods available to interpret results—that is, detailed hydraulic property measurements may be necessary to predict runoff from effluent irrigation whereas estimates of hydraulic properties based on texture may be adequate for broad-scale land evaluation

(e) physical resources available for sampling and testing

(f) the limiting physical processes—for example, waterlogging is unlikely to limit soil use in the arid zone whereas soil strength may do, and therefore it may be more important to measure strength-related properties.

BIBLIOGRAPHY

Aitchison, G.D., Butler, P.F. and Gurr, C.G. (1951), 'Techniques associated with the use of gypsum block soil moisture meters', *Australian Journal of Applied Science* 2, 56–75.

ASTM (1975), *US Standards for Testing Soils*, American Society for Testing and Materials, USA.

Australian Standard, AS 1289.
    3.6.3 Determination of the particle size distribution of a soil-standard method of fine analysis using a hydrometer (1994).
    3.8.1 Determination of the Emerson Aggregate Class of a soil.
    3.8.2 Determination of the per cent dispersion of a soil (1997).

Bowman, G.M. and Hutka, J. (1999), 'Particle Size Analysis' in N.J. McKenzie, K.J. Coughlan and H.P. Cresswell (eds), *Soil Physical Measurement and Interpretation for Land Evaluation*, Australian Soil and Land Survey Handbook Series Volume 5, CSIRO Publishing, Melbourne (in press).

Chepil, W.S. (1951), 'An air elutriator for determining the dry aggregate soil structure in relation to erodibility by wind', *Soil Science* 71, 197–207.

Cochrane, H.R. and Aylmore, L.A.G. (1999), 'Modulus of Rupture' in N.J. McKenzie, K.J. Coughlan and H.P. Cresswell (eds), *Soil Physical Measurement and Interpretation for Land Evaluation*, Australian Soil and Land Survey Handbook Series Volume 5, CSIRO Publishing, Melbourne (in press).

Cresswell, H.P. (1999a), 'Bulk Density and Pore Space' in N.J. McKenzie, K.J. Coughlan and H.P. Cresswell (eds), *Soil Physical Measurement and Interpretation for Land Evaluation*, Australian Soil and Land Survey Handbook Series Volume 5, CSIRO Publishing, Melbourne (in press).

Cresswell, H.P. (1999b), 'Soil Water Characteristic' in N.J. McKenzie, K.J. Coughlan and H.P. Cresswell (eds), *Soil Physical Measurement and Interpretation for Land Evaluation*, Australian Soil and Land Survey Handbook Series Volume 5, CSIRO Publishing, Melbourne (in press).

Cresswell, H.P. and Paydar, Z. (1996), 'Water retention in Australian soils. I. Description and prediction using parametric functions', *Australian Journal of Soil Research*, 34, 195–212.

Decker, R.S. and Dunnigan, L.P. (1977), 'Development and use of the Soil Conservation Service dispersion test', ASTM STP623, *Dispersive Clays, Related Piping and Erosion in Geotechnical Projects*, 94–109.

Emerson, W.W. (1967), 'A classification of soil aggregates based on their coherence in water', *Australian Journal of Soil Research*, 5, 47–57.

Emerson, W.W. (1977), 'Physical properties and structure of semi-arid soils in Australia' in J.S. Russell and E.L. Greacen (eds), *Soil Factors in Crop Production in a Semi-arid Environment*, Aust. Soc. Soil Sci., Queensland.

Emerson, W.W. (1999), 'Emerson Dispersion Class' in N.J. McKenzie, K.J. Coughlan and H.P. Cresswell (eds), *Soil Physical Measurement and Interpretation for Land Evaluation*, Australian Soil and Land Survey Handbook Series Volume 5, CSIRO Publishing, Melbourne (in press).

Eriksson, J., Hakansson, I. and Danfors, B. (1974), 'The effect of soil compaction on soil structure and crop yields', *Swedish Institute of Agricultural Engineering Bulletin No. 354*.

Field, D.J., McKenzie, D.C. and Koppi, A.J. (1997), 'Development of an improved Vertosol stability test for SOILpak', *Australian Journal of Soil Research*, 35, 843–52.

Geeves, G.W., Cresswell, H.P., Murphy, B.W., Gessler, P.E., Chartres, C.J., Little, I.P., and Bowman, G.M. (1995), *The Physical, Chemical and Morphological Prop-*

erties of Soils in the Wheat-belt of Southern NSW and Northern Victoria, CSIRO Division of Soils/NSW CaLM joint report, CSIRO, Adelaide.

Greacen, E.L., Walker, G.R. and Cook, P.G. (1989), *Procedure for Filter Paper Method of Measuring Soil Water Suction*, CSIRO Division of Soils, Australia, Divisional Report No. 108.

Green, W.H. and Ampt, G.A. (1911), 'Studies on soil physics. I. The flow of air and water through soils', *Journal of Agricultural Science*, 4(1), 1–24.

Greenland, D.J. (1981), 'Soil management and degradation', *Journal of Soil Science* 32, 301–22.

Ingles, O.G., Lang, J.G., Richards, B.G. and Wood, C.C. (1969), *Soil and Water Observation in Flagstaff Gully Dam, Tasmania,* Technical Paper 11, Division of Soil Mechanics, CSIRO, Melbourne.

Isbell, R.F. (1996), *The Australian Soil Classification*, CSIRO Publishing, Melbourne.

Kay, B.D. (1990), 'Rates of change of soil structure under different cropping systems', *Advances in Soil Science* 12, 1–51.

Kemper, W.D. and Rosenau, R.C. (1986), 'Aggregate stability and size distribution' in A. Klute (ed.), *Methods of Soil Analysis Part 1: Physical and Mineralogical Method*, Soil. Sci. Soc. Am., Madison, Wisconsin.

Klute, A. (ed.) (1986), *Methods of Soil Analysis Part 1: Physical and Mineralogical Methods*, Soil. Sci. Soc. Am., Madison, Wisconsin.

Loch, R.J. (1994), 'A method for measuring aggregate water stability of dryland soils with relevance to surface seal development', *Australian Journal of Soil Research*, 32, 687–700.

Loveday, J. (1974), 'Recognition of gypsum responsive soils', *Australian Journal of Soil Research,* 12, 87–96.

Loveday, J. and Pyle, J. (1973), *The Emerson Dispersion Test and its Relationship to Hydraulic Conductivity*, Technical Paper 15, CSIRO Division of Soils, CSIRO, Melbourne.

McDonald, R.C., Isbell, R.F, Speight, J.G., Walker, J. and Hopkins, M.S. (1990), *Australian Soil and Land Survey Field Handbook,* Second Edition, Inkata Press, Melbourne.

McKenzie, N.J. and Jacquier, D.W. (1997), 'Improving the field estimation of saturated hydraulic conductivity in soil survey', *Australian Journal of Soil Research*, 35, 803–25.

McKenzie, N.J., Coughlan, K.J. and Cresswell, H.P. (eds) (1999), *Soil Physical Measurement and Interpretation for Land Evaluation*, Australian Soil and Land Survey Handbook Series Volume 5, CSIRO Publishing, Melbourne (in press).

Marshall, T.J. (1947), *Mechanical Composition of Soil in Relation to Field Descriptions of Texture*, CSIRO Aust. Bull. No. 224, CSIRO, Melbourne.

Micrometrics Instrument Corporation USA (1979), *Instruction Manual: Sedigraph 5000D Particle Size Analyser*, 5680 Goshen Springs Rd, Norcross, Georgia 30093, USA.

Moore, P.J., Wrigley, R.J. and Styles, J.R. (1985), 'Identification and amelioration of some dispersive soils',

Civil Engineering Transactions, The Institute of Engineers, Australia, 371–83.

Mullins, C.E., MacLeod, D.A., Northcote, K.H., Tisdall, J.M. and Young, I.M. (1990), 'Hardsetting soils: behaviour, occurrence and management', *Advances in Soil Science,* 11, 37–108.

Murphy, B.W., Koen, T.B., Jones, B.A. and Huxedurp, L.M. (1993), 'Temporal variation of hydraulic properties for some soils with fragile structure', *Australian Journal of Soil Research* 31, 179–97.

North, P.E. (1976), 'Towards an absolute measurement of soil structural stability using ultrasound', *Journal of Soil Science* 27, 431–59.

Northcote, K.H. (1979), *A Factual Key for the Recognition of Australian Soils,* Rellim Technical Publications, Glenside, South Australia.

Perry, E.B. (1975), *Piping in Earth Dams Constructed of Dispersive Clay: Literature Review and Design of Laboratory Tests,* Technical Report S-75-15, US Army Engineers Waterways Experiment Station, US Army, Washington DC.

Philip, J.R. (1957), 'The theory of infiltration. I. The infiltration equation and its solution', *Soil Science,* 83, 345–57.

Pierson, F.B. and Mulla, D.J. (1989), 'An improved method for measuring aggregate stability of a weakly aggregated loessial soil', *Soil Science Society of America Journal* 53, 1825–31.

Ritchie, J.A. (1963), 'Earthwork tunnelling and the application of soil testing procedure', *Journal of the Soil Conservation Service of NSW* 19, 111–29.

Rosewell, C.J. (1970), 'Investigations into the control of earthwork tunnelling', *Journal of the Soil Conservation Service of NSW* 26, 188–203.

Salter, P.J. and Williams, J.B. (1967), 'The influence of texture on the moisture characteristics of soils. IV. A method of estimating the available water capacities of profiles in the field', *Journal of Soil Science* 18, 174–81.

Salter, P.J. and Williams, J.B. (1969), 'The influence of texture on the moisture characteristics of soils. V. Relationships between particle size composition and moisture contents at the upper and lower limits of available water', *Journal of Soil Science.* 20, 126–31.

Sumner, M.E. (1958), 'A simplified technique for the determination of soil aggregation', *South African Journal of Agricultural Science* 1, 301–4.

Utomo, W.H. and Dexter, A.R. (1981), 'Soil friability', *Journal of Soil Science* 32, 203–13.

Verburg, K., Ross, P.J. and Bristow, K.L. (1996), *SWIMv2.1 Users Manual*, Divisional Report No. 130, CSIRO Division of Soils, Australia.

Williams, J. (1983), 'Soil Hydrology', in *Soils: An Australian Viewpoint*, Division of Soils CSIRO, Melbourne; Academic Press, London.

Young, R.A. (1984), 'A method of measuring aggregate stability under water drop impact', *Transactions of the American Society of Agricultural Engineers* 31, 1351–54.

Soil
Engineering
Properties

CHAPTER 11

R.W. Hicks

Soil engineering properties are important to the soil conservationist for a number of reasons. First, it is these properties that affect the success or failure of earthworks, which form such a vital component of erosion control in New South Wales. Second, some soil conservationists are involved in soil mapping for urban land use and, in such cases, some understanding of soil engineering properties is necessary to produce useful soil maps. Lastly, some engineering properties are useful guides to predicting the performance of agricultural soils under some conditions.

Soil materials vary in their engineering properties, dependent on their particle size distribution, clay mineral type(s) and the assemblage patterns of the constituent particles.

To assess the soil materials for their reaction to disturbance and potential use as construction materials, it is necessary to carry out a number of observations and test procedures. The Australian Standard AS 1289 lists methods of testing soils for engineering purposes.

11.1 Soil Composition—Particle Size Distribution

The distribution of the sizes of ultimate particles in a soil is frequently referred to as its grading in an engineering context. Procedures for determining the grading are outlined in Chapter 10. The particle sizes used for engineering properties differ slightly from those used for soil science, with the silt, sand and gravel fraction being subdivided (see Table 11.1).

A brief assessment of soil engineering properties in relation to texture is given in Table 11.2. However, the most useful description of likely soil engineering behaviour in relation to grading is the Unified Soil Classification System, which will be discussed more fully later in this chapter.

Table 11.1 Particle size fractions used for engineering (AS 1289)

Engineering Name	Size	Soil Science Name
Boulders	> 200 mm	
Cobbles	200 mm–60 mm	
Coarse Gravel	60 mm–20 mm	Gravel
Medium Gravel	20 mm–6 mm	
Fine Gravel	6 mm–2 mm	
Coarse Sand	2 mm–600 μm	Coarse Sand
Medium Sand	600 μm–200 μm	
Fine Sand	200 μm–60 μm	Fine Sand
Coarse Silt	60 μm–20 μm	
Medium Silt	20 μm–6 μm	Silt
Fine Silt	6 μm–2 μm	
Clay	< 2 μm	Clay

Table 11.2 Engineering inferences that can be made from soil texture (after Ingles and Metcalfe, 1972)

Observation	Material Behaviour
Sandy soils . . . high sand content	Good engineering properties, but if monosized will need mechanical stabilisation, and if wholly sand may need some clay, cement or bituminous binder.
Silty soils . . . high silt content, some clay	Good when dry, but will lose all bearing capacity and trafficability when wet, no economic treatment except good water shedding.
Silty soils . . . high silt content, little or no clay	Dusty and cohesionless when dry, no bearing capacity when wet, no economic treatment — avoid.
Clay soils . . . high clay content (cracking)	Very troublesome engineering properties unless protected from moisture change. Montmorillonites and illites responsive to lime stabilisation.
Clay soils . . . high clay content (either non-cracking or open-textured)	Reasonably good engineering properties when heavily compacted.
Clay soils . . . clay softens rapidly with water; clay greasiness develops rapidly	Saline clays, poor trafficability when wet.
Clay soils . . . clay softens only slowly with water; greasiness develops slowly	Calcium-rich montmorillonites of poor trafficability and high adhesiveness (sticky) when wet.

11.2 Soil Plasticity

The engineering behaviour of fine-textured soils is very dependent on the types of clays present. To account for this, the soil scientist Atterberg developed a method of describing the effect of varying water content on the consistency of fine-grained soil. He established stages of consistency and defined definite but arbitrary limits for each, as shown in Table 11.3.

The plastic limit (PL) is the water content at which the soil begins to break apart and crumble when rolled by hand into threads 3 mm in diameter (AS 1289 3.2.1, 1995) (see Plate 11.1).

The liquid limit (LL) is defined as the water content at which a trapezoidal groove of specified shape, cut in moist soil held in a special cup, is closed after 25 taps on a hard rubber plate (AS 1289 3.1.1, 1995) (see Plate 11.2).

The shrinkage limit (SL) is the water content at which the soil reaches its theoretical minimum volume as it dries out from a saturated condition.

The limits are customarily expressed as whole numbers of percentage, omitting the per cent sign.

The actual values of these vary from soil to soil. Heavy soils with a montmorillonite clay mineral may have liquid limits of 80 or more; these are the cracking clays (Ug soils). However, other soils including red and yellow duplex soils (Dr and Dy soils) and structured earths (Gn3, Gn4 soils), can also have subsoils with liquid limits of up to 80, but are more commonly in the 40–80 range. The massive earths with loam or clay loam texture usually have liquid limits in the 25–40 range.

In themselves the Atterberg limits mean little, but, as indices to the significant properties of a soil,

Table 11.3   Schematic diagram for soil plasticity

Phase	Solid State	Semi-Solid State	Plastic State	Liquid State	Suspension
Water	← ———————————— Water content decreasing ————————————				
Limits	Dry Soil	Shrinkage Limit SL	Plastic Limit PL	Liquid Limit LL	
			Plasticity Index PI		
Shrinkage	Volume constant	← ———————— Volume decreasing ————————			
Condition	Hard to stiff	Workable	Sticky	Slurry	Water-held suspension
Shear Strength (kN/m²)	← ——— Shear strength increasing ———			Negligible	
Moisture Content	0	SL	PL	LL	

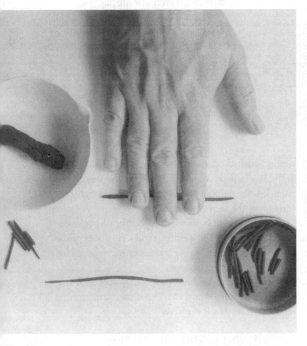

Plate 11.1   The plastic limit test, showing the initial soil condition; the rods into which the soil is rolled; and the finished sample, for moisture content determination, in the tin

Plate 11.2   The liquid limit test, showing the soil paste in the cup which has been grooved ready for the test. The apparatus then jolts the cup regularly and the liquid limit relates to the number of jolts required to close the groove at different moisture contents

they are very useful. The liquid limit has been found to be directly proportional to the compressibility of a soil, and hence its ability to support a load and its trafficability when wet. It is also an indication of the shrink-swell potential of a soil (Mills et al., 1980). The difference between the liquid and plastic limits, termed the 'plasticity index' (PI), represents the range in water contents through which the soil is in the plastic state. Soils with a low index can change from a solid to a liquid state with little change in moisture content. They are, therefore, potential problem soils and may be prone to mass movement.

The Atterberg limits and their associated relationships are simple empirical expressions of the water adsorbing and absorbing ability of soils containing clay. They thus express both the clay-water behaviour and how that is diluted by the non-clay particles. The tests are usually standardised on the portion of the soil finer than 0.42 mm (passing No. 40 sieve: fine sand sizes and smaller). However, if little coarse sand is present, the tests are often made on the total sample, but with some differences in the results.

In the field, hand plasticity tests (see Table 11.4) are useful in assessing soil conditions for construction activities, particularly compaction (Ingles and Metcalfe, 1972).

Table 11.4 Field plasticity tests

Pencil Test	Squeeze Test	Compaction Behaviour
Rolls to pencil before crumbling	Stays together	Approximately at optimum moisture content for compaction
Rolls very fine without crumbling or smears without rolling	Sticks to the hands or extrudes	Too wet for compaction
Crumbles and will not roll	Will not cohere	Too dry for compaction
		If a heavy clay, added water will not be distributed evenly or quickly

11.3 Unified Soil Classification System (USCS)

The most widely used, relatively simple classification of soil materials for engineering purposes is the Unified Soil Classification, developed by Casagrande (1947). The overall system is summarised in Figures 11.1 and 11.2. Much of the material in this section is drawn from *Design of Small Dams* (Anon., 1960), which is a comprehensive basic reference on small earthwork design.

The classification and its interpretation are most applicable in the design of small dams and earthworks, such as soil conservation structures. They take into account soil engineering properties, are descriptive and easy to relate back to actual soils, and have the flexibility of being adaptable both to the field and to the laboratory. A major advantage of the scheme is that a soil can be classified easily by visual and manual examination by experienced persons without the need for laboratory testing. Nevertheless, the classifi-

cation is usually done in the laboratory. The Unified Soil Classification System is based on the size of the particles, the amounts of various sizes and the characteristics of the very fine particles.

It must be kept in mind, though, that no classification is a substitute for specific engineering tests of the soil where site-specific, accurate data are required for design and construction purposes. This is especially the case where there are severe consequences of failure such as severe damage to expensive structures or potential injury or loss of life.

Classifying a Soil Material— Laboratory Method   11.3.1

Classification is usually done in the laboratory because of the need to have information on the particle size distribution of a soil.

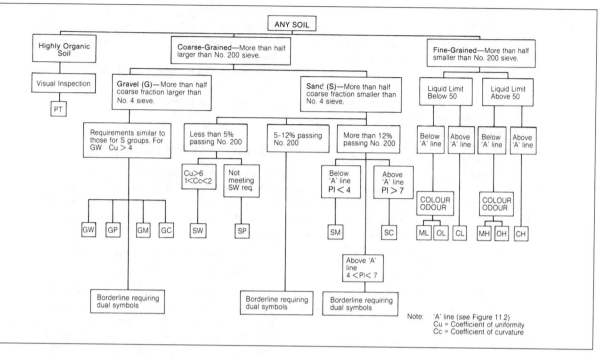

Figure 11.1 Unified Soil Classification System (after Anon., 1960)

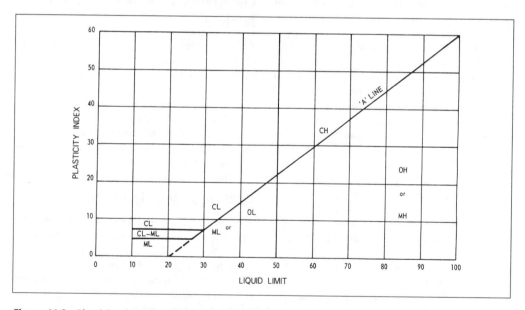

Figure 11.2 Plasticity chart for the laboratory classification of fine-grained soils (after Anon., 1960)

Particle Size

Particles larger than 75 mm diameter are excluded from the Unified Soil Classification System. The amount of such oversized material, however, may be of great importance in the selection of sources for embankment material. Within the size range of the system there are two major divisions, namely, the coarse grains and the fine grains.

Coarse grains are those larger than the No. 200 sieve size (0.074 mm), and they are further divided as follows:

Gravel (G): from 75 mm to No. 4 sieve
(4.76 mm)
Coarse gravel—75 mm to 20 mm
Fine gravel—20 mm to No. 4 sieve
(4.76 mm)
Sand (S): from No. 4 sieve to No. 200 sieve
Coarse sand—No. 4 to No. 10 sieve
(4.76 mm to 2 mm)
Medium sand—No. 10 to No. 40 sieve
(2 mm to 0.42 mm)
Fine sand—No. 40 to No. 200 sieve
(0.42 mm to 0.074 mm)

These size fractions differ from those used in Chapter 10 and in Table 11.1. However, to use the USCS it is necessary to use these size fractions, and care should be taken when interpreting results from a grading curve that the correct size fraction is used. The 0.074 mm/No. 200 sieve is particularly important as this is used to divide the coarse and fine-grained soils. All sieve sizes are to the ASTM standard in this system (American Standard for Testing and Materials).

Character of the Coarse Grains

The amounts of the various sizes of grains present in a soil can be determined in the laboratory by means of sieving, for the coarse grains, and by sedimentation, for the fines.

The laboratory results are usually presented in the form of a cumulative grain-size curve (for an example see Figure 10.2). For soils consisting mainly of coarse grains, the grain-size distribution reveals something of the physical properties of the material. On the other hand, the grain size is much less significant for soils containing a preponderance of fine grains.

Sands and gravels may be defined as:

well graded (W)—good representation of all particle sizes from largest to smallest
poorly graded (P)—uniform, most particles about the same size; or gap gradation—absence of one or more intermediate sizes.

In the field, soil is estimated to be well graded or poorly graded by visual examination. For laboratory purposes, the type of gradation can be determined by the use of criteria based on the range of sizes and on the shape of the grain-size curve. To define the grading, the values Cu, coefficient of uniformity, and Cc, coefficient of curvature, are used. These are defined as:

$$\text{Coefficient of Uniformity (Cu)} = \frac{D60}{D10} \quad \ldots\ldots\ldots\ldots 11.1$$

$$\text{Coefficient of Curvature (Cc)} = \frac{(D30)^2}{D10 \times D60} \quad \ldots\ldots\ldots\ldots 11.2$$

where D60, D30 and D10 are the diameters of the particles corresponding to the 60%, 30% and 10% cumulative frequencies on the grading curve, respectively.

A well-graded soil must meet the following criteria:

Cu > 4 for gravel
Cu > 6 for sand
1 < Cc < 3 for gravel and sand.

Character of the Fine Grains

The degree of plasticity of the fine fraction is determined from the Atterberg limits. The liquid limit (LL), plastic limit (PL) and the plasticity index (PI) are used to allocate soils to a group. The general grouping is as follows:

LL > 50—high plasticity (MH, CH)
LL < 50—low plasticity (ML, CL).

Whether a soil is considered MH or CH, ML or CL is determined by the 'A' line, which is a plot of plasticity index against liquid limit. The more silty soils (ML, MH) have lower plasticity indexes than the corresponding CL and CH soils (see Figure 11.2).

Classifying a Soil Material—Field Method

11.3.2

As Atterberg limits are time consuming and costly to obtain, they are not often available. Therefore, techniques to identify the USCS soil groups without the limits have been developed. These require the use of dilatancy, dry strength and toughness as outlined in the following sections (Anon., 1960).

These procedures are to be performed on the less than No. 40 sieve size, about 0.4 mm. For field classification purposes, the coarse particles that interfere with the tests are simply removed by hand.

Dilatancy (Reaction to Shaking)

After removing particles larger than No. 40 sieve size, prepare a pat of moist soil with a volume of about 10 cm³. Add enough water, if necessary, to make the soil soft but not sticky.

Place the pat in the open palm of one hand and shake horizontally, striking vigorously against the other hand several times. A positive reaction consists of the appearance of water on the surface of the pat, which changes to a livery consistency and becomes glossy. When the sample is squeezed between the fingers, the water and the gloss disappear from the surface, the pat stiffens and finally it cracks or crumbles. The rapidity of appearance of water during shaking and of its disappearance during squeezing assists in identifying the character of the fines in a soil.

Very fine clean sands give the quickest and most distinct reaction whereas a plastic clay has no reaction. Inorganic silts, such as a typical rock flour, show a moderately quick reaction.

Dry Strength (Crushing Characteristics)

After removing particles larger than No. 40 sieve size, mould a pat of soil to the consistency of putty, adding water if necessary. Allow the pat to dry completely by oven, sun or air drying, and then test its strength by breaking and crumbling between the fingers. This strength is a measure of the character and quantity of the colloidal fraction contained in the soil. The dry strength increases with increasing plasticity.

High dry strength is characteristic for clays of the CH group. A typical inorganic silt possesses only very slight dry strength. Silty fine sands and silts have about the same slight dry strength, but can be distinguished by the feel when powdering the dried specimen. Fine sand feels gritty whereas a typical silt has the smooth feel of flour.

Toughness (Consistency near Plastic Limit)

After removing particles larger than No. 40 sieve size, a specimen of soil about 10 cm^3 in size is moulded to the consistency of putty. If too dry, water must be added and if sticky, the specimen should be spread out in a thin layer and allowed to lose some moisture by evaporation. Then the specimen is rolled out by hand on a smooth surface or between the palms into a thread about 3 mm in diameter. The thread is then folded and rerolled repeatedly. During this manipulation the moisture content is gradually reduced and the specimen stiffens, finally loses its plasticity and crumbles when the plastic limit is reached.

After the thread crumbles, the pieces should be lumped together and a slight kneading action continued until the lump crumbles.

The tougher the thread near the plastic limit, and the stiffer the lump when it finally crumbles, the more potent is the colloidal clay fraction in the soil. Weakness of the thread at the plastic limit and quick loss of coherence of the lump below the plastic limit indicate either inorganic clay of low plasticity, or materials such as kaolin-type clays and organic clays which occur below the 'A' line of the plasticity chart (see Figure 11.2).

Highly organic clays have a very weak and spongy feel at the plastic limit. Soils that are typical of the various groups are readily classified by the foregoing procedures. Many natural soils, however, will have property characteristics of two groups, because they are close to the borderline between the groups, either in percentages of the various sizes or in plasticity characteristics. For this substantial number of soils, boundary classifications are used; that is, the group symbols most nearly describing the soil are connected by a hyphen, such as GW-GC.

Proper boundary classification of a soil near the borderline between coarse-grained and fine-grained soils is accomplished by classifying it first as a coarse-grained soil and then as a fine-grained soil. Such classifications as SM-ML and SC-CL are common.

Soil Components and Their Properties

11.3.3

Gravel and Sand

Both of the coarse-grained components of soil have essentially the same engineering properties, differing mainly in degree. The division of gravel and sand sizes by the No. 4 sieve is arbitrary and does not correspond to a sharp change in properties. Well-graded, compacted gravels or sands are stable materials. The coarse-grained soils when devoid of fines are pervious, easy to compact, little affected by moisture and not subject to frost action. Although grain shape and gradation, as well as size, affect these properties, gravels are generally more pervious, more stable and less affected by water or frost than are sands, for the same amount of fines. Poorly graded sands and gravels are generally more pervious than well-graded sands and gravels.

As a sand becomes finer and more uniform, it approaches the characteristics of silt, with a corresponding decrease in permeability and reduction in stability in the presence of water. Very fine, uniform sands are difficult to distinguish visually from silt. Dried sand, however, exhibits no cohesion and feels gritty in contrast to the very slight cohesion and smooth feel of dried silt.

Silt and Clay

Even small amounts of fines may have important effects on engineering properties of the soils in which they are found. As little as 25% of particles smaller than the No. 200 (0.074 mm) sieve in sand and gravel may make the soil virtually impervious, especially when the coarse grains are well graded. Also, serious frost-heaving in well-graded sands and gravels may be caused by less than 10% of fines.

Soils containing moderate amounts of clay and with small amounts of silt are quite suitable for small earth structures, particularly those intended for water storage. These fine materials exhibit marked changes in physical properties with change of water content. A hard, dry clay, for example, may be suitable as a foundation for heavy loads so long as it remains dry, but may turn into a quagmire when wet. Many of the fine soils shrink on drying and expand on wetting, which may adversely affect structures founded upon them or constructed of them. Even when the water content does not change, the properties of fine soils may vary considerably between their natural condition in the ground and their state after being disturbed. Deposits of fine particles that have been subjected to loading in geologic time frequently have a structure that gives the material unique properties in the undisturbed state. When the soil is excavated for use as a construction material, the soil structure is destroyed and the properties of the soil are changed radically.

Silts are different from clays in many important respects but, because of similarity in appearance, they often have been mistaken one for the other, sometimes with unfortunate consequences. Dry, powdered silt and clay are indistinguishable, but they are easily identified by their behaviour in the presence of water. Recognition of fines as being silt or clay is an essential part of the Unified Soil Classification System.

Silts (M) are the non-plastic fines. They are inherently unstable in the presence of water. Silts are fairly impervious, difficult to compact and are highly susceptible to frost-heaving. Silt masses undergo change of volume with change of shape (the property of dilatancy) in contrast to clays, which retain their volume with change of shape (the property of plasticity). When dry, silt can be pulverised easily under finger pressure (indicative of very slight dry strength), and will have a smooth feel between the fingers in contrast to the grittiness of fine sand. Silts differ among themselves in the size and shape of grains, which is reflected mainly in the property of compressibility.

Clays (C) are the plastic fines. They have low resistance to deformation when wet, but they dry to hard, cohesive masses. Clays are virtually impervious, difficult to compact when wet and impossible to drain by ordinary means. Large expansion and contraction with changes in water content are characteristic of some clays. The small size, flat shape and type of mineral composition of clay particles combine to produce a material that is both compressible and plastic. In the Unified Soil Classification System, the liquid limit is used to distinguish between clays of high compressibility (H) and those of low compressibility (L). Differences in plasticity of clays are reflected by their plasticity indexes. For the same liquid limit, the higher the plasticity index, the more cohesive is the clay.

Field differentiation among clays is accomplished by the toughness test in which the moist soil is moulded and rolled into threads until crumbling occurs, and by the dry strength test which measures the resistance of the clay to breaking and pulverising.

Organic Matter

Organic matter in the form of partly decomposed vegetation is the primary constituent of peaty soils. Varying amounts of finely divided vegetable matter are found in plastic and in non-plastic sediments, and often affect their properties sufficiently to influence their classification. Thus, we have organic silts and silt clays of low plasticity, and organic clays of medium to high plasticity. Even small amounts of organic material in colloidal form in a clay will result in an appreciable increase in liquid limit of the material without increasing its plasticity index. Organic soils are dark grey or black in colour and usually have a characteristic odour of decay. Organic clays feel spongy in the plastic range as compared to inorganic clays. The tendency for soils with significant organic content to create voids by decay, or to change the physical characteristics of a soil mass through chemical alteration, makes them undesirable for engineering use.

Allocating a Soil in the USCS

11.3.4

The general procedure for allocating soils in the USCS is shown in Figures 11.1 and 11.2. Laboratory information is used in this procedure. However, soils can be allocated in the field using the following field procedure.

A representative sample of soil (excluding particles larger than 75 mm) is first classified as coarse-grained or fine-grained by estimating whether 50%, by weight, of the particles that can be seen are coarse-grained; soils containing more than 50% of particles smaller than the eye can see are fine-grained soils. If the soil is predominantly coarse-grained, it is then identified as being a gravel or a sand by estimating whether 50% or more, by weight, of the coarse grains are larger or smaller than 5 mm (No. 4 sieve size).

If the soil is a gravel, it is next identified as being 'clean' (containing little or no fines, less than 5%) or 'dirty' (containing an appreciable amount of fines, more than 12%). For clean gravels, final classification is made by estimating the gradation: the well-graded gravels belong to the GW group, and uniform and gap-graded gravels belong to the GP group. Dirty gravels are of two types: those with non-plastic (silty) fines (GM), and those with plastic (clayey) fines (GC). The determination of whether the fines are silty or clayey is made by the three manual tests for fine-grained soils.

If a soil is a sand, the same steps and criteria are used as for gravels in order to determine whether the soil is a well-graded clean sand (SW), poorly graded

Table 11.5 Qualitative comparison of engineering properties for compacted soil and suitability for various purposes (Category 1 is best)

Unified Soil Group	Permeability	Piping Resistance (Non-tunnelling)	Cracking Resistance (Differential Settlement)	Compaction Characteristics	Compaction Equipment	Homogeneous Dam	Core of Zoned Dam	Shell of Zoned Dam	Requirement for Seepage Control
GW	Pervious	High	—	1	Tractor, rubber-tyred roller; flat roller	—	—	1	Positive cut-off
GP	Very pervious	High to moderate	—	2	Tractor, rubber-tyred roller; flat roller	—	—	2	Positive cut-off
GM	Semi-pervious to pervious	High to moderate	3	4	Rubber-tyred roller; sheepsfoot roller Close control required	2	4	—	4 Toe trench to none
GC	Impervious	Very high	4	3	Rubber-tyred roller; sheepsfoot roller Fair ability	1	1	—	1
SW	Pervious	High to moderate	—	1	Tractor Good	—	—	3 if gravelly	Upstream blanket and toe drains or wells
SP	Pervious	Low to very low	—	2	Tractor Good	—	—	4 if gravelly	Upstream blanket and toe drain or wells
SM	Semi-pervious to pervious	Moderate to low	3	4	Rubber-tyred roller; sheepsfoot roller Good with close control	4	5	—	5 erodible Upstream blanket and toe drains or wells
SC	Impervious	High	4	3	Sheepsfoot roller; rubber-tyred roller Fair	3	2	—	2 No controls required
ML	Semi-pervious to pervious	Low to very low	6	4 to 6	Rubber-tyred roller; sheepsfoot roller Good to poor Close control required	6	6	—	6 erodible Toe trench to no controls
CL	Impervious	High	2 to 5	4	Sheepsfoot roller; rubber-tyred roller Fair to good	5	3	—	3 None
MH	Semi-pervious to impervious	Moderate to high	Variable	4 to 8	Sheepsfoot roller Poor to very poor	9	9	—	None
CH	Impervious	Very high	1	4 to 7	Sheepsfoot roller Fair to poor	7	7	—	8 Expansive None

clean sand (SP), sand with silty fines (SM), or sand with clayey fines (SC).

If a material is predominantly (more than 50% by weight) fine-grained, it is classified into one of six groups (ML, CL, OL, MH, CH, OH) by estimating its dilatancy (reaction to shaking), dry strength (crushing characteristics) and toughness (consistency near the plastic limit), and by identifying it as being organic or inorganic.

As with any system of classification, the utility of the USCS is related to the amount of useful information that can be obtained or inferred for the various groups. Tables 11.5 and 11.6 show some important engineering properties of the main soil groups and two frequently occurring boundary groups. The values shown are average values based on more than 1500 soil tests performed by the US Bureau of Reclamation (Anon., 1960).

Table 11.6 Engineering characteristics of soil groups (after Anon., 1960)

USCS	Maximum Compaction		Permeability** at Maximum Compaction (mm/day)	Compressibility		Shearing Strength	
	Max. Dry Density (g/cm³)	Optimum Moisture (%)		at 138 kPa	at 345 kPa	C_0 (kPa)	C_{sat} (kPa)
GW	> 1.90	< 13.3	22464 ± 10800	1.4	(*)	(*)	(*)
GP	> 1.76	< 12.4	53568 ± 28512	0.8	(*)	(*)	(*)
GM	> 1.83	< 14.5	0.25	1.2	3.0	(*)	(*)
GC	> 1.84	< 14.7	0.25	1.2	2.4	(*)	(*)
SW	1.91 ± 0.08	13.3 ± 2.5	(*)	(*)	(*)	39.3 ± 4.1	(*)
SP	1.76 ± 0.03	12.4 ± 1.0	12	0.8 ± 0.3	(*)	22.8 ± 6.2	(*)
SM	1.83 ± 0.02	14.5 ± 0.4	6.3 ± 4.0	1.2 ± 0.1	3.0 ± 0.4	51.0 ± 6.2	20.0 ± 6.9
SM-SC	1.91 ± 0.02	12.8 ± 0.4	0.67 ± 0.50	1.4 ± 1.0	2.9 ± 1.0	50.3 ± 21.4	14.5 ± 5.5
SC	1.84 ± 0.02	14.7 ± 0.4	0.25 ± 0.16	1.2 ± 0.2	2.4 ± 0.5	75.1 ± 15.2	11.0 ± 6.2
ML	1.65 ± 0.02	19.2 ± 0.7	0.49 ± 0.19	1.5 ± 0.2	2.6 ± 0.3	66.9 ± 10.3	9.0 ± (*)
ML-CL	1.75 ± 0.03	16.8 ± 0.7	0.11 ± 0.06	1.0 ± 0.0	2.2 ± 0.0	63.4 ± 16.5	22.0 ± (*)
CL	1.73 ± 0.02	17.3 ± 0.3	0.07 ± 0.03	1.4 ± 0.2	2.6 ± 0.4	86.9 ± 10.3	13.1 ± 2.1
MH	1.31 ± 0.06	36.3 ± 3.2	0.13 ± 0.09	2.0 ± 1.2	3.8 ± 0.8	72.4 ± 29.6	20.0 ± 9.0
CH	1.51 ± 0.03	25.5 ± 1.2	0.04 ± 0.04	2.6 ± 1.3	3.9 ± 1.5	102.7 ± 33.8	11.0 ± 5.9

*	Insufficient data	C_0	Strength of soil placed at maximum dry density
±	90 per cent confidence limits	C_{sat}	Strength of soil placed at maximum dry density then saturated
		**	Provided soils are not highly aggregated

11.4 Soil Shrink-Swell Properties

11.4.1 Expansive Soils in New South Wales

Soil materials can undergo changes in volume as a result of fluctuations in their moisture content. These variations in soil volume occur in both the horizontal and vertical planes, thus leading to soil cracking upon drying, rather than just surface settlement.

The degree of shrinkage and swelling is dependent on the clay content and clay mineral types present in the soil, depth of fluctuation in moisture content, and the rainfall variations both in the short and long terms. Soils that shrink and swell sufficiently to cause problems for buildings and roads are termed expansive or reactive soils.

In New South Wales, such expansive soils are common over much of the semi-arid and subhumid interior, some of the best examples being black, grey and brown cracking clays (Ug5.1, Ug5.2, Ug5.3) found in north-western New South Wales and, to a lesser extent, in mid and south-western New South Wales. Because of the climate, areas of these soils have a high movement potential. Many of these areas show the phenomenon known as gilgai formation, which results in various types of surface humps and hollows formed by massive shrinking and swelling of the soils present.

However, some red and yellow duplex soils (Dr, Dy) or structured gradational soils (Gn3, Gn4) can also have subsoils with expansive clays. These have been noted in the western suburbs of Sydney (Builders Licensing Board, 1984), and the central western tablelands and slopes (Mills et al., 1980).

Basalt-derived soils are probably the most common group of expansive soils and include the cracking clays (Ug) and structured gradational (Gn3, Gn4) soils with expansive clays. Even so, expansive soils have been recorded on a wide range of rock types including shales, andesites and even granodiorites. Soils with moderate to low clay contents can still be expansive, even though their clay minerals are not those conventionally considered to be expansive.

In the more humid coastal areas, the degree of shrinkage and swelling is not as great and movement potential is less. Variations in shrinkage and swelling occur because of differences in rainfall, evaporation demand, soil depth, depth of tree roots and the effect of additional sources of soil water, such as leaking pipes or groundwater seepages.

Shrinkage and Swelling Processes 1.4.2

Soil shrinkage is caused by capillary tension. When a susceptible soil dries, a meniscus develops in each void at the soil particle surface. This produces tension at the soil/water interface and compression in the surrounding soil fabric. During shrinkage, the voids become smaller and the capillary tension increases. The soil continues to shrink until the capillary tension is unable to compress the soil any further. Once a soil is drier than the shrinkage limit, tension is released in the larger voids by crumbling and, if the soil contains fibrous organic matter and/or mica, it will start to expand.

Soil swelling occurs owing to a number of phenomena, including elastic rebound of soil particles, attraction of clay minerals for water, electrical repulsion of clay particles and their adsorbed cations from each other, and the expansion of air trapped in the voids. However, the uptake of water between the clay

plates is the major cause of expansion. Extremely high pressures can develop where a dry soil is wetted while confined. With montmorillonite-type clay minerals, expansion pressures as great as 500 kPa are possible. Thus, swelling can and does have major impacts on structures such as houses, service pipes and road pavements.

Identifying Expansive Soils 11.4.3

Field Identification

Areas that have the following features are likely to have expansive soils:
—soils with surface cracks 6 mm or more wide, 300 mm or more deep and at least one crack per square metre
—soils that develop gilgai microrelief
—soils with obvious medium to heavy clay subsoils
—soil depth greater than about 50 cm.

However, not all expansive soils can be identified by these features, and further testing is required to positively identify a soil as expansive or non-expansive. The preceding features are useful for reconnaissance investigations.

Surface Movement

In construction practice, the total potential movement at the soil surface is critical, particularly in the design of footings and/or slabs for residential buildings and road pavements. Engineers commonly use the term 'reactive soil' to describe soils liable to shrinkage and swelling movement.

Reactive soils are classified on the basis of surface movement. The Australian Standard AS 2870 (1986) (Residential Slabs and Footings) uses the categories outlined in Table 11.7.

Slightly different limits can be placed on these classes where detailed testing is undertaken in a particular region or locality; for example, the *Swelling Soils Study of the Sydney Metropolitan Region* undertaken by the Builders Licensing Board (1984).

Determination of which category best describes a particular soil can be achieved by direct measurement over a number of years or by estimation, using laboratory testing and climatic information.

Table 11.7 Categories of reactive soils

Soil Description	Surface Movement (mm)
Slightly reactive or non-reactive soil	< 20
Moderately reactive	20–40
Highly reactive	40–70
Extremely reactive	> 70

Cores

Laboratory testing can be carried out on undisturbed cores or disturbed samples. Undisturbed cores give the most reliable estimate as the soil structure and chemistry remain close to that occurring in the field. However, due to time and cost considerations, disturbed sample testing is the most commonly used.

Where detailed information is available using direct measurements and/or undisturbed cores in a locality, disturbed sample testing can be used with some confidence in estimating surface movements elsewhere in that locality.

The usual test involves measurements of swell of an undisturbed core with increasing moisture. Interpretation of the test is usually done by engineers with local experience of an area.

Index Tests

Disturbed sample tests are known as index tests and consist of the Atterberg tests for liquid limit and plastic limit, as well as the linear shrinkage test (AS 1289 3.4.1, 1995).

Linear shrinkage is the one-dimensional shrinkage of a soil paste moulded at its liquid limit then oven dried at 105°C for 24 hours. It is expressed as a percentage of the original dimension, and is used as a measure of the shrink-swell potential of a soil. A number of assessment systems are available to assess the soil's potential for movement using these test results such as that of Mills et al. (1980), the Public Works Department (1977), and Holland and Richards (1982) (see Table 11.8).

Other index tests include the Potential Volume Change Meter and the Coefficient of Linear Extensibility (COLE), which are discussed in Mills et al. (1980). These tests have been widely used in the USA to determine shrink-swell potential of soils.

The index tests are useful for the preliminary identification of potential problem soils and hence have been commonly used in soil surveys for urban land use.

Australian Standard for Residential Footing and Slab Construction

An estimate of the likely surface movement can be obtained by this technique.

(i)   The reactivity of the site is assessed by one or more of the following tests:
PI—plastic index
LL—liquid limit
LS—linear shrinkage.

(ii)   The movement is then estimated by multiplying the actual depth of the clay layer in each metre of the profile by the percentage factor in Table 11.9.

(iii) For the purpose of this table, 'temperate and humid' includes Sydney, Hobart and Brisbane, and 'arid and semi-arid' includes Melbourne, Adelaide and Perth. The wet winters and dry summers in these latter three cities lead to a

Table 11.8 Assessment of shrink-swell potential of expansive soils from index tests

(a) MILLS et al. (1980)		(b) PUBLIC WORKS DEPARTMENT (1977)			
Linear Shrinkage (%)	**Expansive Rating**	**Linear Shrinkage**	**Liquid Limit**	**Plasticity Index**	**Shrink-Swell Potential**
0–12	non-critical	0–13	< 45	< 25	low
12–17	marginal	13–17	45–55	25–35	medium
17–22	critical	17–21	55–75	35–45	high
> 22	very critical	> 21	> 75	> 45	very high

	(c) HOLLAND AND RICHARDS (1982)		
	Linear Shrinkage		
	Arid to Semi-arid Climate	**Humid Climate**	**Shrink-Swell Potential**
Non-critical generally refers to soils where the shrink-swell potential is sufficiently low that no problems are expected with building foundations.	0–5	0–12	low
Critical generally refers to soils where the shrink-swell potential is sufficiently high that specific design criteria need to be incorporated in building foundations to prevent movement problems.	5–12	12–18	moderate
	> 12	> 18	high

Table 11.9 Ground movement factors based on index tests expressed as a percentage of clay depth

Test Value			Percentage Swell as a Function of Climate and Depth from the Surface to the Clay Layer					
			Temperate and Humid			Arid and Semi-arid		
PI	LL	LS	0–1 m	1–2 m	> 2 m	0–1 m	1–2 m	> 2 m
< 25	< 50	< 12	1.0	0.4	0	1.0	0.6	0
< 40	< 65	< 20	2.0	0.7	0	2.5	0.4	0
> 40	> 65	> 20	3.0	1.0	0	4.0	2.0	0

wider range of soil moisture conditions during the year.

Shrink-Swell Tests for Earthworks

The successful construction of earthworks for soil conservation purposes requires some knowledge of the shrink-swell potential and expansive qualities of the soil being used. Linear shrinkage is probably the most appropriate test using the interpretation shown in Table 11.8. However, this is a relatively time-consuming test and some evaluation of the soil's expected behaviour can be obtained from the quicker volume expansion test. As Mills et al. (1980) have described, there are problems in interpreting the results of this test if soils are dispersible, but for the very broad categories required to predict soil behaviour in conservation earthworks, it probably gives sufficient information. The interpretation of the volume expansion test is given in Chapter 20.

11.5 Other Soil Properties and Their Effects on Soil Engineering Properties

11.5.1 Acid Sulfate Soils

Acid sulfate soils (described in Chapter 13) have very poor engineering properties, and their disturbance can lead to severe environmental problems.

11.5.2 Saline Soils

Saline soils have high salt levels and this can influence the corrosivity of soils, especially in relation to iron and steel materials. If high sulfate levels are present this may also severely affect concrete materials. High salt levels might also result in high soil permeability, even in clayey materials.

11.5.3 Soil Erodibility

Soil with high erodibility may cause severe problems on construction sites and downstream of construction sites (see Chapter 21). Soil high in sodium (sodic soils) can be especially susceptible to erosion (see Chapters 10 and 12).

BIBLIOGRAPHY

Anon. (1960), *Design of Small Dams*, US Bureau of Reclamation, Dept of the Interior, Washington DC.

Australian Standard, AS 1289.
  3.1.1 Determination of the liquid limit of a soil (1995).
  3.2.1 Determination of the plastic limit of a soil (1995).
  3.4.1 Determination of the linear shrinkage of a soil (1995).
  5.3.2 Determination of the bulk density of a soil by the sand displacement method (1993).
  5.5.1 Soil compaction and density tests. Determination of minimum and maximum dry density of a cohesionless soil (1977).

Australian Standard, AS 2870 (1986), Residential slabs and footings.

Builders Licensing Board (1984), *Swelling Soils Study of the Sydney Metropolitan Region*, NSW Builders Licensing Board, Sydney.

Casagrande, A. (1947), 'Classification and identification of soils', *Proceedings of the American Society of Civil Engineers* 73, 783–810.

Holland, J.E. and Richards, J. (1982), 'Road pavements of expansive clays', *Australian Road Research* 12, 173–9.

Houghton, P.D. and Charman, P.E.V. (1986), *Glossary of Terms Used in Soil Conservation*, Soil Conservation Service of NSW, Sydney.

Ingles, O.G. and Metcalfe, J.B. (1972), *Soil Stabilization: Principles and Practice*, Butterworths, Sydney.

Mills, J.J., Murphy, B.W. and Wickham, H.G. (1980), 'A study of three simple laboratory tests for the prediction of shrink-swell behaviour', *Journal of the Soil Conservation Service of NSW* 36, 77–82.

Public Works Department (1977), *Identification of Expansive Soils in NSW*, Report No. 7, Manly Vale Soils Laboratory, Sydney.

Quilty, J.A., Hunt, J.S. and Hicks, R.W. (1978), *Urban Erosion and Sediment Control*, Technical Handbook No. 2, Soil Conservation Service of NSW, Sydney.

Transport and Road Research Laboratory (1982), *Soil Mechanics for Road Engineers*, Dept. of Environment, Transport and Road Research Laboratory, HMSO, London.

United States Department of the Interior (1960), *Design of Small Dams*, USDI Bureau of Reclamation, Government Printing Office, Washington DC.

Soil Erodibility

CHAPTER 12

G.W. Geeves, J.F. Leys and
G.H. McTainsh

In general, soil erodibility is defined as the susceptibility of a soil to the detachment and transport of soil particles by erosive agents during wind or water erosion (Houghton and Charman, 1986). Soil erodibility can be a significant factor influencing soil erosion hazard. While some long-term erosion prediction methods consider soil erodibility as a constant characteristic inherent to each soil, it is more appropriate to consider it as a characteristic of a soil surface that can vary with time, soil surface management and antecedent soil water conditions. This chapter considers soil erodibility as a factor first influencing water erosion hazard and second influencing wind erosion hazard.

12.1 Soil Erodibility to Water

12.1.1 Effects of Soil Erodibility on Water Erosion Hazard

Water erosion refers to the non-reversible movement of soil by water. It includes inter-rill, rill, gully, tunnel and streambank erosion that are described in detail in Chapter 2. The suite of processes involved in each of these forms of water erosion differs significantly, but each is influenced by the general susceptibility of soil particles to detachment and transport known as soil erodibility. Erosion hazard can be a more useful measure for land managers than soil erosion rate because rainfall intensity and duration are probabilistic. Water erosion hazard refers to the potential for soil movement by water, given the probability of an erosive rainfall event. An area of erodible soil on a steep slope with little surface cover will have a high erosion hazard but may still have a negligible erosion rate if the expected rain does not fall. Conversely, a soil of low erodibility may still exhibit high erosion rates if it occurs on a steep, long slope, has low surface cover and experiences intense rainfall.

12.1.2 Inter-rill Erosion under Rainfall

Inter-rill erosion refers to erosion of the soil surface in between the small, temporary, concentrated flow channels that are known as rills. Erosion on these inter-rill areas is dominated by the impact of energetic raindrops detaching soil particles and raising them into very shallow water flows that carry soil towards rills. Because of the low flow velocities in these shallow, un-concentrated inter-rill flows and the tendency for particles to settle back to the soil surface under gravity, inter-rill erosion requires continual re-detachment and re-entrainment by raindrops. Unless the soil surface is cultivated following every rainfall event, it is common for a layer of detached soil particles to remain on the soil surface from previous erosion events, covering or partially covering the soil surface. It is the resistance of the mosaic of this detached soil layer and the soil surface (where it is not covered) to detachment-entrainment and re-detachment-re-entrainment that will influence inter-rill erodibility (Hairsine and Rose, 1991). This in turn depends on a number of inherent soil factors and transient surface conditions:

—the rate of surface runoff generation, which depends on the rate at which the surface can absorb water, the rate at which subsurface layers allow infiltrated water to drain, and the amount of water each soil horizon or layer is capable of absorbing

—the size distribution of soil aggregates and particles at the soil surface and the ability of these aggregates to withstand disintegration caused by raindrop impact, slaking caused by rapid wetting and dispersion caused by saturation

—the cohesion between the surface aggregates and particles and the rest of the soil surface, which depends on clay content, soil structure and previous sediment deposition as well as consolidation and surface seal formation under rainfall

—the rate at which entrained aggregates and particles settle from overland flow back to the soil surface (known as settling velocity)

—the roughness of the soil surface, which influences the capacity of the surface to trap runoff and sediment, and influences the velocity, and hence the transport capacity, of overland flows.

12.1.3 Rill and Gully Erosion

Rills and gullies differ in scale and their definitions are arbitrary and vary widely. One system defines rills as small, concentrated flow channels that can be obliterated by subsequent tillage. They may reform under subsequent rainfall but not in exactly the same location because their flow paths are controlled by features such as tillage marks and tyre tracks rather than topography. Ephemeral gullies can also be obliterated by tillage but will always reform in the same location because their flow paths are determined by topography. Gullies are too large to be obliterated by tillage and their flow paths are also determined by topography. Despite differences in scale, each of these erosion features results primarily from detachment and transport of soil by concentrated water flow.

Erosion of the side of the channel of a rill or gully may be affected by raindrop impact, diffuse flow over the gully side, and seepage flows through the channel side. However, channel erodibility is primarily controlled by whether the critical shear stress of the soil material is greater than the hydraulic shear applied by the flowing water. If it is, then soil will not be detached. Erosion at the gully head can be influenced by head shape and potential for undercutting and mass movement processes (see Chapter 2), but again erodibility will depend on the hydraulic shear applied by the inflowing water relative to the critical shear stress of the soil surface at the gully head.

The critical shear stress of the gully material is dependent on soil cohesion, which in turn is a function of soil properties such as clay content, clay dispersion, aggregation and bulk density. Another important factor influencing gully shape and erosion rate is the relative erodibility of surface soil and subsurface soil material underlying it. Soils with dispersible subsoils are particularly susceptible to gully erosion.

12.1.4 Tunnel Erosion

Initiation of tunnel erosion generally requires a concentrated lateral flow of water through soil that re-emerges at a soil surface. This can occur when water seeps through an embankment and re-emerges at the toe-slope or when water infiltrates near the head or side of a gully and re-emerges from the gully wall.

Because subsurface seepage flows generally have very low flow velocity, only very fine soil particles can remain in suspension long enough to be transported. For this reason clay dispersibility is the key factor controlling tunnel erodibility. Strong soil cracking and the presence of impermeable subsurface horizons can be subsidiary factors promoting the concentrated lateral subsurface flows required to initiate tunnelling. Subsequent water flows through an eroded tunnel can be of high volume and high flow velocity, especially in cases where a tunnel has formed through an embankment designed to divert or store water such as a diversion bank or farm dam. In these later stages of tunnelling, erodibility will be determined similarly to the flow-driven rill and gully erosion processes referred to earlier.

12.1.5 Inherent Soil Properties Affecting Erodibility

Many soil attributes have been found to affect soil erodibility including soil structure (size and shape of aggregates), permeability, horizonation, particle size analysis, particle density, silica-sesquioxide ratio, organic matter content, texture, aggregate stability, shear strength, infiltration capacity, chemical content (including fertility), mineralogy, aggregate density, antecedent moisture content, cohesiveness, surface roughness and stoniness. Specific studies in New South Wales have also used soil depth, degree of cracking, soil fabric, soil colour and hardsetting characteristics to subjectively assess soil erodibility.

Cohesiveness

Cohesiveness of soils influences the ease with which raindrop impacts or erosive soil water flows can detach soil from the rest of the soil surface. It combines aspects of deformation resistance, and therefore resistance to lateral shear by raindrops, as well as resistance to bedload shear by overland flow. Among other methods, cohesiveness may be estimated from texture assessment, measured by penetrometer or shear vane apparatus (Lloyd and Collis-George, 1982), by compression shear apparatus (Skidmore and Powers, 1982), or by calculation of a consistency index (de Ploey and Mucher, 1981).

Soil Texture

Generally, fine-textured soils are more cohesive and, therefore, less easily detached from the soil mass than coarse-textured soils. Once detached, however, finer materials are more easily transported. Coarse-textured soils are less cohesive, but they can have much higher infiltration rates and thus generate less runoff. Soils least resistant to erosion tend to be those with moderate silt or fine sand contents and limited clay contents. First, silt and fine sand particles are easily detached and transported, and, second, cohesion is not strong as in soils of higher clay content. This picture is complicated for soils that tend to surface seal under rainfall impact, a process that increases runoff but can reduce detachment.

Aggregate Size Distribution

Larger surface aggregates can reduce erodibility because their greater mass increases resistance to raindrop detachment. They can also reduce runoff velocity by increasing surface roughness and reduce runoff by delaying surface sealing under rainfall. The size distribution of stable aggregates also affects sediment deposition and hence transportability. Large aggregates will settle more quickly from overland flows, and hence are more readily deposited than small aggregates but less readily than denser equivalent sand or gravel particles.

Aggregate Stability and Dispersion

The stability of surface aggregates determines how the size distribution of surface aggregates will change under rainfall. It also influences the detachability and transportability of the resulting soil particles. Aggregate stability may be assessed by determining structure grade or may be measured using various wet and dry sieving procedures or a range of other tests. These tests include aggregate wetting tests, which impose large potentials for disruption, such as water-stable aggregate tests; moderately disruptive tests (Emerson Aggregate Test, Childs' [1940] quick wetting test, Middleton's [1930] dispersion ratio, SCS of NSW dispersion percentage [1981]); and restrained disruptive wetting (Collis-George and Figueroa, 1984). The response of aggregates to raindrop impacts is commonly measured by the size or settling velocity of repacked aggregate beds following application of simulated rainfall (for example, Lovell and Rose, 1988; Loch, 1994). Single waterdrop tests using short or long fall paths have also been used (for example, Farres and Cousen, 1985).

Surrogates for aggregate stability tests include organic matter determination and sesquioxide measurement (components of which act as cementing agents for aggregates of primary soil particles), infiltration capacity (uniformly packed aggregates drain more freely the larger the aggregates), and exchangeable sodium percentage combined with electrical conductivity measurements (clays tend to flocculate in high ionic strength, low sodium solutions).

Permeability

Surface runoff, and therefore potential for sheet and rill erosion, is less from permeable, well-drained soil types than from soils with impeded drainage. Impeded infiltration and drainage may result from such factors as soil structural degradation, crusting, presence of an impermeable horizon or layer (plough-pan or illuvial clay horizon), a shallow soil profile, the closure of soil cracks, and high antecedent moisture. A soil that is already wetted will respond very rapidly to an erosive rainfall event and runoff will be generated; in a dry cracking soil, the presence of open cracks will normally dominate the infiltration process.

Subsoil Properties

Erodibility of subsoils, particularly in relation to gully erosion, can be influenced by different factors.

Subsoils in their undisturbed state undergo fewer wetting and drying cycles than topsoils, and the fluctuations in moisture content that do occur are less extreme than in topsoils. Subsoils have an overburden pressure applied to them and tend to have larger aggregate sizes. These circumstances influence the development of aggregates and microstructural features. Subsoils, therefore, have latent structure, and predictions of erodibility of exposed materials made on the basis of their morphology determined during profile description (Wischmeier et al., 1971) may not be accurate. Even when structure is correctly interpreted, the strength of subsoil aggregates may differ from that of topsoil aggregates so that predicted and measured erosion losses differ markedly (Romkens et al., 1975).

With respect to erodibility, subsoils generally differ from topsoils in the following ways.

—Subsoils often have higher clay content. This property affects water uptake, aggregate swelling and the extent of disruption. Cohesion is, therefore, generally greatly reduced by wetting subsoil.

—Subsoils typically have a lower osmotic potential and thus, as critical shear stress has been shown to increase as salinity levels in soils increase, it is likely that rates of erosion in subsoils will be decreased for given flow velocities.

—Subsoils often have a higher sodicity. The degree of clay dispersion is thus likely to be greater in subsoils than in topsoils. While a relatively high external solution concentration will cause a reduction in dispersion, the exposure of subsoil to low ionic strength water (such as rainwater) combined with the effect of adsorbed sodium can produce low levels of soil shear resistance, especially in gullies. For example, Singer et al. (1982) proposed that the USLE (Universal Soil Loss Equation) erodibility factor be increased by 20% for soils with an ESP ((Exchangeable Sodium Percentage) exceeding 2%.

—Subsoils have a lower organic matter content and a higher sesquioxide content than topsoils. The nature of the resistance and disruption by swelling and dispersion forces is, therefore, different between subsoils and topsoils, being influenced by the detachment of brittle oxide cements in subsoils and by mucilaginous organic complexes in topsoils. Greater resistance to detachment is generally correlated with increasing stability of aggregates.

—Subsoil can have a higher stone content, which will result in increased surface armouring. In the USLE the effect of stone cover is considered in the crop and cover management factor, C. The effect of stone armouring is extremely variable, but increases with surface cover (Bergsma, 1985).

—Subsoil particles are denser. For the same particle volume, resistance to transportation increases with density.

—Subsoils tend to form crusts more readily than topsoils. Crusts result in smooth surfaces that restrict development of turbulent flows. They also tend to increase coherence and reduce soil detachment.

2.1.6 Assessing Erodibility Using Field Observations

Soil erodibility assessments and indices are commonly used for the purposes of ranking soil capability or prescribing erosion hazard management recommendations. A range of schemes exists.

Generalised Field-based Erodibility Assessment Schemes

Middleton (1930) proposed the first 'non-empirical' erodibility index. This was a compound index based on dispersion, clay content and moisture retention. Numerous erodibility assessment schemes have been proposed based on convenient observations of the soil properties that are known to be related to soil erodibility, combined with observations of previous erosion at a particular site, but few have been tested widely.

One class ranking scheme proposed by Murphy (1984) for assessment of erodibility of agricultural soils in New South Wales is presented in Table 12.1 as an example of a field-oriented method suitable for use when carrying out soil surveys. The scheme is based on the following generalisations.

Structurally stable and non-dispersible soils are less erodible than unstable, dispersible soils. Aspects of this principle are:

—topsoils high in organic matter are less erodible than those with little or no organic matter (Wischmeier and Smith, 1978)

—subsoils high in sesquioxides are less erodible than those with little or no sesquioxides (Romkens et al., 1975)

—soils that are dispersible owing to high sodium levels are more erodible (for example, Singer et al., 1982)

—water-stable aggregates less than about 2 mm can be readily detached and transported; thus, structurally stable aggregates do not necessarily have low erodibility, and the effect of aggregate stability on soil erodibility can be complex (for example, Wischmeier and Smith, 1978)

—soils with a large number of particles in the very fine sand (0.02–0.10 mm) and silt (0.002–0.02 mm) size range are most erodible (Wischmeier and Smith, 1978)

—soils with high infiltration rates are less erodible owing to the tendency for less runoff to be produced; soils high in coarse sand or with stable, highly structured surface soils tend to have high infiltration rates, whereas hardsetting or unstable fine-textured soils tend to have low infiltration rates.

Assessment of Forest Soil Erodibility

Recent interest in the perceived risk of water pollution from forest operations has led to development of a number of schemes for assessing erodibility of forest soils (for example, Laffan et al., 1996). The scheme currently used to assess soil erosion and water pollution hazard in New South Wales forests (Fogarty and Ryan, 1999) is a two-by-two matrix classification of soil regolith as cohesive or non-cohesive, and soil regolith sediment delivery potential as high or low. Soils with greatest hazard are those that are non-cohesive with high sediment delivery potential. The classification is based on soil mapping and field observations of morphology and erosion behaviour. The type of forest harvesting operation allowed is determined by these two assessments in combination with annual rainfall erosivity and slope class. In addition to this, mass movement hazard is assessed on the basis of soil mapping or field observations, and soil dispersibility is assessed using a modified Emerson Aggregate Test (see Chapter 10).

12.1.7 Erodibility Parameters in Soil Erosion Models

Soil erosion models vary widely from generalised statistical summary models to detailed physically based simulation models. While the concept of soil erodibility is common to them all, the erodibility

Table 12.1 Soil erodibility classes (after Murphy, 1984)

Erodibility	Topsoil	Subsoil
Low	High organic matter (> 3%) (soils have a dark colour and feel greasy when textured)	Cemented layers including silcrete, ortstein and laterite iron, manganese and silicon pans
	High coarse sand	High coarse sand
	Well-structured, non-dispersible clay loams and clays having aggregates that do not slake in water to particles less than 2 mm (Emerson Aggregate Classes 4, 6, 7 and 8), such as red, smooth and rough-ped earths (Gn3, Gn4 soils), some cracking clays (Ug5.1, Ug5.2, Ug5.3 soils), some structured loams (Um6.1 soils) and friable duplex soils (Dr4, Db3 soils)	
Moderate	Moderate organic matter (2–3%)	
	Moderate fine sand and silt, such as hard, pedal red duplex soils (Dr2 soils)	Stable, non-dispersible loams and clay loams, such as red and yellow massive earths (Gn2.1 and Gn2.2 soils)
	Well-structured clay loams and clays that slake in water to particles less than 2 mm (Emerson Aggregate Classes 3 to 6), such as Vertosols (Ug5.1, Ug 5.2 and Ug5.3 soils)	Non-dispersible or slightly dispersible clays with particles that slake to finer than 2 mm Emerson Aggregate Classes 3 to 6), (such as non-sodic, red, brown and yellow duplex soils (Dr, Db and Dy soils)
High	Low (1–2%) to very low (< 1%) organic matter, such as soils with bleached A_2 horizons	Dispersible clays (Emerson Aggregate Classes 1 and 2), such as sodic, yellow and red soils (Dy3.4, Dr3.4, Dr2.3 soils)
	High to very high silt and fine sand (> 65%)	Unstable, dispersible clayey sands and sandy clays, such as yellow and grey massive earths formed on sandstone and some granites (Gn2.3, Gn2.8, Gn2.9, Dy5.8 soils)
		Unstable materials high in silt and fine sand, such as unconsolidated sediments and alluvial materials

parameters and their values vary. Erodibility parameters for two types of models are presented here as illustrations of the range of approaches.

The K Factor of the Universal Soil Loss Equation (USLE) and its Derivatives

The USLE is an empirical model developed by Wischmeier and Smith (1978) using statistical analysis of runoff plot data from the American Midwest. The original model took the form:

$$A = R \times K \times (L \times S) \times C \times P,$$

where

A is the predicted long-term average soil loss (t/ha/yr),

R is a measure of the erosivity of rainfall (the product of storm kinetic energy and maximum 30-minute intensity = EI_{30}),

K is the measure of soil erodibility expressed as long-term average soil loss rate per unit of EI_{30} under standard conditions,

L and S are factors that define the effects of slope length and slope steepness respectively,

C is a crop and cover management factor that describes the effect of crop management in protecting the soil surface from erosion,

P is a land management practice factor describing effects of soil conservation practices such as contour ploughing, minimum tillage, contour banking, strip cropping and so on.

K values for use in the USLE were determined experimentally from short-term plot erosion measurements under natural rainfall for a limited range of soils in the USA. Measurements using simulated rain extended the data to a wider range of soils. They have also been determined experimentally for a small number of soils in conjunction with the computer-based implementation of USLE in New South Wales known as SOILOSS (Rosewell, 1993). A nomograph for estimating K values in metric units based on five attributes (silt + very fine sand content, sand content, organic matter content, structure and permeability) is presented as Figure 12.1. The nomograph provides a reasonable method for determining K when it cannot be experimentally determined. Loch and Rosewell (1992) proposed two modifications to the method of particle size determination for estimating K, one using non-dispersed particle size from 16 hours' end-over-end shaking in water and the other using surface soil particle size after rainfall wetting.

In New South Wales, the bare fallow plots used for experimentally determining USLE K values are cultivated after every significant rainfall event to reduce the effect of surface sealing and rilling on the erosion caused by subsequent rain. The long-term average erosion rates predicted by the USLE are only valid under similar conditions. However, in most applications of the USLE and its derivatives, sites of interest are not cultivated regularly in this way and erodibility and the long-term average erosion rates have to be corrected by modification to the C factor. Experience has shown that with careful selection of the factor values, the USLE can provide accurate estimates of soil loss rates on uniform topography and can provide a useful index of soil erosion hazard.

The Revised Universal Soil Loss Equation (RUSLE) (Renard et al., 1991) retains the same equation structure as the USLE and provides the option of using average soil particle size to estimate K for organic soil and soils derived from recent volcanic ash that are not covered by the original nomograph. It also allows for K values to vary over a season depending on the frost-free period. Most soils in New South Wales do not experience periods of freezing and this and other seasonal effects on soil erodibility have not been evaluated for New South Wales.

The basic structure of both the USLE and RUSLE and the empirical method of determining K dictate that any properties of soil that influence runoff rate will be lumped into the erodibility parameter for that

Figure 12.1   Soil erodibility nomograph (after Rosewell, 1993, in SI units)

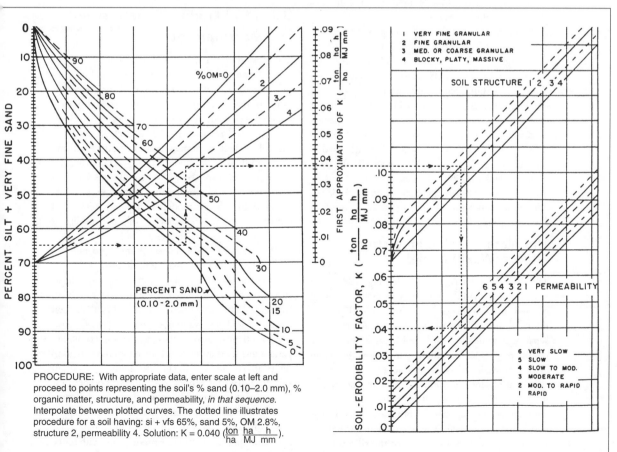

PROCEDURE: With appropriate data, enter scale at left and proceed to points representing the soil's % sand (0.10–2.0 mm), % organic matter, structure, and permeability, *in that sequence.* Interpolate between plotted curves. The dotted line illustrates procedure for a soil having: si + vfs 65%, sand 5%, OM 2.8%, structure 2, permeability 4. Solution: K = 0.040 ($\frac{ton}{ha} \frac{ha}{MJ} \frac{h}{mm}$).

soil. The Modified Universal Soil Loss Equation (MUSLE) (Williams and Berndt, 1977) separates runoff production characteristics from K by replacing the rainfall erosivity term R with a runoff term based on runoff volume and peak runoff rate. This sediment yield model eliminates erroneous predictions made with the USLE and RUSLE when no runoff results from a rainfall event. However, it requires knowledge or prediction of runoff volume and peak runoff and therefore generally needs to be applied with an independent runoff prediction model.

Limitations of using USLE-based models are widely known:

—these models predict only rill and inter-rill erosion; separate account must be made of deposition and of concentrated flow erosion's contribution to sediment production

—these models are parameterised using small plots and should not be applied to larger catchments without taking account of sediment delivery ratios

—the USLE was derived using data from a range of sites dominated by medium-textured soils, and its applicability to soils at either end of the texture scale is less certain

—the approach assumes that soil erodibility is a constant, at least within each event.

Strengths of the approach include its simplicity and its very substantial US database. For New South Wales, the SOILOSS model, which is based on RUSLE, has been validated against long-term plot data and has been shown to give good results (Rosewell, 1992). This approach forms the basis for the soil erodibility characterisation currently used in soil landscape mapping in New South Wales (for example, Jenkins, 1996). It also forms the basis for assessing erosion hazard associated with urban development in New South Wales (NSW Department of Housing, 1998).

The Water Erosion Prediction Project (WEPP) Model

The WEPP model (Lane and Nearing, 1989) is a more process-based simulation model predicting rill and inter-rill erosion. In contrast with USLE, inter-rill and rill erodibility are characterised by separate parameters that do not incorporate effects of soil properties on runoff generation. Inter-rill erosion is predicted as the product of rainfall intensity squared, slope, ground cover, canopy cover and inter-rill erodibility, K_i. Rill erosion is based on continuous prediction of both detachment and deposition within a rill. The detachment capacity of the flowing water is a function of the rill erodibility, K_r, and the excess of hydraulic shear of the flow over the critical shear required for soil detachment. The detachment rate is a function of the detachment capacity and the sediment load being carried relative to the maximum sediment load that could be carried. Deposition rate is calculated from the flow rate, turbulence, settling velocity, V_f, of entrained material, and the sediment load being carried relative to the maximum sediment load that could be carried. The model simulates erodibility parameters varying over a season with soil and crop management.

As with the USLE, implementation of WEPP in the USA involved a very significant field calibration effort. In the case of WEPP, extensive use was made of rainfall simulation experiments rather than plots monitored under natural rainfall. This effort was required to determine the inter-rill and rill erodibility parameters described above. While WEPP has been used and some calibration of erodibility parameters has been carried out in New South Wales, it has yet to be rigorously validated for a wide range of the state's soils, and methods for estimating the erodibility parameters from readily available soil data are lacking.

Using Soil Erodibility Assessments and Measurements

Soil erosion by water results from a complex set of processes that each depend on variable soil properties, climatic variables and antecedent surface conditions. Hence, erosion rates and soil erodibility vary widely over time in response to these factors. Process-based research models of soil erosion, and the soil erodibility parameters in them, have become more complex as our understanding of the processes has improved in an attempt to quantify this variation in erodibility. However, the collection of data required to validate models and parameterise soil erodibility over a wide range of soils and conditions lags behind model development.

Practitioners of soil management, who need to use soil erodibility information to determine erosion hazard or to prescribe soil erosion management measures over large areas, currently still resort to subjective erodibility assessment schemes based on soil maps and field observations, or to USLE-based methods with their known limitations. The soil erodibility parameters used in the USLE-based methods tend to lump all erodibility effects into a single, constant value. Clearly this results in significant potential for error, especially when the methods are used outside their limitations. However, it may be that such applications make the best use of very limited data available and they may provide predictions within the required accuracy for many planning and management purposes. Further, it is difficult to see this situation changing or improving without a significant investment in field-based experimental work to determine refined soil erodibility parameters, and to develop relationships for predicting them from other soil properties.

Soil Erodibility to Wind

Wind soil erodibility can be defined as the susceptibility of a soil to erosion by wind. However, while the above definition sounds simple, soil erodibility is a complex issue. This complexity manifests itself both conceptually and physically. These complex issues are not only a problem for wind erosion researchers, as seen by the number of reviews undertaken by water erosion researchers (see previous section and Loch and Rosewell 1992).

There are two broad aspects to the discussion on soil erodibility. First is the conceptual; that is, what is soil erodibility? when does the concept apply? and what factors influence soil erodibility? Second is the practical; that is, how do you measure and map soil erodibility?

Erodibility—the Concept

Conceptually, soil erodibility is based on the premise that some soils are eroded more often and at a higher rate than other soils. It is also a soil property that is influenced by a number of soil factors and is time dependent. Soil erodibility is not a single value because it changes in time, and is therefore best thought of as a continuum (Leys et al., 1996). The soil properties that influence erodibility also change with time and include:
—soil texture
—soil moisture
—soil binding agents, both mineral ($CaCO_3$) and
     organic (polysaccharides).

As these factors change through time, they in turn influence other factors such as soil aggregation and soil crusts (both physical and biological). Based on this temporal understanding of the processes, it becomes more obvious that soil erodibility is variable and not static.

Historically, erodibility has been represented as a single value. This stems from the practical constraints of wind erosion modelling undertaken in the 1960s and 1970s by the United States Department of Agriculture for both wind and water erosion. Both the US water erosion model Universal Soil Loss Equation (USLE) and wind erosion model Wind Erosion Equa-

tion (WEQ) used average annual factors, the K and I factors, to describe erodibility. Implicit in these factors is the fact that they are averages, which implies that there was some variation about the mean which they represent. However this variability was seldom referred to. So even the early modellers were aware that the erodibility factor was time dependent.

Soil erodibility also has different time scales for the different factors. Some factors, such as soil moisture, soil surface grain size and physical crusts, have short time scales that range from minutes to days. These factors can change with weather conditions or during an erosion event. Other factors like particle size, organic matter, calcium carbonate, soil biota and some biological crusts change over longer time scales of months to years.

Therefore, the concept of an erodibility continuum helps describe the range of susceptibility of a soil to erosion through time. However, when we couple erodibility with a probability analysis, the concept of soil erodibility becomes very useful because both the magnitude and the frequency of the soil susceptibility become apparent.

Figure 12.2 shows diagrammatically the change in magnitude of erodibility for two soils—a sand and a clay. The sand initially has a moderate erodibility which increases to a high level, let's say because of cultivation used to control weeds on a long fallow. Once the crop is planted the erodibility falls again owing to increased aggregation and crusting caused by wetting and drying cycles during the crop growth stage. The clay soil undergoing the same management during this time has adequate aggregation and its erodibility remains low. Then a drought occurs. The sand behaves in much the same way as in other years but the lack of wetting and drying during the drought results in the clay aggregation breaking down and it too is now erodible until the crop is sown, at which point its erodibility falls again. The last year is normal again and the sand is erodible while the clay is stable.

In this way we see how the erodibility of a soil can change along an erosion continuum as a result of climatic and management influences. The utility of the

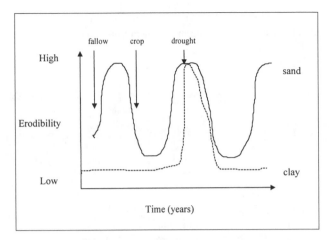

Figure 12.2 Erodibility of two soils to wind through time, showing the repeated high erodibility of the sand and the intermittent high erodibility of the clay

erodibility concept is thus seen. However, the difficulty arises when we try and measure erodibility.

12.2.2 Erodibility—the Practice

When comparisons between soils and sites are made by soil scientists, the tendency is to try and objectively measure differences. This is when difficulties in erodibility assessment occur, because the indices used by scientists are often confused with the conceptual ideas discussed above.

The most utilised measure of erodibility to date has been that of soil transport rate, expressed as either a sediment flux (with units mass/unit area/unit time, for example $g/m^2/s$) or sediment transport (with units mass/unit width/unit time, for example $g/m/s$). Of these indices, the best known are those from the Wind Erosion Equation (WEQ) (Woodruff and Siddoway, 1965). However, in the development of the WEQ there is some confusion about what erodibility is. In the following discussion we will see how the index became removed from the concept of erodibility and confusion began to permeate the literature.

Many types of erodibility indices have been described in the literature. In the works of W.S. Chepil he mentions four types including soil, knoll, wind tunnel and field (Chepil, 1959). Each erodibility type is measured under its own standard conditions and has the following definitions.

The soil erodibility index (I), as used in the wind erosion equation (WEQ) (Chepil and Woodruff, 1963a; Woodruff and Siddoway, 1965), is the potential annual wind erosion for a given soil under a given set of field conditions. The soil erodibility index is expressed as the average annual soil loss in tons per

acre per year for a field that is isolated (incoming saltation is absent), unsheltered (barriers are absent), wide (soil flow has reached its maximum), bare (vegetation cover is absent), smooth (ridge roughness factors are absent), level (knolls are absent), loose (aggregates are not bound together, however clods maybe present), non-crusted (the surface is not sealed), and at a location where the climatic factor (C) is 100. C is 100 for Garden City, Kansas, based on long-term climatic data. The C factor comprises wind velocity and soil moisture parameters (Woodruff and Siddoway, 1965). This is a very precise definition and as such very difficult to replicate.

The soil erodibility index has been determined from a combination of wind tunnel and field measurements (Chepil and Woodruff, 1963a). In Chepil's work on wind erodibility of farm fields (1959, p. 214), erodibility is described in terms of the 'relative quantity of erosion and as a rate of soil flow'. To determine soil erodibility, Chepil used both wind tunnel and field measurements. The wind tunnel method is more fully described below under wind tunnel erodibility. In brief, the rate of soil erosion from the tunnel was expressed as the quantity of soil eroded per unit area before soil removal ceased—that is, soil flux. With field plots, erosion was expressed in terms of rate of soil flow (otherwise known as soil transport). Field plot measurements were standardised to a 65 km/h (40 mph) wind velocity at 15.34 m (50 feet) above the ground where the field was smooth, level and unsheltered (Chepil, 1959). Therefore, we see that Chepil used two measures of soil movement and undertook the research at two scales, which gave him two indices of erodibility. Up until now, the index measured, I, has been a soil property. The next index, knoll erodibility, seems to drift away from the concept of erodibility being a soil property.

The knoll erodibility index (I_s) is used to account for the increased wind erosion potential on topographic features characterised by short, abrupt slopes (knolls). It varies with slope and is applied to windward slopes less than 150 metres long and with a slope gradient greater than 3% from the adjacent landscape (Woodruff and Siddoway, 1965). The knoll erodibility is therefore a slope factor that accounts for the increased erosivity of the wind caused by the acceleration of the wind flow up and over a knoll.

Adjustments to the soil erodibility index (I) are made on the basis that there is increased erodibility near the windward slope of the crest and downwind of the knoll because of the subsequent avalanching effects. However, knoll erodibility is not a true erodibility. The erosivity of the wind is increased owing to increased wind speed and drag and not the erodibility of the soil. As such this is an inappropriate use of the term erodibility.

Wind tunnel erodibility (*i*) is an index of relative erodibility under standard wind tunnel conditions expressed in tons/acre. Field soil samples are placed in trays 5 feet long and 8 inches wide and exposed to wind with a friction velocity (*u**) of 0.61 m/s for five minutes. Erosion usually ceases in that period (Chepil, 1953). Erosion is limited by the length of the tunnel and so *i* does not account for resistance of soil textural classes to abrasion and avalanching (Soil Conservation Service, 1988).

The wind tunnel erodibility (*i*) is similar to the soil erodibility (*I*) in that it has (i) a soil loss rate that is predicted using (ii) a standard climatic erosivity under (iii) standard field and surface conditions. However, the erosivity conditions for *I* are different to those for *i* because of the use of a distinct friction velocity. Similarly, the field conditions are different because the 'wide' field condition is not met for *i* since the soil flow has not reached its maximum. Therefore, *I* is another index of erodibility with a precise definition.

Chepil's next step was to extend the concept of erodibility from a soil to an entire field. With this change in emphasis, the concept of erodibility has been exceeded because more than just soil factors are considered, and as such Chepil's field erodibility (*E*) is not a soil property but rather a site property.

According to Chepil (1959), field erodibility (*E*) is determined by application of the Wind Erosion Equation (WEQ) which uses those factors that influence the potential soil loss. The equation has several forms, however the most widely used is that of Woodruff and Siddoway (1965) and is expressed as:

$$E = f(I', C', K', L', V'),$$

where

> *E*   is the potential soil loss in tons per acre per annum,
> *I'* is soil and knoll erodibility,
> *C'* is the local wind erosion climatic factor,
> *K'* is the soil ridge roughness factor,
> *L'* is the field length,
> *V'* is the equivalent quantity of vegetation cover.

The term field erodibility (*E*) appears to be in contradiction with the other erodibility definitions. Unlike soil erodibility (*I*) and wind tunnel erodibility (*i*), *E* has a *C* factor that changes from place to place (which influences erosivity), *L*-type factors may be variable (which influences whether or not soil flow has reached a maximum), a range of *K*-type factors (which indicates that the surface is not smooth), and *V*-type factors (which indicates that the surface has vegetation cover). In fact *E* is more related to the ero-

sion hazard, which is used in Australia to indicate when all factors such as climate, soil, land use and land management factors are used to predict soil loss (Houghton and Charman, 1986).

From the above discussion it is apparent that erodibility has been measured in a variety of ways and used to describe several erosion conditions. For the practical determination of wind erodibility, a distinct and universal definition can be applied: soil erodibility is a measure of soil movement (expressed in mass per unit width or area, per unit time) under standard erosivity and ground/field conditions.

What also is apparent is that with each of these indices, the scale (plot, field, region), place in the erosion cycle (erosion, transport or deposition zone), the measurement technique (wind tunnel, field dust traps and so on), the type of sampler (active or passive filtration), and the time scale (minutes to months) all influence the measurement and therefore the index. Without a knowledge of where the measurements are in the erodibility continuum, it is difficult to make comparisons between measurements and sites.

Measures of Erodibility　　12.2.3

Despite the difficulties of measuring erodibility, it is still possible to measure and predict the erodibility of a soil *provided* the reader understands:
—the differences between the concept and the practical application
—where they are in the erodibility continuum
—the limitation of the method.

Two things need to be considered before measuring the erodibility. First is the time scale and the second is the method. The time scale can range from instantaneous to years, while the methods can be divided into indirect and direct methods. Indirect methods use a soil property to infer the sediment movement rate, while direct methods measure the sediment movement rate directly.

Suggested indirect methods include:
—soil texture for long time scales (years)
—dry aggregation for moderate time scales (weeks/months)
—a combination of dry aggregation, soil moisture, binding agents and roughness for short time scales (minutes/hours).

Soil Texture

Soil surface texture is a good indicator of erodibility over long time scales. Sandy soils are eroded at higher rates and more often, owing to their inherently low level of soil aggregation, lower soil moisture-holding capacity and lower organic matter levels. This is supported by both Australian and US wind tunnel data. Figure 12.3 shows the sediment transport rate for

Table 12.2 Description of wind erodibility groups (after Soil Conservation Service, 1988)

WEG	Soil Surface Texture Class	Wind Erodibility Index (t/ha)
1	Very fine sand, fine sand or coarse sand	659
2	Loamy very fine sand, loamy fine sand, loamy sand, loamy coarse sand or sapric soil materials	300
3	Very fine sandy loam, fine sandy loam, sandy loam or coarse sandy loam	193
4	Clay, silty clay, non-calcareous clay loam or silty clay loam with more than 35% clay content	193
4L	Calcareous loam and silt loam or calcareous clay loam and silty clay loam	193
5	Non-calcareous loam and silt loam with less than 20% clay content or sandy clay loam, sandy clay and hemic organic soils	126
6	Non-calcareous loam and silt loam with more than 20% clay content or non-calcareous clay loam with less than 35% clay content	108
7	Silt, non-calcareous silty clay loam with less than 35% clay content and fibric organic soil material	85
8	Soils not susceptible to wind erosion	0

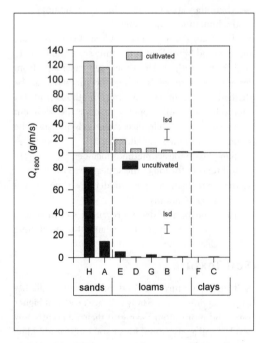

Figure 12.3 Sediment transport rate of nine soils in the cultivated and uncultivated condition (after Leys, 1990)

nine soils in the cultivated and uncultivated state. Soils with less than 13% clay (sandy loam) content have higher erosion rates than soils that have more clay.

From wind tunnel studies in the USA (Chepil, 1953; Zingg, 1951), the United States Department of Agriculture introduced the concept of wind erodibility groupings (WEGs), which used the predominant soil surface texture class to predict the wind erodibility index. The WEGs and wind erodibility index (I, the ratio of the mass of sediment eroded from a soil containing 60% greater than 0.84 mm, to the mass eroded from a soil containing any other proportion of clods greater than 0.84 mm under the same conditions) are given in Table 12.2.

Table 12.2 shows that there is a rapid increase in the erodibility index (I) for sandy loams, loamy sands and sands with I increasing from 193 to 300 to 695 t/ha respectively. Therefore, soils sandier than sandy loams appear to have a high erodibility compared to more loamy and clayey soils. Over longer time scales, texture appears to be a good indicator of soil erodibility. A full discussion of the role of texture in determining sediment movement can be found in Leys (1991, 1999).

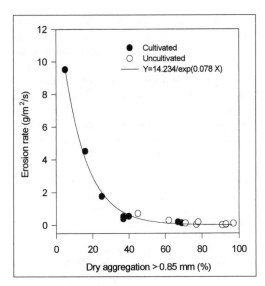

Figure 12.4 Relationship between dry aggregation and erosion rate for nine soils at a wind speed of 65 km/h, in the cultivated and uncultivated condition (after Leys et al., 1996)

Dry Soil Aggregation

The percentage mass of soil aggregation greater than 0.84 mm when it is dry (dry soil aggregation) is a good indicator of the erodibility over moderate time scales like weeks and months. In contrast to soil texture, which changes very slowly with time, dry aggregation changes more rapidly in response to soil texture, organic matter, soil binding agents, climate and management.

Dry aggregation has been adopted for the Revised Wind Erosion Equation (USDA-ARS, undated) as the measure of erodibility in the model (Fryrear et al., 1994). Dry aggregation has also been shown to be a good predictor of erosion rate in Australia (Figure 12.4).

Dry aggregation is a simple field measure that can be quickly done to estimate the erodibility based on Figure 12.4. Dry aggregation can be determined by gentle hand-sieving of 1 to 2 kg of the top 3 cm of dry soil. A fuller discussion of the measurement and use of dry aggregation can be found in Leys et al. (1996) and Leys (1999).

Combination of Dry Aggregation, Soil Moisture, Binding Agents and Roughness

To determine the soil erodibility over very short time periods those soil factors which control the erosion rate and which change rapidly need to be assessed. Currently, this would entail the application of a wind erosion model such as the wind

erosion assessment model (Shao et al., 1996) or some other wind erosion model (USDA-ARS, undated; Gregory, 1991; Marticorena and Bergametti, 1995). The application of such models is generally still in the hands of the model developers and none, with the possible exception of the Revised Wind Erosion Equation, is readily available at this time. These models are generally designed to predict erosion hazard, which is the erosion rate when all factors such as climate, soil, land use and land management factors are used to predict soil loss (Houghton and Charman, 1986). The discussion of erosion hazard is beyond this section and will not be dealt with here.

The above discussion covers indirect methods of measuring erodibility. The other option is to measure it directly such as with field dust traps or a wind tunnel.

Dust Traps

Dust traps are suitable for direct measurement of soil movement for moderate to longer time scales. Several dust traps have been calibrated and used in Australia. The efficiencies of each trap are detailed in Shao et al. (1993).

Wind Tunnels

Wind tunnels are suitable for the direct measurement of soil movement over short time scales (minutes to hours). Extensive use of wind tunnels has been undertaken in Australia and the method for their use is detailed in Leys and Raupach (1991). The major difficulty with wind tunnels is their availability and cost to establish. However, they are very effective in the determination of erodibility and the assessment of the various factors that influence erodibility.

Erodibility Scale

12.2.4

Once the practical measurement of erodibility has been achieved, the next question is, how erodible is the soil relative to other soils? An erodibility scale has been proposed by Leys (1999) based on field data collected with a wind tunnel.

Based on the wind tunnel results presented in Figure 12.3, an erodibility classification is proposed. The erodibility scale can also be used to select an erosion control target. This target can then be used objectively to identify threshold levels, as evaluated in the wind tunnel, for various erosion control factors such as dry aggregation and soil moisture.

Figure 12.4 shows a rapid increase in erosion rate at the lower percentage clay content and aggregation levels. When aggregation levels are greater than 30% and clay contents are greater than 15%, then the erosion rate is low—that is, in the order of 1.19 g/m^2/s (or 5 g/m/s in the tunnel). When

Table 12.3 Erodibility classification according to the sediment transport rate at a wind speed of 65 km/h (Q_W). The corresponding dry aggregation level greater than 0.85 mm (% DA) and percentage clay content (% Clay) that provide effective wind erosion control (that is, less than 5 g/m/s) for a cultivated soil is also included

Erodibility Class	Q_W	% DA	% Clay
Low	< 5 g/m/s	> 31	> 13
Moderate	5–25 g/m/s	11–31	6–13
High	> 25 g/m/s	< 11	< 6

the aggregation levels are low (< 10%) and clay content is low (< 5%), then the erosion rate is high, in the order of 5.95 $g/m^2/s$ (or 25 g/m/s). Therefore, the point on the curve where the erosion rate rapidly starts to increase is an appropriate threshold level.

Similarly, measurements with the wind tunnel over many years indicate that when the soil's surface shows few signs of erosion, the sediment transport rate is less than 5 g/m/s. When the surface looks eroded and ripples are present, the sediment flux rate is greater than 25 g/m/s. Based on the curve in Figure 12.4 and these field observations and measurements, a three-class erodibility scale for use with the wind tunnel is proposed (Table 12.3).

Erodibility levels can also be subjectively assessed by inspecting the surface features of predominantly bare and dry soil after high winds. Soils with low erodibility reveal only the slightest movement of particles. Soils with moderate erodibility have a windswept appearance on the higher surfaces, with deposition behind objects and in surface depressions. Soils with high erodibility have a surface that will look smoothed, and ripples may be evident. These features have been described and photographed by Semple et al. (1988).

In conclusion, soil erodibility is a simple concept with great utility. However, as the above discussion has shown, conceptually there is still some confusion about soil erodibility to wind and its measurement. Soil erodibility is a soil property that is time dependent. As such, the erodibility of soil changes with time and should not be described by a single value. Viewing soil erodibility as a continuum along which a soil might be at any one time helps reinforce the soil erodibility concept.

Measurement of soil erodibility is not as straightforward as it might first seem. Complexities with measurement method, time scale and location on the erodibility continuum all add to the confusion of what the measure of erodibility is providing. Despite this, a number of methods for the determination of erodibility are suggested. Each method is suitable for different time scales, and simple guidelines are presented along with a soil erodibility scale for use with the portable wind tunnel used in this research.

BIBLIOGRAPHY

Anon. (1982), *Soil Erosion by Wind,* Agriculture Canada Publication 1266, Ottawa, Ontario.

Bagnold, R.A. (1941), *The Physics of Blown Sand and Desert Dunes,* Methuen, London.

Bergsma, E. (1985), 'Development of soil erodibility evaluation by simple tests', *Proc. EEC Workshop, Land Degradation due to Hydrological Phenomena in Hilly Areas,* Cesena, Florence.

Bryan, R. B., Govers, G. and Poesen, J. (1989), 'The concept of soil erodibility and some problems of assessment and application', *Catena,* 16, 393–412.

Chepil, W.S. (1945), 'Dynamics of wind erosion. III. The transport capacity of wind', *Soil Science* 60, 475–80.

Chepil, W.S. (1950), 'Properties of soil which influence wind erosion. I. The governing principle of surface roughness', *Soil Science* 69, 149–62.

Chepil, W.S. (1951), 'An air elutriator for determining the dry aggregate soil structure in relation to erodibility by wind', *Soil Science* 71, 197–207.

Chepil, W.S. (1953), 'Factors that influence clod structure and erodibility of soil by wind. II. Water stable structure', *Soil Science* 77, 389–99.

Chepil, W.S. (1959), 'Wind erodibility of farm fields', *Journal of Soil and Water Conservation* 14, 214–19.

Chepil, W.S. and Woodruff, N.P. (1963a), 'Physics of wind erosion and its control', *Advances in Agronomy* 15, 211–302.

Chepil, W.S. and Woodruff, N.P. (1963b), 'Estimations of wind erodibility of field surfaces', *Journal of Soil and Water Conservation,* 257–65.

Childs, E.C. (1940), 'The use of the soil moisture characteristic in soil studies', *Soil Science* 50, 239–52.

Cole, G.W. (1984a), 'A method for determining field wind erosion rates from wind tunnel derived functions', *Transactions of the American Society of Agricultural Engineers* 27, 110–16.

Cole, G.W. (1984b), 'A stochastic formulation of soil erosion caused by wind', *Transactions of the American Society of Agricultural Engineers* 27, 1405–10.

Collis-George, N. and Figueroa, B.S. (1984), 'The use of the high energy moisture characteristic to assess soil stability', *Australian Journal of Soil Research* 22, 349–56.

Emerson, W.W. (1967), 'A classification of soil aggregates based on their coherence in water', *Australian Journal of Soil Research* 5, 47–57.

Farres, P.J. and Cousen, S.M. (1985), 'An improved method of aggregate stability measurement', *Earth*

Surface Processes and Landforms, 10, 321–29.

Fogarty, P.J. and Ryan, P.J. (1999), 'The soil stability factor in erosion hazard assessment: the NSW approach' in *Proc. Second Forest Erosion Workshop, Report 99/6*, Co-operative Centre for Catchment Hydrology, Canberra, 31–2.

Fryrear, D.W. (1986), 'A field dust sampler', *Journal of Soil and Water Conservation* 20, 117–20.

Fryrear, D.W., Krammes, C. A., Williamson, D. L. and Zobeck, T. M. (1994), 'Computing the wind erodible fraction of soils', *Journal of Soil and Water Conservation*, 49(2), 183–8.

Gabriels, D. and Moldenhauer, W.C. (1978), 'Size distribution of eroded material from simulated rainfall: effect over a range of textures', *Soil Science Society of America Journal* 42, 954–8.

Gregory, J.M. (1991), *Wind Erosion: Prediction and Control Procedures*, Report prepared for US Army Corps of Engineers, Waterways Experimental Station, Vicksburg, Mississippi, Texas Tech. University, Lubbock, Texas.

Hack, J.T. (1941), 'Dunes of the western Navajo country', *Geographical Review* 31, 240–63.

Hairsine, P.B. and Rose, C.W. (1991), 'Rainfall detachment and deposition: sediment transport in the absence of flow driven processes', *Soil Science Society of America Journal* 55, 320–4.

Houghton, P.D. and Charman, P.E.V. (1986), *Glossary of Terms Used in Soil Conservation*, Soil Conservation Service of NSW, Sydney.

Jenkins, B.R. (1996), *Soil Landscapes of the Braidwood 1:100 000 Sheet*, Report, Dept. Land and Water Conservation, Sydney.

Laffan, M., Grant, J. and Hill, R. (1996), 'A method for assessing the erodibility of Tasmanian forest soils', *Australian Journal of Soil and Water Conservation* 9(4), 16–22.

Lane, L.J. and Nearing, M.A. (1989), *USDA Water Erosion Prediction Project: Hillslope Profile Model Documentation*, USDA National Soil Erosion Research Laboratory Report No.2, West Lafayette, Indiana.

Lettau, K. and Lettau, H. (eds) (1978), 'Experimental and micrometeorological field studies of dune migration' in *Exploring the World's Driest Climate*, IES Report 101, University of Wisconsin-Madison, Institute for Environmental Studies, 110–47.

Leys, J.F. (1990), 'Blow or grow? A soil conservationist's view to cropping Mallee soils', in J.C. Noble, P.J. Joss and G.K. Jones (eds), *The Mallee Lands, A Conservation Perspective*, Adelaide, 1990, CSIRO, Melbourne, 280–6.

Leys, J.F. (1991), 'The threshold friction velocities and soil flux rates of selected soils in south-west New South Wales, Australia', *Acta Mechanica*, Suppl 2, 103–12.

Leys, J.F. (1999), 'Wind Erosion Processes and Sediments in South-eastern Australia', PhD Thesis, Griffith University, Brisbane.

Leys, J.F. and Raupach, M.R. (1991), 'Soil flux measurements using a portable wind erosion tunnel', *Australian Journal of Soil Research* 29, 533–52.

Leys, J.F. and Semple, W.S. (1984), *Estimating the Weight of Crop Residues for Wind Erosion Control*, Western Area Tech. Bull. No. 24, Soil Conservation Service of NSW, Sydney.

Leys, J.F., Koen, T. and McTainsh, G.H. (1996), 'The effect of dry aggregation and percentage clay on sediment flux as measured by portable wind tunnel', *Australian Journal of Soil Research* 34, 849–61.

Lloyd, J.E. and Collis-George, N. (1982), 'A torsional shear box for determining the shear strength of agricultural soils', *Australian Journal of Soil Research* 20, 203–11.

Loch, R.J. (1994), 'A method for measuring aggregate water stability of dryland soils with relevance to surface seal development', *Australian Journal of Soil Research* 32, 687–700.

Loch, R.J. and Rosewell, C.J. (1992), 'Laboratory methods for measurement of soil erodibilities (K factors) for the Universal Soil Loss Equation', *Australian Journal of Soil Research* 30, 233–48.

Lovell, C.J. and Rose, C.W. (1988), 'Measurement of soil aggregate settling velocities. I. A modified bottom withdrawal tube method', *Australian Journal of Soil Research* 26, 55–71.

Marticorena, B. and Bergametti, G. (1995), 'Modelling the atmospheric dust cycle', *Journal of Geophysical Research, 100(D8)*, 16, 415–30.

Middleton, H.E. (1930), *Properties of Soils Which Influence Soil Erosion*, USDA Tech. Bull. 178, Washington DC.

Murphy, B.W. (1984), *A Scheme for the Field Assessment of Soil Erodibility for Water Erosion*, Technical Paper 19/84, Wellington Research Centre, Soil Conservation Service of NSW, Sydney.

NSW Department of Housing (1998), *Managing Urban Stormwater: Soils and Construction*, Third Edition, NSW Department of Housing, Housing Products Division, Liverpool, NSW.

de Ploey, J. and Mucher, M.J. (1981), 'A consistency index and rainwash mechanisms on Belgian loamy soils', *Earth Surface Processes and Landforms* 6, 319–30.

Renard, K.G., Foster, G.R., Weesies, G.A. and Porter, J.P. (1991), 'RUSLE: Revised universal soil loss equation', *Journal of Soil and Water Conservation* 46, 30–3.

Romkens, M.J.M. (1985), *The soil erodibility factor: A perspective*, S.A. El Swaify, W.C. Moldenhauer and A. Lo (eds), Soil Conservation Society of America, Ankeny, Iowa, 445–61.

Romkens, M.J.M., Nelson, D.W. and Roth, C.B. (1975), 'Soil erosion on selected high clay subsoils', *Journal of Soil and Water Conservation* 30, 173–6.

Rosewell, C.J. (1992), 'The development of land protection technology in Australia', in *Proceedings of 7th ISCO Conference*, International Soil Conservation Organisation, Sydney.

Rosewell, C.J. (1993), *SOILOSS: A Program to Assist in the Selection of Management Practices to Reduce Erosion*, Tech. Handbook No. 11 (Second Edition), Department of Conservation and Land Management, Sydney.

SCS of NSW (1981), *Dispersion Percentage, Test P7a, Laboratory Procedures*, Soil Conservation Service of NSW, Sydney.

Semple, W.S., Leys, J.F. and Speedie, T.W. (1988), 'Assessing the Degree of Wind Erosion', *Soil Conservation Service Technical Report No. 3*, Soil Conservation Service of NSW, Sydney.

Shao, Y., McTainsh, G.H., Leys, J.F. and Raupach, M.R. (1993), 'Efficiencies of sediment samplers for wind erosion measurement', *Australian Journal of Soil Research* 31, 519–32.

Shao, Y., Raupach, M.R. and Leys, J.F. (1996), 'A model for predicting aeolian sand drift and dust entrainment on scales from paddock to region', *Australian Journal of Soil Research* 34(1), 309–42.

Singer, M.J., Janitzky, P. and Blackard, J. (1982), 'The influence of exchangeable sodium percentage on soil erodibility', *Soil Science Society of America Journal* 46, 117–21.

Skidmore, E.L. and Powers, D.H. (1982), 'Dry soil-aggregate stability: energy based index', *Soil Science Society of America Journal* 46, 1274–9.

Thompson, D.F. (1981a), 'Wind flow in western New South Wales', *Journal of the Soil Conservation Service of NSW* 37, 79–90.

Thompson, D.F (1981b), 'Wind erosivity indices for western New South Wales', *Journal of the Soil Conservation Service of NSW* 37, 157–65.

USDA-ARS (undated), *Revised Wind Erosion Equation Version 5.01 Training Manual*, US Department of Agriculture, Agriculture Research Service—Wind Erosion Management Unit, Washington DC.

Williams, J.R. and Berndt, H.D. (1977), 'Sediment yield prediction based on watershed hydrology', *Transactions of the American Society of Agricultural Engineers* 27, 129–44.

Wischmeier, W.H. and Smith, D.D. (1978), *Predicting Rainfall Erosion Losses: A Guide to Conservation Planning*, USDA Agriculture Handbook No. 537, Washington DC.

Wischmeier, W.H., Johnson, C.B. and Cross, B.V. (1971), 'A soil erodibility nomograph for farmland and construction sites', *Journal of Soil and Water Conservation* 26, 189–92.

Woodruff, N.P. and Siddoway, F.H. (1965), 'A wind erosion equation', *Soil Science of America Proceedings* 29, 602–8.

Zingg, A.W. (1951), 'Evaluation of the erodibility of field surfaces with a portable wind tunnel', *Soil Science Society of America Proceedings* 15, 11–17.

Soil Chemical Properties

CHAPTER 13

In this chapter the chemical properties of soil as they can affect sustainable land management are considered. Soil chemical properties that are important in land management include soil acidity, soil salinity, nutrient levels, levels of oxidisable sulfur (acid sulfate soils), levels of toxic compounds in soils, and cation exchange capacity.

The chemical properties of soils are largely dependent on the chemical make-up of the materials from which they were formed and the nature of the soil-forming processes they have been subjected to during their formation. Vegetation has an important role to play in soil chemistry, through the recycling of organic matter and its products in the soil. The use of soils for various purposes may involve changes to their chemical characteristics through the application of amendments such as fertilisers, soil ameliorants such as lime and gypsum, herbicides, and organic materials; or through pollution from waste disposal or industrial contamination. Soil management practices such as tillage, cropping, clearing, pasture growth, irrigation and leaching can also have long-term effects on soil chemical properties.

The chemical make-up of soils has direct effects on plants through the supply of nutrients for plant growth and the presence of elements and compounds in toxic quantities that restrict growth. Indirect effects on plant growth are also important because clay chemistry and the relative amounts of various cations can have a vital bearing on the structural condition of the soil (see Chapter 10). The chemical make-up of the soil can also influence the on-site and off-site environmental impacts that may result from various land uses.

This chapter is presented in several distinct sections. Each section concentrates on a particular soil chemical property and discusses the land management and environmental issues associated with that property in some detail.

13.1 Cation Exchange Capacity P.E.V. Charman

Cation exchange capacity refers to the potential capacity of the soil to interact with and bind elements and compounds in the soil. As such it is a measure of the capability of the soil to store and filter chemicals or other reagents, and buffer the soil chemical properties against changes.

The cation exchange capacity of the soil is largely based on clay particles that are the smallest mineral component of soils, and those generally less than 1 μ in size display colloidal properties. Because of their extremely small size, their specific surface area is large in relation to their mass. The clay particles are mainly layered silicates that have a net negative charge because of an excess of oxygen atoms within the crystal structure of the clay particles. This net negative charge is available to adsorb cations onto the clay particles. There may also be some negatively charged sites on the broken edges of some clay particles. Soil organic matter can also contribute to the cation exchange capacity of the soil, although it does this in a complex way. The cation exchange capacity of the organic matter depends on the type of organic matter and the inherent soil chemical environment within the soil.

If the clay colloid is dispersed in soil water these charges are available to attract oppositely charged ions from the soil solution. The layered silicates forming the clay particles are negatively charged owing to these surface ions, so in soils the attraction for positive ions such as sodium, potassium, calcium, magnesium, aluminium and hydrogen tends to predominate.

This phenomenon is called cation exchange and it has an important bearing on many soil properties, particularly fertility, salinity, acidity and structure. Soils vary widely in their cation exchange capacity

(CEC), according to their mineralogy, clay structure and conditions of formation and management. Soils can show some anion exchange activity (the attraction of negative ions such as chloride and sulfate to clay surfaces), depending on their chemistry, but it is generally not as important as cation exchange.

The cations attracted to clay surfaces in this way are said to be adsorbed. Note that the cations are adsorbed as they are only attached to the surface of the clay particles. They do not become absorbed into the crystal structure of the clay. As they are only adsorbed, they can be readily replaced by other cations in soil solutions. The determination of the CEC of a soil is carried out by leaching it with a concentrated solution with alternative cations to sodium, calcium, potassium and magnesium. The CEC can be estimated from the sum of these cations which gives CECB. The CEC can also be estimated from the total uptake of the alternative cation. This gives CEC_{ex}. Leaching solutions include ammonium salt, barium or silver thiourea. These alternative cations in the leaching solution replace all the cations on the clay. The resulting leachate is analysed for sodium, potassium, calcium and magnesium, and the soil for total cation taken up. The difference between the two totals is attributed to exchangeable hydrogen and aluminium, the chemistry of which is important to soil acidity and the pH level in soils. Aluminium itself becomes toxic to plants in very acid soils. Calcium, on the other hand, has a beneficial effect on acid soil; hence the importance of liming where soil acidification has become a problem.

Calcium is also important for the structure of soils, as this ion tends to replace sodium on the exchange complex and causes the colloidal clay particles to coagulate (hang together), giving rise to a more stable structure. Sodium has the opposite effect, causing clays to disperse when it dominates the exchange complex. Magnesium and potassium have similar, but lesser, effects to calcium and sodium respectively. However, magnesium aggregates swell more and are more easily dispersed than calcium aggregates, and thus may give rise to some instability in subsoils, especially when they are partly sodic. The role of exchangeable sodium is also significant in relation to soil salinity, as the salinisation process involves the soil being affected by this element in the form of sodium chloride. The balance between this salt in the soil solution and sodium on the exchange complex depends largely on rainfall patterns and amounts, and determines the degree of instability induced into the soil profile.

Potassium, calcium and magnesium are all essential plant nutrients, and sodium is of some benefit to particular crops. These elements are held in soils partly in exchangeable form, and as such represent a prime source for plants as the exchange processes between clay surface and soil solution take place.

13.2 Soil Acidification G. Fenton and K.R. Helyar

Soil acidification degrades the soil resource by altering the chemical fertility of the soil. Reduced plant growth is the most visible effect of soil acidity, caused by a reduction in availability of some essential nutrients, an increase of other nutrients to toxic levels, and the effect of the acidity on biological systems.

The effect of soil acidity on the environment, and hence the community as a whole, is not as obvious. When plants do not grow to their climatic potential they have lower water use efficiency leading to excessive recharge of the soil aquifers and, among other things, increased dryland salinity. It also means more bare ground resulting in increased runoff rates. This runoff, in turn, can increase soil erosion and addition of silt and organic matter to waterways. A detailed picture of the off-site impacts of soil acidity is given in Figure 13.1.

Soil acidity reduces the income and increases the costs of farming systems. Failure of acid-sensitive crops such as barley and canola to establish and flourish, or the appearance of yellow patches in cereal crops, are early signs that soil acidity is affecting production. Similarly poor establishment or lack of persistence of acid-sensitive pastures, such as lucerne and acid-sensitive varieties of phalaris, indicates that the soil may be acid. The only practical way to reverse the effects of soil acidity is to use lime, increasing the cost of farming.

Generally the surface soils are the first to show a drop in pH. As the pH in the topsoil drops to below 5.2 (all pH values discussed here are determined in

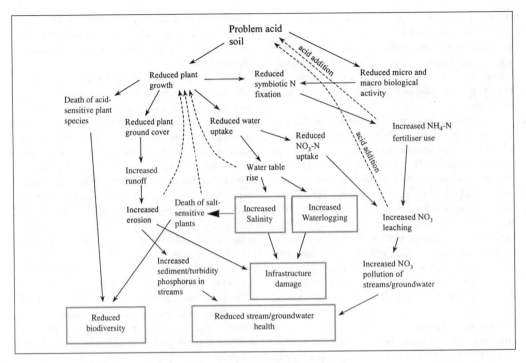

Figure 13.1 Systemic impacts of soil acidification including off-site impacts; components in boxes represent major external impacts (Cregan and Scott, 1998)

0.01 M $CaCl_2$), net movement of acidity down the profile occurs. Both the losses to agriculture and the effect on the environment are increased as the depth of the rooting zone affected by acidity increases. The deeper the soil acidity, the more difficult it is to correct, and thus the more serious the degradation.

The purpose of this section is to assist readers to understand the cause and effect of soil acidity and to know the options for managing acid soils. Both farmers and the community will benefit from better management of acid soils.

13.2.1 The Process of Acidification

In natural ecosystems most soils become acid with time. The older and more weathered soils are generally more acidic than younger soils. This is the natural process. Agricultural production on these soils increases the rates of soil acidification, and is usually called agriculturally induced acidity.

Acidification starts when rock surfaces are first colonised by algae and lichens. Acids, resulting largely from the carbon and nitrogen cycles (Helyar and Porter 1989), are involved in the dissolution of soil and rock minerals during the process of weathering and soil development (see Chapter 1). The 'base' cations, that is, Ca^{2+}, Mg^{2+}, K^+ and Na^+, are released into the soil solution as part of this process and form an association with the nitrate and organic anions.

The pH of the soil fluctuates through the seasons and with changes in soil moisture. Permanent reduction in pH only occurs when there is a loss from the soil (for example, farm produce removed or nitrogen leached). This loss of anions, and their associated cations, entrenches the acidification that has occurred. The removal of the base cations is often referred to in older texts as the cause of acidity in the soil but is, in fact, a consequence of acidification of the soil. A permanent increase in pH only occurs with the addition to the soil of alkaline materials such as plant material, manure or lime.

The Natural Process

The natural processes that affect the acidity of a soil are:

—age, as younger soils have had less time for acid addition

—climate, particularly rainfall, which affects the vegetation produced and nitrate fixation and leaching

—soil parent materials, because of differences in their acid-neutralising capability. Basalt or alluvium are rich in basic minerals, which have a high neutralising capability, compared to previously weathered materials such as a sandstone with a 98% quartz (SiO_2) content.

Plate 1.1 Soil formation factors—climate, parent rock, vegetation, animals, slope
(see 'Soil Formation', p. 3)

Plate 4.3 Subterranean clover showing poor growth and poor ground cover on acid soils
(see 'Soil Acidification, p. 64)

Plate 4.4   Poor crop growth due to soil structural degradation
(see 'Soil Structural Degradation', p. 65)

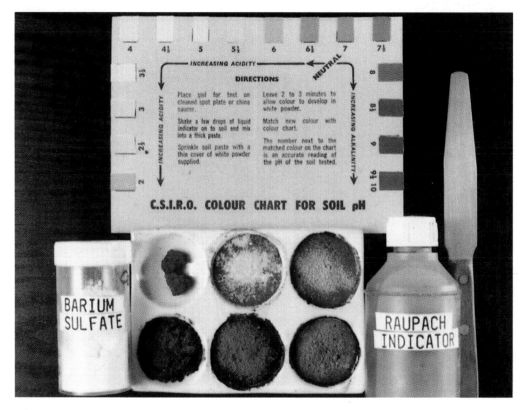

Plate 5.5   The field kit for measuring soil pH
(see 'Field Soil Tests', p. 80)

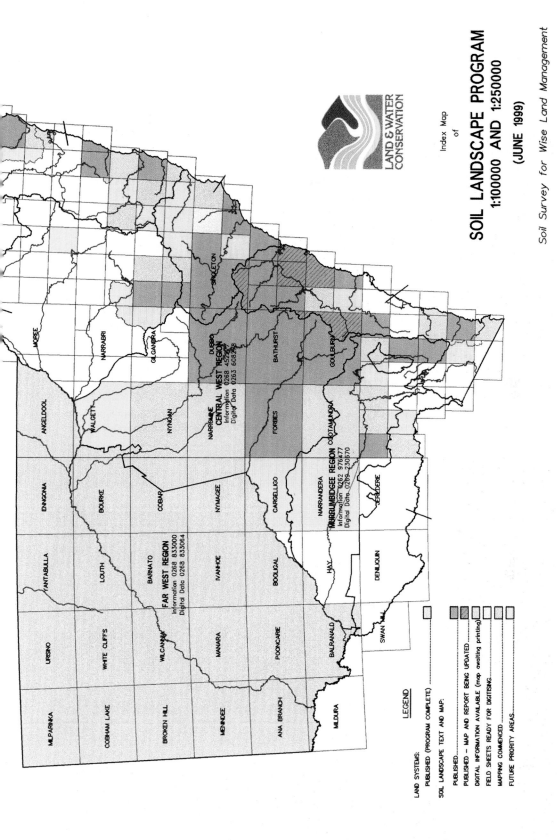

Plate 7.1 Areas covered by the DLWC's soil landscape mapping program

Plate 9.1   A shallow yellow podzolic soil with associated Permian sandstone in the Mount Piper land-scape near Lithgow (see 'Undulating and Rolling Low Hills with Yellow Podzolic Soils', p. 155)

Plate 9.2   The landscape associated with soils similar to the podzolic soil shown in Plate 9.1 (see 'Undulating and Rolling Low Hills with Yellow Podzolic Soils', p. 155)

Plate 9.3   A red earth soil on Paleozoic sediments near Sunny Corner
(see 'Undulating and Rolling Steep Hills with Red Earths', p. 154)

Plate 9.4   The soil landscape associated with the red earth soils similar to that shown in Plate 9.3 (see
'Undulating, Rolling and Steep Hills with Red Earths', p.154)

Plate 9.5 A hillslope black earth showing calcareous material in the lower B horizon (see 'Rolling to Steep Hills with Black Cracking Clays', p.152)

Plate 9.6 A soil landscape near Gunnedah with black earths occurring on both lower hillslopes and valley plains (see 'Undulating Low Hills and Level Plains of Black Cracking Clays', p.152)

Plate 13.1 The effect of acidity on the growth of rapeseed in granitic soils with all nutrients supplied—the pots in the left foreground were not treated with lime (see 'Soil Acidity', p.229)

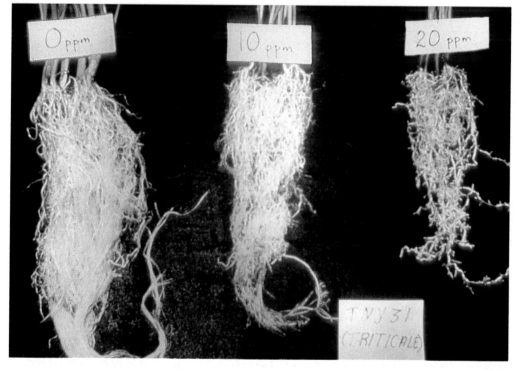

Plate 13.2 Effect of soil acidity on root growth of wheat

Plate 13.3 Lime spreading at a rate of 4 t/ha for the amelioration of soil acidity
(see 'Using Lime', p.234)

Plate 13.4 A badly salinised area showing the effects of salts on vegetative cover
(see 'Salt Toxicity', p.256)

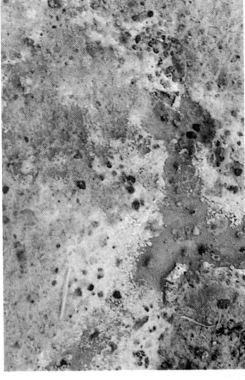

Plate 13.5 Salt encrustations on the soil surface produced by dryland salinisation
(see 'Recognition of Soil Salinity', p.242)

Therefore natural pH varies greatly between soil types (Figure 13.2). Of those soils that have an acid topsoil, weakly weathered soils have increasing pH with depth, reflecting acid production by plant and microbial processes in the surface soil. The pH values often approach 7.5 in the deeper soil layers (weathering classes 1 to 3 in Figure 13.2). This is the pH where precipitation and dissolution of $CaCO_3$ resists or buffers the change in soil pH.

As the soils become more strongly weathered the pH of the surface soil falls and the acid effect is leached deeper down the profile (weathering classes 4 to 6 in Figure 13.2). The pH of the most highly weathered soils is strongly buffered by the dissolution of Al minerals (clays and oxides) at pH 4, so pH values near 4 are the lower limit of pH in most soils. The pH of siliceous sands with very low reserves of minerals containing aluminium or iron can be lower than 4 because of the absence of the aluminium minerals to buffer the pH against acid addition.

A knowledge of the acidity status of a soil down the profile to at least 50 cm is important when making management decisions for an acid soil. Results from a lime trial near Wagga Wagga show that a soil with deep acidity (less than pH 4.5 to 0.5 m) will give a smaller response to liming the top 10 cm than a soil that has a pH greater than 4.5 from 20 cm deep (Helyar et al., 1997).

If a measure of the pH of each layer down the profile is not available when making management decisions, then an estimate of the soil pH profile can be made by selecting the pH profile for the soil type in Figure 13.2 that is comparable to the soil being examined.

Agriculturally Induced Acidity

Normal agricultural practices accelerate the acidification of agricultural lands (Figure 13.3). The relative importance of the factors causing agriculturally induced acidity in two farming systems is given in Table 13.1.

The pH of the surface 20 cm of an agricultural soil can be affected by both the agriculturally induced acidifying processes and use of liming materials.

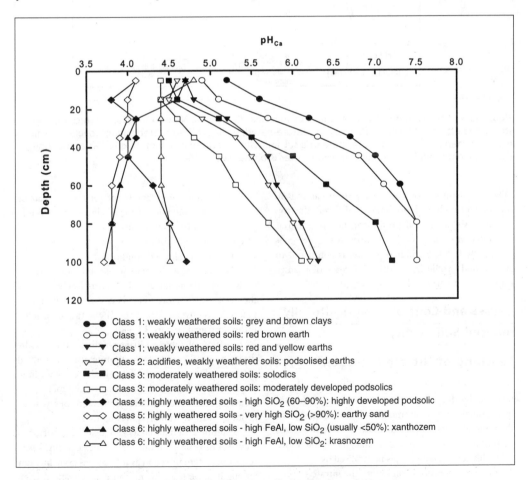

Figure legend:

●—● Class 1: weakly weathered soils: grey and brown clays
○—○ Class 1: weakly weathered soils: red brown earth
▼—▼ Class 1: weakly weathered soils: red and yellow earths
▽—▽ Class 2: acidifies, weakly weathered soils: podsolised earths
■—■ Class 3: moderately weathered soils: solodics
□—□ Class 3: moderately weathered soils: moderately developed podsolics
◆—◆ Class 4: highly weathered soils - high SiO_2 (60–90%): highly developed podsolic
◇—◇ Class 5: highly weathered soils - very high SiO_2 (>90%): earthy sand
▲—▲ Class 6: highly weathered soils - high FeAl, low SiO_2 (usually <50%): xanthozem
△—△ Class 6: highly weathered soils - high FeAl, low SiO_2: krasnozem

Figure 13.2 Soil pH profiles for 10 soils in six soil weathering classes

Table 13.1 The relative importance of factors causing agriculturally induced soil acidity for two farming systems. Both systems require between 200 and 250 kg lime per year to neutralise the acidity.

Cause of acidification	Annual pasture, Southern Tablelands, NSW (%)	Cropping/pasture rotation (1:1), Wagga Wagga (%)
Leaching of nitrate nitrogen	40–60	60–70
Build-up of soil organic matter	30–40	5–10
Removal of product	10–30	10–20
Use of nitrogenous fertilisers	(not applicable)	5–10

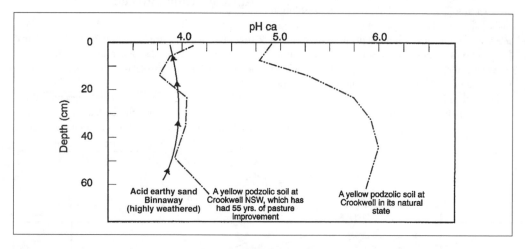

Figure 13.3 The pH to depth of two adjacent soils near Crookwell, New South Wales, which demonstrates the effect of 55 years of improved pasture-based, well-fertilised subterranean clover. This soil has been changed from its natural state of a subsoil with no acidity problems to a soil as acid as one that has weathered for thousands of years (after Bromfield et al, 1983).

Where agricultural practices have been carried on for long periods and no lime used, even the top metre or so of the soil profile can become acid as illustrated in Figure 13.3. The Binnaway soil, which is a naturally formed acidified profile, is compared to an acidified profile induced by 55 years of pasture improvement (Bromfield, et al., 1983).

13.2.2 Causes and Control of Agriculturally Induced Soil Acidity

Leaching of Nitrate Nitrogen

Nitrate nitrogen is the main form of nitrogen that is taken up by the plant. It is made available in the soil by the nitrification of ammonium fertilisers and from the mineralisation of organic matter. These processes of producing nitrate nitrogen make the soil more acid. This acidity, however, is neutralised by:
—plants or microflora converting the nitrate to protein, an alkaline reaction, and either discharging an alkaline compound to the soil (bicarbonate or hydroxyl ions) or storing alkaline compounds in the plant (organic anions)
—conversion of nitrate nitrogen to nitrogen gas under waterlogged conditions.

Nitrate nitrogen is very soluble and therefore easily leached. If the nitrate nitrogen is leached out of the root zone before the plant can take it up, or before it is converted to nitrogen gas, the soil is left a little more acid (Figure 13.4).

Leaching of nitrate nitrogen will be reduced by drying the soil profile over summer using perennial pastures. Experimental data indicate that lime requirement can be reduced using these practices by an average of 50 kg of lime per hectare per year or about 20% of the lime requirement.

In summer rainfall areas (for example, northern New South Wales) summer crops and perennial pastures use nitrate nitrogen as it becomes available, not leaving it to be leached and cause acidification. Even if some nitrate nitrogen is leached, deep-rooted

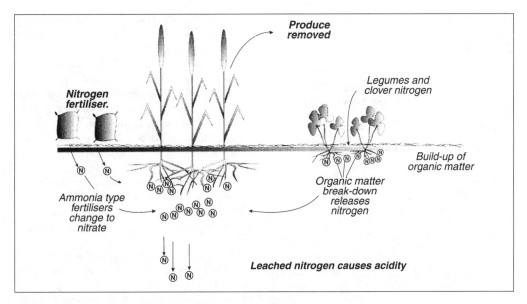

Figure 13.4 The causes of agriculturally induced soil acidity

crops and pastures, including lucerne, can take up nitrate nitrogen from depth.

Efficient use of water and nitrogen will reduce leaching of nitrate nitrogen. Using the correct fertilisers to grow vigorous crops and (perennial) pastures, as well as growing acid-tolerant plants where the soil is acid, will ensure that acidification due to leaching of nitrate nitrogen is kept to a minimum. The amount of nitrate nitrogen produced can be reduced by maintaining a low proportion of clover in pastures, but this restricts productivity.

In cropping paddocks the amount of nitrate nitrogen produced, and hence the potential to acidify the soil, can be reduced by retaining crop residues and using herbicides to control summer-growing weeds, rather than by cultivating the fallow. By sowing a suitable winter cereal crop as soon after the autumn break as possible, the moisture and nitrate nitrogen can be used, or 'caught', before they move too far down the soil profile.

The greatest reduction in acidification caused by nitrate leaching can be made in the winter rainfall areas of southern New South Wales, where annual winter pastures and winter crops predominate. The nitrate nitrogen that accumulates over summer can be reduced by replacing annual pastures with perennial pastures and adopting conservation farming techniques.

Build-up of Soil Organic Matter

A build-up in organic matter has the same effect on the soil acidity as removing produce from the paddock (see below). Over the last 50 to 60 years the regular use of fertiliser and improved pastures, particularly subterranean clover, has increased the amount of organic matter in the soil up to four times. Increasing organic matter has many benefits, for example improving soil structure, but it also makes the soil more acid. Organic matter does not build up in the soil indefinitely and there is no further acidification due to this cause once the organic matter stabilises at a new level. By contrast, burning of stubble is an alkaline process equivalent to reducing organic matter in the soil/plant system.

Higher levels of organic matter (greater than 6%) in the soil will adsorb soluble aluminium minerals resulting in less aluminium being available to the plants. In this situation plants sensitive to aluminium can grow in soils with a low pH.

Increased organic matter levels in the soil are often associated with higher levels of nitrate nitrogen. Therefore, in addition to the acidification from the build-up in organic matter, the potential for further acidification from leaching of nitrate nitrogen is increased.

Removal of Produce

Grain, pasture and animal products are slightly alkaline and their removal from a paddock leaves the soil more acid. If very little produce is removed, such as in wool production, then the system remains almost balanced. If, however, a large quantity of produce is removed, particularly clover or lucerne hay, the soil is left significantly more acid.

If the produce is sold off-farm, regular liming is the only way to maintain pH and production. The rates of lime required to neutralise the acidification caused by removal of produce are given in Table 13.2.

Table 13.2 The amount of lime needed to neutralise the acidification caused by removal of produce (Slattery et al., 1991)

Produce	Lime requirement (kg/t of produce)
Milk[a]	4
Wheat	9
Wool[b]	14
Meat[b]	17
Lupins	20
Grass hay	25
Clover hay	40
Maize silage	40
Lucerne hay	70

[a] There is an additional 25 kg/ha/yr of lime requirement where there is continuous use of the same paddock to hold the herd overnight.

[b] With set stocking there is an additional acidification because pasture is removed to be deposited at stock camps as dung and urine.

The acidification caused by removing hay or silage is neutralised if it is fed to livestock in the paddock where it was made.

Use of Fertilisers

The acidification that results from using fertilisers depends on the fertiliser type. Some nitrogenous fertilisers, for example urea, have no effect on pH if all the nitrogen is utilised by the plant. Acidity will only result if the nitrate nitrogen produced from these fertilisers is leached (Table 13.3).

Other nitrogenous fertilisers, for example sulfate of ammonia, have an acid reaction, and their use acidifies the soil even if fully utilised by plants. Additional acidification results if nitrate nitrogen is formed and leached. Calcium and sodium nitrate neutralise soil acidity if utilised by plants and have no effect if all the nitrate is leached, but they are expensive and rarely used.

Superphosphate has virtually no direct effect on soil pH, but its use stimulates growth of clovers, resulting in increased soil N status and a build-up of soil organic matter, both of which increase soil acidity.

Elemental sulfur is acidifying and requires 3.125 kg of lime for each kilogram of sulfur to neutralise its effect. This effect can be avoided by using products that contain sulfur in the sulfate form such as gypsum, potassium sulfate and superphosphate.

Acidification of the soil can be reduced by avoiding the use of highly acidifying fertilisers such as sulfate of ammonia and mono-ammonium phosphate (MAP). When pre-sowing ammonium forms of nitrogen fertiliser (including urea), it should be placed in narrow bands to slow nitrification and subsequent leaching. Surface application of nitrogenous fertiliser for crops before sowing, even if harrowed, can result in nitrate leaching and consequent acidification. Nitrate nitrogen formed from post-emergent applications is more likely to be utilised by the crop and will cause less acidification.

Table 13.3 Acidifying effect of nitrogenous fertilisers and legume-fixed nitrogen

Source of nitrogen	Lime required to balance acidification (kg lime/kg N)	
	0% nitrate leached	100% nitrate leached
High acidification Sulfate of ammonia Mono-ammonium phosphate (MAP)	3.6	7.1
Medium acidification Di-ammonium phosphate (DAP)	1.8	5.4
Low acidification Urea Ammonium nitrate Aqueous ammonia Anhydrous ammonia Legume-fixed N	0	3.6
Alkaline forms Sodium and calcium nitrate	−3.6*	0

* Equivalent to applying 3.6 kg lime/kg N

Soil Acidity and Plant Growth

As soil becomes more acid the availability of aluminium and manganese to the plant increases to toxic levels, while the availability of other nutrients, mainly molybdenum, phosphorus, magnesium and calcium, decreases. These changes in availability of nutrients cause most of the effects on plant growth attributed to acid soils (see Colour plates 13.1 and 13.2). Other effects of soil acidity include reduced nodulation and nitrogen fixation by some legumes, and decreased mineralisation of organic matter (Robson and Abbott, 1989).

Aluminium (Al)

The principal effect of aluminium toxicity is to restrict root development by disrupting the structure and function of roots. Symptoms of aluminium in the soil are similar to those of phosphorus deficiency as it reduces the plant's ability to fossick for phosphorus. Aluminium also decreases the ability of the plants to extract moisture from the soil, which causes crops to wilt more quickly during dry periods and an overall reduction in water use efficiency of both pastures and crops.

While tolerant plants can survive high levels of aluminium, these levels of aluminium immobilise phosphorus in the soil and reduce utilisation of phosphorus in the plant. High levels of aluminium in the soil also affect the uptake and utilisation of calcium and magnesium, however a serious deficiency of cal-

cium is seen only in the most tolerant plants such as sugar cane.

In cereal crops aluminium toxicity sometimes appears as yellow patches in the field that, on closer inspection, show yellow elongated zones on the leaves.

Plants vary in their tolerance to soluble aluminium. Table 13.4 gives the tolerance of a number of commonly grown agricultural crops and pastures. The plants are arranged into four tolerance classes based on both experimental and anecdotal information.

Two methods are used to measure soluble aluminium in the soil. The first method determines the exchangeable aluminium. The proportion of the Effective Cation Exchange Capacity (ECEC) occupied by aluminium, expressed as a percentage, is then calculated (referred to as $\%Al_{ex}$). This is the most common measure of soluble aluminium and is determined routinely by commercial laboratories.

The second method is to measure the aluminium extracted in the 0.01 M $CaCl_2$ extract used to determine pH (referred to as Al_{Ca}). This measurement will give the best prediction of plant growth.

The $\%Al_{ex}$ and Al_{Ca} cannot be reliably compared. A critical level (say the level where lucerne will suffer a 10% drop in production) in terms of $\%Al$ is dependent on the electrical conductivity of the soil. For example, in a soil with an electrical conductivity of 0.05 dS/m the critical level is about 4%, while in a

Table 13.4 Aluminium sensitivity (tolerance) of some crop[a] and pasture[b] plants

Highly sensitive	Wheat (Durum), barley (Schooner, Yerong), faba beans, chickpeas, lentils, Barrel, Strand and burr medic, lucerne, strawberry, balansa, berseem and Persian clover, buffel grasses, Agropyron and tall wheat (Tyrell) grass.
Sensitive	Canola, wheat (Hybrids, Vulcan, Rosella, Glebe, Janz), barley (O'Connor, Skiff), Albus lupins, red grass (Wagga), wallaby grass (*Danthonia linkii*), phalaris, red clover, snail medic, Caucasian and Kenya white clover and murex.
Tolerant	Brindabella barley, wheat (Sunstate, Swift, Currawong, Hartog, Diamondbird), field peas, tall fescue, Pioneer Rhodes grass, annual rye grass, Haifa white clover and subterranean clover.
Highly tolerant	Narrowleaf lupins, oats, triticale (Tahara), cereal-rye, yellow and slender serradella, cocksfoot, Consol lovegrass, paspalum, kikuyu, *Microlaena stipoides*, *Danthonia racemosa*, *Themeda* species, Maku lotus, common couch and sugar cane.

[a] For a current list of varietal reactions to soil acidity, see *Winter Crop Variety Sowing Guide* published annually by NSW Agriculture.

[b] After Helyar and Conyers (1994)

soil of 0.15 dS/m the critical level is about 1.5%. The critical Al_{Ca} (0.1 to 0.4 mg/L) will be consistent across all soils.

An apparent variation across soil types of the ability of a pasture or crop species to tolerate aluminium measured as $\%Al_{ex}$ may be due to variation in the electrical conductivity of the soil. The variation may also be due to other factors, particularly the level of available manganese or different pH profiles.

In spite of these apparent problems, a measure of available aluminium is more meaningful than just measuring pH as a means of estimating whether acidity is a problem in a given soil. There are clear differences in the solubility of aluminium minerals in different soils at the same pH. Examining results of tests over a range of soil types will show that the level of soluble aluminium at any one pH varies markedly between soil weathering classes (Figure 13.5).

Typical concentrations of Al_{Ca} for the six soil classes are shown in Figure 13.5. The most soluble aluminium minerals are contained in Class 6 soils,

that is, the highly weathered soils low in SiO_2 and high in iron and aluminium where aluminium solubility is controlled by aluminium oxides. The least soluble minerals are contained in the weakly weathered soils (Class 1) in which aluminium solubility is controlled by aluminium silicate clay minerals. Other soils contain minerals intermediate in their solubility. In practice this means that a critical level of aluminium will occur at different pH values for soils in different weathering classes. The pH can vary up to 0.8 units for the same concentration of aluminium in the extract.

In the absence of a measurement of soluble aluminium, the soil pH value and the soil weathering class can be used to predict a potential aluminium toxicity problem.

An aluminium toxicity problem in the soil will be reduced with an application of lime. Alternately using aluminium-tolerant crops and pastures (Table 13.4) will enable a soil high in soluble aluminium to remain productive. Where the subsoil is high in

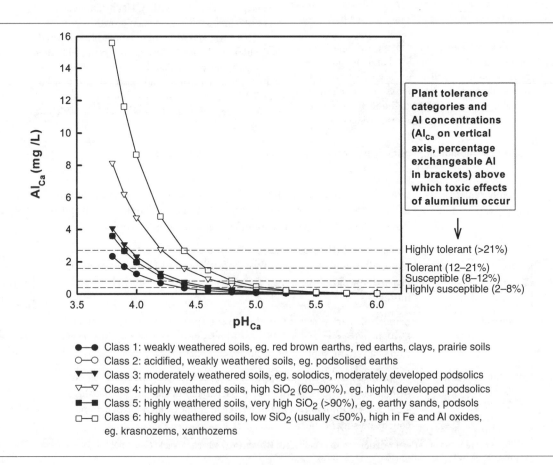

Figure 13.5 Aluminium solubility in soil weathering classes show that a critical aluminium concentration will occur at different pH values for each soil class; classes 2 and 5 are the same line

soluble aluminium, using aluminium-tolerant species may be the only option.

As the available aluminium can vary throughout the year, the time that a soil test has been taken is important when formulating management plans. Research work at Wagga Wagga has shown that aluminium in the soil can be more soluble in winter and early spring (Conyers et al., 1997).

Manganese (Mn)

Crops with a history of being affected by manganese toxicity are canola, French beans, some winter pulse crops and some soya beans. Pasture plants known to be sensitive to manganese are lucerne, medics, serradella and some cultivars of subterranean clover.

Manganese toxicity problems occur in soils with a pH less than 5.5, but only in some soils and then only at certain times of the year. Note that while both toxicities and deficiencies of manganese can occur in agriculture, the main problem in New South Wales is toxicity.

In broadleaf crops, yellowing of the margins and cupping of the leaves can indicate manganese toxicity. In grasses, poor seedling vigour, yellowing at the tips and margins, and some flecking of the older leaves are indicators. However, other nutritional problems can have similar effects, particularly in grasses, and soil testing or tissue analysis is required to diagnose manganese toxicity.

In some situations the presence of manganese can actually reduce the effect of aluminium on root development, resulting in increased crop and pasture growth (Helyar and Conyers 1994).

Available manganese in soil can vary up to five-fold throughout the year (Table 13.5). Soils most likely to develop manganese toxicity are weakly weathered soils, such as red earths and red-brown earths (Class 1, Figure 13.2), and soils that are high in iron and aluminium oxides (Class 6).

Most soil tests give the manganese available at the time of sampling only. Because of the wide variation in levels of available manganese through the year, the date when the soil test was taken must be considered when making recommendations for managing manganese toxicity.

The seasonal variation in manganese availability makes interpretation of soil tests difficult, and analysis of plant tissue can be more useful in determining if there is a toxic manganese problem. Plants vary in their sensitivity to manganese. Some plants are affected by only a small amount of manganese in the tissue, while others are tolerant to quite high tissue concentrations. The concentration of manganese in plants at which a small decline in growth will occur varies from 200 to over 1000 mg/kg of plant dry matter, depending on the tolerance of the plant. Critical manganese concentrations sufficient to cause a 10% decline in growth for a number of species are given in Table 13.6. These concentrations are determined on the youngest, fully developed leaf.

High levels of available manganese in the soil can be lowered by increasing the soil pH to above 5.5

Table 13. 5 The effect of seasons on the balance between manganese being available and not available to plants

SUMMER:	**Manganese is most available.** Hot and dry conditions favour the conditions that make manganese available.
AUTUMN:	**Manganese becomes less available.** Warm (more than 15°C) and moist conditions favour microbial activity that changes manganese to an unavailable form. In autumn it is important to note that: • if the soil pH is less than 5.5, then lime will increase this microbial activity • if the hot and dry conditions return and the soil dries out again, the manganese becomes more available, as in summer • as the soil temperature drops below 15°C, microbial activity slows and the changing of manganese from available to unavailable forms also slows • below 10°C there is very little microbial activity, and the availability of manganese will not change.
WINTER:	**Manganese is least available,** unless the temperature dropped below 10°C before all the manganese was changed to unavailable forms.
SPRING:	**Manganese slowly becomes more available** as conditions become hot and dry. If the soil becomes warm (greater than 15°C) and waterlogged, then manganese becomes available very quickly. In these conditions, however, a lack of available nitrogen and oxygen will usually have a greater effect on plant growth than the excess manganese.

Table 13.6 Tolerance of some crop* and pasture plants to manganese (Mn) and the critical concentrations of manganese in the youngest, fully developed leaf sufficient to cause a 10% decline in growth for each tolerance category

Manganese tolerance category	Plant	Critical leaf Mn level (mg/kg)
Highly sensitive	lucerne, pigeon pea, barrel and burr medics	200–400
Sensitive	white clover, strawberry clover, chickpea, canola	400–700
Tolerant	sub. clover, cotton, cowpea, soya bean, wheat (Matong, Vulcan, Dollarbird), barley (Yerong, Lara, Schooner), triticale (Tahara), oats*	700–1000
Highly tolerant	rice, sugar cane, tobacco, sunflower, most pasture grasses, cereal rye*	> 1000

* See *Winter Cereal Variety Sowing Guide*, published annually by NSW Agriculture, for the current list of varietal differences.

with lime. The effects of manganese toxicity can often be avoided by not sowing manganese-sensitive crops and pastures at the autumn break but waiting until the second autumn rain, or avoiding manganese-sensitive plants altogether.

Molybdenum (Mo)

Where the pH of a soil is below 5.5, an application of 50–100 g molybdenum per hectare may increase pasture production (higher rates on Class 6-type soils, Figure 13.5). Molybdenum is a trace element required in the nitrogen fixation process and the response to the correction of a deficiency is due to increased nitrogen fixation by the clovers.

The response to applying molybdenum will vary from district to district, and local information from a district agronomist or fertiliser outlet should be sought before proceeding. An application every three to five years (depending on soil type) as a spray of sodium molybdate or molybdenum trioxide, or as a component of a fertiliser, is sufficient for all plants. Molybdenum can also induce a copper deficiency in livestock if used on some light soils. A lime application that increases pH by 1, for example from 4.5 to 5.5, increases the available soil molybdenum sufficiently in most soils to overcome most molybdenum deficiencies. However there are some soils that are very low in total soil molybdenum where this response to lime will not occur.

Calcium (Ca)

Most soils in New South Wales contain an adequate supply of calcium for the crops and pastures grown in this state. Where there is a moderate calcium deficiency, it is seen first in parts of the plant that are fur-

thest from the main flow of water within the plant. Examples of the effect of a moderate calcium deficiency are poor seed set in peanuts and subterranean clover, and blossom end rot in tomatoes. More severe calcium deficiency causes death of growing points, for example November leaf in bananas.

Low levels of soil calcium can also adversely affect the nodulation of subterranean clover. Short-term deficiency can cause petiole collapse of young expanding leaves.

Examples of very severe calcium deficiency are most likely in sandy soils with a pH less than 4 that are low in organic matter, or where there has been excessive use of highly acidifying fertilisers. Under these circumstances the exchangeable calcium can drop to concentrations less than 40% of the exchangeable cations other than aluminium, and calcium deficiencies can occur. As these soils have very high levels of soluble aluminium, all but the most acid-tolerant plants, such as sugar cane, are killed before the symptoms of calcium deficiency become apparent. The symptoms are stubby, unbranched and discoloured roots or dead growing points in the shoots. The root symptoms are difficult to distinguish from symptoms of aluminium toxicity.

Magnesium (Mg)

Loss of production in crops and livestock due to a magnesium deficiency is most unusual in New South Wales. While low levels of magnesium have been recorded in the 0 to 10 cm topsoil in parts of the state, there are nearly always adequate amounts in the subsoil. Signs of magnesium deficiency have occurred in wheat on sandy acid soils of the central and south-western slopes of New South Wales while the plants were young and their root development confined to the surface layers. As the plants developed their root system they accessed the sub-surface soil, which is high in magnesium, and yields were not affected by magnesium deficiency.

While grass tetany in cattle is usually attributed to low soil magnesium, recent research has indicated that soil potassium and calcium and climatic conditions, as well as magnesium, play a role in the onset of grass tetany. If the ratio of potassium to calcium plus magnesium in plant tissue rises above 2.2 (that is, K/[Ca + Mg] > 2.2), then the risk of grass tetany increases. Graziers wishing to reduce the incidence of grass tetany by applying lime or dolomite to the soil should seek specialist advice (Elliott, 1995).

Ca:Mg Ratios

Research at Wagga Wagga Agricultural Institute has indicated that the Ca:Mg ratio is not an important factor in crop and pasture nutrition. At the low end of the scale, that is, Ca:Mg < 2 (Mg % CECB >30), clays will tend to disperse. This is a more common problem in the subsoil than in the topsoil.

Research has failed to show a response to correcting a Ca:Mg ratio up to 25:1 (Mg % CECB<2) in paddock situations. This is because there is generally ample magnesium in the subsoil to balance any deficiency of magnesium or excess of calcium in the topsoil.

Reduced Fixation of Nitrogen

Soil acidity will affect the survival of acid-sensitive *Rhizobium* bacteria in the soil. Acid-sensitive rhizobia are usually those that infect the more acid-sensitive plant species and are often more sensitive to soil acidity than the host plant (Agfact P2.2.7).

Practical Options for Managing Acid Soils

13.2.4

The minimum information required to plan the management of an acid soil is:

—pH and availability of aluminium in the top 10 cm of soil
—pH of the 10–20 cm layer
—the time of the year that the soil was taken and analysed
—the intended crops or pasture to be grown.

Soils with a pH (0–10 cm) in CaCl$_2$ greater than 7.0

These soils may endure decades of the most acidifying practices before developing the problems associated with acid soils. These soils are often strongly buffered because of large amounts of undissolved lime, high organic matter, and clay content. However, these soils may cause problems associated with nutrient deficiencies, particularly deficiencies of zinc, boron and, in strongly weathered sandy soils, manganese. In these soils the availability of phosphate is low, and phosphate fertiliser quickly becomes unavailable after application.

Soils with a pH (0–10 cm) between 5.6 and 7.0

These soils are unlikely to cause acidity problems in crops and pastures (Figure 13.6a), and no action is necessary.

Soils with a pH (0–10 cm) below 5.5

These soils may have a molybdenum deficiency (see 'Molybdenum (Mo)' above). As this effect is not universal, checks should be made with an agronomist or

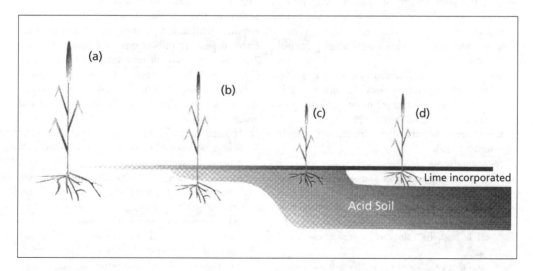

Figure 13.6 The effect of increasing subsoil acidity on an acid-sensitive plant

farm adviser, or spray on test strips of sodium molybdate. Manganese (see 'Manganese (Mn)' above) may, in some situations, become toxic in soils with a pH as high as 5.5.

Soils with a pH (0–10 cm) between 5.1 and 5.5

These soils will slightly reduce production in highly acid-sensitive plants, reduce the effectiveness of some rhizobia, and may cause molybdenum deficiency and manganese toxicity in susceptible plants. These soils can quickly become more acid in the following situations:

—where there is excellent clover growth and little or no perennial grass
—where high rates of nitrogen fertiliser are used
—where the rate of removal of produce is high (for example, in silage, hay or potato production, or in paddocks that are frequently cultivated for dairying, vegetable growing or other intensive industries)
—if the soil is sandy
—if the soil is in southern New South Wales in the 500–800 mm rainfall zone.

Soils with a pH (0–10 cm) between 4.5 and 5.0

These soils are likely to have soluble aluminium (see Figure 13.5). Plants that are sensitive to acidity have trouble establishing in these soils. If they do establish, they do not grow strong surface root systems (see Figure 13.6b), resulting in loss of production. Often the subsoil in this situation is not as acid as the surface, and an application and incorporation of lime neutralises the acidity and fixes the problem.

While liming is the main tool to manage soil acidity, the acidification rate can be reduced by adopting less acidifying farming systems and practices such as:

—using deep-rooted, summer-growing perennial pastures
—using less of the high-acidifying and more of the low-acidifying nitrogenous fertilisers (Table 13.3), taking care that the nitrate nitrogen produced is not leached away
—using chemicals rather than cultivation to fallow farming land
—retaining stubble
—sowing winter crops (other than manganese-sensitive crops) as soon as possible after the first rain
—encouraging maximum growth of crops and pastures by using adequate fertiliser; and where the soil is acid, by using acid-tolerant species of plants.

If a soil that has an acid soil problem is not limed, the pH continues to drop (Figure 13.6c). Once the pH drops below 5.2 there is a net movement of acidity down to the next layer of soil.

Soils with a pH (0–10 cm) below 4.5

These soils in areas with an annual average rainfall over 500 mm will generally be as acid in the 10–20 cm depth, and possibly further down the profile. In areas with rainfall less than 500 mm the subsurface soils are not as acid. To confirm this, testing a sample from 10–20 cm depth will be necessary, and possibly from 20–30 cm, before deciding on management options.

Soils with both pH (0–10 cm) and pH (10–20 cm) below 4.5

These soils will greatly restrict the growth of all acid-sensitive plants. The first option in this situation is to grow acid-tolerant plants. The list of some aluminium-tolerant crops and pastures in Table 13.4 can be used to select plants for their tolerance of acid soils, as aluminium tolerance closely reflects acid soil tolerance in soils this acid.

Figure 13.6d shows that if lime is applied and incorporated, only the surface soil acidity is ameliorated. Acid-sensitive plants will germinate and establish, but they do not develop root systems that are adequate for long-term survival. Acid-tolerant crops, such as triticale, will give increased yield with lime in this situation (Helyar et al., 1997).

If the pH in the limed layer is maintained above 5.5, the pH in the lower levels will improve over one or two years for very light soils and six to 10 years for loams.

Using Lime

13.2.5

Applying lime is the only economical way to reverse acidity in the soil (see Colour plate 13.3). The response of crops and pastures to lime varies considerably depending on soil type and plant tolerance. This has been clearly demonstrated by Geeves et al. (1987), who have prepared a series of response curves describing the effect of lime on the production of wheat, triticale and barley for four soil groups.

Applying sufficient lime to lift the pH of a soil to 5.2 will cure most problems associated with soil acidity. A pH of 5.2 will ensure:

—that the leaching of acidity/alkalinity down the profile is neutral, and therefore will not acidify the subsoil
—there is compensation for the variability of pH in the paddock which is normally more than one pH unit. After lime is applied to lift the pH to an average of 5.2, the proportion of the paddock that still has a pH below this will still be on the

steep part of the lime response curve. This response will balance the parts of the paddock that are now above pH 5.5 where there is a lesser response.

Where the soil below the depth of incorporation of the lime is acid, then a pH of 5.5 or more must be obtained, and maintained, to ensure a movement of the effect of the lime down the profile.

The lime required to change the pH from 4.0, 4.3 and 4.7 to 5.2 is given in Table 13.7 for a range of Effective Cation Exchange Capacities. If the subsoil is acid then the target pH will be 5.5. The additional lime to increase the pH from 5.2 to 5.5 is also given. These rates are based on a standardised titration curve shape developed by Magdoff and Bartlett (1985) and calculating the soil pH buffering capacity using the soil ECEC and pH data (compare Hochman et al., 1989).

To maintain a given pH profile a maintenance lime rate is required to neutralise the acid being added to the system. The maintenance rate for many of the farming systems in New South Wales is shown in Table 13.8. These lime requirements are adjusted for different amounts of clover content of the pastures, different efficiency of grain production, and different

initial soil pH values. The rate of acidification increases at higher soil N status and slows as the pH falls (Williams, 1980). The lime requirement should also be adjusted for hay or silage removed from the paddock (Table 13.2), and for use of medium and highly acidifying fertilisers (Table 13.3). Other produce, such as wheat, has already been accounted for when calculating the lime requirements for the farming systems in Table 13.8.

Generally lime is applied prior to the most acid-sensitive crop or pasture as this gives the best economic return. Lime should not be applied before a wheat or triticale crop unless the root disease 'Take-all' has been controlled with a break crop, 'winter cleaning' of grasses or early fallow.

The quality of the lime, as well as the rate, will affect the amount of change in pH of the soil. The quality of lime will vary between manufacturers, however in New South Wales lime crushers are obliged to register their lime products. The registration requires that they describe their products in terms of Neutralising Value (NV), calcium/magnesium content and fineness. Pure lime has an NV of 100 and a calcium content of 40%. The fineness is described as the percentage of the product that will

Table 13.7 Lime required (t/ha) to lift the pH of the top 10 cm of soil up to 5.2 for a range of Effective Cation Exchange Capacities (ECEC) and pH normally encountered when making liming recommendations. The additional lime required to lift the pH from 5.2 to 5.5 is also given.

ECEC (cmol+/kg)	Lime required (t/ha) to lift the pH of the top 10 cm:			
	from 4.0 to 5.2	from 4.3 to 5.2	from 4.7 to 5.2	from 5.2 to 5.5
1	1.6	0.8*	0.3*	0.2
2	2.4	1.2	0.5*	0.4
3	3.5	1.7	0.7*	0.5
4	3.9	2.1	0.9*	0.6
5	4.7	2.5	1.1*	0.7
6	5.5	3.0	1.2	0.8
7	6.3	3.3	1.4	1.0
8	7.1	3.8	1.6	1.1
9	7.9	4.2	1.8	1.2
10	8.7	4.6	1.9	1.3
15	12.5	6.7	2.8	1.9

Assumptions: bulk density of soil is 1.4; 70% lime dissolves in one year.
Note: this table will give an overestimate of the lime required for a cracking clay.

* It is recognised that low rates are difficult to apply, but over-liming can cause nutrient deficiencies, particularly in these light soils.

Table 13.8 The amount of lime needed to neutralise the acidification caused by farming systems in New South Wales. These recommendations are for a system where sufficient fertiliser is used to maintain a 20% clover balance in all pastures[a] or to produce crop yields that are 60–80% of the maximum possible[b] and the pH is 5.1 or higher[c].

Farming system	Lime requirement (kg/ha/yr)
Plains, less than 500 mm rainfall[d]	
— crop or crop/pasture	75
Coast and tablelands	
— perennial pasture	100
Southern tablelands	
— perennial pasture	150
— 25% crop, 75% perennial pasture	175
— annual pasture	200
— 25% crop, 75% annual pasture	250
Southern slopes, more than 500 mm rainfall[d]	
— perennial pasture	150
— annual pasture	200
— 50% crop, 50% annual pasture	250
— crop	300
Irrigation or more than 1000 mm rainfall[d]	
— less than 300 kg acidifying nitrogenous fertiliser/ha/yr	450
— more than 300 kg acidifying nitrogenous fertiliser/ha/yr	1000

[a] For above 20% clover in the pasture increase the lime requirement by 50% and for below 20% clover reduce the lime requirement to 70% of the figure in the table.
[b] For above 80% of the maximum crop production increase the lime requirement by 50% and for below 60% reduce lime requirement to 70% of the figures shown.
[c] For a soil with pH 4.6–5.0, use 80% of the lime requirement listed in the table.
For a soil with pH 4.2–4.5, use 45% of the lime requirement listed in the table.
For a soil with pH less than 4.2 there is virtually no acidification.
[d] Annual average rainfall.

pass through a 0.25 mm sieve. A fineness over 95% is the accepted industry standard. The finer the lime, the more effective it is in raising the pH of the soil.

Where the lime is applied to the surface and not incorporated, it will take time for the effect of the lime to move down the soil profile. Movement down is favoured by high application rates (that achieve pH levels higher than 5.5) (Scott et al., 1997a). Light texture, higher rainfall associated with good infiltration and some types of worms favour movement down the profile (Baker et al., 1995). For farmers practising reduced tillage or no-tillage, it may take one to three years before lime is fully effective.

In the non-arable grazing zones and where the soils are often very acid to depth, farmers have found that at current returns for livestock products the response to liming is only marginally profitable. Delaying the liming program for too long, however, allows for increases to the acidity in the subsurface soil. Also, converting the small responses by subterranean clover-based pastures into profits from livestock requires careful pasture management (Helyar et al., 1997).

At a conference at Goulburn in 1997 most of the current knowledge on managing soils on the central and southern tablelands of New South Wales was summarised (Scott et al., 1997b). Small responses to lime (20–50%) by tolerant species of annual and perennial pastures have often been measured, with autumn responses more common than spring responses, and responses being delayed by one to two years for unincorporated compared to incorporated lime.

The options for managing soil acidity in a non-arable pasture area proposed at the Goulburn conference were:

1. Modify the grazing system to reduce acidification. The options are:
    — to establish perennial pastures
    — to reduce or eliminate the use of super-

phosphate (but this is counter-productive as clover will die out and overall production will drop)
— to manage stock to encourage dispersal of camp sites
— not to fertilise with elemental sulfur.
2. Plant pastures that are tolerant to acidity (Fenton, 1995b).
3. Correct acidity with lime, but as pointed out above this is not generally seen as economic.
4. Modify land use; for example, establish forests on land less suitable for grazing.

At the time of writing there are no experiments reporting the response of grazing animals to surface-applied lime in replicated experiments in south-eastern Australia. However a number of replicated trials and farmer demonstrations have recently been established, funded by Acid Soil Action. These experiments will enable the response of the whole production system to be measured, including animal responses to effects of lime on pasture quality, quantity, seasonal production patterns and persistence. Results from this work will become available over the next few years.

13.3 Soil Salinisation

P.E.V. Charman and A.C. Wooldridge

Increasing salinisation of soil, land and water resources in many parts of Australia is widely regarded as a serious environmental problem affecting irrigation areas, dryland areas and urban developments. Traditionally soil conservation organisations were mainly concerned about dryland salinisation, as it was seen as a precursor to erosion on grazing lands, particularly in New South Wales. Nowadays, however, as the problem has spread and its ramifications have been more clearly understood, it is seen in a wider catchment context as affecting all types of land and with a strong relationship to all aspects of water quality.

Soil degradation caused by salinisation is generally referred to as secondary (or induced) salinity. This may be defined as the accumulation of free salts in part of the landscape, to an extent that causes degradation of vegetation, water or soil resources, and which has been caused by post-European management of the land (see Colour plate 13.3). Dryland salinisation is generally a result of saline water discharge associated with non-irrigated land although, in semi-arid areas, erosion of surface soil often exposes a saline subsoil, resulting in 'scalding'. This differs from that typically found on the tablelands and slopes, where problems are related to saline water tables surfacing on hillslopes or coming close to the surface on footslopes. Scalding is not the result of water table effects.

Robertson (1996) outlined the seriousness of the problem in different parts of Australia and predicted that almost 2.5 million hectares of saltland would be apparent Australia-wide by the year 2000 and approaching 4 million hectares by 2050. In 1996 the current estimates for seepage salinity were 2.16 million hectares and in 1998, 2.48 million hectares. For New South Wales the estimates have jumped from 25 000 hectares in 1995 to 120 000 hectares in 1998 (National Dryland Salinity Program Management Plan 1998–2003). Areas affected include the upper catchments of the Lachlan, Murrumbidgee, Wollondilly, Shoalhaven, Goulburn and Wollombi Rivers, and areas within the following districts: Narrandera, Cootamundra, Young, Yass, Gulgong, Boorowa, Wellington and Rylstone.

Causes of Soil Salinity

13.3.1

The general cause of dryland (seepage) salinity is change in land management and use since European settlement. It reflects changes in water balance and the way in which groundwater moves through the landscape. Excessive clearing of native vegetation on hillslopes has allowed an increase in the amount of rainfall recharging groundwater systems, thus causing water tables to rise. Rising water levels cause increased mobilisation of soil salts and subsurface seepage flow across the landscape. Saline seepage then emerges lower downslope, or rises close to the surface, in situations determined by landform and subsurface geomorphology.

Causal factors—Vegetation

The process redistributes salts in the regolith, following hydrological disturbance and subsequent discharge of saline water into drainage systems (Peck, 1978). This disturbance normally follows clearing of native trees in the upper parts of catchments.

The most obvious hydrological result of this change in land use is increased groundwater recharge when trees in a catchment are replaced by grass (Boughton, 1970). Different root depths of the two types of plant cover, and the difference in total moisture deficiency produced, lead to contrasts in infiltration rates and flood peaks. Yule and Prineas (1986), in a review of the relationship between trees and dryland salting, show that the greatest percentage of annual precipitation is lost as evapotranspiration in most forested catchments. Only small percentages are lost as streamflow or seepage to groundwater, depending on season. Interception of rainfall by the tree canopy also reduces the amount of water reaching the soil, compared with that in a grassland.

The clearing of trees from catchments is thus the primary catalyst for the hydrological disturbance that leads to soil salinisation. Wagner (1987) noted considerable lag times following clearing before the development of seepage salinity. This may be about 50 years in south-eastern New South Wales, but appears to vary widely, depending on landform, land use and geomorphology.

Causal factors—Salts

All soils in Australia contain variable quantities of salt, with the most common being composed of sodium and chloride ions. These salts, together sometimes with magnesium, calcium and sulfate ions, occur in the lower soil profile or weathered soil mantle. The amount of salt stored and the location of the salt store vary in different landscapes.

The ultimate sources of salts are the ocean and the weathering of rocks and minerals. However, salts from these can end up in soils by a range of pathways. Isbell et al. (1985) maintain that the origins of salts in soils are difficult to identify in a definite way because of the following: salts, particularly sodium and chloride, are very mobile; it is difficult to determine current salt fluxes based on climate and soil permeability; and the large time scale involved in the evolution of many of Australia's landscapes. It is difficult to distinguish the impact of present and past environments on the salt content of soils. However it is useful to have some estimates of the origins of salts in some of the soils of New South Wales.

The immediate sources of salt in the landscape are variable and can be attributed to marine, aerial and terrestrial sources. Terrestrial sources include mineral weathering and dust blown in from naturally saline areas, especially from arid Australia. Salt may also be blown in from the ocean, although Gunn and Richardson (1979) and Isbell et al. (1985) indicate that marine sources of cyclic salt decrease with distance from the coast. Gunn and Richardson estimate that the accumulation of salt in rainwater is less than 2 kg/ha/yr at 240 km inland. The use of Ca/Na ratios indicates that terrestrial sources of salt are important in rainfall, especially as the distance inland increases. In inland areas the Ca/Na ratio in rainfall is much lower than that for seawater, indicating that Ca has been added from other sources than the sea. Other sources of chloride than the sea have been identified by Isbell et al. (1985), and lead to the conclusion that a marine source of chloride is not always required to account for salt levels in soils.

The dry interior of the continent, which currently has significant areas of bare salty deposits exposed to wind and certainly has had these in the past, is a possible source for large quantities of salt. These could have been transported into western New South Wales as aeolian salt, or on saline dust blown from these areas (Isbell et al., 1985; Bowler, 1976). This could be one explanation for the high salt levels in the soils. There is also the possibility that in the past, seawater covered large parts of the plains of south-western New South Wales.

A major likely source of salts in soil is the weathering of parent rock material, with the subsequent accumulation of salts in older, more deeply weathered materials. Gunn and Richardson (1979) and Isbell et al. (1983) consider that weathering of parent rock is one of the major sources of salts and that the local redistribution of salts in drainage water is a major source of salts on a local scale. However, Isbell et al. (1985) emphasise that it is difficult to reach definitive conclusions on the origins of salts in many soils, and that it is unlikely that one explanation for the origin of salts will apply to all landscapes. The source of salts in the riverine plain of New South Wales is not necessarily going to be the same as that for salts in the eastern highlands.

In some areas of the eastern highlands of New South Wales, soluble ions from weathering rocks are a major source of salt, especially if these become concentrated by a hydrological obstruction in the landscape. Gunn and Richardson (1979) concluded that deep weathering of granites and some shales was a major source of salts in some landscapes. An important finding was that salt content was higher in some freshwater sediments than in some of the marine sediments, possibly because of the hydrological and weathering history of the rocks since they were laid down. This indicates the complexity of trying to predict the origins of salts in the landscape.

The relative importance of the various sources depends largely on physiographic environment, weathering history and climate. There is also a strong influence where salts are stored within the regolith, either in the unsaturated zone nearer the land surface or in the groundwater (saturated) zone at greater depth.

Despite some uncertainties about the origins of salts in soils, there are significant quantities of salt stored in many landscapes. Originally, most of the salt was in relatively deep sinks and aquifers, and out of reach of most plants. Unfortunately, in recent years there has been a general trend of rising water tables and the mobilisation of salt stores in the landscape, with saline patches and areas becoming more and more prevalent in the landscapes of New South Wales. This has resulted in a major set of programs to reduce this increasing incidence of salinity.

Causal factors—Geology/Geomorphology

Wagner (1987), after studying dryland salinity on the southern tablelands for many years, asserted that most seepage salinity occurs on sedimentary material. In this region the problem occurs mainly on Ordovician sediments, and also on Silurian and Cainozoic materials. Various other lithologies can be involved, such as granites with sodium-rich minerals or metamorphosed sediments of marine origin.

Geomorphological factors can be critical in determining the occurrence of saline seepage and its effect on land degradation. If intake (recharge) areas have been cleared, and groundwater is tending to move laterally from hillslope to valley floor, it may be caused to surface in the following ways.

The lateral movement of saturation flow in the soil may be impeded by an impermeable layer of clay or rock, causing it to surface higher on the slope than would be expected. In this case a 'perched' water table is formed, and hillside seepage occurs. Other circumstances may give rise to the confining of the saturation flow between two impermeable layers, forming a 'confined aquifer'. Such an aquifer has an intake (recharge area) higher up in the landscape and an outlet (discharge area) lower down. Clearing of the recharge area allows more intake of rainfall, and an increase in hydrostatic pressure and subsequent discharge at the lower level.

In the absence of impermeable layers, infiltrating rainwater percolates down to the regional groundwater body below the valley floor. A rise in this water table level, due to clearing upslope, may bring it to within capillary range of the surface of valley footslopes or flats, or it may be intercepted by drainage depressions. In each case, seepage salinity is a likely result, depending on terrain, slope, slope shape, soil depth and drainage.

The slope of the land can control the extent of saline-affected areas. In relatively flat areas, large areas can be affected by salinity, whereas in more hilly areas, the salinity tends to be more localised and smaller in area.

Causal factors—Soils

Soils associated with seepage salinity typically have a subsoil (B) horizon of low permeability. This horizon is usually sodic and its low permeability leads to the mobilisation and redistribution of soluble salts (Northcote and Skene, 1972). Cope (1958), in describing solodic and solonetzic soils associated with salting in Victoria, indicated that the combination of a permeable A horizon overlying a somewhat impermeable B horizon tends to cause localised accumulations of soluble salts where subsurface drainage is impeded.

Northcote and Skene, in their study of saline and sodic soils, grouped various soil categories according to the Northcote Factual Key—an essentially morphological classification. The results indicate the distinctive range of sodicity, alkalinity and salinity for each grouping. In this way, the various salinity classes have been related to soil landscape mapping units of the *Atlas of Australian Soils*, which uses the Key as a base.

Causal factors—Climate

Seasonal changes in the severity of the effects of dryland salinity relate primarily to rainfall. The amount of precipitation relative to evapotranspiration is the driving force for the mobilisation of salts and the elevation of water tables.

Wagner (1987) studied approximately 100 sites in south-eastern New South Wales, using aerial photo sequences and discussions with landholders. He concluded that:

—many salinised areas became evident in the 1950s as a consequence of a period of above-average rainfall followed by dry years

—two or three years of above-average rainfall followed by a return to drier years appears to exacerbate the surface expression of the problem, whereas a more prolonged period of wet years appears to alleviate it

—annual average rainfalls outside the range 400 to 800 mm are not conducive to this type of salinity development; maximum occurrence appears to be in the range 600 to 700 mm.

The longer wet periods might significantly dilute the salt and allow some regeneration of vegetation before more normal seasons return (Armstrong, 1987). The shorter wet periods do not allow this, but raise water tables, bringing salts to the surface to be concentrated when drier conditions prevail. However, irrespective of the development of transient surface seepage, any rise in groundwater levels increases the risk of salinity.

Causal factors—Land Use

The change in land use from native tree vegetation to grazing land after clearing is generally seen as the cause of development of dryland salinity in susceptible areas. Management of grazing land to maintain a vigorous, improved pasture cover, rather than overgrazed and degraded native pasture, will help alleviate the development of salinity at lower levels, because higher evapotranspiration can be achieved.

Arable land use, particularly with cultivated fallows, can promote salinisation. For much of the season, the land lacks an actively growing and transpiring crop, and soil conditions may favour intake of rainfall due to its fallow or cultivated condition. Arable land use also affects discharge areas, because tillage favours intake of rainfall and the raising of water tables, as well as increasing evaporation.

The problem of secondary salinisation in irrigated soils can be locally more severe in its effects than that in dryland soils, as there is higher potential for intensively used areas to be affected and large amounts of capital infrastructure can be rendered useless. The amount of irrigation water used is critical, as irrigation can raise the level of saline water tables and recycle river water that is already saline. An irrigation scheme must ensure that salts brought into the surface soil are removed from the root zone of the crop just as fast, and that salt content at which this balance occurs is low enough not to harm the crops appreciably. Hence careful water management is the major means of control of this type of salinisation, along with appropriate provision for drainage of irrigated soils. In this way, water applied is restricted to crop needs, and the saline water table is kept below the reach of sensitive crop roots.

Care should also be taken in the selection of soils for growing certain crops, For example, rice should not be grown on soils with high permeability, as this can lead to high rates of water accession to the water table and result in a rapidly rising water table.

13.3.2 Effects of Soil Salinity

High soil salinity has a number of consequences that are of concern in a land and water management context.
(i)   The adverse effect of salts on plant growth.
(ii) The effect on water quality, whether surface water or groundwater, particularly if it is needed for domestic or industrial supplies, stock water, or if it flows into major watercourses or storage bodies.
(iii) Low vegetation cover can expose areas to high erosion rates.
(iv) The saturation of soils with solutions high in sodium can result in the development of sodic soils, as the sodium cations replace calcium and magnesium ions on the clays.
(v)   Highly saline soil solutions have the potential to mobilise heavy metals and other potentially toxic substances in soils.

Soluble salts can have two types of effect on plants, one due to specific toxicities of ions within the saline solution, and the second due to the more general osmotic effect of the saline solution on plant roots. A high salt content around roots greatly reduces their ability to absorb water from the soil, thus reducing plant health and plant growth. Table 13.9 shows the effects of salinity on plant growth in a general way, and indicates the range of salinity levels over which a plant can survive in a saline environment. Some plant species are more tolerant of saline levels than others. For example barley and related grasses (*Hordeum* species) and couch grass (*Cynodon dactylon*) are more tolerant of saline conditions than lucerne or many fruit trees. Taylor (1993) has a more detailed list of the relative tolerance of species to salinity. High salt in the soil can limit the use of soils for crops such as most winter cereals, sorghum, sunflower, peas, beans, potatoes and most fruit crops.

The worst effects of high salinity result in the death of vegetation that would normally have a major role in controlling soil erosion and/or contributing to crop or stock yield. If localised saline areas are left untreated they tend to become bare and compacted, resulting in high runoff. This can lead to serious sheet and gully erosion, yielding high sediment loads to local water watercourses, causing degradation of lower lands and water storages. Water flowing off such areas is likely to be saline and possibly unsuitable for domestic purposes, stock or irrigation. It can also add to the salt load of major watercourses.

Effects on Plant Growth

The primary effect of salts on vegetation is, as mentioned above, through increased osmotic pressure of the soil water, so that plants find it increasingly difficult to extract water from the soil. This is the major cause of vegetative decline on saline areas, leading to many of the adverse environmental consequences of dryland salinisation. The specific ion effects mentioned are thought to be of secondary importance, especially in grazing areas.

Plant species, and to some degree cultivars within species, vary markedly in their ability to extract water from soil under saline conditions, and there is thus a wide range of salt tolerance in plants, which can be exploited for rehabilitation work. However, most normal pastures and crop plants are not highly tolerant, and will eventually die out under saline conditions.

Table 13.9   Effect of salinity on plant growth (Richards, 1954)

Salinity level Electrical conductivity of saturated extract (EC_{sat}) dS/m or mS/cm	Effect on plant growth
0–2	Salinity effects usually minimal
2–4	Yield of very salt-sensitive plants may be restricted
4–8	Yield of salt-sensitive plants restricted
8–16	Only salt-sensitive plants yield satisfactorily
> 16	Few salt-tolerant plants yield satisfactorily

A 1:5 soil:water extract ($EC_{1:5}$) is often used to measure soil salinity. To convert this to EC_{sat}, a conversion factor is necessary based on soil water-holding capacity and other factors, which can be approximated by soil texture. The approximate conversion factors for a range of soil textures are given below (based on Slavich and Pettersen, 1993).

Soil Texture	Approximate conversion factor from $EC_{1:5}$ to EC_{sat}
Loamy sand, clayey sand, sand	23
Sandy loam, fine sandy loam, light sandy clay loam	14
Loam, loam fine sandy, silt loam, sandy clay loam	9.5
Clay loam, silty clay loam, fine sandy clay loam	8.6
Sandy clay, silty clay, light clay	7.5
Light medium clay	5.8
Medium clay	5.8
Heavy clay	5.8

Changes in vegetation, either to dominance of more salt-tolerant species or through reduced growth of existing species, are often the first signs of dryland salinisation problems. This is worsened under grazing if stock have access to the site, as they are attracted to salt on the surface of the soil and tend to graze such sites more heavily than the rest of the paddock. The stress on pasture plants, through both salinity and grazing pressure, means that areas are rapidly denuded of vegetation and become erosion hazards. These effects depend, of course, on seasonal conditions, with plant growth and root zone salt levels varying according to rainfall patterns and the occurrence of periods of drying weather.

Effects on Soils

Salinity affects soils by:

(i) increasing the osmotic pressure of the soil solution, which increases the difficulty with which plants can extract water from the soil

(ii) increasing the soil's content of specific ions which may be toxic to some plant species; this is mainly of concern in irrigation areas, where chloride and sodium ions may be increased to concentrations toxic for some crops

(iii) increasing exchangeable sodium on the soil's exchange complex. This is the main effect on the soils themselves, and may lead to breakdown of soil structure due to the dispersing effect of the sodium ions on clay particles. Such breakdown is more likely to occur after a wet period in which free salt is leached out and the sodic soil remaining then disperses, leading to reduced aggregation, poorer aeration and drainage, and a lower stability of the soil against erosion (see Chapter 10).

Effects on Erosion

The combined effects of saline conditions on soils and vegetation lead, in most cases, to seepage sites becoming focal points for erosion unless preventative measures are taken. The primary cause is loss of vegetative cover, and erosion is often further assisted by the erodible nature of the surface soil, which is typically light-textured and often dispersible. In semi-arid areas, scalds result from the removal of the surface soil by wind or water erosion and subsequent exposure of a saline and impermeable subsoil. In wetter areas, the removal of surface soil is hastened by saline conditions, the surface seals as a result of clay dispersion, and a similar 'scalded' condition develops. These 'wet' scalds also have low permeability, and generate high levels of runoff during heavy rain, often leading to rill and gully erosion. The typical well-developed seepage on the tablelands or slopes is thus often associated with an eroded site, hence traditionally involving soil conservation agencies with this type of salinity.

Effects on Water Resources/Water Quality

This is a major effect of soil salinisation. Dryland salinisation increases the salt content of groundwater and surface water, both in the area concerned and downstream. Both sources of water may eventually enter streams or storages and be used for stock water, irrigation, industrial, domestic, urban or rural purposes. The presence of salts in increasing quantities is undesirable, and imposes excessive costs on users and consumers. Unnaturally high salt levels in major streams and inland rivers create a hostile environment for aquatic plants and animals, leading to reduced populations in some instances.

Through its effects on soils, vegetation and water resources, the overall impact of seepage salinity is reduced value and productivity of agricultural land, adverse effects on trees and other vegetation, reduced aesthetic appeal of the rural environment, and many adverse off-site impacts. These occur through salt transmission in surface and groundwater flow to streams and rivers, and subsequent use of that water for rural, industrial and domestic purposes. In particular, it may contribute to the salinisation problems of irrigation areas by increasing the salt content of both surface and groundwater supplies used for crop irrigation.

Studies within the Murray-Darling Basin show an increase in salinity in many rivers, especially the Murrumbidgee and Lachlan Rivers (Williamson et al., 1997; Walker et al., 1999).

13.3.3 Recognition of Soil Salinity

General observations, such as bare ground, stunted vegetation and salt inflorescences on the surface, can be used to identify areas of saline soils. For three broad categories of salinisation the visual symptoms include:

Incipient Salting (Class 1):
  Wet patches with constant moisture and a greasy surface
  Areas becoming bare and 'scalded'

Changes in pasture or crop health (possible yellowing of leaves)
  Patches of salt-tolerant plants developing.
Moderate Salting (Class 2):
  Pasture and crop losses with replacement by salt-tolerant species
  Bare patches that remain wet all year
  Salt inflorescences appear on surface when dry
  Surface crusts develop (see Colour plate 13.4)
  Stock gather on bare areas to access salt
  Clear water in nearby dams and soakages.
Severe Salting (Class 3):
  Tree decline at or near the site
  Extensive bare patches with little vegetation growing
  Only highly salt-tolerant plant species survive
  Infrastructure damage to roads, fences, buildings and so on
  Associated sheet, rill and gully erosion.

Following recognition of the problem, soil salinity can be more specifically determined in the following ways:

—measuring the conductivity of a soil/water saturation extract in the laboratory (EC_{sat})
—using a conductivity meter to measure the conductivity of a 1:5 soil:water extract made up in the field or in the laboratory; this calibration can be used to estimate EC_{sat} using the conversion factors based on texture in Slavich and Pettersen (1993). See Table 13.9
—using in situ soil salinity sensors or electrode probes
—using piezometers to monitor groundwater salinity and depth; a piezometer is essentially a piece of slotted pipe inserted into a shallow bore-hole (say to 12 metres) into which groundwater rises
—remotely, by electromagnetic induction techniques.

Electromagnetic induction (EMI) technology is also used to assess areas prone to become saline as a result of the presence of salts at depth. The EMI equipment measures the conductivity of the ground by the transmission of a primary electromagnetic field and detection of the secondary field thus generated. The measurements are influenced by salt concentration, water content, clay type and depth, and require careful interpretation.

Conductivity meters measure the concentration of salts in a soil/water extract through direct measurement of the solution's electrical conductivity. The results are expressed in decisiemens per metre (dS/m), and saline soils are generally defined as those with an electrical conductivity greater than 1.5 dS/m for a 1:5 extract, and greater than 4 dS/m for a saturation extract. The 1:5 extract can be measured in the field for practical purposes, whereas the saturation extract is a standardised and more reliable laboratory procedure. Conversion factors, which depend on soil texture, are available with conductivity kits to convert from one measurement to the other, and are based on Slavich and Pettersen (1993). The units used in soil salinity studies and their conversion are discussed fully in Taylor (1993). The most common are decisiemens per metre (dS/m) and microsiemens per centimetre (μS/cm). The unit dS/m can be converted to μS/cm by multiplying by 1000.

3.3.4 Management of Saline Soils

The care and management of saline areas depend on a strategy to increase water use across the landscape, utilising rainfall where it falls. It is important to be able to recognise problem areas and to then investigate and plan appropriate treatment measures. Management can be considered under the headings of control and prevention.

Control of soil salinisation involves the recognition and reclamation of existing and potentially saline sites to contain both on-site and off-site effects, and then ongoing management of both the site and its catchment to ensure that the problem does not recur.

The development of recent technology has enabled better predictions to be made of potentially saline sites and the possible salt stores that may be activated by rising water tables and increased groundwater flows. This has involved the use of electromagnetic induction at various intensities, catchment water flow models such as TOPOG, and hydrogeology investigations. The water flow models usually use digital terrain models and estimated soil hydraulic properties to predict water flows within catchments. Technology has also enabled better monitoring of the levels of water tables and salinity, and of water use by a range of plants. By identifying the main areas contributing water to a saline area, the areas where treatment of a catchment will be most effective can be identified.

Reclamation

The reclamation of saline soils can be divided into two phases: the treatment at the catchment scale; and the treatment of the saline site where salinity is already a problem.

At the catchment scale, the objective is to reduce the amount of throughflow that is leading to rising water tables in lower slope areas. This aims to get a better hydrological balance within the catchment. It is a longer-term policy and requires an increase in water use by plants within the catchment, especially in the areas that have been identified as recharge areas for the saline sites. Tree planting on higher slopes is one option, possibly with trees that can produce an

income in the future. Other potentially beneficial practices include the growth of perennial plants such as grasses and lucerne, which have higher water use than annuals, especially in summer; minimising fallowing of land which can result in large amounts of water being added to water tables; and maximising plant growth by judicial grazing and selection of crop types. There is a range of techniques being developed to improve water use on the landscape.

Reclamation of the saline seepage zone is largely a control situation as the actual soil salinity problem is only likely to be solved by treatment at the catchment scale. However, the degradation of the soil at the site and the environmental effects of the site can be minimised by a range of practices. The treatment aim is to minimise erosion at the site, and the amount of saline water flow entering watercourses downstream. Techniques include diversion of extraneous surface water from the site, sowing of salt-tolerant plant species including *Puccinellia ciliata* and tall wheat grass (*Thinopyron elongatum*), salt bush or some varieties of river red gum (*Eucalyptus camaldulensis*), mulching with hay or straw, building mounds for plants to grow in a less saline environment, use of gypsum, heavy fertilisation to maximise plant growth, and fencing of the site to exclude grazing. Some productive use can be made of these sites, but stabilisation probably should be the main objective. Long-term management could include judicious grazing to prevent plants getting too rank and actually reducing ground cover. Table 13.10 shows grass species suitable for saline areas.

In recent years research has resulted in a widening of interest in tree planting and agroforestry, due to the development of a range of improved tree-planting techniques and study of a wide range of trees and their role in controlling recharge. Recognition of the role of trees in the broader environmental context has also been important.

Australian trees generally use more water than pastures and crops because of their deep rooting, rainfall interception ability and near-continuous and large leaf area. Actively growing trees are the most desirable, so species selection should be based on their suitability to the particular environment concerned and, where necessary, their salt tolerance. A tree-growing program for effective salinity control must take account of the following:

—the amount of water needing to be used by trees
—those parts of the catchment in which tree planting is most needed (recharge zones)
—the proportion of the catchment to be planted
—planting density to achieve the required recharge control in the given rainfall zone
—the time delay before effective recharge control is achieved

Table 13.10 Useful species for revegetation of saline areas

Salt Tolerance	Species	Comment
High	*Puccinellia ciliata* (marsh grass)	Susceptible to competition, poor germinability
	Thinopyron elongatum (tall wheat grass)	Susceptible to heavy grazing
Moderate	*Lolium rigidum* (Wimmera ryegrass)	
	Cynodon dactylon (couch grass)	
	Polypogon spp. (annual beard grass)	Occurs naturally
	Trifolium fragiferum (strawberry clover)	Mulching necessary
Useful	*Phalaris aquatica* (phalaris)	
	Chloris gayana (rhodes grass)	
	Festuca arundinacea (tall fescue)	

—integration of tree planting with other aspects of land use.

Prevention

Prevention of soil salinisation must be on a catchment basis and depends on an understanding of the processes involved and recognition of potential problem areas. Each salinisation site, and the complex association of recharge and discharge zone, is a unique combination of landform, climatic conditions and land use factors, and should be treated accordingly. Prevention of salinisation may require a number of approaches, integrated according to local circumstances. There is no magic formula that can be applied in all situations.

Salinity-prone areas must first be identified using geological, geomorphological and/or hydrological knowledge, or programs of salinity or water table monitoring. The use of EMI techniques for the detection of subsurface salt levels is rapidly developing, and there is an acknowledged need for catchment-wide groundwater measurement and monitoring.

Land use in recharge zones is critical to salinity occurrence. Clearing control is a desirable option where native forest occurs. Where recharge zones have already been cleared of timber, permanent crop or high water-using pasture should be encouraged, to make maximum use of rainfall. Fallowing should be discouraged, as it promotes maximum water intake and minimum transpiration by vegetation. Replanting of trees with deep roots and high rates of water use should be encouraged in appropriate parts of these zones.

The best strategy for salinity prevention is based on the concept of constant water use from a variety of vigorous species of trees, shrubs, grasses and crops. Deep-rooted perennial plants are highly desirable. Large areas with similar water use, based solely on shallow-rooted annual plants, may give rise to problems if the potential for salinisation is present in the catchment.

Trees have an all-important role in recharge zones, and clearing of native vegetation should be avoided where possible. Agroforestry, natural regeneration, interception planting and land retirement may all be considered as supplements to the existence of natural tree cover on hillslopes.

With respect to pastures, perennial types have the potential to optimise water use throughout the year, particularly if deep-rooted species such as lucerne are incorporated. Annual pastures are less effective at drying out the soil in late spring, summer and autumn. Cool season and warm season species have different growth habits, and if local soil and climatic conditions are suitable, grazing paddocks of each in a farm catchment will help maximise overall water utilisation. Management of pastures to maintain a large functioning leaf area depends on a balance between annual and perennial components, and grazing should be managed to ensure that plants become neither too rank nor overgrazed with loss of too much leaf. This will enhance water loss on a continuous basis. A ground cover level of at least 70% should be aimed for through judicious grazing, and this will also be made easier if a mix of species is used in the pasture.

With respect to cropping management to prevent the development of soil salinity, the overall aim should be to minimise deep drainage and evaporative losses of soil water. This is achieved by encouraging the growth of healthy, vigorous crops with a large leaf area to maximise water use by transpiration. Deep-rooted crops reduce drainage losses, and surface mulching reduces evaporative losses and erosion. Soil management to maintain fertility and satisfactory levels of soil acidity and structure is important to encourage growth of high-yielding crops. On erodible soils and steeper slopes, special erosion control measures may be necessary.

Other factors to be considered in the growing of satisfactory crops to resist salinisation include:
—use of rotations to limit crop disease; break crops such as canola, lupins and peas in a wheat rotation also help improve soil structure and fertility
—minimisation of the use of long fallows and tillage
—optimisation of sowing times and seeding rates
—weed control, by herbicides where appropriate
—strip cropping, to maximise water use and control of erosion
—opportunity cropping; depending on the season, this may be a valuable means of utilising surplus soil moisture.

13.4 Soil Nutrient Decline   P.E.V. Charman

The soil's fertility is the soil's ability to support plant life. Fertility in soils is influenced by a number of factors that can be broadly classified under three headings: physical, chemical and biological. Physical fertility is related to the physical features of the soil such as water relations, air space, structure or texture. Chemical features of the soil include factors such as pH, plant nutrient availability and salinity. Biological fertility refers to the population of microorganisms in the soil and its activity in recycling organic matter.

Climate has an overriding part to play in the growth of crops of pasture, and the rainfall/evaporation pattern, in particular, will determine what areas can be developed for agriculture or pasture improvement. The removal of topsoil by erosion, in many parts of the state, will also affect the soil's potential for more intensive use of the land for crop growing. This is because the nutrients of major importance to crop growth tend to be concentrated in the surface soil horizons. The subsoil is of importance in relation to drainage, overall water-holding capacity, growth of deep-rooting plants and the long-term supply of minerals and nutrients.

Chemical fertility is naturally of paramount importance to the growth of vegetation in all soils. One of the main problems in the revegetation of eroded areas is that the more fertile topsoil has often been removed. Exposed subsoil presents a harsh environment for new plant growth, usually lacking in sufficient nutrients or having unfavourable nutrient ratios. It also often lacks a suitable structure to enable penetration of air, water and plant roots. If these problems of chemical and physical infertility can be overcome, there is every reason to expect any revegetation work to succeed (see Chapter 19).

Another soil factor that is of importance in relation to soil fertility is that of soil pH. This is a measure of the acidity or alkalinity of a soil on a scale of 0 to 14, with low figures representing very acid soils and high figures very alkaline soils. Soil pH levels generally fall between 4.0 and 9.0 with extreme figures being rare. A pH of 7.0 means the soil is chemically neutral, and most plants thrive best at a pH between 6.0 and 7.5. Different plants have different tolerances to acidity or alkalinity, some preferring quite acid soil while others do better under alkaline conditions. For the establishment of legumes, certain rhizobia are inhibited under acid conditions, hence the benefits of lime pelleting of legume seeds. Surface soils are normally more acid than subsoils, because of the effects of leaching and organic matter in producing more acid conditions in the surface soil, and the more likely accumulation of calcium carbonate and similar alkaline compounds in subsoil. The availability of various nutrient elements also varies with pH, with the majority being more available on the acid side of neutral, though with some notable exceptions such as molybdenum.

Salinity is another soil factor that may affect chemical fertility. Salinity is due to the accumulation of soluble salts in soil, mainly sodium chloride. As with pH, plants very widely in their tolerance of salts in soils, but most pasture species and crop plants will not grow well when salt contents rise above normal levels. Salinity is assessed by measuring the soil's electrical conductivity and is expressed in decisiemens/metre. A level of 4.0 for a saturated soil extract and 1.5 for a 1:5 soil:water extract are generally accepted as the upper limits of salinity for normal plants.

Soil moisture relations are clearly of paramount importance when plant growth is under consideration. The ability of a soil to absorb and hold water to sustain growth in dry periods characterises the best agricultural soils, and this is no less important in revegetation work.

13.4.1 Chemical Fertility Decline

Many soils in New South Wales have only limited amounts of the nutrients that are required for plant growth. Given that they have been farmed for decades without too much attention being paid to balanced fertiliser programs and the maintenance of organic matter in the soil, it is likely that these nutrient levels will have to be replenished to maintain plant yields and production. The removal of crops and other agricultural products from the land has removed nutrients; the amount of nutrient removed by various crops and agricultural products is shown in Table 4.1. The result has been a decline in organic matter content and nutrients to a level at which fertility is very largely determined by fertiliser input. This particularly applies to cropping soils, but in the more naturally fertile ones, such as the black soils of the Liverpool Plains, the decline has been less significant. Nonetheless, although these soils have generally not been farmed for a very long period, in some instances their organic matter levels are less than half the figure under natural pasture. Even these soils cannot be cropped regularly in the long term without attention to fertiliser programs and the return of plant residues to the soil to keep the organic matter content at a satisfactory level. On many arable soils increases in crop yields have been probably due more to improved farming techniques such as weed control and fallowing and plant breeding, than to a conscious program of fertility maintenance.

Phosphorus

Superphosphate has been the main fertiliser used throughout the agricultural and grazing areas of New South Wales for many years, because of the deficiency of many of our soils in phosphorus. Due to the dramatic increases in production that have been achieved by applying this fertiliser, the other chemical aspects of soil fertility have been neglected to some degree, and the approach to fertiliser and soil management has become rather one-sided. Under these conditions, it was inevitable that, sooner or later, some chemical imbalance would arise. One current sign of

this imbalance is increasing soil acidity, which is particularly evident in parts of the southern tablelands and south-west slopes.

The story with phosphorus is not really one of decline, because most Australian soils are not well endowed with this element, with the exception of the black earths of northern New South Wales and southern Queensland. Sandy soils are particularly deficient. The fertiliser superphosphate, produced from natural rock phosphate by treatment with sulfuric acid, is essentially a calcium hydrogen phosphate. Although soluble in water, much of the applied phosphate reverts to an insoluble form in the soil and thus is protected against leaching. Depending on the chemistry of the particular soil, including its pH level, clay mineralogy, and iron and aluminium oxide levels, the phosphate may become further 'fixed' so that it is relatively unavailable to plants. Krasnozems, for example, are well known for their ability to fix phosphate. This 'fixation', a most complex chemical phenomenon that varies widely between soils, has been the main problem with the phosphate status of many of them. Application of phosphorus fertilisers continues to be a need in both cropping and grazing situations.

Nitrogen and Potassium

While phosphorus tends to have been the main focus of fertiliser management and use because of its non-availability or scarcity in many soils and its importance for legumes, potassium and nitrogen are equally important for plant growth. Their loss from soils under agricultural use has significantly contributed to fertility decline. For wheat cropping, nitrogen is needed to enhance both yield and grain quality (protein content).

Compared with phosphorus, potassium is more mobile in soils and nitrogen is almost totally mobile. Potassium is taken up strongly by crops but most of our soils are reasonably well endowed with this element, particularly the more clayey soils with high cation exchange capacities. Deficiencies are thus normally only experienced where cropping is intensive such as with horticultural production. Potato growing on krasnozem soils is a situation where potassium fertilisers must be regularly applied. However, to arrest the long-term decline in the potassium status of soils, replacement is also required in all cropping situations and under intensive grazing and hay production.

Most nitrogen in soils, and that used by plants, is derived ultimately from nitrogen in the air which has been assimilated by certain soil micro-organisms, many associated with leguminous plants. Nitrogen fixing by legumes is vital to the fertility of many soils. Following fixation, the nitrogen is stored temporarily in the soil in organic forms. These may then be broken down by other soil micro-organisms into ammonium and nitrate forms, via mineralisation and nitrification, for direct uptake by plants in solution. Thus nitrogen is very mobile and is constantly being cycled between atmosphere, plants, animals and soil, subject to changes in climatic, edaphic and environmental conditions. In the nitrate form, it is also leached with ease, which may lead to soil acidification.

The nitrogen status of soils is thus related closely to organic matter levels. In their virgin state it is not of great significance except in the case of soils such as black earths, prairie and chocolate soils which are intrinsically more fertile because of their geology. Other soils such as red-brown earths, non-calcic brown soils and red earths also have good levels of organic matter in their pristine condition, under a grass/woodland type of native vegetation. However they tend to lose it rapidly with cultivation. Under cropping regimes, nitrogen levels may be seriously depleted over periods as short as a few years. Thus for all soils being cropped or intensively grazed, nitrogen replacement in the form of fertilisers is essential to reverse fertility decline. The general aim should be to satisfy the needs of the current crop only, particularly following non-leguminous pasture phases, two or three successive non-leguminous crops, or when incorporating large amounts of stubble as part of a conservation tillage system.

Calcium and Sulfur

The important minor elements essential for plant growth in our soils are calcium and sulfur. Sulfur shortage is not generally a problem, with sufficient being released from soil reserves by cultivation in cropping regions. However, in the longer term, regular cropping is likely to require sulfur applications in the form of specific fertilisers. Canola, in particular, has a high requirement for sulfur, and gypsum is often used as a fertiliser for this crop. One of the problems is that since the reduction in the use of superphosphate, which contains some sulfur, shortages of this element are becoming more frequent in cropping soils. Sulfur-fortified superphosphate has been found to have significant benefits for pasture production in a number of grazing regions.

While calcium is an essential plant nutrient, it also has an important function in maintaining satisfactory pH levels in soils. Unfortunately it is subject to a constant 'draw-down' in all agricultural systems and is regularly exported out 'through the farm gate' in significant quantities in both plant and animal products (grain, milk, hay, meat, bones and so on). This process is partly responsible for the increase in soil acidity which has become a problem over the last twenty years or so on many farms in southern New South Wales. Hence soil acidification is a significant indicator of soil fertility decline or 'nutrient depletion', and liming becomes an important strategy to prevent it.

The other main causes of nutrient depletion, apart from product removal and lack of replacement, are erosion and leaching. Soil erosion is a primary cause as it is directly responsible, not only for the removal of soil from the land, along with absorbed and adsorbed nutrients, but also for the more direct removal of soluble nutrients in runoff. Leaching, whereby rainfall percolates downwards through the soil, also takes with it soluble nutrients, which eventually may find their way into the drainage water and are removed from the soil altogether. Many such substances, whether from erosion or leaching, inevitably end up in streams and rivers and may contribute to an unnatural enrichment of such waters. This often leads to an upsetting of their ecological balance, to the detriment of aquatic flora and fauna. Such is the case with eutrophication and the development of blooms of blue-green and other toxic algae, which received so much attention during the nineties. This decline in water quality may also have serious implications for further irrigation, stock or domestic use.

Nutrient Requirements for Plant Growth

13.4.2

With respect to fertility requirements, it is well understood that different plants have different requirements for and different tolerances of elements that are available to them in the soil. Table 13.11 should be read with this proviso in mind. Responses to added nutrients may be commonly observed if other nutrient concentrations already present in the soil are not limiting (this is usually true of nitrogen fertiliser additions). Provided there are no physical or climatic limitations, the guidelines summarised in the table should describe conditions associated with satisfactory growth and good ground cover protection, especially for species used for the establishment of cover on disturbed areas. Sources for the information in Table 13.11 are listed in the bibliography at the end of this chapter.

Table 13.11 Guidelines for nutrient concentrations necessary for satisfactory vegetative growth

Element	Concentration Range (ppm) Total	Available	Deficiency Limit (ppm)	Toxicity Limit (ppm)	Remarks
Nitrogen	500–3000	500–1800	70	8 (NH$^+_4$aq)	A C:N ratio of less than 32–35 is necessary for significant N mineralisation to occur.
Phosphorus	200–1500	4–20	6	200–300	These values are based on 0.5M NaHCO$_3$ extraction. Phosphorus buffering capacity is known to strongly influence availability of P.
Sulfur	50–400		3	400 (SO$_4$-S aq)	A N:S ratio of 14–16 is desirable. Ratios > 17 are associated with Sulfur deficiency
Aluminium	to 15%	0.01–0.50 (aq)	nap	0.5 (SE)	Soluble aluminium concentration increases with decreasing pH, decreasing organic matter levels and increasing soluble salt levels. Toxicity may also be assessed by measuring the corrected lime potential (e.g. CLP > 3.13 is necessary for legume growth).
Arsenic	0.2–10	0–0.6	nap	2 (aq)	
Boron	10–85	0.1–2	0.3–0.5	0.8–1.5	These values are based on a hot water extract.
Calcium	0.2–20%	2–18 me/100 g	2 me/100 g	10%	Ratios of Ca:Mg in exchangeable form of 5:1 to 1:1 are desirable for good plant production with ex.Mg > 0.2 me/100 g in acid sandy soils or > 1.5 me/100 g in clay soils.
Chromium	200–400	0.02–3 (aq)	nap	5–10 (aq)	Some plants are especially sensitive to quite low levels of chromium, and to chromium in the form Cr(VI).
Cobalt	1–40	0.1–1.0 (HAc)	nap	ndf	
Copper	25–60	2–11 (HCl)	0.5 (HCl) 1 (EDTA)	10 me/100 g	
Iron	2–5%	0.005–0.5	2 (NH$_4$Ac) 0.03 me/100 g	nap	Plants depend on chelation of iron to make this element available and soil extraction techniques are most unreliable indicators of availability.
Magnesium	0.05–0.5%	0.6–1.0 me/100 g	0.2 me/100g sandy soils	nap	A K:Mg ratio of < 0.4 is desirable for adequate magnesium nutrition.
			0.5 me/100 g clay soils		High exchangeable Mg is associated with poor soil structure, particularly in combination with high Na.
Manganese	150–3000	25–100 (Hq)	1.7(aq) 0.01 me/100 g 15 (Hq)	5 (aq)	Manganese toxicity is unlikely above pH 5 and is ameliorated by the presence of Si, Al, Fe and NH$_4$ ions, which occur in acid soils.
Molybdenum	0.5–4	0.01–1.0 (HAc)	0.2 (pH 5)(T) 0.05 (pH 6)(T) < 4 μg/L (LK)	0.8 (pH 5.0)(T) 0.3 (pH 6–8)(T)	Critical molybdenum levels vary with soil pH.
Nickel	15–50	0.1–4 (HAc)	nap	10	
Potassium	0.5–4%	0.05–1.0 me/100 g	0.03 me/100 g	ndf	Potassium buffering capacity is known to strongly influence the availability of K.

Table 13.11   continued

Element	Concentration Range (ppm) Total	Available	Deficiency Limit (ppm)	Toxicity Limit (ppm)	Remarks
Selenium	0.1–6	0.02–0.2	0.45	50	Deficiency levels of 0.45 ppm in soils produce unthrifty plants. Levels of 50 ppm in plants adversely affect consumers of plants.
Sodium	0.2–2%	0.0005–0.05 me/100 g	nap	0.1% 4.34 me/100 g	
Zinc	10–300	1–4	33(total) 1 (HCl)(pH 7.0) 2.5 (HCl)(pH 6)	250–350 (HAc) 100 (HCl)	

nap   not applicable

HAc   mild acetic acid

EDTA   0.05M EDTA

T   Tamms reagent

aq   aqueous extract

ndf   not defined

NH₄Ac   1M NH₄Ac (pH 4.8)

LK   equilibrium soil solution

SE   saturation extract

HCl   0.1M HCl

Hq   1m NH₄Ac + 0.2% hydroquinone

NB   For sources see Bibliography, p. 257

Acid Sulfate Soils   M.D. Melville and I. White

In 'A world perspective on acid sulfate soils', Dent and Pons (1995) open their review with the statement: 'Acid sulfate soils are the nastiest soils in the world.' Awareness of the existence and problems involved with acid sulfate soils (ASS) is rapidly increasing (for example, Dent, 1986; White et al., 1996a). Acid sulfate soils are widespread throughout the world and their problems and description have been known for centuries (Dent and Pons, 1995). Linnaeus (who seems to have classified and named everything), in 1735 coined the term 'Argilla vitriolacea' for clays with sulfuric acid, that is, acid sulfate soils. Nevertheless these soils have only been widely recognised in Australia in the past ten years (there were probably only three papers on acid sulfate soils in Australian scientific journals before 1989: Teakle and Southern, 1937; Walker, 1972; Willett and Walker, 1982). The importance and message of Walker's (1972) paper went unheeded in CSIRO and by coastal land managers. Some of our present environmental problems, arising from acid sulfate soils, could have been avoided if people had noted Walker's message.

What Are Acid Sulfate Soils?   13.5.1

Acid sulfate soil is the name given to soil and sediment containing oxidisable, or already oxidised, sulfides. The main form of sulfides is iron pyrite (cubic FeS_2). Other forms such as monosulfides exist in small concentrations. One of these monosulfides, greigite (Fe_3S_4), has been shown by Bush and Sullivan (1997) to be more common and much more important in the dynamics of pyrite oxidation than previously believed. If the soil remains in a reduced state, so that no oxidation of the sulfides occurs, the material is referred to as a Potential Acid Sulfate Soil (PASS), or a 'sulfidic soil'. Where the soils are exposed to air, so that oxidation can occur, the material is called an Actual Acid Sulfate Soil (AASS), or 'sulfuric soil'. Oxidising conditions usually overlie reducing conditions in the same profile so that AASS and PASS apply to different parts of the same profile. Therefore the generic term Acid Sulfate Soil (ASS) is used for the whole profile.

13.5.2 Origin of Potential Acid Sulfate Soils

The accumulation of sulfides is part of the natural geological global sulfur cycle. During glacial/interglacial periods in the geological timescale, part of the global marine sulfur pool (in the form of dissolved sulfate) is chemically reduced and accumulated in sulfidic sediments of estuaries and coastal lowlands during the rising sea level stage. During high sea level standstill (interglacials), such as the latter part of the Holocene Epoch (less than about 6500 years Before Present), there is still accumulation of sulfides but natural oxidation of them can also occur. During falling and low sea level stages (glacial maxima), much of this sulfidic sediment is oxidised and eroded, and the sulfur returned to the marine pool.

The natural distribution, formation and properties of these soils therefore depend on the balance between the accumulation of sulfides in PASS and the degree of its oxidation to AASS.

In terms of PASS formation, this means a balance between sediment and water inputs from either fluvial or marine sources. Marine sources of sediment and/or water favour sulfide accumulation. In estuaries with small catchment:floodplain area and large tidal exchange:fluvial discharge volumes (for example, Tweed River), estuarine sedimentation will dominate and PASS will occur throughout most of the estuarine floodplain and associated lowlands. On the other hand, in estuaries with a large catchment:floodplain area and large fluvial discharge:tidal exchange volume (for example, Sepik River, Papua New Guinea), the floodplain sediments will be predominantly fluviatile and little, if any, PASS will accumulate. Similarly, in coastlines without estuaries (for example, Great Australian Bight), sediment accumulation does not occur.

The degree of sulfide accumulation in PASS therefore depends on the dissolved sulfate concentration of the water in which estuarine floodplain sedimentation occurs, but it also depends on the prevailing temperature. The biochemical reduction of sulfate depends on the rate of bacterial activity and also on the amount of organic matter in the sediment. Generally, high temperatures in tropical northern Australia favour sulfide accumulation. Nevertheless cool temperate estuarine sedgelands and lake-bottom sediments still can have significant sulfide accumulation (for example, the Baltic coast at 60° N).

The formation of sulfidic sediments can be summarised by the overall, bacteria-driven reaction:

Clearly, the location of PASS relates to present sea level and its position on the coastal margin. In Australia we should therefore anticipate the presence of sulfides in Holocene-age sediments at lower than about 5 m above mean sea level (that is, below about 5 m Australian Height Datum). This rule of thumb allows for variations around the Australian coastline of tidal amplitudes, tectonic uplift and fluvial sedimentation in the Holocene. We are now aware, however, of some instances, such as on the Scott Coastal Plain in Western Australia, where sulfides have been preserved in higher elevation (up to 30 m AHD), mid-Pleistocene-age coastal sediments, despite the intervening low sea levels of glacial maxima.

The preceding understanding of the geomorphic controls on PASS formation were used by the former NSW Soil Conservation Service as the basis for their acid sulfate soil risk mapping program (Naylor et al., 1995). This program relied upon air-photo interpretation of coastal physiography with field checking and some laboratory soil testing. These maps (more than 120 sheets, 1:25 000 scale) have proven the most useful advance for the better management of these problem materials, and provide a model that should be adopted nationally. The maps are now being adapted for planning purposes but overall they show the location of approximately half a million hectares of ASS/PASS.

13.5.3 The Formation of Actual Acid Sulfate Soils

The sulfide minerals in PASS are inert while they remain in a chemically reduced state below the water table. They will begin oxidation to AASS on exposure to air, such as with their excavation, or with the lowering of the water table by periods of prolonged drought (without recharge from the estuary), or by land drainage practices. The formation of AASS and the controls of the water table elevation are therefore a hydrological issue involving the inputs and outputs within the water balance of the estuary floodplain and coastal lowland.

Within the Scott Coastal Plain of Western Australia (previously introduced), there is a seasonal excess of winter rainfall but marked excess of potential evapotranspiration during the Mediterranean-climate summer-time. Thus water tables are at or above the ground surface in winter and lower in summer. Sulfide minerals are preserved in mid-Pleistocene-age sediments within one or two metres

$$4SO_4^{2-} + \quad Fe_2O_3 \quad + \quad 8CH_2O + \tfrac{1}{2}O_2 \quad \xrightarrow[\substack{reduced \\ conditions}]{bacteria} \quad 2FeS_2 \quad + \quad 8HCO_3^- \quad + \quad 4H_2O$$

seawater | ferric iron in sediment | organic matter | | pyrite | bicarbonate

of the ground surface. This demonstrates that while sea levels were much lower during past glacial maxima and rainfall was less than the present, high water tables were maintained because evapotranspiration was the dominant control and this was reduced owing to generally lower temperatures. In the macrotidal rivers of northern Australia the widespread PASS have undergone natural oxidation during the wintertime dry season. The summertime wet season raises water tables to above the ground surface.

The role of land drainage in causing problems with ASS is frequently misunderstood. Not all estuaries and floodplains are the same. In some instances, much of the present-day acidity occurred before drainage. These estuary locations tend to be where well-developed fluviatile natural levees were deposited over PASS materials and these levees effectively cut off backswamp areas from recharge by the estuary. During prolonged periods of drought (perhaps such as the Medieval Warm Period, about AD 1100 to 1300), water tables in the backswamp were markedly lowered by evapotranspiration and AASS were formed (for example, the Rossglen wetland on the Camden Haven River, New South Wales). In other locations (such as the wetlands adjacent to Cobaki Creek, northern New South Wales), minimal fluviatile sediment was available to form levees so that even in the most prolonged droughts the wetlands were continually inundated by tidal waters and the PASS was preserved.

The Pyrite Oxidation Process

The pyrite oxidation process is complex and generally bacteria-driven. However, overall, complete oxidation of 1 tonne of pyrite yields 1.6 tonnes of pure sulfuric acid. Some of this acid may be neutralised by reactions with clay minerals and carbonates in the soil profile. Some acid may also be buffered and stored in partial oxidation products such as jarosite and metal sulfates in the soil profile. Some soluble ferrous iron may be exported to receiving water where its further oxidation to ferric iron in iron oxyhydroxides produces 'acid-at-a-distance'. These iron precipitates can themselves be harmful, smothering photosynthetic surfaces on aquatic plants and blocking the gills of organisms. Some of the soluble acidity can be neutralised by dissolved alkalinity in the receiving waters (marine waters have a neutralising capacity up to about 2 moles H^+/m^3 water). Nevertheless, in many cases the rate and amount of acid production exceeds the neutralising capacity of the environment.

Therefore extreme acidification and its impacts are observed. We have measured pH < 2 in soil profiles and drainwater. Sammut et al. (1996) measured pH 4.5 in a tidal section of the Richmond River in northern New South Wales following a flood-promoted acid discharge of several thousand tonnes of sulfuric acid from one subcatchment (the Tuckean Swamp). Wilson (1995) estimated the discharge of acidity into the Tweed River during 1992 as about 2500 tonnes of sulfuric acid. These discharges make industrial spills trivial and the impacts described by Easton's (1989) important revelation are clearly explained. Impacts from ASS in general are discussed in White et al. (1996b).

The Role of Land Drainage   13.5.4

Flood mitigation and drainage schemes with one-way tidal flap-gates have increased greatly the rates of outflow from estuary floodplains with well-developed natural levees. The period of inundation in the backswamp following floods has been reduced from about 100 days to only five days or less. This is necessary so as to protect pastures and crops. These drained backswamp areas underlain by PASS are potentially very productive and now widely used for agriculture. In fact, most of the sugar cane areas of New South Wales are on ASS, and the all-time state record for a one-year cane crop in New South Wales was grown recently on these materials at McLeods Creek on the Tweed River. It is arguable as to whether or not the intensive drainage systems used for agriculture increase the oxidation of sulfides. Nevertheless there are large amounts of existing acidity (of order 50 tonne H_2SO_4/ha at McLeods Creek and 500 tonne H_2SO_4/ha at Rossglen, Camden Haven River), and drainage systems greatly increase the ability to export acidity from these landscapes. In general, in most locations, the existing acidity in ASS profiles should be managed so as to retain it in the landscape as much as possible. Even if it were physically possible to apply the 3:1 lime:acid necessary to neutralise acidity, it is economically impracticable and may otherwise prove ecologically damaging.

Management of Acid Sulfate Soils   13.5.5

As White et al. (1996b) point out, probably the greatest difficulty in improved management of ASS is that clearly defined environmental goals have not been set. Opinions as to the appropriate goals and procedures to attain them vary among governments, industries, individual land managers and the public. Nevertheless, it would probably be generally agreed that further deterioration of environmental qualities should be avoided in the coastal margins where the Australian population is increasingly living. This requires that further oxidation of sulfides due to human activities should be minimised. This may require that ASS should be avoided. If developments on ASS are unavoidable, then management procedures should be undertaken to ensure the appropriate treatment of all acidity that does develop.

Publications released in August 1998 by the New South Wales Acid Sulfate Soil Management Advisory Committee (ASSMAC) and Queensland Acid Sulfate Soil Investigation Team (QASSIT) provide possible national standards for assessment and management so as to minimise ASS problems. These documents, however, tend to be directed more to avoiding acidification from PASS oxidation that might be associated with new developments in ASS landscapes.

The great remaining problem is identifying the best management of those landscapes where acidification has already occurred, whether or not human activities, such as drainage, have been the cause. Whether or not drainage has increased widespread oxidation of PASS, the drainage systems provide the conduit by which greatly increased quantities of acidity can be transferred from floodplains to estuaries.

In recent times some stakeholders and land managers have advocated changes to tidal flap-gate operations so as to increase fish habitat area, to use tidal waters to neutralise acidity in drains, and to raise floodplain water tables to stop further pyrite oxidation. Such procedures may achieve the first two of these aims, however the neutralising capacity of the tidal waters is finite and may still be exceeded by acid discharges under certain rain events. Fish that have been encouraged to enter drain systems may then be killed or become diseased. In terms of stopping further pyrite oxidation, it must be remembered that much of the water table lowering in ASS landscapes is due to evapotranspiration rather than lateral flow of water into drains.

A major concern for agriculturalists who rely on tidal flap-gated drainage systems is that brackish water ingress into the system may salinise their land. The long-term agricultural activities in estuarine floodplains have, in many cases, reduced natural peat topsoil thicknesses, and caused 'ripening' of PASS which caused irreversible shrinkage. These activities cause surface elevation lowering in the ASS landscapes so that the floodplain is more susceptible to inundation by brackish water during high tides than was the case pre-drainage.

Another new land management option that is being advocated to rehabilitate some severely scalded ASS landscapes is by reflooding with rainwater. Trials have shown greatly increased pasture growth. Nevertheless there are two problems with this reflooding. First, the existing acidity in these landscapes is not greatly neutralised by the rainwater and the oxidation of pyrite is not necessarily stopped by anaerobic conditions beneath the water table if the profile is exceedingly acidic and contains large concentrations of soluble ferric iron (see Nordstrom, 1982). The standing waters in the reflooded areas are therefore likely to be very acidic (pH of 3.5). Second, since there is little storage of additional rainfall, each rain event will discharge acidity into receiving waters. In PASS landscapes where little acidity exists, reflooding will not cause such problems. These areas will provide increased fish habitat, and the fish can provide nuisance insect control (for mosquitoes and midges: Easton, pers. comm.).

Given the large amounts of existing acidity in many ASS landscapes, the only feasible option is to manage the landscape so as to retain as much as possible of the acidity in situ. This will probably involve some re-engineering of existing drainage systems, but this must be based on sound understanding of the geomorphic, soil chemical, and hydrological setting of each location. There are no easy prescriptive solutions.

Land managers, regulators and the public must agree to common environmental goals and achieve these through holistic management. An essential plank in this management is the recognition of the existence and problems of ASS, and therefore avoidance and unnecessary disturbance of ASS.

13.6 Toxicities in Soils P.J. Mulvey and G.L. Elliott

Many potentially toxic elements and compounds can be present in soils, but this does not mean that they are necessarily available for plants to take them up. It is only when these toxic agents are in a form that is readily available to plants that their toxicity becomes apparent. The occurrence of toxic agents in soils is therefore dependent on a range of soil variables and plant type. Soil factors include soil pH, soil texture, mineralogy, organic matter and the nature of the contaminant/soil bond. The assessment of toxicity based on total amounts of a contaminant tends to be excessively conservative as the toxic agent may not be available to the plant. It is the soil pH, salinity levels and other soil chemical factors that control if it is available to the plant. Plants also vary in their capability of extracting agents from the soil, for example capeweed (*Arctotheca calendula*) is very effective at extracting cadmium from the soil.

The occurrence of a high level of a toxic agent in the soil does not mean the soil is contaminated, as some soils are naturally high in heavy metals, especially if the soils are associated with ultrabasic rocks or mineral deposits. Toxicants broadly include metals, organic materials, salts and other inorganic contaminants. The more toxic elements are presented in Table 13.12.

In agricultural applications, substances tend to be more commonly associated with phytotoxicity

Table 13.12 Toxicity levels in soils and related plant symptoms

Component Element	Toxicity Symptom Limit	Remarks	Plant Symptoms
Al (Aluminium)	1 me/100 g (EX) 0.5 ppm (SE)	Per cent Al saturation of CEC and corrected lime potential measurement offer accurate criteria for determining toxicity for individual species (e.g. < 30% satn is required by ryegrass and many clovers; CLP values > 3.13 are required for many legumes). pH levels < 5 likely to induce toxicity.	Dark green leaves, purple stems, dwarfing of plants, death of root tips, clubbing of root ends.
Sb (Antimony)		Limited observations suggest the critical toxic level is at least 600 ppm (EDTA).	Reduced plant growth.
As (Arsenic)	2 ppm (WS) 200 ppm (EDTA)*		Apparent wilting of new leaves, slow growth, late maturity, apical chlorosis, marginal and shot-hole necrosis, premature senescence.
Ba (Barium)		Barium levels are critical if exchangeable barium exceeds exchangeable calcium and magnesium.	Uniform chlorosis (associated with sulphur deficiency).
HCO_3^- (Bicarbonate)	20 me/L (SE) 0.5 me/100 g (WS)*		Calcium, magnesium and iron deficiencies.
B (Boron)	1.5–5 ppm (HWS)	Plants exposed to higher light intensities can tolerate higher levels of boron.	Chlorosis followed by necrosis progressing from margins to leaf base.
Br^- (Bromide)	200 ppm (WS)	Vegetables and flowers are sensitive to lower levels (c.50 ppm).	Marginal necrosis.
Cd (Cadmium)	1.5 ppm (EDTA)	3 ppm in food is toxic to humans.	Plant dysfunction.
Cl^- (Chloride)	0.12% w/w (WS) 3.4 me/100 g (WS)		
Cr (Chromium)	5 ppm (SE) 20 ppm (EDTA)*	At about 50 ppm (SE) concentration, 50% yield reduction may be expected in monocotyledons.	Chlorosis, bronzing in lower leaf blade, narrow form of leaf growth, rolling of leaf blade (monocotyledons).
Co (Cobalt)	1 ppm (SE) 75 ppm (EDTA)*		Reduced vigour, chlorosis, withering of leaves, necrosis.
Cu (Copper)	50–200 ppm (Total) 120 ppm (EDTA) 0.5 ppm (SE)	Higher soil CEC compensates for toxic levels, lower pH aggravates toxicity.	Reduction in vigour, chlorosis, dieback of young tissue. Stunting, thickening, abnormal branching of roots, discolouration of roots.
F^- (Fluoride)	5 ppm (WS)*		Interveinal chlorosis, marginal necrosis.

Table 13.12 Continued

Component Element	Toxicity Symptom Limit	Remarks	Plant Symptoms
Pb (Lead)	100 ppm (EDTA) 7–10 ppm (SE)		Reduction in plant growth, death of roots.
Mn (Manganese)	200 ppm (Total) 740 ppm (HHC) 1.5 ppm (WS)	pH levels < 5 are likely to induce manganese toxicity.	As for copper, but including random chlorotic and necrotic spots followed by marginal chlorosis.
Mo (Molybdenum)	100 ppm (WS)		Strongly coloured leaves (colours variable between species).
Ni (Nickel)	7–10 ppm (EDTA)	Exotic species are more susceptible than native species.	Interveinal chlorosis, leaf striping in monocotyledons. Reduced plant vigour, leaves turn white, marginal necrosis.
P (Phosphorus)	200 ppm (NaHCO$_3$)		Cu and Zn deficiency.
Se (Selenium)	0.5 ppm (total) in susceptible soils, 100 ppm (total) in 'unsuited' soils	5 ppm in feed is toxic to farm animals. Well-aerated neutral-alkaline soils with free CaCO$_3$ or CaSO$_4$ are susceptible soils.	Retarded growth, stunting, chlorosis.
Na (Sodium)	0.1% w/w (WS) 4.3 me/100 g (WS)*		Stunting, dull bluish green appearance of leaves, marginal interveinal and apical necrosis. Induced Ca and K deficiencies.
SO$_4$$^{2-}$ (Sulfate)	3 me/100 g (WS) 5000 ppm (SE)		Retarded growth, dark green leaves.
Salt	< 2.8 pC > 1.5 mS/cm (1:5 extract) > 4.0 mS/cm (SE)	pC is the negative logarithm of the electrical conductivity of a 1:5 soil:water extract in S/cm. At about pC = 2.3, 50% yield reduction of grass occurs.	Reduced growth, early maturation.
Zn (Zinc)	250–350 ppm (HAc) 120 ppm (EDTA) 7–10 ppm (SE)	Lower limit for dicotyledonous crops, upper for monocotyledons.	Reduced root growth, limited leaf expansion.

* Values are based on limited data
WS Water soluble, 1:5 extract
SE Saturation extract
EDTA 0.02M EDTA 1:5 extract 3d equilibration
HWS Hot water extract
HHC Hydroxylamine hydrochloride extract
CEC Cation exchange capacity

HAc Acetic acid extract
CLP Corrected lime potential
EX Exchangeable

NB For sources, see Bibliography, p. **257**

causing reduction in plant yield (for example, soluble aluminium); less commonly associated with poisoning of grazing animals due to plant uptake (for example, selenium), and poisoning of humans by consumption of food (for example, cadmium); and rarely consumption or inhalation of soil by workers causing health problems (for example, very high levels of arsenic or lead). There may be indirect long-term effects of toxicity such as that caused by the death of earthworms and burrowing animals causing soil structural problems.

Metal Toxicities   13.6.1

Threshold criteria for contaminated soil for environmental protection that include most metals have been published by ANZECC (1992). Similarly the NSW EPA (1998) published so-called threshold criteria for phytotoxicity for contaminated soil. These guidelines are based on total concentrations. Published papers and unpublished data at numerous locations have found natural exceedance of these guidelines at Port Macquarie (Cr, Cu, Mn and Ni) (Lottermoser, 1997), Armidale (Cr), Southern Wollongong and Dapto (Cu), and all metalliferous mining areas of the state.

All other states of Australia similarly have large areas exceeding the guidelines. For instance, a large portion of Victoria including parts of Melbourne, all of Bendigo and Ballarat have concentrations of arsenic above most thresholds of investigation. Soil at Port Macquarie exceeds the maximum arsenic levels published by ANZECC (1992) as representative of soil, by over twenty times. Most duplex soils with manganese nodulation will exceed the manganese criteria, particularly if a nodule is included in the analysis.

Metal concentrations are normally elevated in areas with igneous or volcanic parent material, which can include zinc, copper, chromium, nickel and, to a lesser extent, arsenic. Manganese can be enriched in any weathered profile subject to either ancient or current nodulation processes. Generally in areas with high background concentrations the metals are not available for uptake.

Threshold criteria for environmental protection (ANZECC, 1992) included background levels from the Netherlands, and data from the late Kevin Tiller (Cr, Co, Cu, Mn and Ni) then included environmental thresholds mostly derived from the draft Canadian guidelines which were formulated with the protection of groundwater in mind. Copper concentrations were derived from the British leachable threshold and mistakenly interpreted as total.

The problem with these environmental investigation levels in terms of phytotoxicity and reduction in plant yield was first identified when Tiller (1992) noted that the threshold criteria were unnecessarily conservative, particularly for copper, zinc, chromium, arsenic and nickel. Consequently these guidelines are not necessarily widely accepted for all purposes of interpreting soil test results with regard to heavy metals.

Only a minor portion of total metal concentration is available for uptake. If the metal concentration is largely due to previous anthropogenic additions then the labile pool available to plants becomes less with time (Hamon et al., 1997). As metal availability depends on soil and plant species, measures other than total concentration need to be considered. Tiller (1992) suggests that zinc uptake has a strong correlation with EDTA extractability, however other metals do not show such a strong correlation between plant uptake and EDTA extractability. Other metal extractants said to simulate plant uptake include hot water, acetic acid, hydroxylamine hydrochloride, dilute sulfuric and hydrochloric acids, citric acid and a range of combination extractants.

Aluminium is not considered a toxicant for contaminated sites but in an available form is one of the most phytotoxic of all metals, being toxic at very low concentrations in the soluble and/or exchangeable phases. At a soil pH below 5.0, toxicity is likely. Such soil acidity rarely affects plant function directly, but produces the secondary effects of aluminium and manganese toxicity which reduce plant growth.

Metallic and other inorganic contamination usually occurs in agriculture from industry practices. Lead arsenate was used in orchards, arsenic as a dipping agent and an insecticide or termiticide, while bromide and copper are associated with the use of fungicides. Weathering and runoff from galvanised iron will cause elevated zinc concentrations, as will weathering of leaded paint from steel structures. Where such occurrences are suspected, and poor plant vigour is noted, testing is required.

Where soils have formed on mineral-rich or mineralised rocks, toxic levels of certain elements may occur. Some soils developed in the presence of serpentinites (for example, the Great Serpentine Belt, a narrow arc of outcrops from Warialda to Wingham) contain toxic concentrations of nickel, chromium and perhaps cobalt. The Great Serpentine Belt is associated with a range of soil types including chernozems in the east to gravelly red earths and red-brown earths in the west.

Further occurrences of soils that might contain toxic levels of base metals are in areas generally characterised by the presence of metalliferous mines. Copper and zinc toxicities occur in red earths at Captain's Flat, lead is present in duplex soils near Leadville in central western New South Wales, and arsenic is present in soils near Hillgrove. Toxicities also occur in areas not associated with mining, for example manganese toxicity has been identified in soil developed on some marine rocks of Permian age in the Hunter Valley.

Other metal toxicities occur associated with mining, such as selenium from coal, and base metals and arsenic from gold and base metal mining. If these metals have been accumulated as part of geologic weathering processes, they are usually not available and not toxic. Metals and inorganic toxicants released during mining activities can often be highly available and toxic at much lower concentrations than metals accumulated by weathering processes.

Organic Materials
`13.6.2`

Organic contaminants are usually the result of human activities and include contamination by fuels and oils, pesticides, herbicides, termiticides, drenches, dips and other agricultural chemicals.

Fuels and oils are not a direct toxicant to plants but at levels over 10 000 ppm, air and water flow is impeded as too much of the soil void space is occupied by the hydrocarbons. There may be some surfactant effects at the root surface, but nevertheless levels up to 5000 ppm have been shown to have slightly positive effects on plant growth. The aromatic component to hydrocarbons has little effect on plants but is toxic to terrestrial organisms at quite low concentrations (NSW EPA, 1994). The highly volatile nature of aromatics is such that with the exception of leaking underground tanks, toxic concentrations of aromatics in soil from agricultural use is not expected.

Polycyclic aromatic hydrocarbons are found in waste oils and ash (particularly coal ash), and they tend to strongly adhere to soil. Phytotoxic data for these is limited but above several hundred ppm retardation of plant growth is apparent. For plants dependent on mycorrhizae for aspects of plant function, some retardation may be apparent at lower levels. Terrestrial organisms and humans are more sensitive, but levels less than 100 ppm associated with earthworks in agricultural grazing and broadacre farming are likely to have minimal effects on humans, terrestrial organisms and plants.

Organochlorine pesticides have minimal effects on plants. Levels as high as 3000 ppm of total DDT have been recorded with quite luxurious plant growth. But clearly these levels are toxic to insects and the smaller terrestrial organisms. Organochlorines adhere strongly to the soil and break down slowly. They can be found in old and current market gardens, banana plantations and orchards. Levels of total organochlorine pesticides in market gardens rarely exceed 1 ppm (Stuckey and Valley, 1998), while banana plantations reportedly have levels as high as 50 ppm.

There is little need to set phytotoxicity thresholds for organochlorine compounds. Threshold investigation levels in soil for different land uses, based on health hazards for people, have been set by NEHF (1998). All land uses proposed within these guidelines involve intensive occupancy by people and do not cover occasional exposure that would be expected in an agricultural situation. Nevertheless the threshold levels recommended as commercial/industrial would be sufficiently conservative to be appropriate for human exposure in agriculture. It is not expected that these levels would be exceeded under intensive agricultural usage.

Termiticides were required to be sprayed under all building slabs from the early 1960s to the early 1990s. Aldrin, dieldrin and chlordane were used and can be locally very high in areas of former concrete slabs. Based on Australian Standards for application of termiticides the concentrations at the time of application were about 300 ppm or 600 ppm, depending on whether the soil was pre-wetted. After twenty years concentrations have been reported between 1 and 20 ppm. Column F in the NEHF guidelines gives threshold criteria for aldrin/dieldrin as 50 ppm and for chlordane as 250 ppm. Unless the termiticide has been applied in the early nineties or was dumped as a concentrate, toxicities from organochlorine termiticides are not expected.

Current pesticides such as endosulfan and chlorpyrifos are more readily degraded. Criteria from overseas indicate that these are less toxic to animals than aldrin or dieldrin. However, endosulfan is extremely toxic to fish. Levels have not been set in Australia, but for soils with minimal human contact, a threshold criterion below 50 ppm would be considered non-toxic.

Organophosphates break down extremely quickly in soil at commercial application rates. On market gardens, within two years of operations ceasing, organophosphate concentrations are normally nondetectable, or after about five years if used as a termiticide.

Salt Toxicity 13.6.3

Accumulations of soluble and precipitated salts that have pedological origin occur over a wide range of soils. The location of these soils and the effects of the salt on plant growth are discussed in the soil salinity section in this chapter.

Salt accumulation in soils also occurs as a result of overclearing and certain agricultural practices in dryland and irrigated land use situations, as described earlier in this chapter. Water table rises in these circumstances lead to subsoil salts being brought into the root zone of plants, and even to the surface, with consequent toxic effects on plant growth and yield. It is this type of salt toxicity that has the greater effect on agricultural production than that due to the more natural occurrence of salinity.

The 'toxic' effect of soil salinity results mainly from the osmotic potential created due to the presence of soluble salts in soil water. This reduces the availability of water to plants. The presence of elevated levels of soluble salts may also produce specific toxicities due to sodium, chloride and bicarbonate ions.

Land Use and Soil Pollution 13.6.4

Industrial pollution is known to be capable of producing contamination by fluoride, zinc and chromium from smelting; cadmium from the manufacture of tyres, oil and smelting; and lead from fuel additives and smelting. There have been only limited measurements of soil concentrations of these elements in this state. Table 13.12 contains guidelines relevant to the interpretation of levels for these elements.

Land use practices may contribute to the accumulation of elements that are toxic to plant growth. Bromide accumulation associated with soil sterilisation with methyl bromide, arsenic in insecticides, copper and mercury in fungicides, and zinc in bacteriocides are possible examples of this contamination. The use of phosphate fertilisers and gypsum derived from the manufacture of phosphate fertilisers can lead to the build-up of cadmium in soils.

On a much larger scale, land use practices involving nitrate leaching under legume-based pastures and a long history of fertiliser application have been noted to result in soil acidification and toxicities associated with aluminium and sometimes manganese (see soil acidity section). It is not so much the low pH that affects plants as the toxicities of the elements that come into solution with the low pH. Plant roots can only function above pH 3.0 and when the pH is above 9 the roots can no longer absorb phosphorus.

13.6.5   Critical Toxic Limits

The critical toxic limits as they are understood to affect plant species are listed in Table 13.12. Plants vary in their tolerance of toxic levels of elements and agents, and some species can tolerate quite high levels of certain toxic agents. Plants also vary in their capacity to take up different toxic agents. Some plants will accumulate high concentrations of toxic agents, and others will take up little. This variation in plant response should be kept in mind as the table is consulted.

A further complicating factor is that there may be additive effects due to the presence of several toxic agents occurring together, and this additive effect may lower the critical toxic limits. By contrast, the presence of some agents may compensate for the high concentrations of others. For example: phosphorus compensates for high copper levels; sulfate suppresses the uptake of molybdenum; calcium appears to 'detoxify' plants affected by several metals including Al, Co, Pb, Ni, Mn and Zn; and plants grown in soils high in available iron may not readily express chlorosis of the leaves that would be expected with high uptakes of Cu, Co, Ni, or Zn—it is possible that the iron reduces the activity of these elements in the plant.

BIBLIOGRAPHY

Agfact AC.15, *Liming Materials*, NSW Agriculture.

Agfact AC.19, *Soil Acidity and Liming*, NSW Agriculture.

Agfact P2.2.7, *Inoculating and Pelleting Pasture Legume Seeds*, NSW Agriculture.

Agriculture, NSW, *Winter Cereal Management Guide*, NSW Agriculture.

ANZECC and NHMRC (1992), *Australian and New Zealand Guidelines for the Assessment and Management of Contaminated Sites*, Australian and New Zealand Environment and Conservation Council and National Health and Medical Research Council.

Armstrong, J.L. (1987), 'Dryland Salinity', *Annual Conference Proceedings*, Soil Conservation Service of NSW, Sydney.

Arnhem Environmental Impact Assessors (1991), *Soil Survey—Martin Street Area*, unpublished consultancy report, Job No. AE 901101.

Baker, G.H., Barrett, V.J., Carter, P.J., Buckerfield, J.C., Williams, P.M.L. and Kilpín, G.P. (1995), 'Abundance of earth worms in soil used for cereal production in south-eastern Australia and their role in reducing acidity' in R. Date (ed.), *Plant Soil Interactions at Low pH*, Kluwer Academic Publishers.

Bingham, F.T. (1973), 'Boron in cultivated soils and irrigation waters', *Advanced Chemistry Series* 122, 130–8.

Boughton, W.C. (1970), *Effects of Land Management on Quantity and Quality of Available Water*, University of NSW Water Research Laboratory Report No.120.

Bowler, J.M. (1976), 'Aridity in Australia: age origins and expressions in aeolian landforms and sediments', *Earth Science Review* 12, 279–310.

Bromfield, S.M., Cummings, R.W., David, D.J. and Williams, C.H. (1983), 'The assessment of available manganese and aluminium status in acid soils under subterranean clover pastures at various stages', *Australian Journal of Experimental Agriculture and Animal Husbandry* 23, 193–200.

Bush, R.T. and Sullivan, L.A. (1997), 'Morphology and behaviour of greigite from a Holocene sediment in eastern Australia', *Australian Journal of Soil Research* 35, 853–61.

CaLM (1993/4), *Dryland Salinity*, Extension Booklets 1–8, Department of Conservation and Land Management, Sydney.

Charman, P.E.V. and Junor, R.S. (1989), 'Saline seepage and land degradation—a New South Wales perspective', *BMR Journal of Australian Geology and Geophysics* 11, 195–203.

Clayton, P.M. and Tiller, K.G. (1979), *A Chemical Method for the Determination of the Heavy Metal Content of Soils in Environmental Studies*, Divn. Soils Tech. Pap. 41, CSIRO, Melbourne.

Conyers, M.K., Uren, N.C., Helyar, K.R., Poile, G.J. and Cullis, B.R. (1997), 'Temporal variation in soil acidity', *Australian Journal of Soil Research* 35, 1115–29.

Cope, F. (1958), *Catchment Salting in Victoria*, Soil Conservation Authority of Victoria, Melbourne.

Craze, B. (1968), 'Investigations into corrected lime potential relations of fourteen NSW soils and effects of aluminium on plant growth', BSc (Agric.) Thesis, University of Sydney.

Cregan, P.D. and Scott, B.J. (1998), 'Soil acidification—an agricultural and environmental problem' in J. Pratley and A. Robertson (eds), *Agriculture and the*

Environmental Imperative, Australia Agronomy Society, 98–128.

Dent, D.L. (1986), *Acid Sulphate Soils: a Baseline for Research and Development*, Pub. No. 39 ILRI, Wageningen, The Netherlands.

Dent, D.L. and Pons, L.J. (1995), 'A world perspective on acid sulphate soils', *Geoderma* 67, 263–76.

Easton, C. (1989), 'The trouble with the Tweed', *Fishing World*, March, 58–9.

Elliott, M. (1995), 'Grass tetany research' in G. Fenton and P. Orchard (eds), *Making Better Fertiliser and Gypsum Recommendation, a Workshop in Wagga Wagga*, NSW Agriculture.

Environmental and Earth Sciences (1990), *Preliminary Contaminant Investigation for Soils and Groundwater, Kanahooka Road Dapto*, unpublished consultancy report, Job No. 9038.

Environmental and Earth Sciences (1992), *Feasibility of Pesticide in Soil Causing Mental Retardation in a Child*, unpublished consultancy report, Job No. 9213A.

Fenton, G. (1995a), 'Making better recommendations for the management of soil acidity' in G. Fenton and P. Orchard (eds), *Making Better Fertiliser and Gypsum Recommendation, a Workshop in Wagga Wagga*, NSW Agriculture.

Fenton, G. (1995b), 'The acid test: do you need lime?' in M. Casey (ed.), *Pasture Plus. The Complete Guide to Pastures*, Kondinin Group, Rural Research Project.

Fenton, G., Helyar, K., Abbott, T. and Orchard, P. (1996), *Soil Acidity and Liming*, Agfact AC.19, Second Edition, NSW Agriculture.

Geeves, G., Chartres, C., Coventry, D., Slattery, W., Ridley, A., Lindsay, C., Fisher, R., Poile, G., Conyers, M. and Helyar, K. (1987), *Benefits from Identifying and Treating Acid Soils*, CSIRO, Australia.

Gunn, R.H. and Richardson, D.P. (1979), 'The nature and possible origins of soluble salts in deeply weathered landscapes of eastern Australia', *Australian Journal of Soil Research* 17, 197–215.

Hamon, R., Wundke, J., McLaughlin, M. and Naidu, R. (1997), 'Availability of zinc and cadmium to different plant species', *Australian Journal of Soil Research* 35, 1267–77.

Helyar, K.R. and Conyers, M.K. (1994), *Ranking Commercial Pasture Cultivar Sensitivity to Acidity and Allocating Estimated Response Functions to the Cultivars*, Final report, Wool Research and Development Corporation Project DAN 50.

Helyar, K.R. and Porter, W.M. (1989), 'Soil acidity, its measurement and the processes involved' in A.D. Robson (ed.), *Soil Acidity and Plant Growth*, Academic Press, Sydney.

Helyar, K.R., Conyers, M.K. and Li, G. (1997), *MASTER, Managing Acid Soils Through Efficient Rotations*, Final Report DAN 207 International Wool Secretariat, AWRAP Project 1994–97, NSW Agriculture, Wagga Wagga.

Hochman, Z., Godyn, D.L. and Scott, B.J. (1989), 'The integration of data on lime use by modelling', in A.D. Robson (ed.), *Soil Acidity and Plant Growth*, Academic Press, Sydney.

Imray, P. and Langley, A. (1998), *Health-Based Soil Investigation Levels*, National Environmental Health Forum Monographs, Soil Series No. 1, South Australian Health Commission, Adelaide.

Isbell, R.F., Reeve, R., and Hutton, J.T. (1983), 'Salt and sodicity', in *Soils—An Australian Viewpoint*, CSIRO, Melbourne; Academic Press, London.

Linnaeus, C. (1735), *Systema Natural* XII, Vol. 111, Gen. 52, 5, p. 11.

Lottermoser, B.G. (1997), 'Natural enrichment of topsoil with chromium and other heavy metals, Port Macquarie, New South Wales, Australia', *Australian Journal of Soil Research* 35, 1165–76.

Magdoff, F.R. and Bartlett, R.J. (1985), 'Soil pH buffering revisited', *Soil Science Society of America Journal* 49, 145–8.

Mengel, K. and Kirkby, E.A. (1978), *Principles of Plant Nutrition*, Int. Pot. Inst., Berne, Switzerland.

Mortvedt, J.S., Giordano, P.M. and Lindsay, W.L. (1972), 'Micronutrients in Agriculture', *Soil Science Society of America Journal*, Madison, Wisconsin.

Naylor, S.D., Chapman, G.A., Atkinson, G., Murphy, C.L., Tulau, M.J., Flewin, T.C., Milford, H.B. and Morand, D.T. (1995), *Guidelines for the Use of Acid Sulphate Soil Risk Maps*, NSW Soil Conservation Service, Dept. Land and Water Conservation, Sydney.

NEHF (1998), *National Environmental Health Forum Monogram Soil Series No. 1 2nd ed*, Public Health Services, Department of Human Services, South Australia.

Nordstrom, D.K. (1982), 'Aqueous pyrite oxidation and the consequent formation of secondary iron minerals' in *Acid Sulfate Weathering*, SSSA Spec. Pub. No. 10, *Soil Science Society of America Journal*, Madison, Wisconsin, 37–56.

Northcote, K.H. and Skene, J.K.M. (1972), *Australian Soils with Saline and Sodic Properties*, Soil Publication No. 27, CSIRO, Melbourne.

NSW EPA (1994), *Contaminated Sites: Guidelines for Assessing Service Station Sites*, New South Wales Environmental Protection Authority, Sydney.

NSW EPA (1998), *Contaminated Sites: Guidelines for the NSW Site Auditor Scheme*, New South Wales Environmental Protection Authority, Sydney.

Peck, A.J. (1978), 'Salinisation of non-irrigated soils and associated streams—a review', *Australian Journal of Soil Research* 16, 157–68.

Richards, L.A. (ed.) (1954), *Diagnosis and Improvement of Saline and Alkaline Soils*, USDA Handbook No. 60, Washington DC.

Robertson, G.A. (1996), 'Saline land in Australia—its extent and predicted trends', *Australian Journal of Soil and Water Conservation* 9(3), 4–7.

Robson, A.D. (ed.) (1989), *Soil Acidity and Plant Growth*, Academic Press, Sydney.

Robson, A.D. and Abbott L.K. (1989), 'The effect of soil acidity on microbial activity in soils' in A.D. Robson (ed.), *Soil Acidity and Plant Growth*, Academic Press, Sydney.

Russell, E.W. (1973), *Soil Conditions and Plant Growth* (Tenth Edition), Longman Green and Co. Ltd, London.

Sammut, J., White, I. and Melville, M.D. (1996), 'Acidification of an estuarine tributary in eastern Australia due to drainage of acid sulphate soils', *Marine and Freshwater Research* 47, 669–84.

Scott, B.J. (ed.) (1997), *The Amelioration of Acidity in Non-arable Soils of the Tablelands and Slopes*, a workshop held at Goulburn, March 1997, NSW Agriculture, Wagga Wagga.

Scott, B.J., Conyers, M.K., Poile, G.J. and Cullis, B.R. (1997a), 'Subsurface acidity and liming affect yield of cereals', *Australian Journal of Agricultural Research* 48, 843–54.

Scott, B.J., Ridley, A.M. and Conyers, M.K. (1997b), 'Soil acidity in non-arable permanent pastures' in B.J. Scott (ed.), *The Amelioration of Acidity in Non-arable Soils of the Tablelands and Slopes*, NSW Agriculture, Wagga Wagga.

Slattery, W.S., Ridley, A.M. and Windsor, S.M. (1991), 'Ash alkalinity of animal and plant products', *Australian Journal of Experimental Agriculture* 31, 321–4.

Slavich, P.G. and Pettersen, G.H. (1993), 'Estimating the electrical conductivity of soil paste extracts from 1:5 soil water suspensions and texture', *Australian Journal of Soil Research* 31, 73–81.

Stuckey, M. and Valley, L. (1998), 'Organochlorine pesticide concentrations in the surface soil of Western Sydney former market gardens' in P. Mulvey (ed.), *Proceedings of the 1998 National Soils Conference, Environmental Benefits of Soil Management*, Aust. Soc. Soil Sci. Inc.

Taylor, S. (1993), *Dryland Salinity—Introductory Extension Notes* (Second Edition), Department of Conservation and Land Management, Sydney.

Teakle, L.J. and Southern, B.L. (1937), 'The peat soils and related soils of Western Australia. A soil survey of Herdsman Lake', *Journal of the Department of Agriculture, Western Australia* 14, 404–24.

Tiller, K.G. (1992), 'Urban soil contamination in Australia', *Australian Journal of Soil Research* 30, 937–57.

Wagner, R. (1987), *Situation Statement—Dryland Salinity*

Abatement, New South Wales*, Soil Conservation Service of NSW, Sydney.

Walker, G., Gilfedder, M. and Williams, J. (1999), *Effectiveness of Current Farming Systems in the Control of Dryland Salinity*, CSIRO Land and Water, Canberra.

Walker, P.H. (1972), 'Seasonal and stratigraphic controls in the coastal floodplain soils', *Australian Journal of Soil Research* 10, 127–42.

White, I., Melville, M.D., Sammut, J., van Oploo, P., Wilson, B.P. and Yang, X. (1996a), *Acid Sulfate Soils: Facing the Challenges*, Monograph No. 1, Earth Foundation Australia, Millers Point.

White, I., Melville, M.D., Sammut, J., Wilson, B.P. and Bowman, G.M. (1996b), 'Downstream impacts of acid sulfate soils' in H.M. Hunter, A.G. Eyles and G.E. Rayment (eds), *Downstream Effects of Land Use*, 165–72.

Willett, I.R., and Walker, P.H. (1982), 'Soil morphology and distribution of iron and sulphate fractions of a coastal floodplain toposequence', *Australian Journal of Soil Research* 20, 283–94.

Williams, C.H. (1980), 'Soil acidification under clover pasture', *Australian Journal of Experimental Agriculture and Animal Husbandry* 20, 561–7.

Williamson, D.R., Gates, G.W.B., Robinson, G., Linke, G.K., Seker, M.P. and Evans, R. (1997), *Salt Trends—Historic Trend in Salt Concentration and Salt-load of Streamflow in the Murray-Darling Drainage Division*, Dryland Technical Report No. 1, Murray-Darling Basin Commission, Canberra.

Wilson, B.P. (1995), 'Soil and hydrological relations to drainage from sugarcane on acid sulfate soils', PhD Thesis, University of NSW.

Yule, R.A. and Prineas, A. (1986), *The Relationship Between Trees and Secondary Dryland Salting*, Lockyer-Morton Regional Workshop, Department of Primary Industries, Queensland.

Soil Organic Matter

CHAPTER 14

P.E.V. Charman and M.M. Roper

Soil organic matter is that material in a soil which is directly derived from plants and animals. It supports most important microfauna and microflora in the soil, and, through its breakdown and interaction with other soil constituents, is largely responsible for both chemical and physical fertility. Simply expressed, this is a consequence of the direct provision of a range of substances, some of which directly feed plants and some of which bind soil particles together.

There is, however, little scientific evidence to suggest that organic matter has any unique properties that cannot be done without, although true organic farmers believe this to be so. They shun the use of synthetically produced fertilisers, herbicides, pesticides and other like products. Their farming system relies totally on nutrients from organic residues. A team of scientists in the USA conducted a study of organic farming in America and Europe and found that, while it did not generally result in the most productive type of farming, it frequently showed substantial benefits in terms of costs of production and soil, water and energy conservation (USDA, 1980).

In the UK, where soils have been cultivated for centuries in a generally mixed farming system, a major study in the sixties highlighted widespread problems of physical infertility there. In particular, it identified soil structural damage and compaction by agricultural machinery on the more clayey soils, and the decline in organic matter following the omission of grass leys from arable systems of farming (MAFF, 1970). Following the publication of the study report, soil management in the UK has devel-oped, particularly in the areas of reduced tillage and the avoidance of soil compaction.

In New South Wales, the climate is generally not conducive to the production of high levels of organic matter in soils. This is due basically to the dryness and unreliability of the climate and the low inherent chemical fertility of many of our soils. As a result, they are generally low in organic matter content by world standards. In addition, there has been little need for the retention of straw for bedding or feed for stock, and plant residues in cereal-growing areas have traditionally been burnt to facilitate cultivation for weed control and subsequent sowing. We thus have the situation in many cereal-growing areas, which have only been regularly farmed for less than 100 years, where organic matter levels have been drastically reduced over this period (see also Anon., 1986).

In these areas, soils seal rapidly at the surface, aeration and rainfall infiltration are poor, root penetration is reduced, and plant establishment and growth are generally substandard. In addition, hardpans have often formed below cultivation level, soils are more difficult to work and the risk of soil erosion is greatly increased.

These soils have been regularly cultivated without thought being given to the return of organic matter or the adverse effects of years and years of tillage. Their ability to support continued cropping irrespective of the amounts of fertiliser which might be put on, is now in serious doubt. Their problem is one of physical, not chemical, fertility.

14.1 Composition of Soil Organic Matter

The majority of soil organic matter derives from the breakdown of plants after they have died. At this stage it is composed of complex carbohydrates (polysaccharides and celluloses), lignins (complex plant polymers giving plants their woody character), and plant proteins (large amino acid complexes high in nitro-gen) (Schnitzer and Khan, 1978). The carbohydrates and some of the proteins are rapidly broken down by microbial activity in the soil, but the lignins are oxidised more slowly and enriched with nitrogen from the remaining proteins to form a highly complex material of dark colour, generally known as humus.

This, together with dead soil organisms and finely divided plant residues that have lost their structure, forms the soil's organic matter.

Humus is a structureless, amorphous, jelly-like substance that can coat sand, silt and clay particles. It has strongly developed colloidal properties, swelling on wetting and shrinking on drying, forms complexes with clays and has cation exchange capability. It is acidic in nature and on analysis it can be shown to contain about 25% of an insoluble material known as humin, and 75% of a mixture of fulvic and humic acids. These mainly differ with respect to their contents of carbon and oxygen. Overall the composition of humus is very approximately 55% carbon, 35% oxygen, 5% nitrogen, 4% hydrogen and 1% phosphorus and sulfur. However, this can vary depending on soil conditions and the make-up of the plant material from which it was derived.

Soil organic matter can be characterised with respect to plant growth according to its carbon to nitrogen ratio (C/N). This ratio is remarkably constant for a given soil despite changes in cropping management. It is usually around 10 to 12 for an agricultural soil but may range from 5 to 15. High figures indicate that nitrogenous material or fresh green manure crops can be added to the soil with benefit, whereas low figures indicate that mature plant material such as cereal stubble can be incorporated without detriment. Forest and peaty soils have much higher values.

14.2 The Role of Soil Organic Matter

The importance of organic matter in soil relates to its control over fertility in the broadest sense of the term. Fertility is the soil's ability to support production of food or fibre while maintaining a stable soil structure. As such, it has two major components— chemical and physical. Chemical fertility is concerned with the supply of nutrients to the growing crop, whereas physical fertility is concerned with the provision of soil conditions that allow plants to make optimum use of those nutrients. It involves the structure of the soil, or the physical arrangement of soil particles, whereby air and water supply to plants is optimal for the seasonal conditions and root penetration is not impeded. This applies not only to the surface soil, where such conditions are of paramount importance, but also to the subsoil into which roots must grow.

Organic matter is a major determinant of both physical and chemical fertility. A third component of soil fertility—biological fertility—relates to the amount of organic matter in a soil and the activity of the microflora and microfauna that it supports. Through the breakdown of organic matter by micro-organisms, plant nutrients are supplied directly to plants and to other micro-organisms, and substances are also produced which help bind soil particles together to give a stable structure (Degens, 1997). This is in addition to the effects of the many organisms themselves, which also directly assist the aeration and aggregation of soils.

The presence of organic matter in soil also enhances its cation exchange capacity, a property related to clay content, which influences clay aggregation, nutrient availability and water-holding capacity. This property is also important in holding nutrients against leaching, and is thus of particular significance in lighter soils.

The processes of organic matter decay are largely controlled by soil micro-organisms and are therefore influenced by temperature, moisture, pH and soil aeration. Other factors that affect their activity are the supply of carbon, nitrogen and some trace elements, competition from other organisms, the availability of surfaces on which they can grow and, of course, cultivation, through its effect on soil moisture and aeration. A typical soil population might comprise fungi, bacteria, actinomycetes, algae and protozoa, as well as the larger animals such as worms, molluscs, nematodes and small insects. Their importance in this context is their effect on soil properties, both direct and indirect, and this is intricately bound up with the role of organic matter on the fertility of the soil.

Greenhouse Gas Control   14.2.1

Soil organic matter is also important as a possible sink/source for greenhouse gases, which trap re-radiated solar energy and thus keep the earth's surface at a temperature level necessary to support life. These atmospheric gases, mostly carbon dioxide

(CO_2) but also methane (CH_4) and nitrous oxide (N_2O), are being increased by human activity such as burning fossil fuels and land clearing, and many scientists believe that significant global warming is likely, with attendant sea level changes and regional climate change. The consequences of such changes could be socially, economically and environmentally disastrous.

Carbon dioxide is the main gas contributing to this enhanced greenhouse effect, and its small but highly significant concentration in the atmosphere has increased by about 30% in the last two centuries. This has occurred largely because of the burning of coal, oil and natural gas and the clearing and burning of natural vegetation.

The condition of the soil, while being vital for agricultural production, is also important in terms of providing a long-term 'sink' for atmospheric carbon. In the form of carbon dioxide, this carbon is tied up (sequestered) through the process of photosynthesis by plants to give organic compounds that ultimately come to reside in the soil for varying lengths of time.

The soil ecosystem has the potential to sequester carbon for very long time periods (Senanayake, 1993). Depending on soil conditions, these compounds are eventually broken down and the carbon is returned to the atmosphere through decomposition, and respiration of the micro-organisms involved. Methane and nitrous oxide may also be released as a result of the decomposition.

This 'sequestering potential' of soils becomes important in the overall strategy to reduce greenhouse gas emissions and tie up carbon dioxide in the form of soil organic matter. Increasing organic carbon in soils also has benefits in terms of productivity and sustainability. Soils that have lost organic matter (through overcropping, for example), and may be in a degraded condition, have substantial sink capacity for sequestering more carbon dioxide. In a stable ecosystem the emission of the gas from soils (due to decomposition and respiration) is more or less balanced by its input to soils (due to photosynthesis). However, recent quantification of these processes shows that emissions are exceeding inputs as a result of clearing and cultivation of natural vegetation. As the soil carbon pool is approximately twice the size of the atmospheric pool (Wild, 1993), there is considerable

scope for this to continue if land management practices are not improved.

Sustainable practices that enhance the tying up of carbon (dioxide) in the soil can be considered under four groupings:
—Reduced cultivation
—Conservation tillage
—Vegetation enhancement
—Soil conservation.

Cultivation has been shown to increase the breakdown of soil organic matter, mainly through aeration and disturbance of the soil matrix and acceleration of microbial activity. The mouldboard plough is particularly effective in encouraging carbon dioxide loss from soils (Reicosky, 1998). Any system in which tillage is minimised is therefore likely to reduce emissions of carbon to the atmosphere.

Conservation tillage generally involves stubble retention and rotation of grain crops with cover, forage or green manure crops. These increase photosynthesis within the system and foster the retention of soil organic matter, particularly when combined with minimum tillage.

Vegetation enhancement involves a wide range of practices including agroforestry, reduced clearing, reduced stocking, improved pasture systems and rangeland management. These should all increase photosynthetic activity on a landscape scale, thereby significantly boosting the intake of carbon dioxide to plants and soil.

Soil conservation may include all of the above, but also specific erosion control practices and the rehabilitation of degraded land. Many types of degradation may be involved here which include various physical, chemical and biological processes such as repeated wetting/drying of soil, salinisation, and interactions between different types of organic matter. These have been shown to lead to an increase in the volume of carbon dioxide entering the atmosphere (Senanayake, 1993).

Of the four groupings described, it is likely that those practices involving enhancement of vegetative cover, such as improved pasture systems, agroforestry and reduced clearing, will have the greatest effect in reducing greenhouse gases through the sequestering of carbon from the atmosphere into the form of soil organic matter.

14.3 The Effect of Organic Matter on Crop Production

The world's longest-running experiments looking at this question are the classical Rothamsted experiments in England. The Broadbalk experiment, in particular, has been comparing winter wheat yields under a variety of fertiliser treatments continuously since 1843. The effects of nitrogen, phosphorus and potassium (NPK) fertilisers, singly and in various combinations, have been compared with those of farmyard manure (FYM), with the same treatment applied to each plot each year.

During 1970–75 the largest yields of winter wheat were on the FYM plots where soil organic matter had been increased by many applications (35 t/ha/yr). Average yields for the period, as quoted in Boels et al. (1982), appear in Table 14.1.

The annual applications of these amounts of organic manure have more than doubled the organic matter level as compared with that of an unmanured soil, and levels are still increasing. While such yield figures would not be obtained under New South Wales dryland conditions, it is of interest to note the potential of organic matter to lift crop production without the need for other fertilisers.

The majority of Australian long-term experiments of this type have been carried out in southern Australia, but relatively few have related yield of crops directly with organic matter levels. Such data, however, are available from a long-term runoff and soil loss experiment where increased wheat yields were associated with higher soil organic carbon (Figure 14.1). The plots were part of a 30-year experiment comparing runoff and soil loss from a fallow/wheat rotation and a fallow/wheat/pasture rotation at Wagga Wagga. The higher organic carbon

levels reflect reduced erosion as well as increased organic matter in the plots where pasture was part of the rotation.

Other experiments that examined the effects of different crop sequences on yield were reported by Russell and Greacen (1977), and are summarised in Table 14.2.

The figures clearly show the effect of an alternate fallow year in giving rise to greater wheat yields. This

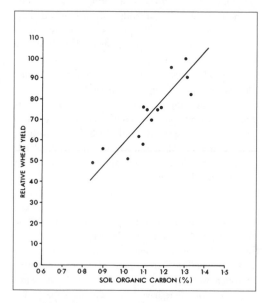

Figure 14.1 Relationship between wheat yield and soil organic carbon on old runoff and soil loss plots at Wagga Wagga Soil Conservation Research Centre, 1981 (r = 0.65)

Table 14.1 Wheat yields (1970–75), Broadbalk, Rothamsted

Wheat Grown Continuously	Wheat Grown in Rotation	t/ha
with FYM (35 t/ha/yr)		5.8
with NPK (48 kg/ha/yr N, + P + K)		2.1
	with FYM (as above)	6.8
	with NPK (as above)	3.2

Table 14.2 Wheat yields and rotations, southern Australia

Location	Years	Mean Wheat Yield (kg/ha) Wheat/Wheat	Fallow/Wheat	Fallow/Wheat/Pasture
Rutherglen, Vic	(29)	460	1540	1450
Longerenong, Vic	(31)	640	2110	2310
Booborowie, SA	(13)	—	1800	1850
Waite Institute, SA	(46)	770	1870	2400
Chapman, WA	(16)	360	540	770
Wongan Hills, WA	(16)	310	870	1160
Walpeup, Vic	(27)	—	1510	1710
Werribee, Vic	(35)	400	1100	1210

NB Figures in brackets show period of trial in years.

is probably due to better weed control and moisture accumulation for the subsequent crops.

Generally, there is additional benefit from the inclusion of pasture in the rotation, and this is attributed to the improved organic matter status of the soil and its likely effects in improving soil physical conditions. This is particularly evident in the long-running Waite Institute experiment, in which there has been a dramatic decline in crop yields where there is no pasture ley in the rotation.

Soil Conservation Service experiments in the conservation tillage area, started in the early seventies, have been primarily concerned with soil stability, runoff and soil loss rather than crop yield. The general trend emerging is that, by following conservation tillage practices involving reduced tillage and the retention of crop residues, yields of winter cereals can be maintained at, or somewhat above, previous levels which were achieved using more traditional cultural practices. Only in favourable seasons can consistent and significant yield increases be expected. Some seasons will strongly favour reduced tillage systems, and some will not. On present evidence, major increases in winter cereal crop yields cannot be expected on a consistent basis just by using conservation tillage.

This view does not, of course, take account of the clear soil conservation advantages of these practices, and it is on these advantages that conservation tillage is primarily justified. As farmers adapt to the changing technology, and research and product development achieves solutions to the associated problems such as stubble handling, herbicide costs and changes in pest and disease patterns, yield increases may well be achieved under a wide range of conditions, consistent with the maintenance of soil stability and gradual increase in soil organic matter (see also Chapter 16).

14.4 Soil Improvement

14.4.1 Increasing Soil Organic Matter

Organic matter levels of uncultivated Australian topsoils range from about 1% for desert loams in the dry inland up to extreme figures of about 50% for some alpine humus soils in the cold, well-watered parts of the southern uplands. A selection of cereal-growing soils in five states quoted by Hamblin (1980) shows a narrower range, which would be more typical of agricultural soils, from 1.6% to 4.6%.

Harte (1982) studied a range of soils in northern New South Wales and compared organic matter levels, from the top 5 cm of soil, of virgin sites with adjacent sites cultivated for varying periods of time (Table 14.3).

These figures indicate that many soils have had their organic matter levels halved in a period of 15 to 20 years of traditional cultivation practices. Some soils deteriorate rapidly after cultivation, with a sharp drop in organic content and aggregation after only

Table 14.3 Soil organic matter levels, northern New South Wales (after Harte, 1982)

| Soil Classification | Organic Matter % | | Cultivated Period |
(Northcote Code)	Virgin	Cultivated	(Years)
Dr2.23	2.4	1.6	15
Gn2.13	3.4	1.6	15
Dr4.23	5.1	3.0	12
Gn3.43	3.9	2.6	1
Gn3.43	3.9	2.0	10
Dd1.13	4.2	4.1	2
Dr2.23	4.6	2.1	20

one year, whereas others react only slowly. For example, Hunt (1980) showed the dramatic drop in soil aggregate stability in mallee soils after only one year's cultivation, and demonstrated the importance of at least a five-year pasture phase after a period of cropping to restore the levels to those of uncleared land.

The amount of organic matter in soils is affected by soil type, previous cropping history, climate, tillage practices, the amount and kind of plant or animal material returned or removed, and methods of destroying litter (such as fire). The most significant of these in the long term is the amount of organic material returned to, or removed from, the soil. In this context, plant roots have an important role since, when they decompose, not only are humic substances added to the soil, but nutrients are also released throughout the rooting zone, thus benefiting both soil structure and the uptake of nutrients by the following crop.

The rate of oxidation, breakdown and utilisation of organic matter depends on the microbial population present and the suitability of the soil environment for their effective operation, particularly in terms of moisture content, temperature and aeration. The intensive cultivation of a soil and the growing of high-yielding crops will reduce the level of organic matter, often to an equilibrium at a much lower level in the long term.

To boost humus levels in cropping soils, the amount of plant material can be manipulated by leaving crop residues on the soil surface, incorporating them by cultivation, by fertilisation to increase biomass, by growing a green manure crop or by adding some other plant or animal waste materials from elsewhere. Under New South Wales broadacre farming conditions, the last three of these are likely to be of doubtful viability. Reliance must, therefore, be placed on the return of crop residues *in situ*, and, in the typical cereal cropping situation, this means stubble retention. This practice has marked soil conservation benefits, in addition to the value of added organic material, through enhanced infiltration, reduced soil splash and runoff velocity. Soil erosion under retained stubble is greatly reduced.

Crop Residue Breakdown

14.4.2

Crop residues are an enormous source of carbon and provide energy for a range of soil micro-organisms that are responsible for many of the nutrient transformations in soils. For example, stimulation of nitrogen fixation by free-living bacteria in soils has been observed by a number of people. In a field study on three sites in the wheat-belt of New South Wales, Roper (1983) measured substantial rates of nitrogen fixation when wheat straw was added to the soil.

Crop residues contain complex celluloses, hemi-celluloses and lignin. A few free-living nitrogen-fixing bacteria can use some of these complexes directly (Halsall et al., 1985), but others rely on decomposition first. Soils contain a wide range of microflora capable of breaking down crop residues to smaller units, which are then available for use by other micro-organisms. Free-living nitrogen-fixing bacteria occur naturally in most soils, although their composition and number vary. Nitrogen fixation by such bacteria-utilising residues is best when the soil is moist and warm, and in soils with a high clay content (Roper, 1983, 1985; Roper and Halsall, 1986). Clays are beneficial because they protect bacteria from predation and parasitism, and from desiccation, and this probably explains the larger population of nitrogen-fixing bacteria found in soils with a high clay content (Bushby and Marshall, 1977; Roper and Marshall, 1978). It is likely that such soils also support larger populations of other groups of micro-organisms responsible for different nutrient transformations in the soil.

Indications are that nitrogen fixation by bacteria using stubble may lead to a reduction in the need for nitrogenous fertilisers, and this has been observed in some areas following the adoption of straw retention practices (for example, Goldsmith, 1980). However, there are many examples in the literature where depressions in growth of winter cereal crops have occurred as a result of stubble retention. For example, White (1984) reports such depressions, and attributes them to lowered levels of nitrate nitrogen (or immobilisation of nitrogen) in the soil as a result of the wide carbon:nitrogen ratios associated with the retained stubble. Many workers feel that a period of a few years' adjustment of the soil microflora may be required when a soil is changed from a stubble-burnt to a stubble retention system. In the interim, some additional nitrogen fertiliser may be required to prevent available nitrogen shortage to the establishing crop as a result of stubble retention.

Immobilisation of nitrogen by stubble is usually only temporary and nitrogen becomes remineralised, particularly in spring when a crop has a greater need for it (Lynch, 1983). It may even be an advantage to have some nitrogen immobilisation in the winter, particularly in winter rainfall areas, because losses by leaching can be reduced.

There are other potential problems with retaining crop residues. There is a possibility of disease carry-over from one crop to another, and difficulties may arise with sowing into heavy crop residue. These problems are more evident in northern New South Wales, where cropping is the dominant land use, stubble is needed for erosion protection and summer fallows are important for moisture conservation, and where stock are often not available to reduce stubble levels prior to sowing. The stubble handling problem is being addressed with the development of new sowing equipment in Australia.

14.4.3 Crop Rotations

In a mixed farming situation, such as the southern cropping areas of New South Wales, there is far greater scope for the improvement of soil organic matter levels than when the land is continuously cropped. This is because of the recycling of plant material through the animal, and return to the land as dung and urine. This process returns a greater proportion of organic material from plants growing in the soil, in a condition in which it can be broken down more easily, and with enhanced nitrogen recycling which can greatly assist the further breakdown of plant residues.

Perhaps the greatest scope for improvement of soil organic matter levels lies in the return of cropland to a regular pasture phase to break up periods of continuous cropping. Soil aggregates from pastureland are not only more porous, but are also more stable. This enhanced stability has been shown by many studies in different parts of the world to be associated with higher levels of organic matter in pastureland and the more intense root proliferation of pasture grasses. The use of legumes in pasture mixtures is seen as particularly valuable in improving nitrogen levels in the soil and in generating more humus which leads to enhanced soil aggregation. An example is shown in Figure 14.2 (Tisdall and Oades, 1980).

Whereas tillage operations can be used to create a range of pores in surface soils, the stability of aggregates is largely dependent on organic materials added to them. Tillage breaks down soil aggregates into smaller units and often reduces overall porosity by breaking down large pores such as wormholes and root channels as well. There is, thus, a strong argument for the concept of reducing tillage to the minimum necessary to allow sowing and growing of the crop satisfactorily.

Addition of organic matter and enhanced biological activity are both characteristic of the changes that occur in the soil under a pasture. Hence, management of the soil to develop optimum physical conditions for arable crop production is unlikely to be achieved by tillage alone, but will require that the soil be managed in such a way that conditions are maintained which are similar to those developed under pasture. For many soils, this requires a pasture/arable rotation system, but there are also indications that on some soils the direct drilling system, similar to that increasingly used in southern New South Wales, may give rise to similar conditions after a few years of operation.

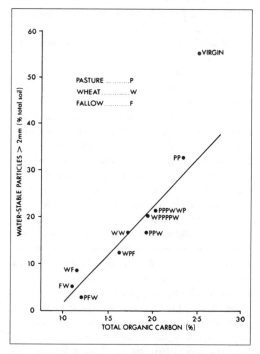

Figure 14.2 Relationship between percentage organic carbon and water-stable aggregation in Urrbrae fine sandy loam after 50 years of crop rotations (Tisdall and Oades, 1980)

14.5 Agricultural Practices and Soil Organic Matter

Generally, the effect of breaking up pasture, whether it be natural, introduced or improved, or of clearing and cultivating vegetated areas of any sort, is to reduce the organic matter level in the soil. Often this occurs rapidly within one year, as in the case of the mallee.

Following this initial break-up, traditional cultural practices have aerated and disturbed the soil to such an extent that levels of organic matter have been more than halved in a relatively short space of time. While the short-term effect of this breakdown of organic material may be increased mineralisation and availability of nutrients, the longer-term effect is often reduced soil structure.

The practice of burning stubble has aggravated this situation by removing a large proportion of cereal crop residues from the land altogether. The regular cropping of such land, in many instances continuous cropping without much added fertiliser, has assisted this depletion process by removing sources of nutrients in the soil.

The ongoing interest in conservation tillage gives some indication of farmers' concern for the soils of New South Wales. The aim of this technology is to maintain or improve crop yields while reducing soil degradation, conserving energy and offering the farmer a more flexible and, hopefully, lower cost system of cropping.

The two key features of conservation tillage are reduction in the intensity of tillage and the retention of crop residues. The first of these acknowledges the fact that traditional cultivation practices are generally harmful to soil structure. Crops can generally be grown satisfactorily in seedbeds much rougher than traditional practice has required. Reduced cultivation also slows up the oxidation of organic matter and reduces losses of moisture as a result of soil disturbance. The retention of crop residues aims to provide soil protection in the short term and enhancement of organic matter levels in the long term, as the soil microflora adjusts to the practice.

As studies of these practices show, in general we can be cautiously optimistic about their potential for improving soil organic matter contents and aggregation. Hamblin (1980) studied changes in aggregate stability in five wheat-growing soils in different states after three to eight years of continuous cropping using conventional cultivation and direct drilling.

Only small increases in organic carbon and aggregate stability were found for the direct drilling treatments and her conclusion was that structural improvement appeared to take place at a slower pace than in wetter environments.

Packer and Craze (1982) have measured changes to soil properties in an experiment at Cowra comparing direct drilling with reduced and conventional tillage. They found that while organic matter levels had remained fairly constant over three seasons in the reduced tillage and conventional treatments, there was a steady increase in the direct drill treatments—25% for ungrazed and 17% for grazed. These improvements have shown themselves in increased percentage water-stable aggregation in the size range larger than 0.25 mm.

Tisdall and Oades (1980) reported the effect of 50 years of crop rotations on roots, hyphae, organic matter and water-stable aggregation in red-brown earths. They found that aggregation larger than 0.25 mm was reduced by rotations that included fallowing and multiple cultivation. The role of roots and fungal hyphae was important in this size range. Aggregation in the range 0.05 to 0.25 mm occurred as a result of organic stabilisation, but was unaffected by management. The general conclusions were that soil stability in red-brown earths can be improved by growing plants with extensive root systems with minimum cultivation.

Ferris et al. (1983), reporting the results of some tillage studies at Tamworth, New South Wales, documented the inputs and outputs of organic carbon relating to both conservation tillage and conventional tillage systems for the growing of sorghum. They found that under conventional tillage there was a gradual net loss of organic material from the soil. However, under the conservation system, which involved no-tillage using herbicides for weed control, the balance of organic matter in the soil was maintained.

These experiments summarise our expectations in a cropping situation under New South Wales dryland conditions. Using carefully managed conservation farming systems, we can expect to maintain soil organic matter levels or gradually improve them in the more reliable rainfall areas given favourable seasons. CSIRO work in this field generally confirms this view (Anon., 1986). In marginal cropping areas, such improvement is likely to be extremely difficult, and hence there is considerable concern about the possibility of soil degradation in such areas. The desirability for the introduction of pasture phases to areas in which continuous cropping has become established must be very seriously considered.

14.6 Soil Organic Matter and Soil Conservation

The significance of soil organic matter in a soil conservation context lies in its vital contribution to the stability of soil structure. While there are soils that rely on such substances as calcium carbonate and complex iron or aluminium compounds to provide the binding agents for soil particles, as well as the interparticle bonding found in clayey soils, it is likely that organic substances and the action of roots and micro-organisms are of major importance in many Australian soils. The build-up of soil structure, due to organic materials, and the particularly stable structure provided by certain types of soil/humus interactions are of the utmost importance when erosive forces occur. Soils with stable structure absorb rainfall at a higher rate and to a greater extent than poorly structured soils, and their aggregates also tend to resist the disruptive forces of raindrop impact, overland flow or wind action.

The surface of soils is especially subjected to disruption in an erosion event, particularly when there is a lack of surface vegetation. Stable surface aggregates resist crusting and surface sealing which tend to increase runoff during heavy rainstorms. Infiltration is increased as a result of good surface aggregation and if subsurface soil materials are also aggregated through past soil management, percolation and drainage will occur to an extent which will ensure that only a small proportion of rainfall is available to cause erosion, depending on rainfall intensity.

Soil conservation measures, therefore, aim not only to restrict the effect of excessive rainfall or wind, but also to reduce the size of the excesses in the first place. Soil management thus complements structural soil conservation works by reducing actual soil erosion and degradation *in situ*. The structural works control and restrict the effects of erosion when the inevitable climatic excesses occur. This is the thrust of soil conservation technology in New South Wales, particularly in the control of soil degradation on arable lands. Key features are organic matter maintenance and enhancement, reduction in cultivation, rotation of crops and/or pastures where feasible, stubble retention, structural works where necessary and the use of the contour principle in all farming operations.

Stubble mulching, or leaving crop residues on the surface of the soil, is the preferred system of stubble retention. Stubble burning is regarded as a strictly strategic operation, to be avoided where possible. Cultivation is regarded in the same light, as a strategic operation, to be carried out only as necessary.

Further and more detailed coverage of agricultural management practices and their relation to soil conservation and the improvement of soil organic matter levels will be found in Chapter 16.

BIBLIOGRAPHY

Allison, F.E. (1973), *Soil Organic Matter and Its Role in Crop Production*, Elsevier, Amsterdam.

Anon. (1986), 'Cultivation and soil organic matter', *Rural Research* 131, 13–18.

Bauer, A. and Black, A.L. (1981), 'Soil carbon, nitrogen and bulk density comparisons in two cropland tillage systems after 25 years and in virgin grassland', *Soil Science Society of America Journal* 45, 1166–70.

Boels, D., Davies, D.B. and Johnston, A.E. (eds) (1982), *Soil Degradation*, Balkema, Rotterdam.

Bushby, H.V.A. and Marshall, K.C. (1977), 'Some factors affecting the survival of root-nodule bacteria on desiccation', *Soil Biology and Biochemistry* 9, 143–7.

Degens, B.P. (1997), 'Macro-aggregation of soils by biological bonding and binding, and the mechanisms and factors affecting these—a review', *Australian Journal of Soil Research* 35, 431–60.

Donaldson, S.G. and Marston, D. (1981), 'Structural stability of black cracking clays under different tillage systems', *Symposium on Cracking Clays, Armidale, NSW*.

Ferris, I.G., Holland, J.F. and Felton, W.L. (1983), 'Effect of tillage and herbicides on carbon cycling', *Proc. of Northern NSW No-Tillage Project, Tamworth* 64-73.

Goldsmith, S.W. (1980), 'Erosion control on a West Wyalong district farm', *Journal of the Soil Conservation of NSW* 36, 48–52.

Greenland, D.J. (1981), 'Soil management and soil degradation', *Journal of Soil Science* 32, 301–22.

Halsall, D.M., Turner, G.L. and Gibson, A.J. (1985), 'Straw and xylan utilisation by pure cultures of nitrogen-fixing *Azospirillum* spp.', *Applied Environmental Microbiology* 49, 423–8.

Hamblin, A.P. (1980), 'Changes in aggregate stability and associated organic matter properties after direct drilling and ploughing on some Australian soils', *Australian Journal of Soil Research* 18, 27–36.

Hamblin, A.P. and Davies, D.B. (1977), 'Influence of soil organic matter on the physical properties of some East Anglian soils of high silt content', *Soil Science* 28, 11–23.

Handreck, K.A. (ed.) (1979), *Organic Matter and Soils*, Discovering Soils Series No. 7, CSIRO Division of Soils, Adelaide, SA.

Harte, A. (1982), *Annual Research Report*, Soil Conservation Service of NSW, Sydney.

Hunt, J.S. (1980), 'Structural stability of mallee soils under cultivation', *Journal of the Soil Conservation Service of NSW* 36, 16–22.

Lynch, J.M. (1983), *Soil Biotechnology: Microbiological Factors in Crop Productivity*, Blackwell Scientific Publications, Oxford.

MAFF (1970), *Modern Farming and the Soil*, HMSO, London.

Marston, D. and Hird, C. (1978), 'Effect of stubble management on the structure of black cracking clays' in W.W. Emerson, R.D. Bond and A.R. Dexter (eds), *Modification of Soil Structure*, John Wiley and Sons, Chichester, UK.

Oades, J.M., Lewis, D.G. and Norrish, K. (eds) (1981), *Red-Brown Earths of Australia*, CSIRO/Waite Inst., Adelaide, SA.

Packer, I.J. and Craze, B. (1982), *Annual Research Report*, Soil Conservation Service of NSW, Sydney.

Reed, J.B. and Goss, M.J. (1981), 'Effect of living roots of different plant species on the aggregate stability of two arable soils', *Soil Science* 32, 521–41.

Reicosky, D.C. (1998), 'Tillage intensity and carbon dioxide emission from soils', *Symposium Working Papers on Conservation Tillage: Can it Assist in Mitigating the Greenhouse Gas Problem?* ASSSI, AIAST, ISTRO and University of Queensland, St Lucia.

Roper, M.M. (1983), 'Field measurements of nitrogenase activity in soils amended with wheat straw', *Australian Journal of Agricultural Research* 34, 725–39.

Roper, M.M. (1985), 'Straw decomposition and nitrogenase activity (C_2H_2 reduction): effects of soil moisture and temperature', *Soil Biology and Biochemistry* 17, 65–71.

Roper, M.M. and Halsall, D.M. (1986), 'Use of products of straw decomposition by N_2-fixing (C_2H_2-reducing) populations of bacteria in three soils from wheat-growing areas', *Australian Journal of Agricultural Research* 37, 1–9.

Roper, M.M. and Marshall, K.C. (1978), 'Effects of a clay mineral on microbial predation and parasitism of *Escherichia coli*', *Microbial Ecology* 4, 279–89.

Russell, E.W. (1971), 'Soil structure: its maintenance and improvement', *Soil Science* 22, 137–51.

Russell, J.S. and Greacen, E.L. (eds) (1977), *Soil Factors in Crop Production in a Semi-arid Environment*, ASSSI with Queensland University Press.

Schnitzer, M. and Khan, S.U. (1978), *Soil Organic Matter*, Elsevier Scientific Publishing Company, Amsterdam, Oxford, New York.

Senanayake, R. (1993), 'Soil ecology, agriculture and the greenhouse effect', *Australian Journal of Soil and Water Conservation* 6(1), 27–30.

Somoni, L.L. and Saxena, S.M. (1975), 'Effect of some organic matter sources on nutrient availability, humus build-up, soil physical properties and wheat yields under field conditions', *Annals of the Arid Zone* 14, 149–58.

Tisdall, J.M. and Oades, J.M. (1980), 'The effect of crop rotation on aggregation in a red-brown earth', *Australian Journal of Soil Research* 18, 423–33.

Tisdall, J.M. and Oades, J.M. (1982), 'Organic matter and water-stable aggregates in soils', *Soil Science* 33, 141–63.

USDA (1980), *Report and Recommendations on Organic Farming*, US Department of Agriculture, US Govt Printing Office 0-310-944/SEA-139.

White, P.J. (1984), 'Effects of crop residue incorporation on soil properties and growth of subsequent crops', *Australian Journal of Experimental Agriculture and Animal Husbandry* 24, 219–35.

Wild, A. (1993), *Soils and the Environment : an Introduction*, University Press, Cambridge.

Soils, Vegetation and Land Use

C H A P T E R   1 5

R.O. Sonter, W.S. Semple
and J.W. Lawrie

In this chapter we examine the natural capability of land to grow vegetation and provide for human needs under agricultural use. Native vegetation reflects both the nature of the soil it is growing in and the other factors such as drainage, aspect and slope determined by the landscape, and may often be used as an indicator of these various factors which are in turn also determinants of land use.

Individual species or types of vegetation are sometimes associated with specific soil attributes such as pH, texture or nutrient deficiency. In other cases the presence or absence of species is due to land form or climate attributes such as drainage, frost frequency, slope or runon water. All these factors, whether soil-related or otherwise, have relevance to the capability of the land to support various uses. Over large areas vegetation types are more likely to be associated with climate than soils, but in smaller areas, associations with parent material, relief or soil characteristics are more likely.

15.1 Soils and Vegetation

15.1.1 Soil Types and Vegetation Types

There are many ways to describe soil types (see Chapters 6 and 7). Soil types can be broad as great soil groups or soil orders (Isbell, 1996), or more detailed as soil series, soil families or soil phases, which can take into account soil parent material and geology, slope position and drainage. Vegetation types can also be defined in many ways, either as assemblages of species (associations, alliances and so on), growth forms (mallee, forest, grassland) or hardness of leaves (sclerophyll forest, rainforest), or growth form and crown separation (see McDonald et al., 1990). Therefore before investigating relationships between vegetation and soils, it needs to be made clear which soil types and vegetation types are being used to define the relationships.

15.1.2 Relationship Between Soils and Vegetation

Some relationships between soils and vegetation may be quite strong and definite at the local level (Plate 15.1) but become less definite over larger areas because of changes in climate or other factors. A vegetation type may be associated with a particular soil property or group of soil properties such as fertility, drainage or water-holding capacity. Therefore it will not necessarily be associated with a soil type defined

a   b

Plate 15.1a and b Two very different types of vegetation are shown in these photos taken west of Sydney on the Cumberland Plain. While the photos were taken only 50 m apart, the vegetation communities are obviously different despite their close proximity. The tall black trees with small amounts of shrub understorey are from an ironbark community (*Eucalyptus fibrosa*), on soil with a dense clay subsoil over shale. Shale was often within 1 m of the surface. The community with contorted trunks and higher amounts of undergrowth (applebark community *Angophora bakeri*) is on deep sandier soil.

in soil classification such as great soil group or soil order (Isbell, 1996). Often relationships between vegetation and soil types occur at classification levels lower than great soil group or soil order, and are more likely between soil series, soil phase or soil family levels of classification. This is often represented by a great soil group on a particular parent material (see Chapters 6 and 7). As soil landscapes largely define soil classification at a more detailed level than great soil group or soil order, they have the potential to have stronger relationships with vegetation types.

A further complication is that the soil properties that influence the distribution of vegetation are not necessarily the same ones that are used for classifying soils. The distribution of each plant species can also be influenced by different soil properties. This leads to the proposal that the degree of change in vegetation between two soil types will be related to the degree of change in soil properties between two soil types. Two soils with many similar properties will have more plant species in common than two soils with widely differing soil properties. If an association between soils and vegetation can be demonstrated, it can be particularly useful to soil or vegetation surveyors in delineating mapping units.

In order to understand the limitations of the relationship, it is necessary to understand why plants grow where they do. For example, the availability of moisture during the growing season is critical for plant survival and is one where soils would be expected to play a major role—but so too do climate and topography. The distribution of plants is also affected by temperature and in this respect, soils would be expected to play a lesser role than climate and its modification by topographic location (for example, 'frost hollows'). Further, some plants (for example tropical types) may not occur in temperate areas so could not be expected to be consistently associated with a soil type that has a broad distribution. These issues are further explored below.

Factors Controlling the Development and Distribution of Vegetation

As used here, 'vegetation' refers to an assemblage of populations of plant species. At a broad level, such assemblages can be classified into formations such as forest, woodland and grassland. Each of these can be further subdivided into open forests, closed forests (rainforests), and so on and further still according to which species is dominant. A dominant species is one that, by virtue of large numbers of individuals, controls many of the resources, including space, available in a layer (for example, the groundstorey) at a particular site. In a relatively undisturbed yellow box –Blakely's red gum woodland, for example, yellow box may be dominant and Blakely's red gum subdominant or equally dominant in the uppermost layer, and the understorey dominated by kangaroo grass. This vegetation type could be described as a 'grassy yellow box–Blakely's red gum woodland' or a *Eucalyptus melliodora–Eucalyptus blakelyi–Themeda australis* woodland.

For reasons described later, it should not be assumed that the type of vegetation present at a particular site has been continuously present since European settlement unless there is evidence such as an accurate description by an explorer, consistency of pollen types in stratigraphic samples, or perhaps a known long period of minimal disturbance. In semiarid New South Wales, for example, there is considerable evidence of a change, without deliberate human intervention, from grassy box woodlands to shrubby box woodlands or even to white cypress pine (*Callitris glaucophylla*) forests, since settlement of that area.

It was once believed that vegetation types were directly related to climatic zones. According to this theory, the vegetation at a site experiencing gradual or sudden disturbance would, by a process of one species replacing another (succession), ultimately develop into a characteristic vegetation type, or 'climax', for that particular climatic zone—regardless of soil type, slope, geology, and so on. Vegetation types in any climatic zone could therefore be classified as 'climax' or 'seral' (that is, immature and moving towards or away from the climax, depending on the frequency and type of disturbance).

Some of our early vegetation surveyors such as Beadle (1948) and Costin (1954) found it difficult to reconcile this theory with what they saw on the ground and adopted a variant, which proposed that the development of vegetation types could be limited by other factors as well as climate. Beadle, for example, considered the plains grass (*Stipa aristiglumis*) vegetation type to be limited by soil type (an 'edaphic climax') in western areas, and Costin considered topography, particularly drainage, as limiting further development of sub-alpine fen and bog vegetation types ('physiographic climaxes') on the Monaro. A vegetation type that was a consequence of human activities (for example due to clearing, regular burning or grazing by domestic livestock), was considered to be an 'anthropogenic climax' or 'disclimax', and many grasslands were considered to be in this category. Though this model of vegetation dynamics still has some adherents, particularly in areas with a reliable and benign climate, vegetation has been classified and mapped independently of it, according to inherent features of the vegetation itself, for many years (for example, by Moore, 1953; Biddiscombe, 1963; Portners, 1993). Recent, but not very successful, attempts have also been made to classify both vegetation and topography, such as the work of Sivertsen and Metcalfe (1995).

A number of other models of vegetation dynamics have been put forward (for example, see review by Whalley, 1994). One of these, the 'functional–factorial' model, proposed that the type of vegetation present at any particular site is dependent on five independent factors: parent material (for example rock type, aeolian or alluvial deposits), climate, relief, time since disturbance and the availability of organisms—similar factors that determine soil type (Chapter 1). This suggests that soils and vegetation at any site should be closely related but this is not always the case. In part, this is due to human activity: humans can change soils to some extent directly (for example by applying fertiliser) or indirectly (for example by increasing the rate of erosion), but their capacity to change vegetation is much greater. For example, a grassland can be maintained across a variety of soil types by clearing trees, introducing fertilisers and/or pasture species and grazing. Even without deliberate human intervention, differences in vegetation between sites with the same soil type (or similarities of vegetation across different soil types) can be enhanced by natural disturbances such as droughts, cyclones or wildfire, which are more disturbing to vegetation than to soils. As vegetation must re-establish after each disturbance, there is an opportunity for a new type of vegetation to emerge. As described in the next paragraph, all species of organisms are not equally available to all sites, and their availability changes over time.

Many plant species are unavailable owing to their limited distributions (due to climatic or other factors) and/or limited means of dispersal. If seed from all native species were available to all sites, then vegetation types could be quite different to those that are now present. We know that if plants or seeds of some native species are deliberately moved to a new site, they can thrive and even out-compete local plants. Cootamundra wattle (*Acacia baileyana*), for example, has been spread from its original habitat on the south-western slopes to many other areas (and many introduced plants have been even more successful and are now considered weeds). Even within the one area, seed of all local species may not be continuously available. Seeds of some species can remain viable in the soil for many years but this is not always the case. Seeds of eucalypts, for example, have a short life in the soil and most seed is stored in woody capsules in the canopy of the trees. A bushfire in a forest dominated by a fire-sensitive species such as mountain ash (*Eucalyptus regnans*) or blue mountains ash (*E. oreades*) can result in replacement by a forest

with different dominant species or even another vegetation type if no seed is present in the canopy at the time of the fire. Acacia seed, on the other hand, may remain viable in the soil for many years. Following a bushfire, seed may germinate and if conditions are favourable, acacias may come to dominate a site where very few were present immediately prior to the fire. Birds and animals can also move seed around the landscape.

Even if all species were equally available at a site, it is unlikely that most would be able to establish. In order to establish, a seed of a species has to be available at the time when conditions are most suitable for germination and early growth. These conditions could include the presence of light, bare soil, high soil temperature and moisture, and the absence of some herbivores and pathogens (for example, species of *Phytophthora*). As most plant species have specific requirements, the opening up of a 'regeneration niche' (Grubb, 1977), which usually means the creation of an unoccupied space, will be favourable to only some of the many species that may be available at the time. The regeneration niche may be created by the death of a plant due to old age, a fire, uprooting, burial by sediment, and so on. The surface conditions created by any one of these disturbances differ as do the weather conditions present when it occurred, thereby favouring some species over others. To describe plant colonisation of an unoccupied space in terms of 'first in—best dressed' is only partially correct as it is really the beginning of a competition. Many species will be out-competed in days or weeks. The winner may not be known for years, but a species has to be in the race to win it. For plants to persist in a particular environment, they must survive grazing, fire, flooding and so on to reach maturity, and if they are to continue to dominate a site, they must reproduce.

The presence of a particular vegetation type at a site is therefore the result of coincidental interactions between the independent site factors at the time the regeneration niche was created and during the subsequent development of the young plants. As already noted, the 'climax' model of vegetation dynamics proposed that vegetation types changed along a continuum from a bare surface to a particular vegetation type, defined by the climate, soil type or other limiting factor for a particular area. An alternative view, the 'state and transition' model, which incorporates the regeneration niche concept, proposes that the type of vegetation at any site changes from one 'state' to another by means of 'transitions', which usually occur following an infrequent event, or combination of events, such as wildfire, unusually high seasonal rainfall or extended drought. The model appears to be particularly applicable to arid and semi-arid areas where climatic conditions are very variable. Vegetation states and transitions for a number of rangeland types have been described by Whalley (1994) and Hall et al. (1994).

Site Factors and Vegetation

Most plants require something to grow in, a substrate or medium, which provides nutrients, anchorage and most importantly, moisture. Sometimes the substrate is rock, as on a hill where the erosion rate exceeds the rate of soil formation, or on waste dumps created by underground mining. Where conditions are favourable for soil development, the influence of parent material may be obscured; but more commonly and particularly in areas with sub-optimal rainfall, the properties of parent material are often evident, for example in the presence of lime, nutrients, sands or clays. Even in areas with subhumid or humid climates, the effect of parent material on soil types is sometimes clear and also of economic significance. For example, soils derived from Hawkesbury Sandstone near Sydney are preferred for horticulture to those derived from Wianamatta Shales. Soils derived from parna and possibly basalt are preferred for cherry orchards at Young to those derived solely from granite. The limestone-derived 'terra rossa' soils are favoured for vineyards in parts of South Australia.

Where plant growth is limited by low rainfall, differences across substrates may be evident in the area of low rainfall. Two examples from semi-arid New South Wales illustrate the interaction between climate and parent material:

—In far south-west New South Wales, mallee (which is a growth form rather than an assemblage of species) is often confined to deep sands on dunes—presumably due to the volume of moisture-containing substrate available for root exploration—whereas the swales, which usually have more clayey substrates, support grasslands or shrublands. Further east, however, mallee may occur on both dunes and flats. It should also be noted that mallee does not occur on similar dunes in north-west of New South Wales—presumably owing to the higher incidence of summer rainfall.

—On the semi-arid part of the Riverine Plain of south-east Australia, a chenopod shrubland dominated by saltbush (*Atriplex vesicaria*) usually occurs on the heavy, saline clay alluvium and shallow sands overlying the clay, whereas deeper sands of sandhills may support woodlands of stunted white cypress pine. The latter vegetation type indicates sandy substrates in this environment but further east, where rainfall is higher, it also occurs on more clayey soils, such as on the Bland Riverine Plain around Quandialla.

Table 15.1   Relationship between slope position and aspect for an area on the south-eastern Riverina (adapted from Moore, 1953)

Slope position	Soil development on north and west slopes	Plant species on north and west slopes	Soil development on south and east slopes	Plant species on south and east slopes
Near crest	No profile development	*E. dealbata* and *Acacia* spp.	No profile development	*E. dealbata* and *E. macrorhyncha*
High on slope	No profile development	*E. dealbata* and *Callitris* spp.	Some profile development	*E. macrorhyncha* and *E. goniocalyx*
Midslope	Some profile development	*E. dealbata* and *E. sideroxylon*	Development of surface and subsoils	*E. macrorhyncha* and *E. rossii/E. mannifera*
Mid to lower slope	Development of surface and subsoils	*E. sideroxylon*	Development of surface and subsoils	*E. macrorhyncha* and *E. rossii/E. mannifera*
Lower slope	Development of surface and subsoils	*E. sideroxylon* and *E. polyanthemos/E. microcarpa*	Development of surface and subsoils	*E. macrorhyncha* and *E. polyanthemos*

In other cases, it is the availability of runon water, an aspect of relief, that determines the more westerly distributions of vegetation types, and relationships between substrate and vegetation may be low, for example:

—bimble box (*Eucalyptus populnea*) woodlands, which are common on a variety of well-drained soil types along the eastern edge of the semi-arid zone, sometimes extend further to the west when additional runon water along ephemeral streams or in small terminal drainage depressions is available

—the distribution of river red gum (*Eucalyptus camaldulensis*) forests and woodlands shows a similar pattern; they may occur on hills in high-rainfall areas, as in the Western District of Victoria, but are restricted to watercourses over a large part of Australia where climate ranges from subhumid to arid.

However as relief, which includes altitude, aspect, position on slope as well as drainage, affects both soil and vegetation types, some relationships between the two would be expected. In some parts of the western slopes, the following sequence occurs: yellow box, to Blakely's red gum woodland on flats and lower slopes, to white box (*Eucalyptus albens*) woodland on upper slopes, with tumbledown gum (*Eucalyptus dealbata*) open forest on crests. A toposequence of soil types may also occur in these situations giving differences in fertility, depth, runoff or drainage, which together with other factors such as the presence of pathogens, temperature or frost frequency, determines which species becomes dominant or out-competes the others (see also Table 15.1). In order to use a model to predict vegetation distribution, it is necessary for those site attributes responsible for the distribution of native vegetation to be experimentally determined.

In his survey of the vegetation of arid and semi-arid New South Wales, Beadle (1948) reported that climate (mean annual rainfall and season of incidence) was often associated with broad vegetation types but soil types, except for grasslands on clays, were not. Rock types too were rarely associated with vegetation types, though in western New South Wales, rocks are not parent materials for many soils (alluvial and aeolian deposits more commonly are). This is not surprising over such a large area where the availability of moisture is particularly limiting to plant growth. More recent surveys—the Land Systems mapping program of the New South Wales Department of Land and Water Conservation (see Chapter 7)—have shown that relationships between vegetation and soils, geomorphology and geology are common in western New South Wales.

In surveys of smaller areas further east, where rainfall is less limiting, Moore (1953; for the south-western slopes) and Biddiscombe (1963; for the area north of Dubbo) both noted that vegetation types and soil types were often correlated. However, it would be incorrect to conclude that one was dependent on the other as relationships between parent material, relief, rainfall (sometimes) and soil and/or vegetation type were also evident. An illustration of the effects of position on slope (which also reflects drainage and depth of colluvium) and aspect on the soils and vegetation of hills on the south-western slopes was provided by Moore (Table 15.1). This model could be further modified by adding variations due to parent material. Further north, for example, open forests dominated by mugga ironbark (*E. sideroxylon*) rarely occur on granite parent material. In other areas too, the change from Palaeozoic sediments to granite is often associated with a change in vegetation type (for example see Semple, 1997).

North of Dubbo, white box woodlands occur mainly on soils derived from basalt, whereas on the south-western slopes in wetter areas they occur on podsolic soils derived from a wide variety of parent materials. This example reinforces the proposal that the relationship between soils and vegetation is often complex and cannot be expected to be constant over large areas, including several climatic zones. For these reasons they are more likely to be related (that is, one may be a good indicator of the other) over relatively small, rather than large, areas.

At the level of individual species or groups of species (compare vegetation type), correlations between properties of soil/substrate are relatively common. Again, these associations may only occur over limited areas, which may or may not coincide with the natural distribution of the species. Some of these are described below.

Site Factors and Plant Species

Soils that are salinised, highly acidic/alkaline or low in chemical fertility may be associated with particular species. Where chemical limitations occur, the range of plant species that can establish and survive under those conditions is relatively narrow, and it is likely that the presence of particular species will reflect that limitation. There are even some instances where particular plant species have been used for mineral exploration purposes, especially when looking for heavy metals associated with ultrabasic rocks.

Saline sites of humid and subhumid areas, for example, commonly support a restricted range of species, sometimes called 'indicator species', which may include couch grass (*Cynodon dactylon*) and the introduced species sea barley (*Hordeum marinum*), annual beardgrass (*Polypogon monspeliensis*), spiny rush

(*Juncus acutus*), or curly ryegrass (*Parapholis incurva*). These species do not occur at all saline sites owing to other restrictions such as climate (for example sea barley is restricted to winter rainfall areas), degree of salinity (for example strawberry clover, *Trifolium fragiferum*, does not survive on highly saline sites), pH and/or the types of salts present. Couch grass is probably the most common and widespread species, occurring on saline sites from Victoria to Queensland, but it also occurs on non-saline sites as do most of the other so-called 'indicator species'. Annual weeds may also be common on saline sites in seasons when salt concentrations are diluted by high rainfall.

In humid and subhumid areas, infertile sites (for example naturally acid, nutrient deficient or with shallow soils) are often dominated by native rather than introduced species; though exceptions occur on sheep camps—usually the most elevated parts of paddocks—due to nutrient concentration. On the infertile sites, herbaceous plants such as red-anthered wallaby grass (*Chionochloa pallida*) and nodding blue lily (*Stypandra glauca*) and native shrubs are often common. Shrubs of the heath family (Epacridaceae), which have hard, often prickly, leaves, are particularly characteristic of infertile sites. Shrubs that can thrive on low-fertility sites can often grow equally well on high-fertility sites. It is their presence in high numbers that is likely to be associated with low fertility. Hence in this case, low fertility may be associated with a particular vegetation type, such as a shrubland or a shrubby woodland, as well as with particular species.

When woody plants are cleared from fertile, humid sites, the maintenance of a grassland is relatively easy because herbaceous plants are able to out-compete the relatively slow-growing seedlings of any woody plants that germinate. The likelihood of woody plants 'reclaiming the site', assuming their seeds were still available, is low unless a major disturbance such as drought or flooding reduces herbaceous competition. However on sites with low fertility or with shallow soils, opportunities for woody plants to re-establish occur more frequently as herbaceous plants rarely dominate all of the site. Hence, caution needs to be exercised when contemplating tree clearing on these areas as, instead of a grassland, a shrubland, such as sifton bush (*Cassinia arcuata*), may result if fertility is not improved at the same time.

Despite Beadle's reporting little correlation between vegetation and soils in western areas, he did find that some species were associated with a particular site attribute such as pH or drainage (Table 15.2). Some species thought to be 'indicators', like kurrajong (*Brachychiton populneus*) of limestone, and wilga (*Geijera parviflora*) of clay soils, were relatively effective on a localised basis in higher-rainfall country, but were not so effective as indicators in western New South Wales. Other relationships are shown in the land system reports (see Chapter 7). A further example of relationships between soil and vegetation is shown in Murphy and Lawrie (1999).

Conclusion 15.1.3

Although soil and vegetation types are dependent on the same independent factors—time, parent material, relief, climate and availability of organisms—the two may not be highly correlated in any one area. In part, this is due to the different processes involved in the development of each. Over time, vegetation is

Table 15.2 Some plant species that are reasonably reliable indicators of some aspect of soil or drainage in western New South Wales (from Beadle 1948, pp. 49–52)

Species	Indicator of:
Acacia aneura (mulga)	sandy soils
Triodia scariosa (porcupine grass)	sandy soils
Maireana aphylla (cotton bush)	clay soils
Eragrostis setifolia (neverfail)	clay soils
Casuarina pauper/C. cristata (belah)	alkaline subsoils
Maireana sedifolia (pearl bluebush)	sandy soils with alkaline subsoils
Eucalyptus intertexta (inland red box)	sandy soils with acid subsoils
Acacia pendula (myall or boree)	clay soils with alkaline subsoils
Astrebla lappacea (curly mitchell grass)	clay soils with alkaline subsoils
Acacia stenophylla (river cooba)	sites subject to flooding
Eucalyptus coolabah (coolibah)	sites subject to flooding
Acacia doratoxylon (currawang)	sites with excessive drainage
Eucalyptus sideroxylon (mugga ironbark)	sites with excessive drainage

more likely to be severely disturbed, for example by wildfire or cultivation, than is soil. On each occasion there is the opportunity for a new type of vegetation to develop as a result of new organisms being available for recolonisation. Even small disturbances such as the death of an individual plant and its replacement by another species (often weedy types) can slowly result in a change from one vegetation type to another. Human activities have been particularly important in creating disturbances, introducing new organisms and changing vegetation types. Over a large area, vegetation types are more likely to be associated with climate than with soils but in smaller areas, associations with soil type, parent material or relief are more likely.

Sometimes individual species are associated with a particular soil attribute, such as pH, texture or nutrient deficiency, though in other cases the presence or absence of species is due to non-soil attributes such as drainage, frost frequency, runon water or the availability of seeds in any particular locality.

15.2 Rural Land Capability

While most areas of land can be used at various levels of intensity, the physical resources of each dictate that some will be capable of more intensive uses than others. Land use decisions should be guided by the capability of the land (see Plate 15.2). This will avoid both the penalties of lower returns through underuse and long-term degradation resulting from overuse. Land capability can be described as the maximum intensity of use sustainable by an area of land without causing permanent degradation or requiring unacceptably high inputs in maintaining that use.

Soil is a key element in determining land capability because practically all forms of land utilisation depend ultimately on the soil as a medium for plant growth or as an engineering material. An understanding of soils is thus essential for maintaining stable surfaces in the landscape and for selecting appropriate land uses.

Rural land capability is usually assessed at two levels:

1. on a large area or small-scale level—based on 1:100 000 topographic maps

Plate 15.2 Melonhole gilgai typical of high shrink-swell soils and a constraint to land use, particularly arable agriculture

2. on a farm or large-scale level—using enlarged aerial photographs, at scales ranging between about 1:5000 and 1:20 000, the 1:20 000 being for the Western Division.

Rural land capability has been mapped at a scale of 1:100 000 for the whole of the Eastern and Central Divisions of New South Wales. This provides a valuable information base for land use planners and managers working at a district or regional level, such as for local government planning.

At the individual farm level, land capability assessments are used as a basis for property management plans. Such plans provide detailed recommendations for development and management of farms.

Land capability assessments at a property scale require more detailed knowledge of the soils than those carried out at a district scale. Property planning also requires a greater level of soil information than is required simply for the capability assessment. Capability classification is primarily concerned with the risk of soil degradation, including erosion, associated with different forms of land use. This may result in individual classifications that contain a range of variables, any of which, in their own right, may be of significance for farming and grazing operations. For example, changes in soil type may be significant to farm operations, but will not necessarily require changes in the land capability classification. In addition, the physical and chemical properties of subsoils may be very important in planning soil conservation earthworks and selecting appropriate soil management practices and crops, but are of less importance in assessing capability, where topsoils are usually of greater relevance.

15.2.1 Rural Land Capability Assessment

Rural land capability assessment involves consideration of all the environmental factors that affect land use. Soil-related factors are the primary concern of this publication and will be treated separately in a following section. However, to maintain a balanced viewpoint, the other factors involved are outlined here. These are climate, landslope and landform.

Relevant climatic factors are those affecting plant growth, such as moisture availability and temperature, and rainfall or wind erosivity. Moisture availability is affected by the amount of rainfall, and its seasonal incidence and intensity, temperature, humidity, winds and evaporation. Aspects of temperature affecting plant growth include diurnal and seasonal variation and extremes, including frost incidence. The erosive potential of rainfall can be related to its intensity, amount and seasonal incidence, while that of wind depends on its strength and seasonal incidence. The incidence of salinity waterlogging is also affected by rainfall, evaporation and plant growth.

Landslope is a primary determinant of land capability because erosion hazard increases with slope steepness and because slope steepness imposes physical limits on many types of land usage. Erosion hazard increases with landslope because steeper slopes produce greater volumes and velocities of runoff water. These more effectively detach and transport soil particles. Greater proportions of rainfall tend to run off steeper slopes because surface water runs away more rapidly, allowing less opportunity for infiltration. The effects of gravity also become more pronounced on steeper slopes, with increasing tendency for detached soil particles to be moved downslope. The physical constraints imposed by steep slopes are quite marked in large-scale mechanised crop production, where the successful operation of most machinery is restricted to relatively low slopes, and the absence of rock outcrop. In addition, land uses requiring soil conservation earthworks can only be considered for slopes low enough to allow the use of earthmoving machinery and the effective functioning of the earthworks.

Various aspects of landform may also have an influence on capability. The position of the area within the landscape can affect the presence and amount of surface and subsurface runon water flows. These can affect soil erosion, flooding, salinity and waterlogging (see Plate 15.3). The intensity of the drainage pattern, dissection of the landscape, and rock outcrop and microrelief, where more intense and exaggerated, can also affect machinery operations and the design of soil conservation earthworks. Particularly in areas of greater local relief which are marginal to a particular form of use, slope aspect can be of importance in determining capability because of differing exposure to sun and winds.

Each of these factors is considered separately in a limiting factor type of analysis, that is, either one or a combination of factors may limit the capability of an area of land. Based on this type of analysis, land is classified into one of eight classes. The limitation to use, the risk of erosion and salinity, and the need for soil conservation measures become greater from Class I to Class VIII.

The basic assessment guidelines and class definitions are set out here.

Classification Guidelines

(i) Land classification takes account of present technology only. Future developments in agricultural technology may necessitate that land capability classifications be altered. For example, conservation farming techniques and a realisation of the causes of salinity have

Plate 15.3 Land degradation results not only from aggressive climate and poor soils, but also from inadequate management and using land beyond its capability

changed earlier land use recommendations based on land classification.

(ii) The scheme is based on the interpretation of the effects exerted on dryland agricultural usage by a combination of permanent environmental characteristics.

(iii) Lands with the same classification are necessarily similar only in respect of the limitations placed on their use and in terms of the intensity of soil management measures required for the maintenance of that use.

(iv) Permanent limitations are those considered to be either unalterable or not economically alterable.

(v) Lands are classified according to the most intensive use of which they are capable, irrespective of whether or not that capability is being, or will be, realised.

Class Definitions

In rural land capability classification, the primary emphasis is placed on the intensity of land use that a soil is capable of sustaining without degradation. Five categories of use have been defined.

(i) The first represents land suitable mainly for cropping. The land must be capable of sustaining at least two successive seasonal or annual tillage phases for crop production in which the tillage layer is inverted or shattered, without producing either a significant increase in soil erosion susceptibility or a significant deterioration in soil structure. It includes land where the soils are: sufficiently deep with a structure that will not readily break down under tillage; free of excessive salts; relatively free of large stones so as not to restrict the use of farm machinery; and which have efficient internal drainage but sufficient moisture-holding capacity to suit the requirements of the crop to be grown.

(ii) The second category includes land suitable mainly for grazing, which is capable of the infrequent growing of crops utilising practices involving a series of soil workings. It is land suitable for grazing that can be occasionally tilled for pasture establishment or renewal, but because of site factors such as soil types, slope, topographic location or drainage is not suited to regular cultivation.

(iii) The third category represents grazing lands unsuited to tillage operations and includes excessively steep or stony land, areas with high erosion potential and/or soils which limit productivity owing to their depth or physical fertility.

(iv) The fourth category is best left protected by trees and ungrazed.

(v) The final category includes land unsuitable for the cropping or grazing enterprises mentioned previously because of the land's physical limitations to conventional rural production, which result in an extreme erosion hazard if general land clearing occurs. Includes land suitable for tree cover or unsuitable for agriculture.

These five categories are further subdivided depending on the soil conservation measures required to sustain permanent production. The resulting eight classes are defined as follows.

15.2.2 Land Capability Classes

Land Suitable Mainly for Cropping

Class I—Wide variety of uses, such as vegetables and fruit production, grain crops, energy crops and fodder, sugar cane. No special soil conservation works or practices necessary.

Class II—Soil conservation practices such as strip cropping, conservation tillage and adequate crop rotations are necessary.

Class III—Structural soil conservation works such as graded banks and waterways are necessary, together with all the soil conservation practices as in Class II.

Land Suitable Mainly for Grazing

Class IV—Occasional cultivation, better grazing land. Soil conservation practices such as pasture improvement, stock control, application of fertiliser and minimal cultivation for the establishment or re-establishment of permanent pasture, maintenance of good ground cover.

Class V—Similar to IV; soil conservation works such as diversion banks and contour ripping, together with the practices in Class IV, like the maintenance of good ground cover.

Land Suitable for Grazing

Class VI—Not capable of cultivation, less productive grazing, can have saline areas. Soil conservation practices including limitation of stock, broadcasting of seed and fertiliser, promotion of native pasture regeneration, prevention of fire and destruction of vermin. This may require some structural works and maintenance of good ground cover.

Land Suitable for Tree Cover

Class VII—Land best protected by trees. Very important habitat areas for protecting biodiversity.

Land Unsuitable for Agriculture

Class VIII—Cliffs, lakes or swamps and other lands where it is impractical to grow crops or graze pastures.

Soils and Rural Land Capability 15.2.3

The soils information required for rural land capability assessment is that which determines the ability of the soil to support viable agricultural and pastoral pursuits on a sustainable basis. This, along with the effects of climate and topography, forms the basis for land capability assessment.

The capability of the soil to support these pursuits is related to its capacity to sustain economically valuable plant growth and its ability to withstand cultivation, grazing and other associated practices without degradation.

Soil characteristics that have a bearing on these capabilities include texture, structure and structural stability, depth, fertility, water-holding capacity, water infiltration and internal drainage capabilities, the occurrence of impeded drainage, rocks and stones; or of salinity, alkalinity or acidity, erodibility and the status of any soil erosion which has already occurred. Each of these characteristics is now further discussed.

Soil Texture

Soil texture is largely determined by the proportions of different-sized soil particles present in a soil. It is a very important factor in determining land capability.

Soils with topsoils having well-graded textures have a relatively even distribution of particle sizes from clay through to sand, and tend to be better able to support agricultural and pastoral activities than either very sandy or very clayey soils. They are better able to withstand cultivation and stock trampling and are more resistant to soil erosion.

Soil texture is closely related to water-holding capacity. Clayey soils hold more water than sandy soils and so can maintain plant growth for longer periods after wetting. Texture is also an important determinant in soil infiltration and internal drainage, with sandier soils tending to have greater infiltration rates and better internal drainage.

Soils with fine texture (the clay soils) tend to be more suitable for grazing than for agriculture. Except where they are composed of very well-structured or self-mulching clays, they may be very difficult to cultivate in either the wet or dry states. On the other hand, soils with coarse texture (the sands) are very unstable and easily eroded, and may need maximum protection against erosion, which can only be provided by an undisturbed cover of vegetation.

Soil Structure and Structural Stability

Soil structure describes the way in which primary soil particles are arranged or combined into soil aggregates. Structural stability refers to the ability of these aggregates to withstand wetting, drying and mechanical disruption. These are important properties in determining soil stability and resistance to mechanical breakdown and soil erosion. Soil structure is also very important in controlling water infiltration into the soil, soil aeration, porosity, permeability and internal drainage.

Well-structured soils tend to be more resistant to erosion owing to their ability to absorb rainfall more freely and over longer periods, and because of the resistance of their aggregates to detachment and transport by raindrop splash and/or overland flow. They also have good soil/water/air relationships for the growth of plants. Poorly structured soils have unstable aggregates and low infiltration rates. They tend to break down quickly under heavy rainfall, which leads to soil detachment and erosion. Under certain conditions, surface sealing occurs and this gives rise to rapid and excessive runoff.

Soil structure and structural stability, therefore, have an important bearing on land capability because of their strong influence on physical fertility and on the capability of the soil to withstand cultivation and other forms of land use.

Soil Depth

Depth, both of the topsoil and of the total soil profile, plays an important role in determining land capability. It is an important co-factor, along with soil texture, governing soil water-holding capacity. It also determines the volume of soil available for plant root penetration, and is important in controlling the available store of plant nutrients.

At least a moderate depth of soil is needed to allow arable agricultural use (75 cm or deeper). Shallow soils are only capable of use for grazing, while very shallow soils need the protection of a permanent undisturbed cover of vegetation.

Shallow soils and soils with shallow topsoils are inherently more fragile in respect to soil erosion. Any level of soil loss must represent a greater proportional loss of the resource and thus is of more significance in shallow soils than in deeper soils.

Soil Fertility

Soil fertility in this context refers to the capacity of the soil to provide suitable physical and biological conditions and adequate and well-balanced supplies of plant nutrients for the growth of economically useful plants. It has important implications for rural land capability as it is a major factor determining the economic usability of land for agriculture.

Soil fertility also influences soil erosion hazard because it affects the suitability of the soil for the establishment and maintenance of a vigorous cover of vegetation for soil protection.

While highly fertile soils may be capable of a wide range of uses, soils of lower fertility will sustain only pastoral uses or timber production. Soils of very low fertility are often not capable of any economic use and should be left in, or returned to, their natural states.

Water-holding Capacity

This refers to the amount of water that can be held in a soil after any excess has drained away following saturation. Together with infiltration rates, this has a large influence on the effects of the soil on runoff. It is, therefore, a factor in assessing the susceptibility of land to soil erosion and so its capability for different uses.

Its significance for land capability, however, is its relation to available water-holding capacity, that is, the amount of water held in the soil that is available for plant growth and survival between rewetting episodes. This differs from water-holding capacity in that a proportion of the water held in soil is unavailable to plants, and this proportion is greater in finer-textured soils.

Soils with a greater available water-holding capacity have a greater capability for agricultural use and, particularly in areas of low and erratic rainfall, for maintaining an effective cover of vegetation that ensures protection of the soil.

The three factors controlling soil water-holding capacity are soil depth, soil texture and soil structure. Deeper soils simply have more volume (per unit area at the surface) in which to store water. Finer-textured, more clayey soils have greater capacity to store water than do coarser-textured, sandy soils because of the far greater surface area (and, hence, absorptive capacity) of their particles per unit volume of soil. Well-structured soils, because of their aggregated state, possess greater pore volume both within and between aggregates for the storage of water.

Greater soil depth can help offset the effects of coarser soil textures, particularly with deep-rooting plants. Combined with the greater proportions of rainfall absorption and soil water availability found in coarser-textured soils, this often results in rapid and prolific response of vegetation to rainfalls on such soils. On the other hand, vegetation on finer-textured soils tends to persist longer into dry periods.

In assessing rural land capability on the basis of available water-holding capacity, deep, moderate- to fine-textured soils have few limitations to use, while shallow, coarse-textured soils will have low productivity and require a high degree of protection.

Water Infiltration and Internal Drainage

These are two related soil characteristics. Water infiltration refers to the rate at which water moves into the soil from the surface. Internal drainage refers to the rate at which water, once in the soil, moves through it. This may be affected by site characteristics as well as by soil properties.

The main soil-related factors governing these characteristics are soil structure and its stability, which determine soil porosity, and soil texture.

Soils with a good stable structure and a high porosity have the highest rates of water infiltration and internal drainage, provided that site characteristics are also satisfactory. Soils whose structure is weak, or liable to break down under wetting and mechanical disruption, are prone to the blocking of soil pores and surface sealing. This greatly reduces infiltration and increases the likelihood of surface runoff.

Coarse-textured soils tend to have higher rates of infiltration and better internal drainage than fine-textured soils. This, however, can be greatly modified by structural differences. For example, well-structured, fine-textured krasnozem soils can have much greater rates of infiltration and drainage than some poorly structured, coarse-textured soils.

In addition to the above soil-related factors, soil cover and general biological activity near the surface can play a big role in determining infiltration.

Higher rates of water infiltration allow for more rapid replenishment of soil and groundwater reserves, and reduce rates of surface runoff. Free internal drainage reduces the likelihood of damage by waterlogging, allows rapid development of suitable conditions for plant growth following heavy rainfall, and increases the soil's capacity for absorption of further rainfall. On the other hand, soils that drain too freely, particularly coarse-textured soils with their associated low water-holding capacity, dry out too rapidly and only provide for short growing periods following rains.

Land capability is affected by water infiltration and internal drainage in two ways. First, these characteristics affect rates of surface runoff and hence the liability of the land to soil erosion. This in turn affects the intensity of land use possible and the degree of protection required. Second, these characteristics affect plant growth conditions provided by the soil, which help determine the economic usefulness of the land and the vigour of the plant cover. The growth of plant cover again governs the level of protection provided for the soil surface.

Impeded Drainage and Regular Inundations

Impeded drainage, high water tables, seepage flows and regular or seasonal inundation greatly restrict the economic utility of affected areas. Such areas are, therefore, relegated to the lower land capability classifications. Less severely affected lands may be included in Class IV, while those more severely affected would be classified as Class VI or VIII.

Rocks and Stones

Rocks and stones in the soil have an effect on land capability that depends on the degree of their occurrence. Obviously rock outcrops do not support economically useful plant growth; shallow rock beds result in shallow soils with their associated limitations; the amount of rock and stone may be sufficient to restrict or prevent cultivation and present other management problems in respect to accessibility and pest control; and high levels of rock exposure may result in high rates of stormwater runoff onto lower lands.

The effect of rock and stone occurrence, then, is to cause soils so affected to be relegated to lower capability classes than would otherwise have applied, depending on the severity of the occurrence.

Soil Salinity, Sodicity, Alkalinity or Acidity

Where the levels of soil salinity, soil alkalinity (particularly when associated with excess sodium) or soil acidity are such as to adversely affect plant growth or soil structure, they will reduce productivity and increase erosion hazard. They therefore must be reflected in the land capability assessment.

Where measures are available to economically relieve these effects, such as drainage or the use of gypsum or lime, the effects on capability may not be great. But in more severe cases, the limitations on potential use can be very restrictive and the needs for soil protection very high. In such cases, the land will be placed in a capability class such as Class VI or even Class VIII.

Soil Erodibility

Soil erodibility is the susceptibility of the soil to the detachment and transportation of soil particles by erosive agents. It results from the interaction of various soil properties, including soil texture, soil structure and structural stability, chemical composition, infiltration and drainage ability, and water-holding capacity. Soil erodibility is a soil characteristic that can be considered independently of other non-soil factors influencing soil erosion, although it can be modified by management.

Soil erodibility has a large influence on land capability, particularly where soils are highly erodible. Soils of low and moderate erodibility may be capable of a wide range of uses. With higher levels of erodibility, however, the range of sustainable uses is reduced, and the costs incurred in maintaining soils in more intensive uses become prohibitive.

Existing Soil Erosion

Soil erosion, depending on its severity, has the potential to greatly modify soil texture, structure, fertility, water-holding capacity, drainage and local topography. Whole soil horizons may be removed, deep draining gullies developed and previously normal forms of land use made very difficult or impossible.

It can, therefore, have a marked effect on land capability.

The extent of the effects on capability depends on the economics of reclaiming the affected land and the amount, condition and potential productivity of the soil resource that is left. In very deep fertile soils, even large amounts of soil loss may not greatly affect land capability, although it may result in high reclamation costs. However, where soils, and particularly topsoils, are only shallow, even moderate soil losses may greatly downgrade productivity and land capability. This becomes more significant when, because of lower land values in poorer areas, even moderate reclamation costs may not be economically justifiable.

15.3 The Effects of Soils on Land Use

At least some of the more economically important effects of soil on possible land uses have long been recognised by land users. Soil-related land use patterns have developed as users have found from experience the types of use that will provide economic returns on particular types of soil. The effects of soil on land use are, therefore, quite dramatically illustrated in many parts of New South Wales.

However, it is important to differentiate between current land use and long-term land capability. To achieve permanently sustainable use, it is essential that land use and management be based on long-term capability rather than short-term economic return. Too often the longer-term deleterious effects of land uses that initially proved to be attractive, have not been recognised until the resulting soil degradation became well advanced.

Some of the more economically important differences in soils that lead to differences in land use are associated with differences in parent materials and differences in the local topographic situations in which the soils have developed. Parent material differences tend to affect belts of country, while topographic factors produce more local effects.

Examples of land use patterns in New South Wales that are controlled by differences in soils are now briefly described. Also mentioned are areas in which soil degradation has developed because of uses that in the longer term have proved to be inappropriate to particular soils.

There are, in New South Wales, widespread examples of differences in soils and resultant land use that are primarily due to differences in soil parent materials. Some of the more striking of these occur where soils derived from basaltic rocks and sediments lie adjacent to those developed from a range of other parent materials. Basaltic soils are generally chemically fertile, have high clay content, are well structured and internally well drained, and have relatively deep topsoils and high water-holding capacity. They withstand cultivation well, and are much sought after for arable agriculture where slopes, climate and rock content are suitable. Where not suited to arable agriculture, they often provide high class grazing land. In contrast, soils derived from parent materials such as granites and sandstones may be relatively infertile and have shallow, sandy topsoils overlying clay subsoils (duplex soils and massive earths). Their water-holding capacity is generally much lower, and they may be poorly structured and drained, and have a low capacity to withstand cultivation. Consequently, these soils, even though having similar climate and slopes to the adjoining basaltic soils, are used much less intensively, often only for low quality grazing.

A number of examples of such contrasts between the uses of soils of basaltic origin and those of poorer adjacent soils exist throughout eastern New South Wales. Examples include the basaltic belts running through Guyra, Inverell and Yallaroi and through the

Liverpool Ranges and Plains, where intensive agriculture, including areas of double cropping and better class grazing land, occurs on basaltic soils such as black cracking clays. Adjoining soils support less intensive agriculture, and sometimes only extensive grazing. The Dorrigo-Ebor area, the Comboyne area, the Canobolas-Millthorpe area, the Robertson-Bowral area and the Crookwell-Taralga area have basaltic soils such as krasnozems and euchrozems which are used for intensive grazing and horticulture, including potato growing, while surrounding soils are used for less intensive grazing. The basaltic areas around Cooma on the southern tablelands and the Monaro have black cracking clays, which provide better quality grazing lands than soils on surrounding parent materials.

Other examples of land use differences associated with differing soil parent materials occur in soils developed on deep alluvium in contrast to those formed on local country rock in major river valleys. Alluvial soils are often deeper, more fertile and generally much better suited to intensive agriculture than are those formed on local country rocks. Such contrasts can be seen in the Hawkesbury Valley around Richmond and Windsor, in the Hunter Valley around Singleton, Muswellbrook and Maitland, and in the Clarence Valley around Grafton and Maclean. In each of these areas, the deep, fertile alluvial soils are used for intensive agriculture, horticulture or dairying, while surrounding soils developed on local country rock are used for grazing and other less intensive uses. Similar, if not always so marked, contrasts in soils and land use related to alluvium are also found in the middle valleys of many of our western-flowing river systems.

A further example of soil-controlled land use differences associated with river systems is found on the Riverine Plain of southern New South Wales. Clay soils on the plain absorb larger amounts of water but make lower proportions of this available to plants than do the sandier or lighter-textured soils developed on surrounding country rock and aeolian deposits. As annual rainfall drops to around 400 mm and below, dryland crop production tends to become uneconomic on the clay soils, while still remaining a viable land use on the sandier or lighter soils. This is because the clay soils tie up moisture received from lighter falls, making it unavailable to crops. Crops on lighter soils can respond to such falls, enabling them to complete growth cycles and produce economic yields.

Examples of this effect can be seen in the area north-west of Griffith to Goolgowi. Within the irrigation areas of the Riverine Plain, soil-controlled land use patterns are also obvious. Rice and pastures are grown on the heavy clay soils, mainly the grey cracking clays, because of their good water-holding capability, while horticultural production is confined to the lighter soils associated with ancient stream courses and sandhills or sand sheets, because these crops require better drainage.

Examples of these land use patterns can also be seen throughout the Murrumbidgee Irrigation Areas, and increasingly in the central western irrigation areas along the Lachlan and Macquarie Rivers.

In the north, along the Namoi River, irrigation and a suitable climate allow the growing of cotton on the well-structured, deep and fertile black cracking clays of basaltic origin in the Narrabri and Wee Waa areas, as well as a range of other crops. Further west, the greyer clays of the Namoi and Barwon river systems are less well structured, sometimes sodic, and generally more difficult to manage. However, their fertility and water-holding capacity allow good crops to be grown, given good seasons and careful management.

Soil differences that may be reflected in the uses made of the land may also result from the relative topographic situations in which soils occur. Some of these differences may be due to the different soil-forming processes that have been at work over time in the different parts of the landscape. These processes, which include differential leaching, both horizontal and vertical, and differential erosion and deposition, produce soils with different characteristics at different points along the soil catena or toposequence. Other differences in land utility can also result from differing soil drainage conditions experienced in different parts of the landscape.

Topographically related soil differences tend to be repeated across soils developed on a given parent material. These soil differences are often reinforced by slope differences, and the combined result is the topographically related land use patterns that are commonly seen in most undulating to hilly landscapes. The shallower, sometimes coarser-textured and sometimes rockier soils of the higher part of the landscape have lower water-holding capacities and are less favoured for agriculture and more suited to lower intensity grazing. The middle and lower slopes have greater soil development and are generally reasonably well drained, and so are often best suited to intensive grazing with occasional cultivation, or may require the provision of drainage for use in arable agriculture. In drier areas, of course, the deeper soils and greater water availability in these low-lying areas may make them the best agricultural lands.

As stated earlier, however, not all land use and management practices that initially were attractive on particular soils have proved to be sustainable over the long term. The use of superphosphate and clovers to produce improved pastures in the better-watered agricultural and pastoral areas of New South Wales, particularly on soils of low clay content, is leading, after many years, to widespread soil acidification problems. These problems are now demanding at least a considerable modification in land management practices. The overclearing of catchments for grazing and agriculture has, in many part of Australia, and again after many years, led to rising valley water tables and soil degradation due to salinity. A similar effect is seen in irrigation areas where rising water tables can also eventually give rise to salinity problems. In many areas of the wheat-belt, long, continued cultivation associated with stubble burning has led, particularly on some red-brown earths, to soil structure decline. This has resulted in serious soil crusting and erosion problems, necessitating reappraisal of the capability of some of these soils and considerable modifications to management practices.

BIBLIOGRAPHY

Beadle, N.C.W. (1948), *The Vegetation and Pastures of Western New South Wales with Special Reference to Soil Erosion*, Government Printer, Sydney.

Biddiscombe, E.F. (1963), *A Vegetation Survey in the Macquarie Region, New South Wales*, CSIRO Division of Plant Industry Technical Paper No. 18.

Costin, A.B. (1954), *A Study of the Ecosystems of the Monaro Region of New South Wales with Special Reference to Soil Erosion*, Government Printer, Sydney.

Flannery, T.F. (1994), *The Future Eaters*, Reed Books, Port Melbourne.

Grubb, P.J. (1977), 'The maintenance of species-richness in plant communities: the importance of the regeneration niche', *Biological Reviews* 52, 107–45.

Hall, T.J., Filet, P.G., Banks, B. and Silcock, R.G. (1994), 'State and transition models for rangelands. II. A state and transition model of the Aristida–Bothriochloa pasture community of central and southern Queensland', *Tropical Grasslands* 28, 270–3.

Isbell, R.F. (1996), *The Australian Soil Classification*, CSIRO, Melbourne.

Mcdonald, R.C., Isbell, R.F., Speight, J.G., Walker, J., and Hopkins, M.S. (1990), *Australian Soil and Land Survey Field Handbook*, Second Edition, Inkata Press, Melbourne.

Moore, C.W.E. (1953), 'The vegetation of the south-eastern Riverina, New South Wales. I. The climax communities', *Australian Journal of Botany* 1, 458–547.

Murphy, B.W. and Lawrie, J.L. (1999), *Soil Landscapes of the Dubbo 1: 250 000 Sheet*, New South Wales Department of Land and Water Conservation, Sydney.

Portners, M. (1993), 'The natural vegetation of the Hay Plain: Booligal–Hay and Deniliquin–Bendigo 1:250 000 maps', *Cunninghamia* 3, 1–122.

Semple, W.S. (1997), 'Native and naturalised shrubs of the Bathurst granites: past and present', *Cunninghamia* 5, 49–80.

Sivertsen, D. and Metcalfe, L. (1995), 'Natural vegetation of the southern wheat-belt (Forbes and Cargelligo 1: 250 000 map sheets)', *Cunninghamia* 4, 103–28.

Whalley, R.D.B. (1994), 'State and transition models for rangelands. I. Successional theory and vegetation change', *Tropical Grasslands* 28, 195–205.

Soils and Sustainable Farming Systems

C H A P T E R   1 6

J.W. Lawrie, B.W. Murphy,
I.J. Packer and A.J. Harte

Given the changing community attitudes and expectations, the evaluation of the suitability of farming systems today requires that some account be taken of the sustainability of those systems, and their potential effects on the environment and on the quality of our natural resources. The most acceptable and better farming practices today need to be sustainable and meet the following three key objectives identified by the Standing Committee on Agriculture (SCA, 1991):

—maximise economic returns by maximising productivity and minimising costs to ensure economic viability

—maintain the quality of soil, land, water and air on the farm

—maintain or improve ecosystems that are influenced by agricultural activities; this means maintaining such things as water quality in river systems, soil and air quality off-site, and climatic systems.

Five principles of sustainable agriculture have been identified, which form the basis of sustainable agriculture (SCA, 1991). These are:

—to sustain and increase farm productivity

—to minimise, ameliorate or avoid adverse impacts on the agricultural natural resource base and associated ecosystems

—to minimise residues resulting from the use of chemicals in agriculture

—to maximise net social benefits derived from agriculture

—to ensure that farming systems are sufficiently flexible to manage risks associated with the vagaries of climate and markets.

The main resource management issues for agriculture in Australia are land and soil degradation, water use and quality, chemical use in agriculture, vegetation degradation, feral and native animals, biodiversity decline, the increase of greenhouse gases and the potential for climate change, and plant and animal health issues (SCA, 1991). Particular aspects of soil degradation identified are erosion, nutrient decline including acidification, structure decline, salinisation and biological decline. Farming practices that contribute to this degradation include excessive cultivation, bare soil and fallowing practices, overgrazing and loss of ground cover, animal/machinery traffic especially on wet soils, poor matching of enterprises to land capability, excessive clearing of deep-rooted perennial native species, and the use of acidifying pastures and fertilisers. This chapter addresses some of these issues, especially those related to dryland cropping practices and their potential effects on land and soil resources and the ecosystem. The focus here is on the soil's physical property of structure—its importance for crop growth, its susceptibility to tillage practices and its vital role in sustaining soil ecosystems. Discussion of relevant chemical and biological properties related to tillage is to be found in Chapters 13 and 14 respectively.

In this chapter conservation farming systems include cropping and grazing practices, which are often combined in a rotation, as part of a whole farm management system. Conservation tillage practices refer to specific cropping practices such as no-tillage (zero tillage), direct drilling or minimum (reduced) tillage.

16.1 Development of Sustainable Farming Systems

16.1.1 Historical Review

The history of dryland farming practices in New South Wales is closely related to the history of wheat growing, which has always been our principal dryland crop. The history of wheat growing in Australia is often divided into four key eras based on average yields of wheat for Australia (Hamblin and Kyneur, 1993):

1. 1860 to 1900, the period of nutrient exhaustion when average yields dropped from about 910 kg/ha to about 420 kg/ha.

2. 1900 to 1950, when wheat was grown as a monoculture using superphosphate and fallowing practices to conserve moisture, and average yields increased from 420 kg/ha to about 860 kg/ha. During this period there was widespread soil and land degradation as long fallows greatly increased the number of tillage operations and left the soil bare for long periods (Kaleski, 1945; Trotman, 1965; Sims, 1977; Breckwoldt, 1988). Some soils were tilled after every rainfall event for up to 12 months prior to sowing a crop (NSW Department of Agriculture, 1937).

3. 1950 to 1978, when the introduction of legume-based pastures into cropping cycles saw increased nitrogen fertility, some improvement in soil structure conditions and reduced erosion, and yields increased from 860 kg/ha to 1229 kg/ha.

4. 1978 to 1993, when yields increased only slightly from 1229 kg/ha to 1375 kg/ha, with improved wheat varieties and the development and adoption of improved farming practices involving reduced amounts of tillage, increased herbicide use, increased fertiliser use including nitrogen- and sulfur-based fertilisers, improved sowing machinery, stubble retention and improved rotations using a range of oil-seed crops (canolas, sunflowers and safflowers) and pulses (lupins, faba beans, field peas and chickpeas) as well as pastures.

This last era has also reduced or at least to some degree reversed soil and land degradation, particularly erosion, nutrient decline and soil structure degradation. Since 1993 this improvement has continued with reduced cultivation, increased fertiliser and herbicide use, a wider range of crops grown, new crop varieties and improved tillage equipment, together with a move towards whole farm planning. Some of these changes have been reasonably rapid. For example, there was a reduction in the average number of cultivations from 3.2 workings to 2.0 in the central west between 1993 and 1996 (Vanclay and Hely, 1997).

A conservation ethic to many aspects of farming practices was introduced with the advent of the Soil Conservation Service in the late 1930s (Trotman, 1965). During the 1970s and early 1980s NSW Agriculture, the Soil Conservation Service and agribusiness commenced the development and promotion of conservation tillage practices with a strong emphasis on stubble retention, minimum tillage and direct drilling. By the 1990s, the concept of conservation farming and the practices and techniques that it encompasses had rapidly expanded. The need to develop and promote farming systems that can meet the principles of sustainable agriculture in a wide range of circumstances has become especially important at the end of the twentieth century.

The Current Situation

16.1.2

The rural community is now aware that land and soil degradation is an important issue (Vanclay and Hely, 1997), but the SCA in 1991 identified the following barriers, which are still preventing some farmers from adopting conservation farming systems:
—difficulties in financing and making a conversion to a new system
—risks associated with change
—lack of a conservation system that is technically and economically viable
—lack of an incentive to change, especially for older farmers
—insufficient local information to make a change
—lack of awareness of the problems, and the benefits or need to change
—lack of awareness of the means of achieving a change in farming practices
—presentation of information in an isolated form and not as part of a whole viable system.

The Department of Land and Water Conservation is now working in cooperation with NSW Agriculture, CSIRO, Landcare and private industry to address many of the above issues and encourage the development and adoption of conservation farming systems. They are targeting some of the specific difficulties farmers are experiencing with conservation tillage practices. These include agronomic problems such as disease and weed control, the availability of suitable machinery for sowing and establishing crops in difficult soils and heavy stubble, and the economic limitations due to lower yields and higher costs, or at least perceived higher costs (that is, up-front purchases of items such as specialised machinery and herbicides). The development and adoption of farming systems that meet all of the key sustainability objectives has not been easy, because a good understanding and working knowledge of how these new practices and techniques interact is required.

Conservation farming associations have formed in New South Wales since 1994 to cater for this need. These include the Moree Conservation Farmers Association, the Central West Conservation Farming Association and Stipa Native Grasses Association.

This chapter discusses the effects that dryland farming practices have on soil and land degradation, their potential effect on ecosystems, and it also presents some of the conservation tillage practices that are being used to achieve long-term sustainability for dryland agriculture in this state.

16.1.3 The Areas Involved

The potential for farming practices to cause environmental degradation is high considering the large areas of land cropped in New South Wales. During 1993/94 about three million hectares of land was under crop (Bray, 1996). This is nearly 4% of the state's surface area. In some regions of the state the proportion is much higher, with 9% being under cultivation in the northern region and over 10% in the central west. In the south of the state 5% of the Murrumbidgee Region and 2.5% of the Murray Region is under cultivation. However the total area of land cropped within a cropping rotation could be at least three to four times greater than the above figures.

Only small areas are irrigated for more intensive production, but the potential for environmental impacts on this irrigated land is very high. So the management of these irrigated farming areas is also critical to the environmental management of the state.

16.1.4 Dryland Cropping Systems in New South Wales

There is a wide variety of dryland cropping systems in New South Wales. These can vary in all of the following operations.

(a) When and how often one-way disc cultivators (see Plate 16.1), two-way disc cultivators and tined cultivators are used.

Plate 16.1 One way disc cultivation, often used in traditional cropping systems

(b) How stubble is managed, that is, how long it is retained standing and when and how it is reduced by burning, grazing or with machinery.

(c) When and how often weeds are controlled using different types and quantities of herbicides and/or with grazing and burning.

(d) How much and what type of soil disturbance is applied at sowing and what type of machinery is used for sowing.

(e) Crop selection and varieties of crop sown and the type and length of different crop rotations.

(f) Amount and type of fertiliser added to the soil before, during and after sowing.

(g) Use of pasture phases, including grazing intensity and what type and amount of plant growth is allowed in the pasture phase.

The condition of the soil, the plant growth and the environmental effects resulting from these operations will depend on the following:

(a) The condition of the soil when the operation is done, that is, soil moisture content, soil friability and bulk and type of plant growth.

(b) Weather conditions before, during and after the operations, including rainfall amount and intensity, evaporation, wetting and drying cycles, wind velocity and frequency.

(c) Landform characteristics, that is, degree and length of slope, concentration of water by the catchment and size of wind fetch areas.

(d) Nutrient levels in the soil materials, that is, natural nutrients plus added fertiliser and organic materials on the surface and in the soil.

(e) Levels of agricultural chemicals in the soil and plant materials including previous history of agricultural chemicals added to the soil and plants; time since adding chemicals, residual amounts of chemicals in soils.

It is apparent from the above lists that there are a large number of permutations and combinations of cropping operations that can be applied to an area of land, and hence a large number of systems. There is also a wide range of consequences from applying these operations to an area of land, which will depend on the soil conditions, rainfall characteristics, temperature regime and landform of an area, as well as the types of crops grown.

While land managers have a wide range of operations to choose from, they can be broadly grouped into a few cropping systems in the two major climatic zones of the dryland cropping areas of New South Wales. While there is no definite line forming the boundary between the northern and southern zones, a broad line from Warren through Dubbo to Coolah is an approximate boundary.

The Northern Dryland Cropping Zone

This area is characterised by predominantly summer rainfall, with relatively high intensity storms being common during the summer. This puts an emphasis on the need to fallow and store water in the soil from the summer rainfall for the winter crops. Fallowing for up to 12 to 15 months is sometimes practised in drier areas. Grazing is less frequently used in conjunction with cropping, and in some wetter areas the land is continuously cropped with both summer and winter crops. This presents a very high erosion hazard, especially during the summer. It is critical to prevent water erosion in the summer period when fallowing is practised or summer crops are grown and the surface is left bare, to prevent excessive soil losses.

Soils with self-mulching surfaces are widespread in the north and these make an ideal seedbed that only requires minimum disturbance to obtain good soil/seed contact and minimal resistance to germination, emergence and establishment of seedlings.

The Southern Dryland Cropping Zone

The southern cropping areas in New South Wales have predominantly winter rainfall, although summer storms do occur, particularly in the centre of the state near Cowra and Wellington. There is less emphasis on summer fallowing to store water in the soil, although there is often a benefit from storing autumn rainfall in drier winters, especially in the more marginal cropping areas to the west in locations such as Condobolin, Nyngan and Balranald. Erosion is a major problem on cultivated soils in the autumn/early winter period when the rainfall can be high, if not intense.

The widespread occurrence of hardsetting or fragile surface soils has also resulted in an emphasis on reducing the effects of cultivation and loss of organic matter which has resulted in structural decline. This area also has a higher percentage of hardsetting or fragile surface soils and fewer self-mulching surface soils than in the north. There are also substantial areas of very hardsetting sodic surface soils, which can pose significant problems when implementing minimum tillage practices.

Cropping Systems

The farming systems in New South Wales can be placed into the following three broad groups:

No-Tillage/Zero Tillage/Direct Drill is where stubble is retained for as long as possible, weeds are controlled with herbicides and rotations, there is limited or no grazing and minimal ground disturbance at sowing. Sowing is carried out through the remaining stubble on the surface with modified or specialised sowing equipment with either some or all of the following features: parallelogram sowing systems for ground following ability, split fertiliser and seeding tubes to allow deep banding of fertiliser, presswheels or a rolling harrow attachment for seed/soil contact, and so on. Sowing machinery designed for no-tillage generally has high breakout forces to handle the harder soil conditions.

A modified system of no-till farming is direct drilling where there is complete ground disturbance at sowing to help control weed, disease, pest or soil structure problems. Sheep grazing is not necessarily a factor in direct drilling but is more related to areas that have a grazing enterprise on the farm and the stock are used to take advantage of extra feed and for weed control. There are many farmers in the south who graze and practise no-tillage.

Minimum Tillage/Reduced Tillage is the most popular cropping system in New South Wales. It addresses the hard soil conditions that have been induced by grazing or tillage compaction, the incorporation of fertilisers or soil ameliorants, and weed, pest and disease control problems. The practice of stubble incorporation with discs is now being replaced by tined cultivation and the retention of as much stubble on the surface as possible for easier management, better erosion control, better moisture retention for biological activity and better options for weed, pest and/or disease control with burning. Farmers cannot always rely on full ground disturbance at sowing to address hardsetting soils. When complete ground disturbance is carried out before sowing using some tined tillage implement, this is called minimum tillage.

Another form of minimum tillage is one-pass tillage, which is also being promoted and adopted for hardsetting soils prior to sowing. One-pass tillage, which uses a combination of wide and narrow points and heavy rollers, is environmentally acceptable because the soil surface is left in a condition less prone to erosion than with traditional cultivation techniques, because more organic matter (both soil and plant) is left on the surface. However, any form of a complete ground disturbance system depletes soil organic matter to some extent.

Traditional Tillage/Conventional Tillage systems rely on burning the stubble as soon as possible after harvest and weed control is achieved with one-way disc cultivators. The effective control of weeds might require up to seven cultivations prior to sowing. The soil is left exposed to summer storms and erosion losses can be very high. Grazing is commonly used to help control weeds in the south.

No-tillage and traditional tillage cropping systems are at the ends of a continuous spectrum of cropping systems based on the many permutations and combinations that land managers have to choose from to suit the resources they have and the condition of their soils.

No-tillage is the best option environmentally but under some conditions traditional farming produces the highest yields. However with the rising costs of fuel and labour and the falling costs of herbicides, the gross returns from the no-till systems are equal if not in front when done properly. Getting the agronomy right in no-till is the key to its success because the rules are a bit different from traditional systems.

Many farmers are still opting for reduced/minimum tillage systems because they have not yet solved problems related to stubble handling, weeds (especially herbicide resistance), pests, diseases, soil nutrition and soil strength. Unfortunately these systems often have more workings than the perception of one operation before sowing. There are operations such as harrowing, wide-lining and complete soil disturbance at sowing plus a further soil disturbance with harrows which are often not considered to be a cultivation operation. The end result is a poorly structured soil very much like a traditional seedbed, with depleted organic matter, which is prone to erosion.

16.2 Conservation Farming—Field Practice

The successful application of conservation farming in the field requires a good working knowledge of the agronomic and soil management principles on which it is based. In some ways it requires a rethink of many of the techniques and principles upon which farming was traditionally based. Conservation farming comprises two main systems—conservation tillage and conservation grazing. This chapter concentrates on conservation tillage or cropping, and its application in the field.

The concept of conservation tillage began in the 1970s with the introduction of the technique of direct drilling to establish crops with minimum disturbance. The introduction of herbicides like paraquat, diquat, hoegrass and glyphosate made this system possible, and it was combined with grazing to help control weeds. It challenged many conventional practices. Despite the noted benefits of erosion control, reduced runoff, improved soil structure and improved friability, there were some initial problems, including difficulties in weed control, difficulties in the sowing operation, poor early growth in minimally disturbed seedbeds, and increase in some diseases. This often led to lower yields, although some farmers were able to achieve comparable yields between direct drill and traditional crops. One of the problems was that direct drilling was often used as a salvage operation which meant late sowing into poorly prepared paddocks, or was applied to paddocks that were in poor condition, that is, heavily grazed and trampled surfaces

with compacted soils or soils that had a long history of excessive cultivation.

Successful conservation tillage including reduced/minimum tillage or no-tillage systems requires early application of the correct agronomic management advice. Management aspects include selecting the right paddock and using the most suitable machinery, plus an awareness of soil nutrient levels, effective weed control with the appropriate chemicals and/or grazing. Diseases can be controlled with carefully selected rotations and with suitable tillage at sowing. Effective stubble and stock management is essential. More information is now available on these aspects and this has substantially improved agronomic productivity of conservation farming systems, in conjunction with a better understanding of the implementation of conservation tillage.

It is now well known that conservation tillage cannot be applied readily over all soils under all conditions, and some flexibility is required to successfully implement conservation tillage under difficult soil conditions.

Practical Issues in Applying Conservation Tillage

16.2.1

To be effective, conservation tillage has to be approached as a conservation farming package. It should be noted that the issues and techniques raised here are subject to rapid development, as new techniques are tried and adopted.

Weed Control

For the successful adoption of conservation tillage, weed control and boomspray technology have become critical, because cultivation is not used to control weeds. The boomspray becomes the major means of weed control and this is even more important when the distance between sowing rows increases to more than 300 mm. Some essential points that need to be considered are outlined below.

(i)   Weeds need to sprayed at the correct stage. This requires innovative techniques to avoid problems associated with high wind velocity and/or poor trafficability due to wet soils. Air-injected nozzles and infra-red and video cameras that allow spot spraying of weeds are now available. However, the overriding factor with any new techniques is that they must be effective and economical.

(ii) Weeds should be sprayed as early as possible because less herbicide is then required to kill young weeds and this reduces costs and minimises potential damage to the environment.

(iii) It is important to understand the mode of action of different types of herbicides to maximise their effectiveness and minimise environmental damage. It may mean that sometimes it is necessary to strategically burn or cultivate to gain complete weed control.

(iv) Herbicides should be rotated to avoid the build-up of herbicide resistance.

(v)   There are other options besides herbicides for weed control including grazing, burning and the retention of dense stubble layers. Pastures may also be used to control weeds in the period before cropping.

(vi) Concern about the potential environmental effects of herbicides is a growing issue. The contamination of produce, soils and waterways with herbicides and their breakdown products (metabolites) is of considerable concern, especially in the use of residual herbicides. Issues that need to be addressed include the consideration of breakdown rates in soils, closeness of spray applications to waterways and other crops, and the potential for movement of herbicides in dust and wind drift.

Stubble Management

The retention of crop residues or stubble for erosion protection is an essential part of any conservation tillage system as it also increases organic matter and soil fauna activity. There are many machinery options for handling stubbles, which begin at harvest and finish at sowing. The system adopted by each farmer will depend on available finances, machinery available,

presence of stock and the extent of the cropping area.

Stubble management needs to consider the capabilities of available machinery, potential agronomic problems associated with stubble, and the climate, as well as the stubble level required for erosion control, moisture retention and increased biological activity in the soil. This is considered to be greater than 4 t/ha, and at least 2–3 t/ha of stubble to significantly increase organic matter. From a soil structure viewpoint it is better to maintain the stubble on the surface and burn it late to remove it for sowing, rather than excessively cultivate the soil to remove it.

Unfortunately, stubble retention may create some of the following practical agronomic problems:

(i)   Stubble can interfere with the action of tines and cultivators, especially at sowing. Specialised machinery can handle larger amounts of stubble. Sometimes, soil and stubble clumps left after the sowing operation can significantly reduce crop germination and establishment. In many areas, stock grazing, burning, and/or stubble-modifying equipment is needed to reduce stubble levels. As a general rule, stubble needs to be less than 4 t/ha for most tillage operations, even with specialised sowing machinery.

(ii) Allelopathy from fresh stubble is a potential problem. The stubble of some crops produces chemicals that can actually reduce the growth of some other crops. Fortunately, this allelopathic effect is reduced if there has been sufficient rain and microbial activity to leach out these chemicals. Allelopathy by some stubbles has the potential to control some weeds, and research is still continuing in this area.

(iii) It has been known for a long time that stubble ties up available nitrogen, and that this is caused by micro-organisms using nitrogen to break down stubble with high carbon contents or high carbon/nitrogen ratios (see Chapter 14). Until microbial populations build up to cope with breakdown of high-carbon stubble, it is recommended to increase nitrogen fertiliser input by 10–20%.

(iv) Weed control options particularly with pre-emergent herbicides may be limited.

(v)   Some diseases, such as yellow spot in wheat and black leg in canola, are thought to be carried over by stubble from one crop to the next.

Crop Establishment

Traditional cultivation systems are very effective for crop establishment and although problems of surface crusting and waterlogging are sometimes associated with traditional tillage, most farmers are confident in this system. Confidence in no-tillage or direct drilling systems is not as high. This is a major concern of farmers when adopting conservation tillage practices.

They are concerned about suitable seedbed formation, accurate seed and fertiliser placement, and seed/soil contact. Selection of suitable tillage equipment, especially points, is critical to achieving these aims. Complete ground disturbance is not necessary to achieve this; rather emphasis should be given to where the seed is placed in the soil. The interaction between soil moisture, soil friability and soil response to the shape of the tine makes recommendations for the ideal tillage point and breakout pressure for each paddock difficult. This is even more difficult when a farmer has a paddock with two or more different soil types.

Sowing Equipment

Point wear on sowing equipment can be more rapid when soils are direct drilled, and tungsten carbide protection may be required for tine points in some soil types.

Also a zone of deeper tillage below the seed (30 to 60 mm) in many soils is advantageous for good establishment and growth. This is particularly the case on structurally degraded soils or very hardsetting soils, which restrict infiltration causing waterlogging, and restrict growth due to high soil strength. Other advantages of deeper tillage are disease control and the application of fertiliser at sowing below the seed (deep banding).

Seedbed formation can only be judged by farmers using a point type on the tines that suits their sowing machinery, speed of travel and soil type. However, important considerations are:

(i) Conventional seeding points are not usually suited to direct drilling. Edge-on (knife points) and inverted T points and spear points are more suited to the system. Problems to look out for in point selection are smearing, particularly in clayey soils, sideways soil throw which exposes the seed, and draught requirements to pull the tine through the soil.

(ii) Generally points give more reliable results than discs for disease control and provide better soil conditions for seedling establishment and root growth.

(iii) To be most effective and maintain a constant sowing depth, tillage points need to be kept at a constant depth and angle, which requires an adequate tine breakout pressure. The depth that a point will sow at will vary with tine point design and soil conditions.

(iv) Consider the condition of the soil into which the crop is to be sown. It may be obvious that direct drilling or no-tillage is not suitable for crop establishment. Such conditions include paddocks coming out of pasture with many well-developed plant tussocks; severe hardsetting (massive) soil surfaces with an obvious crust; and paddocks with a bad infestation of weeds with long, wiry stems, for example wireweed and couch grass.

Presswheels are often recommended for good seed/soil contact and to aid in seed placement using conservation tillage (see Plate 16.2). A range of presswheels is available, but it is essential to match the presswheel width with the sowing point to ensure that they track and press the soil around the seed. Narrow points and presswheels facilitate moisture seeking for sowing (Wylie, 1993). Scrapers can be mounted to remove soil from the presswheels, especially in more clayey soils.

Ideally presswheels can be used in a parallelogram system, which controls the depth of sowing. This system ensures ground-following capability on undulating country or with wide sowing equipment. The ability to adjust presswheel pressure is desirable, as it appears from current studies that as the soil dries out, the pressure should be increased. Presswheel placement is also worth considering. Good germination has been achieved by wheels that press to the side of the cultivated row. This reduces the possible effects of soil crusting and smearing, which may result by the action of the presswheel directly over the seed.

Alternatives to presswheels, that increase seed/soil contact, include loose ring rollers, prickle chains, rolling harrows and soil rollers, but these systems can have the problem of dragging loose soil into the sowing row, which makes the maintenance of sowing depth difficult.

Disease Control

Soil and plant diseases are not new to farmers but conservation tillage, with its minimal soil disturbance and stubble retention, has increased the incidence of some diseases. Some commonly observed problems are:

(i) Reduced early growth of cereals under conservation tillage systems (Chan et al., 1989). Diseases are thought to be a major cause of this and they tend to be more prevalent after a long pasture phase that has allowed the disease organisms to build up. The incidence of disease is reduced after a few years of minimum disturbance cropping. It can be avoided by sowing early, cultivating below the seed when sowing and by giving the soil one cultivation prior to sowing. Reduced early growth does not always lead to reduced yields, as in years that have a

Plate 16.2   Presswheels on sowing equipment to ensure good seed–soil contact.

dry finish the reduced vegetative mass uses less moisture, and more moisture is available to produce grain at the end of the plant's life cycle.

(ii)   Several soil-borne diseases are encouraged by continuous farming using no-till practices. These are Pythium, Take-all and Rhizoctonia. The first two are specific to cereals and it is possible to control them by rotation with a legume or broadleaf crop. Recent work from NSW Agriculture has developed a seed dressing with the potential to control Take-all. Rhizoctonia is one disease that is increased by minimum tillage and no-tillage, and because it is not crop specific, cannot be controlled by crop rotation. Soil cultivation is the most recognised means of control, but this is not compatible with no-tillage or direct drilling. Deep disturbance (40 to 60 mm) below the seed using extended points has given good control in several trials. These soil-borne disease problems highlight the relevance of increased biological activity in the soil under conservation tillage systems, and the potential for interactions between micro-organisms as biological activity increases. Unfortunately, not all the increases in biological activity are beneficial for crop yields.

(iii)   Plant foliage diseases have become prevalent with stubble retention and minimum tillage. Yellow leaf spot, *Septaria triticea*, blotch and eyespot are examples of such diseases. These diseases are more common in the north of the state, particularly yellow leaf spot. There appears to have been little damage of economic significance in the south of the state. These diseases can be controlled by selecting resistant varieties, grazing or burning stubble prior to sowing, and rotations with crops other than cereals.

Deep Tillage

Deep tillage has the aim of breaking up dense, compacted soil layers or plough-pans at 80 to 150 mm depth. The depth of the compacted layer needs to be identified and the tine depth set at a depth below this layer. The soil should be drier than field capacity when the operation is done to maximise the amount of soil shattering that occurs. The effects of the tillage may be only temporary if the soil is unstable to wetting, because when the soil is rewetted it collapses down to its prior condition. This would be expected to happen in the case of bleached A_2 horizons or sodic, dispersible soils. Strategies to prevent this happening in these soils include active root growth, reduced tillage, avoiding stocking in wet conditions and controlled traffic. Deep tillage should be done with tines that minimise the amount of soil inversion, and several types of deep tillage machinery are available that meet this requirement. Non-inversion allows the organic matter-rich surface soil to remain at the surface.

Deep tillage should only be done where there is a clearly definable compacted layer. It is a costly operation and results can be variable. Dann (1986) and Harte (1985b) found that responses were very variable, and could not be relied on in many cases. Dann found positive responses to deep ripping on 16 of 23 sites in southern New South Wales, and negative responses on seven sites showed moderate to large decreases in yield. Deep ripping was found to be detrimental to pastures, and was found to result in rapid drying of the soil profile in some cases. The option of improving compacted soils by plant root activity, minimal tillage and encouraging soil fauna activity may sometimes be the best option. Once soil pores have been formed by root and soil fauna activity, they should be kept intact by using no-tillage practices. A further complication is that nutrient levels may interact with plant responses to deep tillage, and that increased nutrient levels may be required to achieve yield increases (Osborne et al., 1978). The addition of lime with deep tillage gave large increases in yield in Victoria on soils with sodic subsoils (Ellington and Fung, 1984). Yield responses associated with deep tillage will depend on soil type, seasonal conditions, time of tillage, fertiliser history, crop and pasture type, and tillage orientation.

16.2.2 Conservation Farming in the Whole Farm Operation

An important factor in conservation tillage is that the practice must be approached as a whole farming system rather than just growing crops. This is particularly the case where grazing and farming systems are included in the farming enterprise. Particular aspects to be considered in a conservation farming system are outlined below.

Grasses should be controlled in the pasture phase for two years prior to the crop to ensure good weed and disease control for the cropping phase. This can be done by spray topping (this kills the seed in the grass and not the plant) and winter cleaning techniques. Ideally the amount of dry matter in the pasture phase should be maintained to at least 2 t/ha to give surface protection from raindrop impact and erosion, as well provide some protection from compaction by stock. This plant growth will also give sufficient root activity to develop soil structure and increase aggregate stability. An added advantage of a

good pasture phase, especially with perennial grasses, is the high level of water use, which can reduce the amount of recharge and help control the development of salinity. A period of five to 10 years for a pasture phase will maximise the benefits to the soil, but the pasture needs to be kept in good condition by well-managed grazing and weed control.

16.2.3 Future Trends in Farming Practice

New farming practices are being considered, and this is a dynamic area of change. Certainly the options for farming practices are now becoming wider as new technologies and ideas begin to be applied in the field. Developments are rapid, but the principles will still be relevant. Some of these new practices are listed below.

Controlled Traffic

The recognition of the effects of compaction on soils when they are wet or moist, particularly on those with higher clay contents, has led to the concept of controlled traffic farming. In this practice, paddocks are surveyed so that tramlines or traffic paths are marked out for machinery to follow. Ideally, sowing, spraying and harvesting are all carried out using these marked paths, and machinery is set up so that they can use these lanes or multiples of these paths. Sowing and spraying equipment is mainly being set up at this stage, but not headers, even though these are potentially the most damaging.

Potential benefits of this practice are less compaction of the soil, less overlap during sowing and spraying operations, less chemical use, and the potential for night spraying to be done. Large paddocks on low slopes are the best suited to this practice. There is also the potential for shielded spraying of crops with combinations of herbicides and/or insecticides.

Several basic design problems to be solved in implementing controlled traffic are controlling overland flow and erosion associated with the traffic paths, matching equipment wheel widths to the traffic paths (especially headers), and improving traction in wet conditions.

Raised Bed Farming

Waterlogging is frequently a problem in some soils, especially in the south, and this has been identified as a major constraint to crop growth. The coarse-structured clay surface soils, sodic clay soils, and soils that have a coarse-textured surface soil with a sodic subsoil are all affected by waterlogging. Beds are formed to raise the plants above the saturated zone of the soil, and the roots are less affected by waterlogging. Increased evapotranspiration, and biomass accumulation on the beds encourages higher levels of biological activity, increases organic matter levels, and leads to the development of macroporosity.

For dryland farming it is not economical to carry out the expensive land-planing and filling that is done for irrigation areas. However, surface draining raised bed schemes can be successful in dryland areas. Disadvantages of raised beds are the establishment costs, the maintenance of the raised beds throughout the cropping cycle, and the need to modify or purchase machinery to handle sowing and other operations. Raised beds are only useful for soils that have severe waterlogging problems. The advantages are less waterlogging, increased land value, improved timeliness of sowing, and better water use for production. It has been estimated that an area of at least 50 ha is required to give a net return on the costs (Bakker et al., 1998; Bakker and Hamilton, 1998).

Precision Farming

Precision farming combines the use of satellite imagery, infra-red photography, yield mapping using Global Positioning Systems (GPS), and crop monitoring to help farmers make decisions about growing their crops. It helps the farmer to monitor season and to plot maps of crop yield using GPS technology fitted to the harvester. The value of precision farming has not been fully evaluated in Australia but in the future it should enable the development of detailed yield maps of paddocks at a fine scale (paddock footprints), which can be used to determine different soil types, areas of soil chemical and physical constraints such as waterlogging, nutrient deficiencies and acidity. These may be clearly identified, as well as the potential damage of weed and disease to the crop.

The ultimate use of this technology will be to determine the constraints on a paddock at a fine scale, and use this information to apply variable rates of fertilisers and herbicides using GPS and computer-controlled application equipment. This should result in reduced costs, and will also be a gain for the environment.

Climate Modelling

Climate modelling is being used to predict long-term trends in rainfall patterns to assist in the selection of long-term management strategies, and also to assess the results of short-term experimental trials. The major model being used is Agricultural Production Systems Simulator (APSIM), developed in Queensland. The idea is based on rainfall being the main driving force in dryland agriculture. The models look at soil water balance, solute transport, soil nitrogen and the effects of tillage and surface residue. Ideally they can predict long-term crop yields.

Other Cropping Issues

Other issues that may concern farmers and the community in the future include organic farming, biotechnology and expansion of cropping into locations such as the tablelands. Organic farming has not been widely adopted in Australia because of the problems of soils with low nutrients, difficulty of weed control, and lower yields. However, there are a range of farmers who are trying to apply this farming system to dryland farming in New South Wales.

The most immediate use of biotechnology is the development of herbicide-resistant varieties of canola. Ideally, biotechnology will provide some plant varieties that are more efficient water users that will more actively compete against weeds, reducing the need for herbicides and pesticides. The potential cash flow from cropping has led to a possible expansion of cropping into new areas such as the tablelands and in more marginal land further west. Crop varieties for this are available, but the potential environmental consequences from increased soil erosion, structural and nutrient decline need to evaluated and avoided.

16.3 The Effects of Dryland Cropping on Soil Degradation

A wide range of research has shown that increasing tillage reduces aggregate stability and surface infiltration, and increases bulk density (Clarke and Russell, 1977; Hamblin, 1987; and Mullins et al., 1990). The soil physical properties discussed in this section are described in more detail in Chapter 10.

While the immediate effects of tillage on structure appear favourable, the longer-term effects of frequent tillage on soil structure are usually not so. In fact, despite the obvious loosening of some of the soil, the long-term negative effects may be:

(i)   reduction in the overall stability of soil aggregates due to the severing of bonds between aggregates and soil particles, oxidation and loss of organic matter, and pulverisation of aggregates, breaking them into individual soil particles

(ii)   reduction in the effectiveness of existing macroporosity by cutting pores off from the surface or by destroying them

(iii) development of smearing and compaction layers, especially when the soil is too wet for tillage, or tillage tools are incorrectly set

(iv)   removal of surface cover exposing the soil to raindrop impact

(v)   reduction in the amount of biological activity in the soil.

The beneficial effects of pasture on soil physical and chemical properties have often been reported. Clarke and Russell (1977) record that pastures (ryegrass/clover) increase aggregate stability in proportion to the length of the pasture phase. Long-term pasture (10-year phalaris) had the highest aggregate stability improvement.

Tisdall and Oades (1979, 1980a, 1980b, 1982), in a series of papers, showed that aggregate stability of red-brown earths is influenced by crop rotations and the amount of pasture growth. They showed that pasture growth (ryegrass) increased aggregate stability, especially the stability of aggregates larger than 0.25 mm. They concluded that actively growing pasture is necessary for the stability of these larger aggregates. Tisdall (1991) also confirmed that fungal hyphae, which can grow in soils with a high level of active roots or high organic matter, can improve soil structure.

Hamblin (1980) reported that the increase in aggregate stability is slower under direct drilling than under pasture, and in referring to a paper by Tisdall and Oades (1980b), concluded that the most rapid method to build up aggregate stability is frequent watering and clipping of pastures. Plants that were mainly in a wilted condition or cut before reaching a certain size contributed little to aggregate stability.

Land use practices that involve frequent tillage, low vegetative production or low cover levels are always likely to increase erosion. This is especially the case for the 'vulnerable soils', which show rapid structure degradation with losses of organic matter and do not have sufficient clay activity to regenerate soil structure (Lal, 1986; El-Swaify et al., 1985). These soils have a high potential for erosion as shown by Edwards (1987), Murphy and Flewin (1993) and Hairsine et al. (1993). As Murphy and Flewin explain, this potential is extremely high should an erosion event occur when soils are in a poor physical and cultivated condition. In these circumstances surface roughness is lost rapidly as the soil settles

because of instability to wetting. The surface seal greatly increases runoff but below the surface seal, the soil may still be in a loose, easily detached and entrained condition. This combination of high runoff and high detachability and entrainment can lead to large erosion losses. Harte (1984) and Packer et al. (1984) showed that after a period of no-tillage or direct drilling practices (three to five years), runoff and erosion were significantly less than that from traditional tillage practices. Soils under direct drill practices are less erodible than those under traditional tillage (Packer et al., 1992).

The adoption of practices that reduce the amount of tillage, and maintain cover on the soil surface, aims to prevent surface crusting and sealing, erosion and soil structure decline. These practices are now generally referred to as conservation tillage. The development of these practices is described by McNeill and Aveyard (1978), Charman (1985), Poole (1987) and Hamblin (1987). The details of these earlier practices have been significantly modified over the years to overcome problems of weed control, disease, poor seedbed conditions, slow early growth and the need for specialised machinery. However the basic principles of reducing the amount of tillage and soil disturbance, maximising plant growth and maintaining surface cover still form the basis for conservation farming practices.

Changes in Soil Organic Matter   16.3.1

Cultivation depletes the soil organic matter store by increasing its oxidation as well as by exposing inaccessible organic matter to organisms that decompose it. Soil organic matter is also lost by erosion. Cultivation during fallowing removes all plant growth and prevents the replacement of organic matter by further plant growth. Fallowing periods in the northern wheat-belt last as long as the period between crops, which may be as long as six to 18 months. For a red-brown earth in northern New South Wales, 15 consecutive years of cropping reduced soil organic matter levels in the top 20 cm of soil by 41% (Harte, 1984).

Where stubble is completely removed by burning, significant declines in the soil's organic matter reserve have been demonstrated (Loch and Coughlan, 1984; Marston, 1978). Where residues are burnt, organic matter levels are usually higher in untilled soils than in tilled soils. This infers that tillage alone can reduce organic carbon levels.

The adoption of no-tillage cropping featuring full stubble retention has in some cases shown a reversal of the losses of organic matter. Conservation tillage trials on a red-brown earth at Cowra have shown significant increases in organic matter when compared to traditional tillage. However, for the soils in the northern wheat-belt the adoption of no-tillage systems does not seem to have had the same effect. For a range of soils there, including self-mulching black earths, three to five years of no-tillage has produced only slight increases in organic carbon (Harte and Marschke, 1985). This apparent disparity between northern and southern cropping systems with respect to their organic matter response under no-tillage may be related to differences in soil texture and climate. Northern soils are finer textured and so provide more sites to protect organic carbon from breakdown due to cultivation. Also it is more difficult to increase organic matter in the north because more moisture in summer results in higher breakdown rates of organic matter in the soil.

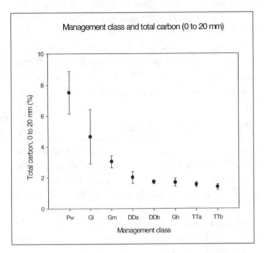

Figure 16.1a   Effect of land management practices on total carbon in the 0–20mm layer for 75 sites in the dryland cropping area of south eastern Australia

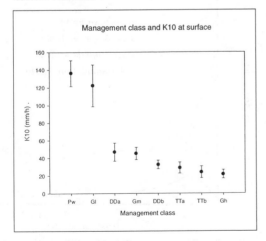

Figure 16.1b   Effect of land management practices on surface infiltration for 75 sites in the dryland cropping area of south eastern Australia. Surface infiltration measured at 10mm tension

Hamblin (1980) reports an increase in organic carbon in four out of five soils studied nationwide, after three to eight years of no-tillage cropping. However, these changes in organic carbon were not large, nor highly correlated with the duration of the treatment. Hamblin summarises by commenting that increases in soil carbon and structural stability resulting from no-tillage practices in Australian soils will be slow in the early years, largely because the Australian climate is not conducive to the retention or rapid build-up of organic matter in the soil.

The loss of soil organic matter as tillage increases has been shown by Geeves et al. (1995), and their data has been graphed in Figure 16.1a. The change in the level of soil organic matter shows a general decrease as the land management systems grade from woodland through to traditional tillage with stubble burning and numerous cultivations. The apparent rapid reduction in organic matter with the introduction of a cropping system or heavy grazing is noteworthy. Note also that pastures do not automatically result in high organic matter levels, as the direct drill cropping systems are shown to have higher organic matter levels than the pastures under heavy grazing.

It has recently been suggested that we should be looking at labile carbon, which would provide a more sensitive measurement for assessing the sustainability of agricultural systems (Blair et al., 1995). Labile carbon does not include the large amount of inert organic matter like charcoal, which does not change under different land management practices.

Soil Physical Properties   16.3.2

The change in soil physical properties under tillage and grazing studied by Geeves et al. (1995) is shown in Figure 16.1b. The farming systems are as for Figure 16.1a. Note the rapid reduction in infiltration with the introduction of any cropping system, even for direct drilling. Note also that pastures do not automatically result in higher levels of infiltration, as the direct drill cropping systems are shown to have higher infiltration levels than the pastures under heavy grazing, which have the lowest average values. The reason for this can be explained by a combination of the high compactive force applied by stock and the low level of plant growth under heavy grazing. Plant growth can provide cover to protect the soil against raindrop impact, and lead to the redevelopment of soil structure.

In Figure 16.2 the chance of the hydraulic conductivity being less than 10 mm/h is shown for each of the farming systems. It is clear that the traditional tillage with burning, which is the most degrading system, has the highest chance of a hydraulic conductivity less than 10 mm/h.

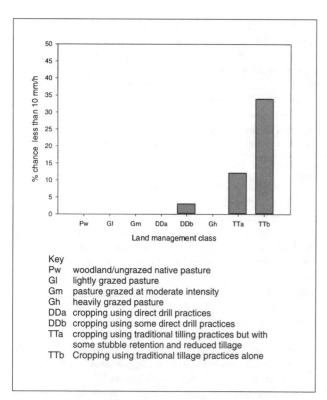

Key
Pw   woodland/ungrazed native pasture
Gl   lightly grazed pasture
Gm   pasture grazed at moderate intensity
Gh   heavily grazed pasture
DDa   cropping using direct drill practices
DDb   cropping using some direct drill practices
TTa   cropping using traditional tilling practices but with
      some stubble retention and reduced tillage
TTb   Cropping using traditional tillage practices alone

Figure 16.2 Effect of land management practices on the likelihood of surface infiltration of less than 10mm/h. Surface infiltration was measured at 10mm tension

Several other instances where the effects on soils of traditional tillage are compared to direct drill systems are reviewed by Hamblin (1987). Moran et al. (1988) used micromorphology to show the large changes in soil structure under the different cropping systems, and Murphy et al. (1993), who showed large changes in infiltration under the different farming systems, also demonstrated considerable seasonal variability in infiltration. Macks et al. (1996) demonstrated changes in soil friability and soil organic matter under different cropping systems, Chan and Mead (1988, 1989, 1992) also showed changes in the physical properties of soils under the different cropping systems, and Chan et al. (1992) showed changes in organic carbon under different tillage systems.

All these studies demonstrated that relative to direct drill systems, soils under traditional systems had one or more of the following: lower organic matter, lower aggregate stability, lower friability, lower infiltration, lower macroporosity, or higher runoff. Carter and Steed (1992) showed an increased water penetration in soils with a history of direct drilling compared to those under traditional tillage. An increase in worm activity under direct drilling was shown by Haines and Uren (1990).

The aggregate stability of the soil is a measure of the soil structure, as the higher the aggregate stability, the better the soil structure (see Chapter 10). The amount of aggregate stability decreases with time that tillage has been applied to the soil, according to Harte (1984). His data show that for cropping soils in northern New South Wales there is a significant reduction in water-stable aggregation, especially during the first eight years of tillage (see Figure 16.3). The reduction is most pronounced in the larger 2 to 6 mm size aggregates, while there is only a very small change in the amount of water-stable aggregates in the 0.5 to 0.25 mm size range. This loss in aggregate stability is very much associated with corresponding losses in organic matter, which are also demonstrated in Figure 16.1b. The effect of tillage on aggregate stability for different soil types is also shown in Figure 16.4. The biggest effect appears to be on the black self-mulching clays, but these clays have a greater capacity to redevelop soil structure through wetting and drying cycles, while the red-brown earths require plant and root growth to redevelop soil structure.

Runoff and Soil Loss

The degree to which cropping practices determine runoff and soil loss depends on a combination of factors such as soil type, slope, surface roughness, surface cover and the physical state of the soil when it rains. A summary of some studies showing this is provided in Packer et al. (1992) and in Plate 16.3, which shows how deep water penetrated into the soil under each tillage system.

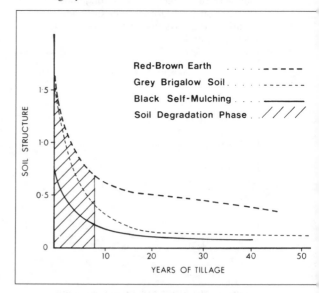

Figure 16.3 Proposed relationship between soil structure (measured as aggregate stability) and period of tillage

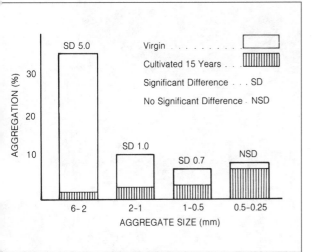

Figure 16.4 Effect of tillage on the size ranges of water-stable aggregates for a red-brown earth

Table 16.1 Differences in soil loss under different tillage practices (after Packer et al., 1984)

Location	Tillage practice	Soil loss (kg/ha)
Ginninderra		
	Direct drill	15
	Reduced tillage	200
	Traditional tillage	290
Wagga		
	Direct drill	190
	Reduced tillage	300
	Traditional tillage	400

Note that these soil losses are based on results from a rainfall simulator with a one square metre plot, and will not necessarily represent the order of magnitude of soil losses in a paddock under a storm.

The amounts of infiltration under different cropping systems on the fragile sandy loams ranged from 4 to 10 mm/h using simulated rainfall. The reduction in runoff under the direct drill system is obvious (see Figure 16.5).

The reduction in erosion under direct drill cropping systems compared to traditional tillage systems is also shown by Packer et al. (1984, 1992) (see Table 16.1).

A significant factor in these results is that measurements were made on a settled seedbed after har-

vest. Soil loss can be expected to be less from a settled seedbed, as the soil detachability and susceptibility to entrainment are much less.

Studies of two major erosion events that occurred at Cowra in autumn, on cultivated soils, indicate that differences in soil loss between direct drill and traditional tillage systems are even greater, given that soils under the traditional tillage system are in a loose cultivated condition, while under the direct system the soil is in a more stable, settled condition. Hairsine et al. (1993) showed that erosion losses for a major

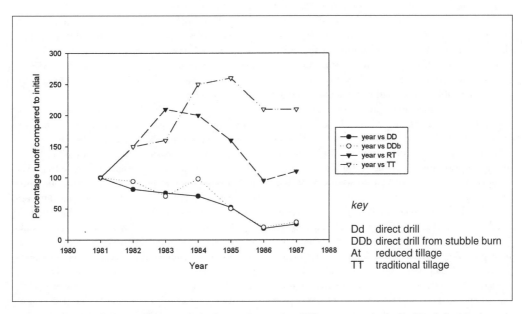

Figure 16.5 Trend in runoff with years of cropping under different cropping practices (after Packer et al., 1992)

Plate 16.3 The difference in infiltration between direct drilling (left) and traditional tillage after 30mm of rain

storm of 81 mm in 45 minutes giving an average intensity of 108 mm/h, with short periods over 300 mm/h, were 342 t/ha for the traditional tillage system compared with 65 t/ha for the direct drill system. The difference in soil loss between the two systems is demonstrated in Plate 16.4, with an obvious step between the two cropping systems. The high erodibility of the soil when it is in a cultivated condition has been demonstrated by Murphy and Flewin (1993), who showed the very high soil erodibility of the cultivated soil. Soil loss, even under relatively low intensity rainfall, was 78 t/ha. They concluded that the high erodibility of the cultivated soil was a consequence of the low aggregate stability of the soil and its tendency to readily form a crust at the surface which produced high amounts of runoff. This combined with the high detachability and low energy of entrainment of the soil under the crust resulted in rapid rill formation and a very high erosion rate.

On a black, self-mulching clay in northern New South Wales, Marston (1978) provided some of the early comparisons of the effect of direct drilling cropping practices on runoff and soil loss (see Figure 16.6). Long fallow with stubble burn had the highest amounts of runoff and soil loss, and short fallow stubble incorporated had the least erosion of the cropping systems. Sown pasture had the lowest runoff and soil loss.

For a similar soil, Harte (1985a) found that the soil structure developed under no-tillage management practice reduced runoff by 30% and soil loss by 50% during 30 minutes of simulated rainfall. The intensity of the simulated rainfall was 90 mm/h. Retention of stubble to give 50% surface cover on the no-tillage treatment reduced runoff and soil loss to negligible levels.

At Gunnedah, on another self-mulching clay, Marschke and Thompson (1983) found that during a single natural storm on an already saturated soil profile, bays with no-tillage cropping practices yielded 4% of the total runoff and 2% of the soil loss of that for bare cultivated bays.

As soil conservation banks were a feature of the areas studied in these experiments, and are commonly used to control erosion, an important conclusion from these erosion studies is that contour banks alone are insufficient to prevent erosion. In fact Wylie (1993) suggests that the presence of contour banks can make farmers complacent about the need to control erosion using cropping practices. One result of significant amounts of erosion between contour banks is that farmers can quickly be faced with the costly operation of cleaning out the banks of deposited sediment. The message is that erosion control requires a range of practices and structures to be effective.

Studies by Lang (1979) and Wylie (1993) both showed that about 70% surface cover is required to prevent serious erosion.

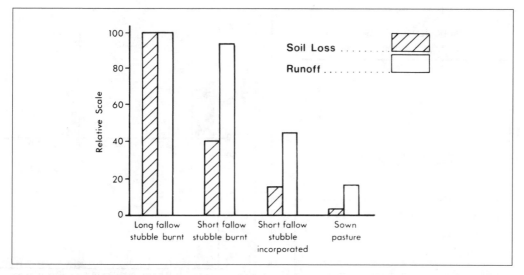

Figure 16.6 Runoff and soil loss data from stubble burnt and stubble incorporated fallows on a black earth at Gunnedah in north western NSW (after Marston, 1978)

6.3.4 Soil Compaction

Increases in bulk density under traditional tillage systems are demonstrated by Harte (1984). The changes in bulk density were measured after harvest in a settled seedbed. The effect of tillage on soil bulk density can be assessed by comparing cultivated sites to adjacent sites that are in an uncultivated, woodland or virgin condition. In northern New South Wales increases in bulk density for the top 20 cm have been recorded after relatively short periods of cultivation in red-brown earths or red earths, which have fragile or hardsetting surface soils. These increases are shown in Figure 16.7 and are based on Harte (1984). The increases in bulk density are attributed to compaction by machinery and cultivators, and to the reduced aggregate stability.

Modern systems of cropping have seen an increase in the mass, power and traction requirements of machinery. The average weight of tractors in the USA has increased from 2.7 tonnes in 1948 to 4.5 tonnes in 1968. During this period, there has also been an average power increase of 5–7% per year (McKibbon, 1971). Hakansson (1994) estimated that the average weight of tractors in the USA in 1968 was 45 kN, but by 1985 had increased to 67 kN, with some large four-wheel-drive units reaching 220 kN. One response to this was to increase tyre size to reduce the pressure on the soil or to have four tyres. Note that 1 kN is the force exerted by 0.102 tonnes at the earth's surface. Therefore a tractor exerting a force of 220 kN at the earth's surface weighs about 22 tonnes.

The total coverage of tractor and plant wheels over the soil surface during a crop preparation period can be surprisingly high. A traditional tillage system will cover 82%, a no-till system 46% and a controlled traffic system 14% of the paddock in one year (Walsh, 1998). Given that these operations are often at a time when the soil is in an optimum condition for tillage, being slightly moist and probably soft, with a low resistance to compaction, there is a large potential for compaction to occur. This has led to interest in controlled traffic systems for cropping.

Another form of soil compaction often occurs in soils because of low aggregate stability associated with low organic matter levels and/or high levels of sodium. Under the stress of raindrop impact or wetting to saturation, the soil aggregates disperse or slake, and the soil collapses to a consolidated layer to form a hard surface crust on drying or a dense massive surface soil. Cotton farmers have been some of the first to recognise the problem of compaction (Mortiss and McGarry, 1993).

Changing Soil Structure 16.3.5

The fibrous root systems of certain grasses tend to be more efficient binders of fine soil particles than are legumes. Grasses have many fine roots that have a better influence on soil aggregation than fewer larger roots. Just three years of pasture (grasses and medics) on a previously well-cultivated and degraded sandy clay loam in northern New South Wales was sufficient to significantly improve the soil's aggregate stability (Figure 16.8). The grasses were more effective than legumes and the most significant improvements occurred in the 2 to 6 mm size range of aggregates.

Pasture leys in old tillage country have the potential to dramatically improve the number of water-stable aggregates (see Tisdall and Oades, 1980). This

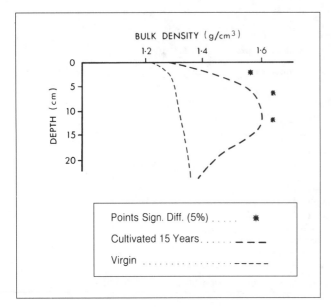

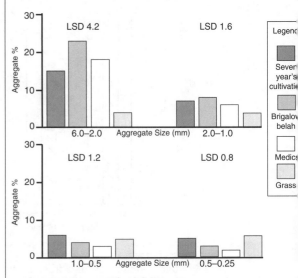

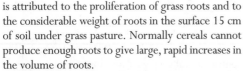

Figure 16.7 Effect of tillage on the bulk density of a red-brown earth in northern NSW (after Harte, 1984)

Figure 16.8 Effect of different land management practices on aggregate stability, showing relative effects of grass and medics

is attributed to the proliferation of grass roots and to the considerable weight of roots in the surface 15 cm of soil under grass pasture. Normally cereals cannot produce enough roots to give large, rapid increases in the volume of roots.

Exudates from plants can also affect aggregate stability (Chan and Heenan, 1991). They showed that canola roots increased aggregate stability considerably and they attributed this to the exudates that are produced by the canola plant.

Improvements in soil structure under plant growth may also be a consequence of shrinkage of soil aggregates as the soil dries out and from the development of pores or channels when old roots decay. Soil fauna activity such as that of earthworms, dung beetles and ants can also redevelop structure in massive soils, but usually need some plant regrowth or cover to maximise their activity (Lee and Foster, 1991).

Some soils redevelop soil structure naturally with wetting and drying cycles. The most notable of these are the self-mulching soils which are in the north of the state.

16.3.6 Consequences of Soil Structure Degradation

Soil Erosion

As discussed earlier, areas of land under traditional tillage systems are likely to have much higher erosion rates than areas under conservation tillage systems.

This erosion can have severe effects on soil quality in the long term, especially on shallow soils and texture contrast soils with subsoils that are unfavourable to plant growth.

The environmental consequences of erosion are also of importance, as discussed in Chapter 3. Consequences include sedimentation on roads and/or washing away of roads, which can cost many millions of dollars. Sedimentation of water storages is also a potential cost of erosion. Erosion can also lead to the deterioration of water quality as sediment is washed into rivers and water storages. The sediment can take nutrients into the rivers and storages leading to development of blue-green algae and other water quality problems, or it can carry agricultural chemicals into water bodies creating further deterioration of water quality.

Crop Yields

It is not always easy to demonstrate reductions in yield due to soil structure degradation because yield effects can be very seasonally dependent (Hamblin, 1991; White, 1988). Causes of loss of yield include loss of useable water from rainfall because of low surface infiltration, poor seedbed conditions which reduce seedling establishment, poor aeration because of high bulk densities and waterlogging, high soil strength restricting the distribution of root growth, and poor trafficability of soils following wet periods which affects the timeliness of sowing, weed spraying and harvest.

The potential for conservation cropping practices to give equivalent yields was demonstrated by Murphy et al. (1989), but in some circumstances direct drilling resulted in lower yields than traditional tillage because of poor weed control, poor seedbeds, lack of tine penetration with the sowing equipment in a hardsetting soil, and lack of a fallowing period in a low-rainfall year.

Wylie (1993) found that there was a 5–20% increase in yield in southern Queensland under zero-tillage, and this was attributed to better moisture storage under these cropping systems. Steed et al. (1993) suggested that yields increased after three to four years of direct drilling in northern Victoria. A major improvement under direct drilling cropping practices was the improved timeliness of sowing and weed control because of better trafficability of the soils. Hamblin (1987) also noted a general improvement in yields in soils under direct drilling practices after a period of time.

However, yield reductions have commonly been observed owing to difficulties with weed control and plant diseases under direct drilling practices. Difficulties with seedbeds and inappropriate sowing equipment have also resulted in lower yields, and is sometimes related to paddock and soil selection. Severely degraded paddocks with hardsetting surface soils, or paddocks with severe weed or disease problems, are not the best options for commencing a conversion to direct drilling or no-tillage practices.

Greenhouse Gases

The reduction in organic matter in soils as a consequence of tillage is one factor contributing to greenhouse gas emissions. One hectare of soil can hold 7 to 12 tonnes of carbon, which is equivalent to 25 to 40 tonnes of carbon dioxide. Changing land use from woodland or pasture can result in the emission of up to 6 t/ha of carbon dioxide. Conversely, changing land use from traditional tillage farming systems to low-grazing pasture or woodland can result in the take-up (sequestration) of 4 t/ha of carbon, which is equivalent to about 12 t/ha of carbon dioxide. Estimates indicate that changing from a traditional tillage system to a direct drilling system could sequester about 0.7 t/ha of carbon, which is equivalent to about 2.5 t/ha of carbon dioxide. However, the changes in carbon levels between soils under direct drilling practices in comparison to traditional cropping practices can be very variable, and depend on the amount of plant growth in the period before the crop, the amount of biomass in the crop sown by direct drilling, and the rate of breakdown of organic matter in the soil. Further work is required to accurately predict the amounts of carbon changes under different land management practices.

16.4 Management of Soil Structure

Soil structure has two key roles in sustainable land use:
—ensuring the productivity of plants
—maintenance of resources, ecosystems and the environment.

For plant productivity in agriculture, the function of soil structure is to provide plants with water, air, space for roots to grow, shelter for the germination of seeds and the establishment of plants, and a trafficable base for machinery operations and stock.

For the maintenance of resources, ecosystems and the environment, the role of soil structure is to accept and retain water to prevent runoff and deep drainage, minimise the dispersal of nutrients and agricultural chemicals, minimise the dispersal of industrial chemicals where soils are used for disposal of liquid waste, maintain a resistance to the entrainment of soil particles by flowing water, and maintain a resistance to wind erosion.

A degraded soil no longer has a structural condition that can perform these functions adequately. Examples are soils that have a surface crust that prevents infiltration and emergence of seedlings; soils with a compacted layer or plough-pan below the surface that inhibits water transmission and root growth; and soils with a hard, dense subsoil that restricts root distribution.

Depth Zones for Soil Structure Management   16.4.1

To describe the structural form of a soil profile, it is necessary to consider the structural form of soil material at different depth zones within the profile (Kay et al., 1994). The depth zones designated here follow those outlined by Northcote (1983). They also relate to those defined by McGuiness (1991) and Murphy and Allworth (1991). It is necessary to consider these depth zones in the soil profile when evaluating the impact of soil structural degradation and in planning management strategies for ameliorating it.

The Soil Surface

This is the air/soil interface and it is characterised by observations on the soil surface, by measurements at the soil surface (for example, infiltration, penetrometer resistance), or by soil measurements on the 0–20 mm layer (for example, organic carbon, exchangeable sodium percentage, aggregate stability and electrical conductivity). This is a critical zone of soil structure as it may severely restrict infiltration and/or plant emergence and growth if it is in poor condition. Poesen and Nearing (1993) describe the importance of the soil surface:

The boundary between soil and atmosphere is an important and active place. It controls to a large extent the flux of nutrients, water, gases and heat to and from the underlying soil. So it has a direct and dramatic influence on the conditions of the soil below the interface, the microbes, animals and plants that live in the soil, and the amount of surface runoff generated from excess rainfall.

A poor structural condition at the surface is indicated by the presence of surface crusting and by the absence of a vegetative cover at the soil surface. Reduced infiltration can greatly increase runoff under rainfall or irrigation with an adverse effect on the use made of rainfall and irrigation by plants, as well as leading to increased erosion rates. The presence of a strong crust may also inhibit or prevent the emergence of seedlings. The surface is usually the least costly to ameliorate by the application of chemical ameliorants such as gypsum or lime, or perhaps temporarily by tillage.

The Subsurface

The subsurface is defined as either the depth of ploughing, usually about 80–100 mm, or the A_2 horizon, which is paler in colour than the surface and the subsoil and usually more poorly structured. This can be a critical zone of soil structure if it is:
—severely compacted, layered or affected by smearing
—strongly cemented by secondary minerals of silicon, iron and manganese, and/or
—strongly weathered or bleached, often giving an adverse chemical environment for plant growth.

The cost of ameliorating this zone is more than that required for the surface soil but in some cases may be possible and economically justified.

The Subsoil

The subsoil or B horizon is beneath the A horizon and is usually finer in texture (more clayey) and more strongly structured and coloured. In dryland agriculture there are not many options for ameliorating the subsoil conditions other than by biological means such as growing specific plant species, encouraging the activity of soil fauna and controlling traffic. Under intensive agricultural production there may be some economic management options to ameliorate the soil structural form of the subsoil, by deep tillage or specific practices such as gypsum slotting (Jayawardane and Chan, 1994; Malinda, 1995).

Specific Functions of Soil Structure 16.4.2

In the growing of commercial farm crops the structure of the soil is vital for the transmission and retention of air and water to allow plants to germinate and grow, and for the provision of pore space and anchorage to allow roots to develop. It also relates strongly to the ability of the soil to support the machinery necessary for tillage, sowing, crop management and harvesting. This is termed the soil's trafficability.

Water Movement and Retention

The structure of the soil largely controls the rate at which water infiltrates into the soil surface, the rate at which it flows through the soil, and the amount of water retained in the soil for plant growth. Hence it affects how rainfall or water added to the soil in irrigation is partitioned between runoff, which causes erosion, water retained in the soil for plant growth, and water that flows through the soil into the water table or back into watercourses. How rapidly the water flows through the soil influences the amount of time that effluent, chemicals and nutrients have to interact with the soil particles, and so how much filtering activity the soil can provide to restrict the movement of these potential pollutants.

Root Growth

The non-limiting water range (NLWR) of Letey (1985) shows how soil structure affects root growth. This concept was expanded by Jayawadene and Chan (1994) and da Silva and Kay (1997a, b). The NLWR is based on the relationship between moisture content and aeration and the relationship between moisture and soil strength. At the dry end of the NLWR, soil strength limits root growth and at the wet end, aeration limits root growth. The difference between these two defines the NLWR, which for a soil with good structure may be quite wide, but for a soil with poor structure (with a clayey, sodic subsoil) may be zero.

The maximum soil bulk densities which permit root penetration are generally thought to be in the range of 1.5 g/cm³ for clayey soils and 1.8 g/cm³ for sandy soils. It is likely that root penetration will be severely restricted above a level of 1.9 g/cm³ in any soil. However, bulk density is only a guide to root growth, and does not take into account the continuity and distribution of pores or planes of weakness or cracks in the soil. Roots can grow into these parts of the soil, but if the bulk density becomes high, the volume of the soil that can be utilised by the roots becomes increasingly small.

The size of soil structural units can have a large effect on the root distribution in the soil, as was noted by Greacen in 1977. In coarse-structured soils, roots are often restricted to channels and cracks. Whether a root can enter a ped is determined by the mechanical state of the soil material in the ped, and in many coarse-structured soils the mechanical state of the soil material will prevent the entry of roots. As a result the coarse-structured soils often have a limited volume of soil available as sources for water and nutrients.

Soil Trafficability

Trafficability of soils has been considered a critical factor in land use, particularly in some environments and on some soil types (Spoor, 1979) where it affects the timeliness of most cropping operations. Soil structure is again important, particularly in more clayey soils, and under wet conditions. The soil properties that control trafficability such as texture, structure, strength and penetration resistance are discussed fully in Chapter 10.

6.4.3 ## Cropping Systems and Soil Structure

The effect of cropping practices and farming systems on soil structure can be estimated by considering the level of activity of the agents causing change in soil structure, as in Table 16.2. Some farming systems will have a high activity of agents leading to the development of soil structure (for example, low-intensity grazing), others will have a very high activity of agent causing soil structure degradation (for example, traditional tillage). Other farming systems will have a wide range of activities of agents (for example, direct drilling). For example, a paddock under a direct drill cropping system may be subjected to a high amount of stress from raindrop impact and compactive force. This occurs when the stubble from the previous crop is heavily grazed, and any weed growth is strictly controlled using herbicides. The soil surface is likely to be bare in the period before sowing. If the crop grows poorly, there is little plant growth to develop soil structure. Of course if a direct drill paddock has a good cover of stubble and then grows a crop with a high biomass, the stress on the soil structure will be low and the plant growth will improve the structure.

In a heavily grazed pasture there is a high compactive force on the soil and only a small amount of plant growth, so the soil structure cannot be expected to improve greatly.

Strategic tillage refers to a carefully timed and applied operation, often to break up a massive or compacted soil layer. Its net effect should be to improve soil structure. However, in general cultivation depletes the store of organic matter in the soil, also breaking up soil aggregates.

The general level of activity of mechanisms causing changes in soil structure within a range of land management systems is shown in Table 16.2. The expected order of the amount of soil structure degradation and loss of soil carbon can be predicted from this table. The general order predicted is:

Irrigation > Traditional Tillage
> Reduced Tillage > Direct Drill > Pasture

The traditional system is most likely to place a high stress on soil structure, both from tillage and from exposure to raindrop impact, and so is most likely to have a low potential to develop soil structure by plant growth. However this order may vary depending on the relative activities of the agents causing soil structure decline, and the development of soil structure within each farming system. As discussed earlier (see Figure 16.1), it is possible for a soil under pasture or direct drill to be more structurally degraded than a soil under a well-managed traditional cropping system. This will occur if the direct drill and the pasture systems have had to sustain large effects of raindrop impact and compaction, and have only a small amount of plant growth. Certainly under irrigation, the potential for the large production of biomass may mean that these soils are not as structurally degraded as expected. Any soil management system needs to balance the stresses on soil structure with the potential for soil structure development. In some circumstances, direct drilling may not be the optimum system to maintain it.

Soil Types in Relation to Soil Structure Management 16.4.4

Land managers at meetings and field days frequently comment that soil type has a large influence on deciding what they feel are the 'best' or most effective land management options. Many times it is stated that land management practices that work well for one soil are not successful on another soil. A simple grouping of soils useful for land management practices can be based on the following:

Table 16.2 Causes of changes in soil structure and the time for them to cause these changes

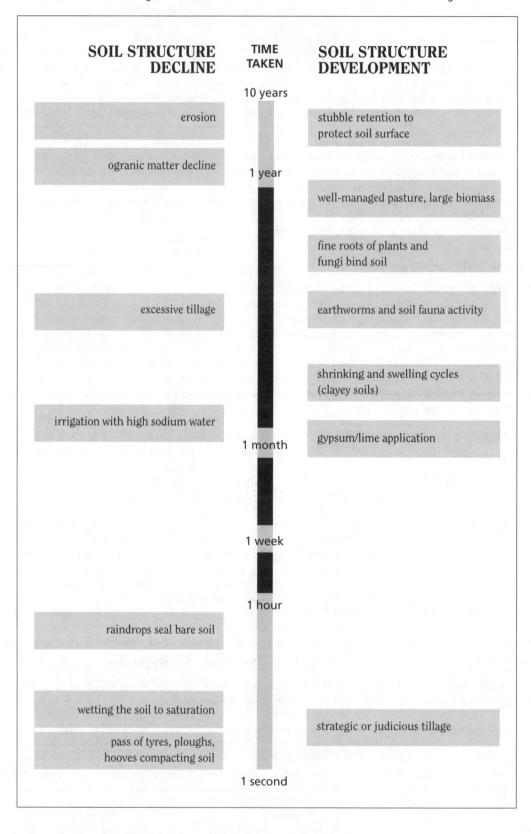

(i)   mechanisms operating to change soil structure in a given soil type
(ii) the effectiveness of the mechanisms based on the activity of the agents causing changes in soil structure and the response of the soil to this activity
(iii) the properties of the soil at various degrees of structure degradation and in particular when the soil is in the most structurally degraded condition possible.

El-Swaify et al. (1985) identify a group of vulnerable soils and say they are characterised by: surface layers enriched with coarse material; low contents of fine material especially at the surface; inactivity of clay minerals; stabilisation of only low amounts of organic carbon; and increasing clay content with depth. These soils are very susceptible to surface sealing.

General soil classification does not relate specifically to surface soil properties, but some general trends are apparent (Murphy and Lawrie, 1996). These trends include:

(i)   hardsetting is most commonly associated with red-brown earths, red earths and non-calcic brown soils (Red Chromosols and Red Dermosols)
(ii) self-mulching surfaces are usually associated with black earths and some grey and red clays (some self-mulching Vertosols)
(iii) friable surfaces are most commonly associated with the krasnozems, euchrozems and prairie soils (Red Ferrosols, Red Dermosols and Black Dermosols)
(iv) clay surfaces that do not self-mulch are usually associated with the grey, brown and red clays (some Vertosols)
(v)   sodic surfaces occur in several great soil groups including the red-brown earths (Red Sodosols) and the grey, brown and red clays (some Vertosols).

The proposed soil groupings are described below with the reasons for choosing these soil types explained. The soil types outlined in this section are considered useful for the management of soil structure and are based largely on groups defined in the literature.

Fragile Surface Soils

The fragile surface soils have low clay activity and usually have weakly pedal or apedal massive to single grain structure (see McDonald et al., 1994, for definitions of structure terms). They are somewhat analogous to the 'vulnerable' soils defined by El-Swaify et al. (1985) and Lal (1986) as having low clay activity, and they have low structure stability and resilience (Kay et al., 1994). Butler's (1955)

description of friability including pulverescence and coalescence can be used to identify these soils, as they tend to be 'brittle' when dry. It is useful to divide these soils into coarse and medium texture groups, which is justified on the basis of the different properties of medium- and lighter-textured soils. The coarser-textured soils include sands, loamy sands, sandy loams and fine sandy loams, while the medium-textured soils include loams, clay loams and sandy clay loams. This is especially the case when the soils are in a structurally degraded condition with the medium-textured soil having higher strengths and poorer aeration and lower hydraulic conductivity. The medium-textured soils (loams and clay loams) are generally more difficult soils to manage (Spoor, 1979). Oades (1993, Table 2) also recognises the need to separate sandy soils (< 15% clay) from loamy soils (15–35% clay). He bases this on the fact that the structure of loamy soils is influenced by shrink-swell capacity and the abiotic development of aggregation, whereas shrink-swell has little influence on the sandy soils. The aggregation and development of structure in sandy soils is almost solely influenced by biotic processes and the level of organic matter. The surface layers of red-brown earths, often with high silt and fine sand, are particularly prone to structural breakdown when cultivated (Greenland, 1971).

Friable Surface Soils

The friable surface soils are usually clay loams or light clays with moderate to strong pedality, usually showing fine or moderate-sized polyhedral structure. They are frequently associated with high levels of aluminium and iron sesquioxides, and are usually derived from basic to intermediate parent materials such as basalts, diorites and andesites and their associated tuffaceous rocks and sediments. Soil types to which they are commonly linked include the krasnozem and euchrozem great soil groups (Stace et al., 1968) and the Oxisol and Dermosol soil orders (Isbell, 1996).

Self-Mulching Surface Soils

The self-mulching surface soils are a discrete group of surface soils showing a natural tendency to form a loose surface made from small (< 10 mm) subangular to angular blocky peds at the surface. This gives the surface soils many advantageous physical properties for tillage, sowing and emergence of crops. In particular, the tendency for the soil to reform this structure following wetting and drying cycles shows that the soils have very high structure resilience. They are associated with soils having high clay contents (> 35%), active clays such as smectites, low levels of exchangeable sodium and high levels of exchangeable calcium. Some of the processes of self-mulching are described by Grant and Blackmore (1991) and Pillai-McGarry and Collis-George (1990a, b). These kinds of surface soils are confined to the black earth and less frequently the red clay and grey clay great soil groups (Stace et al., 1968). In the Australian Soil Classification (Isbell, 1996) they are confined to the Vertosols.

Sodic Surface Soils

Sodic surface soils are defined as having an exchangeable sodium percentage (ESP) greater than 5–6% (Northcote and Skene, 1972; McKenzie et al., 1993). Once ESP levels reach 5–6% the structural stability of the soil to raindrop impact and wetting becomes severely affected, resulting in severe surface crusting, very low infiltration rates, severe cloddiness and poor workability, high soil strength and poor aeration. The sodic soils cause these problems because the high sodium levels make the soils very unstable and susceptible to dispersion on wetting (see Emerson, 1991). Whether dispersion occurs depends on:

(i) the total cation concentration, the amount of total dissolved solids (TDS) in the soil or the electrical conductivity (EC), and

(ii) the amount of sodium in the soil, usually measured as the sodium adsorption ratio (SAR) of the soluble cations or the percentage of sodium in the exchangeable cations (ESP) (Rengasamy et al., 1984); and

(iii) the clay mineralogy of the soils and the amount and type of organic matter in them (see Sumner, 1993).

Sodic soils may be further subdivided as described in Isbell (1996), taking into account factors such as pH, salinity levels, texture, sodium and magnesium levels.

Sodic surface soils are not confined to any one soil group because of the mobility of sodium in the landscape. However sodic surfaces are most commonly associated with red-brown earths, some cracking clays and the solodic soils. In the new Australian Classification (Isbell, 1996) they are most commonly associated with Sodosols and some of the Vertosols, especially the grey and yellow suborders.

When considering the options for the management of sodic soils it is useful to divide them on the basis of soil texture and the pH of the surface soil. This yields the following subgroups within the sodic surface soils:

—heavy texture, alkaline to neutral

—heavy texture, acidic

—medium texture, alkaline to neutral

—medium texture, acidic

—coarse texture, alkaline to neutral

—coarse texture, acidic.

Some consideration also needs to be given to magnesium levels as magnesium can increase the instability to wetting. The coarse- and medium-texture subgroups overlap with the fragile surface soils.

Coarse-Structured Clayey Surface Soils

The coarse-structured clayey surface soils include those clayey surface soils that do not fit into the self-mulching and sodic surface soils. They often have a coarse blocky structure with peds greater than 10 mm in size. A very thin (< 2 mm) sandy crust is sometimes present. The properties of these soils are not as adverse as those of the sodic surface soils with similar texture, nor are they as favourable as those of the self-mulching soils. Their distribution in south-eastern New South Wales is largely confined to the ancient alluvial and colluvial plains west of the Great Divide and the Riverina. Many of these surface soils are marginally sodic with ESP values of 4–5%, and the clay minerals present appear to be less active (lower shrink-swell capacity) than those found in self-mulching soils. These can be difficult to manage because of problems of cloddiness, surface crusting and the ponding of surface water. They are associated with the red, brown and grey clays or transitional red-brown earths in the great soil groups of Stace et al. (1968). In the Australian Soil Classification (Isbell, 1996), they are associated most commonly with the massive great groups of the Vertosols.

Soils with High Levels of Organic Matter

These soils are characterised by having levels of organic matter sufficient to have a large effect on soil properties. They are in the new Australian Soil Classification (Isbell, 1996) as the *melanic* and *melacic* soil materials, having the following features:

—moist colour is black (value < 3; chroma < 5), and dry colour has value < 5

—the structure grade (pedality) is more than weak and the common ped size is 10–20 mm or less

—pH > 5.5 for melanic soils; pH < 5.5 for melacic soils.

To qualify as a melanic or melacic horizon in the classification a layer must be at least 0.20 m thick, but it is intended here that where layers are less than this, the soil materials may be called melanic or melacic. These soil materials are common in the surface soils of many forests and grasslands in native condition, or in grasslands that have had low stocking rates or no cropping for extended periods.

The key features of *humose* soils as defined by Isbell (1996) are that if the soil is low in clay then the organic carbon is between 4% and 12%, or if the soil is high in clay (> 60%) then the organic carbon is between 6% and 18%. These levels of organic carbon are based on loss on ignition, but can be approximated from the Walkley-Black method by multiplying the Walkley-Black values by 1.3 (Isbell, 1996; Rayment and Higginson, 1992). The critical values of organic carbon for humose soils are proportional between these different levels of clay. The high organic carbon levels in these soils tend to dominate the soil properties and these soils are usually very stable to wetting and raindrop impact, but may show strong water repellence when dry. The occurrence of these soils is similar to the melanic and melacic soils but these represent areas where organic carbon accumulation has been greater. Again it is necessary to distinguish between humose horizons as defined by Isbell (1996) and the humose soil materials.

Organic materials are defined by Isbell (1996) as being plant-derived organic accumulations that either:

—are saturated with water for long periods or are artificially drained and, excluding live plant tissue, have 18% organic carbon if the mineral fraction of the soil has greater than 60% clay; or have 12% organic carbon if the mineral fraction has no clay; if the clay content is between 0 and 60% they have a proportional level of carbon between 12% and 18% (see note about determining organic carbon levels above); or

—are saturated with water for no more than a few days and have 20% organic carbon.

These soil materials occur in poorly drained areas of very cold climates, or in wet swampy areas in warmer areas.

The use of these highly organic soils for agriculture is likely to cause a decrease in the level of organic matter, and should this occur to a sufficient degree, they become the surface soil type that is related to the inherent properties of the mineral fraction of the soil.

Note on the Use of the Term 'Hardsetting'

The commonly used term 'hardsetting' has not been included in these groupings because as Mullins et al. (1990) point out, it covers a wide range of soils from loamy sands to clays (see their Figure 1 and their Tables 1 and 2). Northcote (1979) and Northcote et al. (1975) imply that hardsetting is a term that should only apply to duplex soils. However, in Mullins et al. (1990) the term is applied to a range of Australian soils including pale sands (Uc4); loam soils (Um); non-cracking clays (Uf6.4, Uf6.5, Uf6.6, Uf6.7); massive cracking clays (Ug5.4, Ug5.5); and many massive earths (Gn2). Hardsetting also occurs in some smooth-ped structured earths (Gn3) after clearing and continued cultivation.

Some workers have redefined hardsetting on the basis of a specific soil strength. Northcote (1979) originally suggested that hardsetting soils be those that offer a resistance to a 6.35 mm diameter penetrometer of 5 kg/m^2 (= 50 N/m^2 = 0.05 kPa). This is a relatively small pressure. Chartres and Mullins (1994) suggested that a hardsetting soil has a tensile strength of 90 kN/m^2 (= 90 kPa), a much larger resistance to fracture. Alternatively, the value for hardsetting could relate to specific problems such as the strength that will prevent emergence of seedlings, or tine penetration, or the value when the draught of a tine becomes limiting.

Since hardsetting applies to such a wide range of soils and soil conditions the term is probably too broad to be useful when making specific recommendations about the management of soil structure (Chartres and Mullins, 1994; Harper and Gilkes, 1994). Harper and Gilkes (1994) concluded that surface soils should be grouped on the basis of texture rather than on hardsetting. They based this conclusion on the lack of a precise definition for hardsetting and the strong relationship they found between soil strength and clay content.

16.5 Conclusion

Farming practices have changed greatly since the early 1970s and emphasis is placed now not only on productivity but also on the potential environmental effects of farming practices and the sustainability of the farming systems being used. Conservation farming, by controlling the movement of soil and nutrients in catchments, is vital for sustainable agriculture, catchment health, water quality and water utilisation.

It is clear that cropping systems that minimise tillage and soil disturbance, maximise plant growth and maximise the amount of stubble retention lead to improvement of soil structure, reduce runoff and reduce erosion. Therefore the adoption of conservation tillage systems such as no-tillage and minimum tillage are required to prevent soil degradation by soil structure decline and erosion in the long term. However, there are agronomic and economic limitations to the adoption of these systems, as discussed earlier in this chapter.

Conservation farming is a dynamic field for research and experimentation. New ideas are continuously being tried by farmers, and some of these new practices will require some years of trialling under a wide range of conditions before they can be fully evaluated.

BIBLIOGRAPHY

Bakker, D. and Hamilton, G. (1998), 'Production performance of permanent raised beds in widely differing soil and climate regimes in Western Australia', in J.N. Tulberg and D.F. Yule (eds), *Proceedings of Second National Controlled Traffic Conference, August 1998*, The University of Queensland, Gatton College.

Bakker, D., Hamilton, G. and Bathgate, A. (1998), 'Economic assessment and requirements of permanent raised beds in Western Australia', in J.N. Tulberg and D.F. Yule (eds), *Proceedings of Second National Controlled Traffic Conference, August 1998*, The University of Queensland, Gatton College.

Bathke, G.R., Cassel, K.K., Hargrove, W.L. and Porter, P.M. (1992), 'Modification of soil physical properties and root growth response', *Soil Science* 154, 316–29.

Blair, G.J., Lefroy, R.D.B., and Lisle, L. (1995), 'Soil carbon fractions based on their degree of oxidation, and the development of a Carbon Management Index for agricultural systems', *Australian Journal of Agricultural Research* 46, 1459–66.

Bray, G.W. (1996), *Regional Statistics—New South Wales 1996*, Australian Bureau of Statistics, Catalogue No. 1304.1.

Breckwoldt, R. (1988), *The Dirt Doctors—A Jubilee History of the Soil Conservation Service of NSW*, Government Printer of NSW, Sydney.

Bresson, L.M. (1995), 'A review of physical management for crusting control in Australian cropping systems. Research opportunities', *Australian Journal of Soil Research* 33, 195–209.

Brussard, L. (1994), 'Interrelationships between biological activities, soil properties and soil management', in D.J. Greenland and I. Szabolcs (eds), *Soil Resilience and Sustainable Land Use*, CAB International, UK.

Brussard, L. and Kooistra, M.J. (eds) (1993), 'Soil Structure/Soil Biota Interrelationships', *Geoderma* 56, Nos 1–4 and 57, Nos 1–2, Elsevier, Amsterdam.

Buckerfield, J.C. and Auhl, L.H. (1994), 'Earthworms as indicators of sustainable production in intensive cereal cropping', in C.E. Pankhurst, B.M. Daube, V.V.S.R. Gupta and P.R. Grace (eds), *Soil Biota Management in Sustainable Farming Systems*, CSIRO, Australia.

Butler, B.E. (1955), 'A system for the description of soil consistence in the field', *Journal of the Australian Institute of Agricultural Science* 4, 241–9.

Cannell, R.Q., Davies, D.B., Mackney, D. and Pidgeon, J.D. (1979), 'The suitability of soils for sequential direct drilling of combine-harvested crops in Britain: a provisional classification', in M.G. Jarvis and D. Mackney (eds), *Soil Survey Applications*, Soil Survey Tech. Monograph No. 13. Soil Survey of England and Wales, Harpenden, UK.

Carter, M.R. and Mele, P.M. (1992), 'Changes in microbial biomass and structural stability at the surface of a duplex soil under direct drilling and stubble retention in north-eastern Victoria', *Australian Journal of Soil Research* 30, 493–503.

Carter, M.R. and Steed, G.R. (1992), 'The effects of direct-drilling and stubble retention on hydraulic properties at the surface of duplex soils in north-eastern Victoria', *Australian Journal of Soil Research* 30, 505–16.

Chan, K.Y. and Heenan, D.P. (1991), 'Differences in surface soil aggregation under six different crops', *Australian Journal of Experimental Agriculture* 31, 683–6.

Chan, K.Y and Mead, A. (1988), 'Physical properties of a sandy loam under different tillage practices', *Australian Journal of Soil Research* 26, 549–59.

Chan, K.Y and Mead, J.A. (1989), 'Water movement and macroporosity of an Australian alfisol under different tillage and pasture conditions', *Soil and Tillage Research* 14, 301–10.

Chan, K.Y. and Mead, J.A. (1992), 'Tillage induced difference in the growth and distribution of wheat roots', *Australian Journal of Agricultural Research* 43, 19–28.

Chan, K.Y., Mead, J.A., Roberts, W.P. and Wong, P.T.W. (1989), 'The effect of soil compaction and fumigation on poor early growth of wheat under direct drilling', *Australian Journal of Soil Research* 35, 631–44.

Chan, K.Y., Roberts, W.P. and Heenan, D.P. (1992), 'Organic carbon and associated soil properties of a red earth after 10 years of a rotation under different stubble and tillage practices', *Australian Journal of Soil Research* 30, 71–84.

Charman, P.E.V. (ed.) (1985), *Conservation Farming— Extending the Principles of Soil Conservation to Cropping*, Soil Conservation Service of NSW, Sydney.

Chartres, C.J. and Mullins, C. (1994), 'Definition of hardsetting', *ACLEP Newsletter 3*, CSIRO, Division of Soils, Canberra.

Churchman, G.J., Skjemstad, J.O. and Oades, J.M. (1993), 'Influence of clay minerals and organic matter on effects of sodicity in soils', *Australian Journal of Soil Research* 31, 779–800.

Clarke, A.L. (1986), 'Cultivation', in J.S. Russell and R.F. Isbell (eds), *Australian Soils: The Human Input*, Queensland University Press, St Lucia.

Clarke, A.L. and Russell, J.S. (1977), 'Crop sequential practices', in J.S. Russell and E.S. Greacen (eds), *Soil Factors in Crop Production in a Semi-arid Environment*, University of Queensland, St Lucia.

Clarke, A.L., Greenland, D.J. and Quirk, J.P. (1967), 'Changes in some physical properties of the surface of an impoverished red-brown earth under pasture', *Australian Journal of Soil Research* 5, 59–68.

Clarke, G.B. and Marshall, T.J. (1947), 'Influence of cultivation on soil structure and its assessment in soils of variable mechanical composition', *Journal of Commonwealth Scientific and Industrial Research, Australia (CSIRO)* 20, 162.

Clayton, E.S. (1931), 'The control of erosion on wheatlands', *Agricultural Gazette of NSW* 42, Part II. *1931 in SCS History.*

Cornish, P.S. and Pratley, J.E. (eds) (1987), *Tillage—New Directions in Australian Agriculture*, Inkata Press, Melbourne.

Cotching, B. (1995), 'Long term management of Krasnozems in Australia', *Australian Journal of Soil and Water Conservation* 8(1), 18–27.

Creswell, H.P., Smiles, D.E. and Williams, J. (1992), 'Soil structure, soil hydraulic properties and the soil water balance', *Australian Journal of Soil Research* 30, 265–83.

Currie, J.P. and Good, J.A. (1992), 'Soil faunal degradation and restoration', *Advances in Soil Science* 17, 171–215.

Da Silva, Alvaro Pires and Kay, B.D. (1997a), 'Estimating the least limiting water range of soils from properties and management', *Soil Science Society of America Journal* 61, 877–83.

Da Silva, Alvaro Pires and Kay, B.D. (1997b), 'Effect of soil water content variation on least limiting water range', *Soil Science Society of America Journal* 61, 884–88.

Dann, P.R. (1986), 'Deep tillage research on the Southern Tablelands of NSW', *Proceedings of the Southern Conservation Farming Group, Wagga.*

Degens, B.P. (1997), 'Macro-aggregation of soils by biological bonding and binding, and the mechanisms and factors affecting these: a review', *Australian Journal of Soil Research* 35, 431–60.

Edwards, K. (1987), *Runoff and Soil Loss Studies in New South Wales*, Tech. Handbook No. 10, Soil Conservation Service of NSW, Sydney.

El-Swaify, S.A., Pathak, P., Rego, T.J. and Singh, S. (1985), 'Soil management for optimised productivity under rainfall condition in the semi-arid tropics', *Advances in Soil Science* 1, 1–64.

Ellington, A. and Fung, K.K.H. (1984), 'Effects of deep ripping, direct drilling, gypsum and lime on wheat', *Proceedings of the National Soils Conference, Brisbane.*

Emerson, W.W. (1991), 'Structural decline of soils, assessment and prevention', *Australian Journal of Soil Research* 29, 905–21.

Emerson, W.W. (1994), 'Aggregate slaking and dispersion class, bulk properties of soil', *Australian Journal of Soil Research* 32, 173–84.

Fraser, P.M. (1994), 'The impact of soil and crop management practices on soil macrofauna', in C.E. Pankhurst, B.M. Daube, V.V.S.R. Gupta and P.R. Grace (eds), *Soil Biota Management in Sustainable Farming Systems*, CSIRO, Australia.

French, R. and Schultz, J. (1984), 'Water use efficiency of wheat in a Mediterranean climate. I. The relationship between yield, water use, and climate', *Australian Journal of Agricultural Research* 35, 743–64.

Geeves, G.W., Cresswell, H.P., Murphy, B.W., Gessler, P.E., Chartres, C.J., Little, I.P., Bowman, G.M. (1995), *The Physical, Chemical and Morphological Properties of Soils in the Wheat-belt of Southern NSW and Northern Victoria*, CSIRO Division of Soils, Canberra and NSW Department of Conservation and Land Management.

Graham, O.P. (1989), *Land Degradation Survey of NSW 1987–88: Methodology*, Technical Report No. 7, Soil Conservation Service of NSW, Sydney.

Grant, C.D. and Blackmore, A.V. (1991), 'Self-mulching behaviour in clay soils: its definition and measurement', *Australian Journal of Soil Research* 29, 155–73.

Greacen, E.L. (1977), 'Mechanisms and models of water transfer', in J. S. Russell and E. L. Greacen (eds), *Soil Factors in Crop Production in a Semi-arid Environment*, University of Queensland Press, St Lucia.

Greenland, D.J. (1969), 'Soil management and degradation', *Journal of Soil Science* 32, 301–22.

Greenland, D.J. (1971), 'Changes in the nitrogen status and physical condition of soils under pasture with special reference to the maintenance of Australian soils used for growing wheat', *Soils and Fertilisers* 34, 237–51.

Greenland, D.J. (1977), 'Soil damage by intensive cultivation: temporary or permanent?', *Philosophical Transactions of the Royal Society London*, B, 281, 193–208.

Haines, P.J. and Uren, N.C. (1990), 'Effects of conservation tillage farming on soil microbial biomass, organic matter and earthworm populations in north-eastern Victoria', *Australian Journal of Experimental Agriculture* 30, 365–71.

Hairsine, P., Murphy, B., Packer, I. and Rosewell, C. (1993), 'Profile of erosion from a major storm in the south-east cropping zone', *Australian Journal of Soil and Water Conservation* 6(4), 50–5.

Hairsine, P.B., Moran, C.J. and Rose, C.W. (1992), 'Recent developments regarding the influence of soil surface characteristics on overland flow and erosion', *Australian Journal of Soil Research* 30, 249–64.

Hakansson, I. (1994), 'Subsoil compaction caused by heavy vehicles—a long-term threat to productivity', *Soil and Tillage Research* 29, 105–10.

Hamblin, A.P. (1980), 'Changes in aggregate stability and associated organic matter properties after direct drilling and ploughing on some Australian soils', *Australian Journal of Soil Research* 18, 27–36.

Hamblin, A.P. (1984), 'The effect of tillage on soil surface properties and water balance of a xeralfic alfisol', *Soil and Tillage Research* 4, 543–59.

Hamblin, A.P. (1985), 'The influence of soil structure on water movement, crop growth and water uptake', *Advances in Agronomy* 38, 95–155.

Hamblin, A.P. (1987), 'The effect of tillage on soil physical conditions', in P.S. Cornish and J.E. Pratley (eds), *Tillage—New Directions in Australian Agriculture*, Inkata Press, Melbourne.

Hamblin, A.P. (1991), 'Sustainable Agricultural Systems: What are the appropriate measures for soil structure?', *Australian Journal of Soil Research* 29, 709–15.

Hamblin, A. and Kyneur, G. (1993), *Trends in Wheat Yields and Fertility in Australia*, Department of Primary Industries and Energy, Bureau of Resource Sciences, Australian Government Publishing Service, Canberra.

Harper, R. J. and Gilkes, R.J. (1994), 'Hardsetting in the surface horizons of sandy soils and its implications for soil classification and management', *Australian Journal of Soil Research* 32, 603–19.

Harte, A.J. (1984), 'The effects of tillage on the stability of three red soils of the northern wheat belt', *Journal of Soil Conservation Service of NSW* 40, 94–101.

Harte, A.J. (1985a), 'Research on dryland cropping areas, northern NSW', *Proceedings Queensland Crop Production Conference, DPI, Queensland*.

Harte, A.J. (1985b), 'Yield data and some comments on the viability of some land management options from investigations at "Clearview", North Star', Soil Conservation Research Centre, Inverell, Technical Bulletin No 14, NSW Soil Conservation Service, Sydney.

Harte, A. J. and Marschke, G.W. (1985), 'Tillage practices and soil structure at the northern no-tillage regional sites, observation 1984', *Proceedings of Northern No-tillage Team, Tamworth*, NSW Soil Conservation Service.

Isbell, R.F. (1996), *The Australian Soil Classification*, CSIRO Australia, Melbourne.

Jayawardane, N.S. and Chan, K.Y. (1994), 'The management of soil physical properties limiting crop production in Australian sodic soils—a review', *Australian Journal of Soil Research* 32, 13–44.

Jones, A.A. (1983), 'Effect of soil texture on critical bulk densities for root growth', *Soil Science Society of America Journal* 47, 1208–11.

Kaleski, L.G. (1945), 'The erosion survey of NSW (Eastern and Central Divisions)', *Journal of Soil Conservation Service of NSW* 1, 12–20.

Kaleski, L.G. (1963), 'The erosion survey of NSW (Eastern and Central Divisions)', *Journal of Soil Conservation Service of NSW* 19, 171–83.

Kay, B.D. (1990), 'Rates of change of soil structure under different cropping systems', *Advances in Soil Science* 12, 1–52.

Kay, B.D., Rasiah, V. and Perfect, E. (1994), 'Structural aspects of soil resilience', in D.J. Greenland and I. Szabolcs (eds), *Soil Resilience and Sustainable Land Use*, CAB International, UK.

Kirkegaard, J.A., Angus, J.F., Gardner, P.A. and Muller, W. (1994), 'Reduced growth and yield of wheat with conservation cropping. I. Field studies in the first year of the cropping phase', *Australian Journal of Soil Research* 45, 511–28.

Lal, R. (1976), 'Soil erosion on Alfisols in Western Nigeria. II. Effects of slope, crop rotation and residue management', *Geoderma* 16, 363–75.

Lal, R. (1986), 'Soil surface management in the tropics for intensive land use and high sustained production', *Advances in Soil Science* 5, 1–109.

Lal, R. (1994), 'Sustainable land use systems and soil resilience' in D.J. Greenland and I. Szabolcs (eds), *Soil Resilience and Sustainable Land Use*, CAB International, UK.

Lang, R.D. (1979), 'The effect of ground cover on surface runoff from experimental plots', *Journal of Soil Conservation Service of NSW* 35, 108–14.

Lavelle, P., Gilot, C., Fragoso, C. and Pashansii, B. (1994), 'Soil fauna on sustainable land use in the humid tropics' in D.J. Greenland and I. Szabolcs (eds), *Soil Resilience and Sustainable Land Use*, CAB International, UK.

Lee, K.E and Foster, R.C. (1991), 'Soil fauna and soil structure', *Australian Journal of Soil Research* 29, 745–75.

Letey, J. (1985), 'Relationships between soil physical properties and crop production', *Advances in Soil Science* 1, 277–94.

Loch, R.J. (1994b), 'Effects of fallow management and cropping history on aggregate breakdown under rainfall wetting for a range of Queensland soils', *Australian Journal of Soil Research* 32, 1125–39.

Loch, R.J. and Coughlan, K.J. (1984), 'Effects of zero tillage and stubble retention on some properties of a cracking clay', *Australian Journal of Soil Research* 22, 91–8.

Loch, R.J. and Donnollan, T.E. (1988), 'Effects of the amount of stubble mulch and overland flow on erosion of a cracking clay soil under simulated rainfall', *Australian Journal of Soil Research* 26, 661–72.

Loveday, J. (1980), 'Soil management and amelioration', in T.S. Abbott, C.A. Hawkins and P.G.E. Searle (eds), *National Soils Conference 1980—Review Paper*, Australian Soc. Soil Sci. Inc.

Loveday, J. and Bridge, B.J. (1983), 'Management of salt-affected soils' in *Soils—An Australian Viewpoint*, CSIRO, Melbourne/Academic Press, London.

McDonald, R.C., Isbell, R.F., Speight, J.G., Walker, J. and Hopkins, M.S. (1994), *Australian Soil and Land Survey Field Handbook* (Second Edition), Inkata Press, Melbourne.

McGuiness, S. (1991), *Soil Structure Assessment Kit—a guide to assessing the structure of red duplex soil*, Department of Conservation and Environment, Victoria (Bendigo).

McKenzie, D.C., Abbott, T.S., Chan, K.Y., Slavich, P.G. and Hall, D.J.M. (1993), 'The nature, distribution and management of sodic soils in New South Wales', *Australian Journal of Soil Research* 31, 839–68.

McKibbon, G.G. (ed.) (1971), *Compaction of Agricultural Soils*, American Society of Agricultural Engineers, St Joseph, Michigan.

McNeill, A.A. and Aveyard, J.M. (1978), 'Reduced tillage systems—their potential for soil conservation in southern NSW', *Journal of Soil Conservation Service of NSW* 34, 207–9.

MacRae, R.J. and Mehuys, G.R. (1985), 'The effect of green manuring on the physical properties of temperate area soils', *Advances in Soil Science* 3, 71–94.

Macks, S.P., Murphy, B.W., Cresswell, H.P. and Koen, T.B. (1996), 'Soil friability in relation to management history and suitability for direct drilling', *Australian Journal of Soil Research* 34, 343–60.

Malinda, D. (1995), 'Ameliorating a soil with a non-wetting sandy surface and a sodic subsoil', *Australian Journal of Soil and Water Conservation* 8(1), 36–41.

Marschke, G.W. and Thompson, D.F. (1983), 'The influence of conservation tillage practice on runoff and soil loss from a high intensity convective storm', in *No-Tillage Crop Production in Northern NSW*, Proc. No-Till Project Team, Tamworth.

Marston, D. (1978), 'Conventional tillage systems as they affect soil erosion', *Journal of the Soil Conservation Service of NSW* 34, 194–8.

Moran, C.J., Koppi, A.J., Murphy, B.W. and McBratney, A.B. (1988), 'Comparison of the macropore structure of sandy loam surface soil subjected to two tillage treatments', *Soil Use Management* 4, 96–102.

Mortiss, P. and McGarry, D. (1993), 'Farmer consultations on soil compaction—their implications for extension and research', *Australian Journal of Soil and Water Conservation* 6(2), 9–14.

Mullins, C.E., MacLeod, D.A., Northcote, K.H., Tisdall, J.M. and Young, I.M. (1990), 'Hardsetting soils: behaviour, occurrence and management', *Advances in Soil Science* 11, 37–108.

Murphy, B.W. and Allworth, D. (1991), *Detecting Soil Structure Decline*, Pamphlet, Soil Conservation Service of NSW, Wagga Wagga.

Murphy, B.W. and Flewin, T.C. (1993), 'Rill erosion on a structurally degraded sandy loam surface soil', *Australian Journal of Soil Research* 31, 419–36.

Murphy, B.W. and Lawrie, J. (1996), *What surface soil is that?*, Extension Brochure, Department of Land and Water Conservation Research Centre, Cowra.

Murphy, B.W., Koen, T.B., Jones, B.A. and Huxedurp, L.M. (1993), 'Temporal variation of hydraulic properties for some soils with fragile structure', *Australian Journal of Soil Research* 31, 179–97.

Murphy, B.W., Lynch, L.G., and Dwyer, P.J. (1989), 'Agronomic aspects of conservation tillage trials at Wellington, NSW', *Australian Journal of Soil and Water Conservation* 2(2), 39–42.

Northcote, K.H. (1979), *A Factual Key for the Recognition of Australian Soils*, Rellim Technical Publications, Glenside, South Australia.

Northcote, K.H. (1983), *Soils, Soil Morphology and Soil Classification: Training Course Lectures*, CSIRO Division of Soils, Rellim Technical Publications, Glenside, South Australia.

Northcote, K.H. and Skene, J.K.M. (1972), *Australian Soils with Saline and Sodic Properties*, Soil Publication 27, CSIRO Australia, Melbourne.

Northcote, K.H., Hubble, G.D., Isbell, R.F., Thompson, C.H. and Bettenay, E. (1975), *A Description of Australian Soils*, CSIRO Division of Soils, Wilke and Co. Ltd, Clayton, Victoria.

NSW Department of Agriculture (1937), *The Farmers Handbook*, Fifth Edition, Government Printer, Sydney.

Oades, J.M. (1984), 'Soil organic matter and structural stability: mechanisms and implication for management', *Plant and Soil* 76, 319–67.

Oades, J.M. (1993), 'The role of biology in the formation, stabilisation and degradation of soil structure', *Geoderma* 56, 377–400.

Osborne, G.J., Rowell, D.L. and Matthews, P.G. (1978), 'Value and measurement of soil structure under systems of reduced and conventional cultivation', in W.W. Emerson, R.D. Bond and A.R. Dexter (eds), *Modification of Soil Structure*, John Wiley and Sons, Chichester, UK.

Packer, I.J., Hamilton, G.J. and Koen, T.B. (1992), 'Runoff, soil loss and soil physical property changes of light textured surface soils from long-term tillage trials', *Australian Journal of Soil Research* 30, 789–806.

Packer, I.J., Hamilton, G.J. and White, I. (1984), 'Tillage practices to conserve soil and improve soil conditions', *Journal of Soil Conservation Service of NSW* 40, 78–87.

Pankhurst, C.E., Doube, B.M., Gupta, V.V.S.P. Gupta and Grace, P.R. (eds) (1994), *Soil Biota: Management in Sustainable Farming Systems*, CSIRO, Australia.

Passioura, J.B. (1991), 'Soil structure and plant growth', *Australian Journal of Soil Research* 29, 717–28.

Petersson, H., Messing, I. and Steen, E. (1987), 'Influence of root mass on saturated hydraulic conductivity in arid soils of central Tunisia', *Arid Soil Research and Rehabilitation* 1, 149–60.

Pillai-McGarry, U.P.P. and Collis-George, N. (1990a), 'Laboratory simulation of the surface morphology of self-mulching and non self-mulching Vertosols. I. Materials, methods and preliminary results', *Australian Journal of Soil Research* 28, 129–39.

Pillai-McGarry, U.P.P. and Collis-George, N. (1990b), 'Laboratory simulation of the surface morphology of self-mulching and non self-mulching Vertosols. II. Quantification of visual features', *Australian Journal of Soil Research* 28, 148–52.

Poesen, J, and Nearing, M. (1993), 'Soil surface crusting and sealing', *Catena* Supplement 24.

Poole, M.L. (1987), 'Tillage practices for crop production in winter rainfall areas', in P.S. Cornish and J.E. Pratley (eds), *Tillage: New Directions in Australian Agriculture*, Inkata Press, Melbourne.

Pratley, J. (1999), 'Allelopathy—exploring its agricultural potential', Proceedings of CONFARM 21, Second National Conservation Farming and Minimum Tillage Conference, Toowoomba 1999.

Proffitt, A.P., Bendetti, S., Howell, M.R. and Eastham, J. (1993), 'The effect of sheep trampling and grazing and soil physical properties and pasture growth for a red-brown earth', *Australian Journal of Agricultural Research* 44, 317–31.

Prove, B.G., Loch, R.J., Foley, J.L., Anderson, V.J. and Yowger, D.R. (1990), 'Improvements in aggregation and infiltration characteristics of a krasnozem under maize with direct drill and stubble retention', *Australian Journal of Soil Research* 28, 577–90.

Quirk, J.P. and Murray, R.S. (1991), 'Towards a model of soil structural behaviour', *Australian Journal of Soil Research* 29, 829–67.

Quirk, J.P. and Schofield, R.K. (1955), 'The effect of electrolyte concentration on soil permeability', *Journal of Soil Science* 6, 163–78.

Quisenberry, V.L., Smith, B.R., Phillips, R.E., Scott, H.D. and Nortcliff, S. (1993), 'A soil classification system for describing water and chemical transport', *Soil Science* 156, 306–15.

Rayment, G.E. and Higginson, F.R. (1992), *Australian Laboratory Handbook of Soil and Water Chemical Methods*, Inkata Press, Melbourne.

Rengasamy, P., Greene, R.S.B., Ford, G.W. and Mechanni, A.H. (1984), 'Identification of dispersive behaviour and the management of red-brown earths', *Australian Journal of Soil Research* 22, 413–31.

Roper, M.M. and Gupta, V.V.S.R. (1995), 'Management practices and soil biota', *Australian Journal of Soil Research* 32, 321–39.

Rosewell, C.J. and Marston, D. (1978), 'The erosion process as it occurs within cropping systems', *Journal of Soil Conservation Service of NSW* 34, 186–93.

Rovira, A.D. and Greacen, E.L. (1957), 'The effect of aggregate disruption on the activity of microorganisms in the soil', *Australian Journal of Agricultural Research* 8, 659–73.

Rovira, A.D., Smettem, K.R.J. and Lee, K.E. (1987), 'Effect of rotation and conservation tillage on earthworms in a red-brown earth under wheat', *Australian Journal of Agricultural Research* 38, 829–34.

SCA—Standing Committee on Agriculture, Australian (1991), *Sustainable Agriculture*, Report of the Working Group on Sustainable Agriculture, SCA Technical Report, 36, CSIRO Australia, East Melbourne.

Schafer, R.L. and Johnson, C.E. (1982), 'Changing soil condition—the soil dynamics of tillage', in P.W. Unger, D.M. Van Doren, F.D. Whisler and E.L. Skidmore (eds), *Predicting Tillage Effects on Soil Physical Properties and Processes*, ASA Special Publication No. 44, Madison, Wisconsin.

Schultz, J.E. (1974), 'Root development of wheat at the flowering stage under different cultural practices', *Agricultural Record (South Australia)*, January 1974, 12–17.

Scotter, D.R. (1970), 'Soil temperatures under grass fires', *Australian Journal of Soil Research* 18, 273–80.

Scotter, D.R. and Loveday, J. (1966), 'Physical changes in seedbed material resulting from the application of dissolved gypsum', *Australian Journal of Soil Research* 14, 69–75.

Shiel, R.S., Adey, M.A. and Lodden, M. (1988b), 'The effect of successive wet/dry cycles on aggregate size distribution in a clay texture soil', *Journal of Soil Science* 39, 71–80.

Sims, G.K. (1990), 'Biological degradation of soil', in R. Lal and B.A. Stewart (eds), 'Soil Degradation', *Advances in Soil Science* 11, Springer-Verlag, New York, Berlin.

Sims, H.J. (1977), 'Cultivation and fallowing practices', in J.S. Russell and E.L. Greacen (eds), *Soil Factors in Crop Production in a Semi-arid Environment*, University of Queensland Press, St Lucia.

Smettem, K.R.J., Rovira, A.D., Wace, S.A., Wilson, B.R. and Simon, A. (1992), 'Effect of tillage and crop rotation on the surface stability and chemical properties of a red-brown earth (Alfisol) under wheat', *Soil and Tillage Research* 22, 27–40.

Smiles, D.E. (1977), 'Air-heat-water relationships', in J.S. Russell and E.L. Greacen (eds), *Soil Factors in Crop Production in a Semi-Arid Environment*, University of Queensland Press, St Lucia.

Soane, B.D. (1975), 'Studies on some soil physical properties in relation to cultivations and traffic', in Ministry of Agriculture, Fisheries and Food, *Soil Physical Conditions and Crop Production*, Technical Bulletin 29, Her Majesty's Stationery Office, London.

Soil Conservation Service of NSW (1989), *Land Degradation Survey*, NSW.

Spoor, G. (1979), 'Soil type and workability', in M.G. Jarvis and D. Mackney (eds), *Soil Survey Application*,. Soil Survey Technical Monograph No. 13, Soil Survey of England and Wales, Harpenden, UK.

Stace, H.C.T., Hubble, G.D., Brewer, R., Northcote, K.H., Sleeman, J.R., Mulcahy, M.J. and Hallsworth, E.G. (1968), *A Handbook of Australian Soils*, Rellim Technical Publications, Glenside, SA.

Steed, G., Ellington, T. and Pratley, J. (1993), 'Conservation tillage in south-eastern Australia', *Australian Journal of Soil and Water Conservation* 6(2), 34–41.

Stoneman, T.C. (1962), 'Loss of structure in wheat belt soils', *Journal of the Department of Agriculture Western Australia* 3, 493–5.

Sumner, M.E. (1993), 'Sodic soils: new perspectives', *Australian Journal of Soil Research* 31, 683–750.

Tisdall, J.M. (1985), 'Earthworm activity in irrigated red-brown earths used for annual crops in Victoria', *Australian Journal of Soil Research* 23, 291–9.

Tisdall, J.M. (1991), 'Fungal hyphae and structural stability of soil', *Australian Journal of Soil Research* 29, 729–43.

Tisdall, J.M. and Oades, J.M. (1979), 'Stabilisation of soil aggregates by the root systems of ryegrass', *Australian Journal of Soil Research* 17, 429–41.

Tisdall, J.M. and Oades, J.M. (1980a), 'The effect of crop rotation on aggregation is a red-brown earth', *Australian Journal of Soil Research* 18, 423–33.

Tisdall, J.M. and Oades, J.M. (1980b), 'The management of ryegrass to stabilise aggregates of a red-brown earth', *Australian Journal of Soil Research* 18, 415–22.

Tisdall, J.M. and Oades, J.M. (1982), 'Organic matter and water stable aggregates in soils', *Journal of Soil Science* 33, 141–63.

Tisdall, J.M., Cockcroft, B. and Uren, N.C. (1978), 'The stability of soil aggregates as affected by organic materials, microbial activity and physical disruption', *Australian Journal of Soil Research* 18, 423–33.

Trotman, R.C. (1965), 'Conservation farming—the road back to stability', *Journal of Soil Conservation Service of NSW* 21, 13–19.

Valentin, C. (1995), 'Sealing, crusting and hardsetting soils in Sahelian agriculture', in H.B. So, G.D. Smith, S.R. Raine, B.M. Schafer, and R.J. Loach, (eds), *Proceedings of International Symposium on Sealing, Crusting, and Hardsetting Soils: Productivity and Conservation*, Australian Society of Soil Science and University of Queensland.

Vanclay, F. and Hely, A. (1997), *Land Degradation and Land Management in Central NSW: Changes in Farmers' Perceptions, Knowledge and Practices*, Charles Sturt University, Riverina, Wagga Wagga.

Vanclay, F. and Lockie, S. (1993), *Barriers to the Adoption of Sustainable Crop Rotations*, Charles Sturt University, Riverina, Wagga Wagga.

Walker, J. and Reuter, D.J. (1996), Indicators of Catchment Health—A Technical Perspective, CSIRO, Australia.

Walsh, P. (1998), 'Controlled traffic reduces overlap to save on crop inputs', *Farming Ahead* 83, Kondinin Group.

Wenke, J.F. and Grant, C.D. (1994), 'The indexing of self-mulching behaviour in soils', *Australian Journal of Soil Research* 32, 201–11.

White, I. (1988), 'Tillage practices and soil hydraulic properties: why quantify the obvious?', in J. Loveday (ed.), *Review Papers: National Soils Conference, 1988*, Koomarii Printers, ACT, Australia.

Wylie, P. (1993), 'Conservation farming systems for the summer rainfall cereal belt', *Australian Journal of Soil and Water Conservation* 6(2), 28–33.

Soils and Rangeland Management

CHAPTER 17

D.J. Eldridge

Rangelands can generally be defined as areas that are unsuitable for cultivation and are generally used for grazing. They may be unsuitable because of physical limitations such as low or erratic rainfall, steep slopes, shallow soils, poor drainage, or even cold temperatures. About 70% of the area of Australia and 40% of New South Wales can be considered as rangelands. While most rangeland in New South Wales is found in the arid and semi-arid areas, many parts of the northern and southern tablelands can be regarded as rangeland because they support native (unimproved) pastures and are used primarily for grazing. Rangelands also support mining, parks and reserves, and large areas are set aside for traditional ownership.

17.1 Introduction to Rangeland Soils

Rangeland soils are typically shallow and infertile and formed on highly weathered parent materials. In New South Wales they are generally found on the western plains in areas receiving between 180 and 500 mm average annual rainfall. The soils vary widely in their physical and chemical properties and their ability to resist erosion and to recover once eroded. Many of the soils have been subjected to continued reworking and redistribution by wind and/or water. A summary of the soil properties of the main groups of soils found in western New South Wales is given in Table 17.1.

Soil moisture availability is the most important factor influencing pastoral productivity in rangelands. Available moisture is a function of rainfall, soil physical properties such as texture and depth, and location in the landscape. As discussed below, some sites accumulate moisture while others lose it. In western New South Wales there are strong relationships between rainfall amount and biomass of pasture. Rainfall is therefore critical in determining safe carrying capacity for grazing animals in the drier areas of the state.

Soils low in silt and clay have little or no structure, low levels of organic matter in the surface horizon and are not prone to surface sealing. Consequently, sandy soils respond well to small falls of rain and allow rapid germination of seeds. Clay soils have a large water-holding capacity but a low saturated hydraulic conductivity, that is, they do not conduct water very well. They are able to take in large amounts of water until any cracks seal up and infiltration is restricted. Although much of the moisture in a clay is unavailable for plant growth and germination after small rainfall events, these soils will produce a large amount of pasture when saturated. Pastoral production is further complicated on these clayey soils in that they readily crack on drying out and often, after prolonged droughts, perennial shrubs may die when their roots become separated from the soil. Duplex or texture contrast soils have characteristic sandy to loamy surface horizons that are underlain by distinct clayey subsoils. The sandy topsoil responds rapidly to small falls of rain owing to low water-holding capacity and high infiltration characteristics of the surface soil.

Table 17.1 Properties of the main great soil groups in western New South Wales

Great Soil Groups*	Subscript	Northcote Codings	Surface Textures	Landscapes	Erodibility
Siliceous, earthy and calcareous sands	a	Ucl, Uc5	Sands and loamy sands	Dunefields and plains	Highly susceptible (wind erosion)
Skeletal soils (shallow loamy earths)	b	Um1, Um4, Um5	Loams	Ridges and footslopes	Moderate (water erosion)
Grey, brown and red clays	c	Ug5.24, Ug5.28, Ug5.3	Light to medium clay	Plains and depressions	Low to nil
Calcareous red earths	d	Gn2.13	Loamy to clay loam	Plains	Low to moderate (water erosion)
Solonised brown soils	e	Gc1.2, Gc1.22, Gc2.12	Sandy loam to loam	Plains	Moderate (wind erosion)
Desert loams	f	Dr1.13	Sandy loam to clay loam	Slightly undulating plains	Moderate (wind and water erosion)
Red-brown earths Non-calcic brown soils	g	Dr2, Dr4	Sandy loam to sandy clay loam	Slightly undulating plains	Moderate (wind and water erosion)\

*See 'Soils of the Western Zone', Chapter 8.

Table 17.2 Direct and indirect effects of livestock on soil (after Johns et al., 1984)

Effect	Processes	Potential Consequences
	Direct effects	
Compaction (i.e. soil is trodden into a more dense state) Most likely to occur when moist	Lower pore space with smaller pores; reduced infiltration; lower water storage; increased runoff	Initiation of water erosion; animal paths form gullies; less soil water available for plant growth
Surface Pulverisation (i.e. soil aggregates are reduced in size)	Production of 'fines' susceptible to erosion by wind or water; formation of physical crusts; reduced micro-topography; release of sand particles from the soil matrix	Clay, containing nutrients, may be removed differentially from the soil; less soil water available; increased runoff; impaired seedling emergence; no surface water stored; increased runoff; vegetation blasting by wind; formation of scalds
	Indirect effects	
Compaction Reduction of plant density and cover (e.g. trimming the size of perennials, trampling of litter)	Greater area for raindrop impact and hence crust formation; increased soil insulation; greater opportunity for unobstructed overland flow; increased wind velocity at surface; fewer roots to physically bind the soil	Increased runoff; less stored water; oxidation of soil organic matter and consequent reduced aggregate stability; lower populations of soil biota; harsh conditions for seedlings; greater wind erosion hazard, a self-reinforcing cycle; increased sheet, rill and gully erosion

Table 17.3 Mechanical reclamation techniques (after Noble et al., 1984)

Technique	Where	When	How
Pitting	Bare areas with slopes less than 2%	Best results if carried out prior to rainfall, especially if combined with reseeding	Disc or tine pitter
Furrowing	Slopes greater than 2% —furrows staggered on steepest slopes	Best results if prior to expected rainfall	Three-point linkage mounted tine and two opposed discs, one either side of ripper— also mouldboard plough
Waterspreading	Where catchment areas are adjacent to runon area with soils of high infiltration and water-holding capacities	Sites need to be cleared of any problem shrubs before bank construction	Banks constructed with road grader
Waterponding	Bare, scalded areas where the ratio of catchment to ponded area is between 1:1 and 1:2	Bank construction best during dry times	Banks constructed with road grader, size of bank dependent on ponding depth which depends on soil permeability

17.2 Soil–Vegetation Relations

17.2.1 Patchiness and Resource Redistribution

Experimental and anecdotal evidence from rangelands worldwide has shown that stable and productive ecosystems are ones where essential resources such as water, nutrients, organic matter and topsoil are concentrated in specific zones in the landscape (Tongway and Ludwig, 1990). These favoured zones, known as 'fertile patches' or 'resource islands' (Garner and Steinberger, 1989), occupy up to 40% of the landscape. Within these resource-rich zones the amount and availability of resources are regulated by processes such as runoff/runon and erosion/deposition. Associated with resource-rich zones are zones of resource depletion, which are relatively constant through time and comprise about 60–80% of the landscape in stable, functional ecosystems.

Patchiness in the rangelands exists at a number of scales. This range of scales is illustrated by the mulga (*Acacia aneura*) woodlands in eastern Australia. At the broad landscape scale, these woodlands are organised into topographic sequences of alternating mulga groves and intergroves (Ludwig and Tongway, 1995). Maximum soil moisture and nutrients, organic matter, biological activity and hence productivity occur in the timbered groves, which depend on a redistribution of resources from the intergroves to function. At a finer scale, this redistribution and hence patterning occurs at the scale of individual trees and shrubs, fallen logs and termite pavements, and even at the scale of individual grass tussocks.

17.2.2 Vegetation Distribution in Relation to Soils

Broad regional differences in soil types, a consequence of both climate and parent material, have a marked impact on the distribution of vegetation communities. Within New South Wales, rainfall increases from west to east, and is least in the extreme north-west of the state where it is also very unreliable, that is, changes markedly from year to year. Rainfall distribution also changes from north to south. Northern areas receive about 30% more rainfall in the warmer months, while those in the south receive 30% more rainfall in the cooler months. This rainfall distribution means that summer-growing grasses such as *Enneapogon*, *Dichanthium* and *Astrebla* species tend to be dominant in the north while winter-growing grasses such as *Stipa* and *Aristida* tend to be dominant in the south. Higher rainfall in the east means generally that the soils are more fertile and support generally more plants with greater cover and productivity. Consequently, many of these areas of former rangeland in the east are coming under increasing pressure for clearing and cultivation.

In south-western New South Wales the dominant landforms are aeolian in origin (Eldridge, 1985). The soils and landscapes have developed during a period of aridity during the Quaternary period and the soils are predominantly coarse-textured sandy loams and loamy sands. Stunted mallee (*Eucalyptus socialis*, *E. leptophylla*, *E. oleosa*, *E. gracilis*) dominate the sandy areas while belah (*Casuarina cristata*) and rosewood (*Alectryon oleifolius*) are generally restricted to the finer-textured swales between the dunes. Other species such as the shrubs pituri (*Duboisia hopwoodii*) and narrow-leaved hopbush (*Dodonaea viscosa*) occur in areas of deep sand, often covering entire sand dunes.

On the Riverina Plain between Narrandera and Balranald in the south-west, the soils on the plains are dominated by a mosaic of grey cracking clays, interspersed by ancient stream beds ('prior streams') with scalded red duplex soils. The plains are dominated by chenopod shrubs, particularly *Atriplex vesicaria*, which has the ability to extract water from the upper layers of these clay soils which generally have low water availabilities (Eldridge, 1988). The prior streams, however, support bluebushes (*Maireana pyramidata*, *M. sedifolia*) and often small trees such as yarran (*Acacia homalophylla*) and boree (*A. pendula*). Watercourses, which receive intermittent flooding and have slightly deeper clay soils, often support groves of blackbox (*Eucalyptus largiflorens*) or river red gum (*E. camaldulensis*). Vegetation distribution therefore is strongly determined by soil texture, which influences both infiltration and water-holding capacity. For some plants, soil type and depth and even carbonate content are important. In north-western New South Wales a catenary sequence is evident on the hills and ranges, with *Maireana sedifolia* and *M. astrotricha* preferring the shallower, better-drained soils of the higher footslopes and ridges, while *M. pyramidata* and *Atriplex vesicaria* occur lower in the topographic sequence (Eldridge, 1988).

As discussed above, the soils and vegetation of many rangeland systems are strongly patterned. Ephemerals or annuals tend to dominate zones of nutrient-poor sites that act as erodible, water-shedding sites, while the long-lived perennials tend to dominate areas of resource accumulation. Thus fertile patches are dominated by either trees or shrubs, while runoff zones are dominated by ephemerals which have the capacity to rapidly convert accumulated reserves into the production of seeds (Westoby, 1979/80). Short-lived perennials such as grasses are often found in the intermediate areas between runoff and runon zones.

17.3 Soil Biota in Rangelands

Soil biota, that is, organisms living close to or within the soil profile, are important components of rangeland soils. While we know very little about microscopic soil organisms such as protozoa, bacteria and fungi, a considerable amount of research has been done on the macro-invertebrates (particularly ants and termites), and those organisms that make up biological soil crusts, that is, the lichens, bryophytes and cyanobacteria.

Microfauna such as protozoa and nematodes are found in areas where soil moisture is highest, and therefore have a patchy distribution in rangelands. The larger and more dominant mesofauna are dominated by the mites (Acari) and springtails (Collembola) which, along with microfauna, tend to be concentrated in moister, more favourable patches that provide suitable food, habitat and microclimatic conditions (Whitford and Herrick, 1996). Termites and ants are the dominant surface-active fauna in rangelands, particularly in the subtropics. Soil animals actively modify their physical and chemical environment by influencing the movement of soil and water, and the supply of nutrients through processes of decomposition as well as by harvesting organic material and moving it through the landscape (Lobry de Bruyn and Conacher, 1990).

Biological soil crusts, sometimes called cryptogamic crusts, comprise a rich assortment of both macro- and micro-organisms present within the top few millimetres of the soil. These organisms include lichens, mosses, liverworts, cyanobacteria (blue-green algae), bacteria, fungi, procaryotic bacteria and eucaryotic bacteria. True soil crusts result from the intimate association of these small plants with the thin surface soil layer. Crusts are common components of arid and semi-arid rangelands, and are widely distributed in the central and south-west of the state. They provide food and shelter for soil fauna, regulate the flow of water and nutrients through the soil (Eldridge and Tozer, 1997), provide insolation to the soil surface, fix nitrogen and convert carbon dioxide into soil carbon, and bind the soil particles together near the surface, providing a barrier against raindrops and wind erosion (Eldridge and Greene, 1994). Their roles in soil and ecological processes make them useful indicators of soil health in rangelands (Eldridge and Koen, 1998), and the stability of some systems is intimately linked with the health of these organisms.

17.4 Measuring the Productive Potential of Soils

The traditional method of measuring the productive potential of rangeland soils is to measure the vegetation. While this might be appropriate in good seasons when plants are plentiful, it is often not feasible during poor seasons when most of the vegetation is either degraded or absent. A more effective technique is to measure the productive potential of the soil, that is, its capacity to capture and store water and nutrients and therefore to produce pasture and to maintain the stability of the rangeland.

Research during the past few years has shown that there are strong links between the physical characteristics of the soil surface, which can easily be recognised in the field, and its productive potential (Tongway, 1995). Important soil surface attributes include soil surface roughness, the coherence and degree of cracking of any physical surface crusting, the stability of soil surface aggregates in water, the amount and type of water and wind erosion such as sheeting and rilling, the type of eroded material (for example, sand, stone or gravel), and soil texture. Other biological factors such as biomass and cover of plants, cryptogamic crusts and the amount, cover and incorporation of litter are also measured, along with the cover of grass butts and obstructions on the soil surface that are likely to trap sediments which are the result of erosion.

Observations of the soil surface can be condensed into three categories. These indices give a measure of: (1) the stability of the surface or its capacity to withstand erosion; (2) the infiltration capacity of the soil, that is, how it conducts water and resists runoff; and (3) the nutrient capacity or how organic matter is recycled into the soil. These categories can be monitored over time to see how a site is changing in relation to its management, or the success of a reclamation program.

17.5 Erosion Processes in Rangelands

Rangeland soils are ancient, highly sorted and low in nutrients and minerals, and the landscapes from which these soils were derived are often flat and highly weathered (Stafford Smith and Morton, 1990). Grazing animals influence the vegetation and soil directly by disturbing the surface, and indirectly by reducing vegetative cover, making the soil susceptible to wind and water erosion. The main effects are outlined in Table 17.2. Overgrazing leads to a reduction in surface roughness, and an increase in compaction, crusting and hardsetting of the soil surface. There is little opportunity for water to pond on the surface, and ponded water is either lost through evaporation or is shed as runoff (Eldridge et al., 1995). Hardsetting and crusted soils are often associated with reduced levels of soil nutrients such as nitrogen and phosphorus (Tongway and Ludwig, 1990).

17.5.1 Water and Wind Erosion

Both wind and water erosion are a common feature of rangelands. Erosion may be human-induced or natural. An example of natural erosion is the 'Walls of China' at Lake Mungo in south-western New South Wales. Wind erosion is common in areas where the soils are sandy or poorly structured (see Chapters 2 and 12), particularly where plant cover is sparse. Soil erodibility is related to soil physical and chemical

Plate 17.1 View along a fence line showing two areas with markedly different levels of pasture use

properties such as texture, structure, aggregate stability and soil surface characteristics. Well-structured soils such as clays, and massive (no structure) soils such as sands, are resistant to water erosion. It is the intermediate group, the duplex or texture-contrast soils, which are the most susceptible. The amount of wind erosion is also dependent on the type and cover of vegetation, and the erosivity of the wind (see Chapter 2).

Water erosion is often the result of two common types of processes: splash and flow transport. Splash erosion occurs when raindrops dislodge particles from the soil surface. These particles are then moved by a process of overland flow and redeposited in other areas (Rose, 1993). Soils such as red earths are susceptible to splash erosion. Fine, detached soil particles clog up the surface soils resulting in a surface seal of low permeability, reducing infiltration and increasing runoff. Much of the rangelands is flat, and water moves slowly across the landscape. Particles detached by splash erosion are transported in a thin film of overland flow (Kinnell et al., 1990). In areas of higher slope, water often has little time to infiltrate and moves rapidly downslope. This creates rills and gullies that are difficult and costly to reclaim. Eroded soil is often deposited in rivers and streams.

Vegetation: the Key to Erosion Prevention 17.5.2

The key to protecting the soil against erosion is to ensure adequate vegetation cover on the soil (see Plate 17.1). In the higher-rainfall areas on the slopes and the coast, 70% cover is recommended for protection against water erosion (Lang, 1990). In the rangelands, however, these levels are rarely attained. Threshold cover levels for water erosion protection have not been identified despite considerable research (Eldridge and Koen, 1998; Greene et al., 1994). As a rough guide, however, land managers should aim to attain 30–40% cover in the far west and 40–50% on the eastern edge of the rangelands. This cover includes all living and dead plants as well as biological crusts.

Increasing the cover of plants, particularly perennials, on the soil surface increases the likelihood of water percolating into the soil. This is because perennial plants are often present during dry times when the more ephemeral (short-lived) plants have disappeared. Perennial grasses, for example, have well-developed root systems that allow the passage of water through the soil, and also encourage termite activity which further increases infiltration.

17.6 Reclaiming Rangeland Soils

Conventional methods of reclaiming rangeland soils have been based on an engineering approach using a range of agricultural implements to change the environment of the soil surface to make it more suitable for seeds and seedlings. Land degradation often results from unsustainable stocking rates, whether they are caused by domestic, native or feral animals.

17.6.1 Reclamation by Reducing Total Grazing Pressure

Reclamation may be achieved by manipulating grazing rates of domestic, native and feral animals. The decision to manipulate stocking rates, or use mechanical methods or a combination of both will depend on the landscape, the soil and vegetation type, seasonal conditions, and the degree of degradation. Reducing the total grazing pressure, however, often only results in an increase in weedy annual species or woody perennial shrubs that adapt well to heavy grazing. Overgrazing often alters the soil surface so substantially that the vegetation community shifts from desirable to undesirable species. For example around Cobar and Bourke, large areas of red earth soils have become invaded by woody shrubs of the genera *Cassia*, *Eremophila* and *Dodonaea* through inappropriate grazing management and possibly the suppression of wildfires.

17.6.2 Reclamation by Mechanical Methods

In many areas the soil surface has changed so dramatically that reducing grazing rates will have little effect on the landscape in the short to medium term. This is evident in a number of badly eroded exclosures set up by the Soil Conservation Service in the mid-1950s in western New South Wales. Despite complete removal of livestock over the past forty years, erosion is still evident in many areas. The objective of mechanical manipulation is to alter the surface characteristics of the soil to increase water penetration and entrapment of seed, thereby producing a favoured niche for developing plants. For seeds to germinate and plants to grow, essential resources such as water and nitrogen must occur at threshold levels both at the same time and in the same place (Tongway, 1990). Many reclamation techniques have failed because they have been undertaken in a season when one or more of the essential nutrients was absent.

A range of mechanical techniques has been trialled in the rangelands for reclaiming degraded soils (Table 17.3). Early work concentrated on chequerboarding and furrowing on scalded duplex soils in the Riverina (Jones, 1966) and on hard red ridges near Cobar (Cunningham, 1978) with moderate success. Around Broken Hill, contour furrowing has been used successfully using a single tine with attached mouldboard, mounted either on a tractor or on a trailer (Tatnell, 1992). Furrowing has proven useful for regenerating degraded pasture, particularly where dense stands of *Atriplex* or *Maireana* occur nearby. Sometimes seed and fertiliser are added during ripping. The furrows are designed to be long-lived (> 10 years) in order to increase water storage and therefore pasture regeneration and carrying capacity.

Waterponding has been increasingly popular in the eastern margins of the rangelands to reclaim large areas of scalded soils, that is, where the topsoil has been removed exposing the hard subsoil. Ponding banks are usually constructed with a grader to a height of 30–40 cm in order to pond water to a depth of up to 10 cm over areas of 0.5 to 1.0 ha (Rhodes, 1987). In areas near Nyngan, ponding has resulted in large increases in the amount of forage compared with scalded sites. Ponding works by leaching out salts from the normally saline soils and improving the structure of the subsoil, thereby increasing water-holding capacity.

Tine pitting and land imprinting have been used with little long-term success to reclaim eroded soils or rejuvenate degraded pastures. Tine pitting produces boat-shaped pits about 180 cm long, 30 cm wide and 20 cm deep over about 10–20% of the surface. The areas between the pits act as a catchment and the pits fill with water, trap seeds and revegetate the surface. While early successes were reported with tine pitting (Stanley, 1978), success is usually short-lived, with the pits either filling up with sand or disappearing after a few years. Because of the short-term nature of pitting, it has tended to be replaced by more permanent methods. Land imprinting is a similar technique, which pushes pits into the soil into which seeds and water are designed to be stored. Various combinations of pitters and furrowers have been adapted to construct implements to regenerate degraded rangelands, for example the Mallen Niche Seeder and the Crocodile Seeder.

Management of Rangeland Soils

Extensive pastoralism, predominantly the grazing of sheep or cattle on native pastures, is the principal land use over much of the rangelands. Predominant land uses over smaller areas include national parks, forestry, tourism and mining. The two major impacts upon the soil are disturbance (generally trampling by grazing animals) and in some areas, fire.

17.7.1 Grazing and Fire

Grazing animals influence a number of soil processes directly by either compacting or destroying the soil surface, or by removing pasture. Indirectly this leads to reduction in vigour of plants, a replacement of desirable species with less desirable ones, lower populations of soil organisms and eventually a decrease in the productive potential of the soil (Table 17.2). The impact of overgrazing and trampling is most marked close to water points (dams and troughs). Soil compaction occurs where sheep activity is greatest, and this is usually on the narrow tracks radiating from the water point (Andrew and Lange, 1986). Here, trampling and excretion by sheep lead to increases in soil nutrients and the dominance of grazing-tolerant species of poor nutritive value. Destruction of the biological crust removes a source of biologically fixed nitrogen (Eldridge and Greene, 1994), and once the protective lichen crust has been removed, transport of detached soil through overland flow (Eldridge, 1998) or wind erosion increases. The destruction of the soil surface by trampling is greatest in soils with hardsetting surfaces such as red and brown duplex

Plate 17.2 Old man saltbush is frequently used to reclaim saline sites and provide valuable forage for livestock

soils. Heavy clay soils are the least susceptible as they are highly aggregated and able to recover rapidly after rain. Although they are quite resistant to water erosion, they are susceptible to wind erosion, particularly when trampled during droughts.

Fire is a regular feature of many arid and semi-arid landscapes, and occurs either as wildfires or as a result of prescribed burning. In the semi-arid woodlands around Cobar and Bourke, fire has been used economically to control the invasion of woody perennial shrubs (Burgess, 1988), effectively reducing shrub biomass and promoting pasture growth (Hodgkinson and Harrington, 1985). The frequent use of fire destroys biological soil crusts and changes the morphology of the soil surface, reducing organic matter and making the surface unsuitable for germination. Frequent use of fire (annually) has been shown to result eventually in lower infiltration rates (Greene et al., 1990)

Management Strategies to Prevent Degradation

Experience has shown that low-risk or conservative stocking is the most appropriate strategy to prevent land degradation in rangelands. Economic analyses have shown that in many situations, pastoralists can maintain their incomes, and even increase profits, by reducing the number of grazing animals per unit area. This comes about by using better stock and reducing some of the fixed costs associated with managing a grazing enterprise. Strategies that reduce grazing intensity, by either providing more watering points or dispersing stock over larger paddocks, will reduce the risk of damage to the soil and the vegetation. Other useful strategies include the use of 'sacrifice' paddocks within which drought feeding of stock can be carried where necessary so that degradation is confined to small areas. When fire is used as a management tool, it is important to consider fire intensity and frequency, as these may affect the long-term consequences of a fire. Consideration also needs to be given to the combined effects of other stresses such as drought and grazing, which may occur following fire.

Finally, the key to managing soils is to manage the vegetation, and a proven technique is for landholders to undertake a property planning process. Property planning allows property managers to identify risks and opportunities in their enterprises, to plan those areas that require remediation and which will gain the maximum benefit from a revegetation program (see Plate 17.2). Planning enables landholders to determine appropriate stocking rates to maintain the long-term sustainability of the landscape.

BIBLIOGRAPHY

Andrew, M.H. and Lange, R.T. (1986), 'Developments of a new piosphere in arid chenopod shrubland grazed by sheep. I. Changes to the soil surface', *Australian Journal of Ecology* 11, 395–410.

Burgess, D.M.N. (1988), 'The economics of prescribed burning for shrub control in the semi-arid woodlands of north-west New South Wales', *Australian Rangeland Journal* 10, 48–59.

Cunningham, G.M. (1978), 'Waterponding and waterponding on Australian rangelands', in K.M.W. Howes (ed.), *Studies of the Australian Arid Zone. III. Water in Rangelands*, CSIRO, Perth.

Eldridge, D.J. (1985), *Aeolian Soils of South-western New South Wales*, Soil Conservation Service of NSW, Sydney.

Eldridge, D.J. (1988), 'Soil-landform and vegetation relations in the chenopod shrublands of western New South Wales', *Earth Science Review* 25, 493–9.

Eldridge, D.J. (1998), 'Trampling of microphytic crusts on calcareous soils and its impact on erosion under rain-impacted flow', *Catena* 33, 221–39.

Eldridge, D.J. and Greene, R.S.B. (1994), 'Microbiotic soil crusts: a review of their roles in soil and ecological processes in the rangelands of Australia', *Australian Journal of Soil Research* 32, 389–415.

Eldridge, D.J. and Koen, T.B. (1998), 'Cover and floristics of microphytic soil crusts in relation to indices of landscape health', *Plant Ecology* 137, 101–44.

Eldridge, D.J. and Tozer, M.E. (1997), *A Practical Guide to Soil Lichens and Bryophytes of Australia's Dry Country*, Department of Land and Water Conservation, Sydney.

Eldridge, D.J., Chartres, C.J., Greene, R.S.B. and Mott, J.J. (1995), 'Management of crusting and hardsetting soils under rangeland conditions', in H.B. So, G.D. Smith, S.R. Raine, B.M. Schafer and R.J. Loch (eds), *Crusting, Sealing and Hardsetting Soils, Productivity and Conservation*', Australian Society of Soil Science, Brisbane.

Garner, W. and Steinberger, Y. (1989), 'A proposed mechanism for the formation of "Fertile Islands" in the desert ecosystem', *Journal of Arid Environments* 16, 257–62.

Greene, R.S.B., Chartres, C.J. and Hodgkinson, K.H. (1990), 'The effect of fire on the soil in a degraded semi-arid woodland. I. Physical and micromorphological properties', *Australian Journal of Soil Research* 28, 755–77.

Greene, R.S.B., Kinnell, P.I.A. and Wood, J.T (1994), 'Role of plant cover and stock trampling on runoff and soil erosion from semi-arid wooded rangelands', *Australian Journal of Soil Research* 32, 953–73.

Hodgkinson, K.C. and Harrington, G.N. (1985), 'The case for prescribed burning to control shrubs in eastern semi-arid woodlands', *Australian Rangeland Journal* 5, 3–12.

Johns, G.G., Tongway, D.J. and Pickup, G. (1984), 'Land and water processes', in G.N. Harrington, A.D. Wilson and M.D. Young (eds), *Management of Australian Rangelands*, CSIRO, Australia.

Jones, R.M. (1966), 'Scald reclamation studies in the Hay district: natural reclamation of scalds', *Journal of the Soil Conservation Service of NSW* 22, 147–60 (I), 213–30 (II).

Kinnell, P.I.A., Chartres, C.J. and Watson, C.L. (1990), 'The effect of fire on the soil in a degraded semi-arid woodland. II. Susceptibility of the soil to erosion by shallow rain-impacted flow', *Australian Journal of Soil Research* 28, 779–94.

Lang, R.D. (1990), 'The effect of ground cover on runoff and erosion from plots at Scone, New South Wales', MSc Thesis, Macquarie University.

Lange, R.T. (1969), 'The piosphere: sheep tracks and dung patterns', *Journal of Range Management* 22, 396–400.

Lobry de Bruyn, L.A. and Conacher, A.J. (1990), 'The role of termites and ants in soil modification: a review', *Australian Journal of Soil Research* 28, 55–93.

Ludwig, J.A. and Tongway, D.J. (1995), 'Spatial organisation of landscapes and its function in semi-arid woodlands, Australia', *Landscape Ecology.* 10, 51–63.

Noble, J.C., Cunningham, G.J. and Mulham, W.E. (1984), 'Rehabilitation of degraded rangelands', in G.N. Harrington, A.D. Wilson and M.D. Young (eds), *Management of Australian Rangelands*, CSIRO, Australia.

Rhodes, D. (1987), 'Waterponding banks—design, layout and construction', *Journal of the Soil Conservation Service of NSW* 43, 80–3.

Rose, C.W. (1993), 'Soil erosion by water', in G.H. McTainsh and W.C. Boughton (eds), *Land Degradation Processes in Australia*, Longman, Melbourne.

Stafford Smith, D.M. and Morton, S.R. (1990), 'A framework for the ecology of arid Australia', *Journal of Arid Environments* 18, 255–78.

Stanley, R.J. (1978), 'Establishment of chenopod shrubs by tyne pitting on hardpan soils in western New South Wales', *Proceedings of the 7th Biennal Conference of the Australian Rangelands Society*, Cobar, 84–91

Tatnell, W. (1992), 'The effect of site condition, exclosure and contour furrowing on pasture changes over a five year period', in *Proceedings of the 7th Biennal Conference of the Australian Rangelands Society, Cobar*.

Tongway, D.J. (1990), 'Soil and landscape processes in the restoration of rangelands', *Australian Rangeland Journal* 12, 54–7.

Tongway, D.J. (1995), 'Monitoring soil productive potential', *Environmental Monitoring and Assessment* 37, 303–18.

Tongway, D.J. and Ludwig, J.A. (1990), 'Vegetation and soil patterning in semi-arid mulga lands of eastern Australia', *Australian Journal of Ecology* 15, 23–34.

Westoby, M. (1979/80), 'Elements of a theory of vegetation dynamics in arid rangelands', *Israel Journal of Botany* 28, 169–94.

Whitford, W.G. and Herrick, J.E. (1996), 'Maintaining soil processes for plant productivity and community dynamics', in N.E. West (ed.), *Proceedings of Vth International Rangelands Congress*, Salt Lake City, Utah.

Soils and Coastal Dune Management

C H A P T E R   1 8

P.A. Conacher and R.J. Stanley

The essential soil resource of the coastal strip is the sand, which comprises the basic material of the beach and dune areas along our coastline. Management of these areas to ensure appropriate long-term use of the coastline thus depends on a knowledge of the interaction of this material with the vegetation necessary to ensure its stability.

Concepts in coastal dune management rely on the reduction in transport of wind-blown sand when it encounters the ridged form of the frontal dune. The uniformly elevated form of this sand dune, interacting with zoned vegetation, is recognised as one of the basic requirements for coastal stability, and is used as a basis for reducing the amplitude of foreshore fluctuations. These concepts are used in conjunction with an understanding of coastal processes and must account for the supply and demand for sand arising from ocean changes (see also Chapter 2).

The basic principle in dune management is to maintain a satisfactory vegetative cover on the frontal dune, which prevents sand blowing inland where it can damage developments and where it is lost from the beach system.

Management of coastal dunes may involve the application of sound land use principles, control of recreational activity or rehabilitation of disturbed dunes. Management plans are generally site specific and should be formulated and treated as such.

18.1 Coastal Dune Soils

Soils developed from dune sands may be quite complex and fragile. Despite a relative uniformity in parent material, soils formed from unconsolidated littoral sands have developed under different conditions of climate and colonising vegetation to become as varied as the plant communities themselves. Just as vegetation needs time to establish and protection to develop to a climax community, dune sands require time and vegetative protection to endure and develop to a mature soil.

At the start of any dune rehabilitation program, an examination of soil types, conditions and properties is essential to identify any problems that may be present. Techniques used in revegetation can then account for soil problems once the relationship between plant and soil is understood.

Soils that have developed on coastal sands reflect the various soil-forming processes that operate near the coast (see Chapter 1). These processes are not independent and interact closely to provide the resultant soil. Their influence is seriously interrupted by outside influences. Poor land use and management, traffic, deforestation, mining and repeated burning are among the practices affecting soil-plant relationships. Reinstatement of affected areas needs to re-establish, as far as possible, the predisturbance situation if revegetation is to be permanent and maintenance-free.

Soil Characteristics 18.1.1

(a) Particle Size

Most coastal sands along the New South Wales coastline are of relatively uniform particle size. This indicates that the coastal sands have been thoroughly sorted. Because of this high degree of sorting, the pore space between the grains of sand is a large percentage of the total sand volume, a factor that has an important bearing on their water-holding capacity.

(b) Moisture Retention and Drainage

In the deep sandy soils of foredunes, the high percentage of pore space provides an excellent medium for free drainage. The pore space between grains may be too large for the water film to adhere and, as a result, water freely flows through the profile, leaching salt and nutrients that may be present on the surface.

As vegetation establishes on bare sand, organic matter from the vegetation and associated fauna is incorporated into the soil. Water-holding capacity of the soil is closely dependent on organic matter content; as time progresses and organic matter builds up, water-holding capacity increases. A general rule to follow is that soils low in organic matter will generally dry out much quicker than soils that have a high organic matter component. This drying behaviour has important implications when it comes to revegetating dune soils.

Old dune soils have a density approaching 50% of that of young dune soils. Old dune soils have developed under vegetation over a longer period of time than younger, more seaward sands, and have a higher water-holding capacity. It may vary from 7% (by volume) in young dune sand to 33% in old dune sand.

(c) Soil Fertility

Soil fertility is governed by two factors:
—*physical factors*: related to features such as soil structure, air porosity and structural strength
—*chemical factors*: related to pH, nutrient availability and salinity.

Dune soils are generally considered to have low fertility, which decreases as the depth from surface organic layers increases. This confines plant roots to the upper horizons of the profile and increases the plant's reliance on frequent rainfall stored by organic matter near the surface.

Revegetation programs implemented on foredunes require the addition of artificial fertiliser to assist in plant establishment and growth.

(d) Erodibility

Erodibility is an intrinsic property of soils used to describe their susceptibility to erosion. It is generally related to the detachability of their particles and their infiltration characteristics with respect to water erosion, and their detachability and aerodynamics with respect to wind erosion (see Chapter 12).

18.1.2 Soil Types

The major soil types present in dune systems along the New South Wales coastline are:
—siliceous beach sands
—siliceous dune sands
—sand podzols
—acid peats
—dredged material.

Within each of these soil types, there may be minor variations in soil physical and chemical characteristics, resulting from minor differences and gradual changes in the soil-forming processes. More importantly, between soil types there can be significant differences in soil characteristics, including particle size, water-holding capacity, drainage, fertility, pH and erodibility.

Soil types and soil characteristics generally vary according to distance from the sea, topographical position, vegetation and landform drainage. The location of the major soil types within the coastal dune system is shown schematically in Figure 18.1. However, the actual location of these soils can vary significantly in relation to sequence and inland distance, depending on a number of variables. On a receding coastline, for example, sand podzols may adjoin siliceous beach sands.

When developing dune rehabilitation or management strategies, an understanding of the main characteristics of each soil type is important. These characteristics are outlined in the following paragraphs.

(a) Siliceous Beach Sands

These sands are loose fawn-coloured sands with no soil structure and no profile development. They are deep well-graded sands that are free draining and have low water-holding capacity.

Siliceous beach sands are found on the beach berm and incipient foredunes. Quite often these sands are utilised in dune construction. It is important, when they are used for this purpose, that they be allowed time to settle before revegetation. Time to settle, and rainfall or irrigation, will reduce the pore space between the sand grains (thus reducing free drainage and increasing water-holding capacity), and reduce salt levels in the soil by leaching.

These soils are very low in nutrients, and fertilisers must be added when bare areas are revegetated.

(b) Siliceous Dune Sands

With the natural establishment of dune vegetation on siliceous beach sands, these soils gradually develop into siliceous dune sands.

Dune sands are fawn to pale grey sands showing little profile development except for some accumulation of organic matter (producing a typically dark brownish grey colour) in the surface 30 cm. These are deep profiles with no soil structure developed. Permeability is high and they are very well drained to a deep-water table.

Like siliceous beach sands, these soils are low in nutrients and require fertiliser application to assist plant establishment.

Both siliceous beach sands and siliceous dune sands are relatively 'young' soils comprising Holocene sediments.

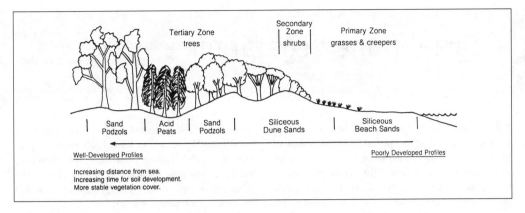

Figure 18.1 A schematic representation of soils and vegetation in a coastal dune system

(c) Sand Podzols

Podzols are acid sandy soils with strongly differentiated profiles distinguished by: a grey, organic A_1 horizon; a thick, whitish, sand A_2 horizon; and with a reddish or yellowish brown to black B horizon dominated by the accumulation of organic matter and/or sesquioxides. Podzols form as a result of strong leaching.

Often a hardpan (known as 'coffee rock' or indurated sand) forms at the top of the B horizon where organic and iron deposits are cemented together by the action of a fluctuating water table. The hardpan inhibits water infiltration, restricts the penetration of roots and air, and is relatively infertile.

Quite often the A horizons of this soil type may have been eroded, exposing the hardpan which is then extremely difficult to revegetate. Plants can be established by deep ripping the hardpan and covering it with at least 30 cm of siliceous beach or siliceous dune sands.

Sand podzols are generally found landward from the frontal dune, under vegetation dominated by heath communities and sometimes by stands of eucalypt forest. They are much 'older' soils comprising Pleistocene sediments.

(d) Acid Peats

This soil type commonly occurs in the swale areas behind the foredune where impeded drainage and a shallow water table occur. Acid peats show little horizon differentiation and the dominant characteristic is a surface accumulation of black, peaty organic matter, which is strongly acid and has a high water-holding capacity.

Acid peats form by the accumulation of organic matter from the associated vegetation under predominantly anaerobic conditions. They are often associated with swamp vegetation in swales and also occur in estuarine areas dominated by *Casuarina*

species. Because of their acid nature and the presence of a high water table, it is essential to select plant species for revegetation that are adapted to these soil characteristics. Alternatively, these soils can, if necessary to obtain desirable dune shape, be covered with several metres of another sandy soil type and revegetated with more conventional dune plants.

(e) Dredged Material

The use of sand dredged from bays and river mouths as foundation material has grown in recent years. The Port Botany Development Project and the Kooragang Island Project at Newcastle are two examples of large-scale development involving the use of dredged material for land reclamation. Sand dunes may also be artificially constructed from dredged material. Frequently, dredged material needs to be stabilised by vegetation.

Experience has shown that dredged material may have a very low pH, may harden when dried, may be so well sorted as to not retain any moisture and may be chemically infertile. It is recommended that any such material be carefully analysed for heavy metal concentrations, nutrient levels, water-holding capacity, salinity and pH. Analysis will provide invaluable assistance when planning and implementing revegetation programs. Problem areas can be isolated and solutions to these problems may be found before the costly step of revegetation is commenced. Dredged material should also be tested for its potential to produce an acid sulfate soil.

The salinity levels of recently dredged sand do not pose many problems because salt is leached out of the top few centimetres of sand after about 50 mm of rain has fallen (or the equivalent in irrigation), provided that infiltration and drainage are normal. The rapid leaching of salts from the soils, however, also means that rapid leaching of applied fertiliser may occur, making frequent and regular applications of fertiliser necessary to support plant growth.

18.2 Coastal Dune Vegetation

Native coastal dune vegetation can have a profound impact on several of the pathways of the sediment cycle already illustrated in Figure 2.12. It can, therefore, exercise considerable influence on the rate of net coastal erosion or accretion. Dune vegetation is highly adapted and, combined with the dune ridge, will provide protection for other terrestrial species; dune species are essential to reduce the exposure which would quickly kill other species. They do this by a high degree of adaptation, zonation and interdependence (see Figure 18.1).

In a well-vegetated dune system, a foundation of primary zone grasses colonises lower parts of the beach and traps abrasive sand particles. This action initiates the development of a higher dune, which can be colonised by secondary zone shrub species to provide a wind-deflecting wall near the shoreline. Finally, a tree zone, which will further elevate the wind and provide shelter to vegetation, is provided by tertiary zone coastal species.

Native coastal vegetation is well adapted to withstand harsh conditions, such as strong wind, salt spray, sand blast and low nutrient levels. It is at the same time, however, a fragile community that is easily damaged by disturbance. Dune stability relies on keeping the three zones intact. The loss of any component can cause deterioration in other parts.

18.3 Coastal Land Use

At present, three-quarters of the Australian population lives within 40 km of the coast. Beach areas are very popular for recreation and holidays. Industries are also attracted to the coastal region by the availability of water from coastal lakes, heavy minerals and sand from coastal dunes and estuaries. Associated with these developments is a range of infrastructure—roads, harbours, airports, railways and waste disposal sites. Much of this usage and development conflicts with the fragile environment of the coastal zone.

18.3.1 Urban and Residential Development

Much of the early development in coastal areas allowed for a beachfront buffer zone between private property and the active beach area. This buffer zone accommodated the short-term fluctuations of the shoreline and provided a measure of natural resistance to the erosive forces of the sea. With the passage of time, the buffer zone has been eroded in many places, leaving some development under threat from wave attack during storms. More recent develop-

ment on the foreshore often did not allow for a buffer zone and it, too, is under threat from wave attack during storms.

To avoid this situation recurring, it is important to understand the coastal process as detailed in Chapter 2. Simply put, the natural recuperative processes which restore a beach after an erosion event rely on sand being transported back to the beach by wave and wind action during calm periods, thus replenishing the beach and dunal system. If this is interfered with, for example by removal of the dune, then developments may have to be abandoned to the sea or protected by special structures. The construction of protective works such as rock or concrete walls significantly degrades the beach amenity that originally attracted people to the area.

In undeveloped areas, the opportunity exists to plan development and thus avoid future problems. In such areas, no permanent development should be allowed within an area likely to affect or be affected by coastal processes within, say, 100 years. However, on an unstable shoreline, the developed area will finally come under threat. Therefore, it is desirable to be as generous as possible in defining that area and to set aside as great a depth of land as possible for low-intensity uses that can economically be abandoned when necessary.

Careful attention must be paid to the siting of stormwater discharge pipes from urban areas. Incorrect location may result in erosion of the dune and beach and deposition of rubbish and litter which detracts from the aesthetics of the area. Where possible, stormwater pipes should discharge onto rock shelf areas where storm flows can dissipate without causing erosion.

18.3.2 Recreational Use

Recreational use of beaches has increased and diversified markedly over recent years, resulting in use throughout most of the year. On the more popular beaches, pedestrian and vehicular access can lead to the destruction of dunes and their vegetation. This can be avoided to a large extent by fencing off sensitive areas, providing formal accessways for pedestrians and vehicles, and allocating less-sensitive areas for use by vehicles. Formal parking areas and amenities can be located as far as possible behind the frontal dune to prevent damage to the structures from storms.

18.3.3 Heavy Mineral Mining and Sand Extraction

Heavy mineral mining removes only a small portion of the total sand mass. However, it changes the geomorphology and soil structure of the dune system.

Effects may include:
—broadening and lowering of the frontal dune
—seaward or landward displacement of the frontal dune
—where indurated sand beds (that is, sediments cemented together to form soft rock) have been removed, stability of the frontal dune is decreased and it thus becomes susceptible to wave erosion
—disturbance to groundwater flow patterns
—destruction of dune vegetation. Although mined areas are commonly revegetated, it takes many years of careful management for a mature sequence of stable vegetation types to develop.

The consequences of these changes should be given careful consideration prior to any mining operation.

Sand extraction for building materials has far greater impacts on the dunal area than does heavy mineral mining, as all of the material is removed. It must be recognised that sand extraction results in:
—permanent loss of sand from the beach system
—reduced prospects for natural regeneration of vegetation due to the time taken for the mining operation
—limitations on subsequent land use options due to the drastic modifications of the landscape.

Considering these factors, extraction sites should be restricted to relatively small sites in non-sensitive areas that can be satisfactorily restored, and the volumes removed should not have a significant effect on the local coastal processes.

18.3.4 Grazing

Grazing is a completely unsatisfactory use of dunes. Stock consume large quantities of dune grasses and trample large amounts while grazing. Past grazing activity has left a legacy of uncontrolled sand drift in some areas, due not only to the loss of vegetation but also to the direct trampling and disturbance of the dunal material by hooved stock.

18.3.5 Effluent Disposal

Disposal of sewage effluent from coastal settlements has become a matter of major concern in recent years due to:
—rapid growth in permanent populations accompanied by intermittent heavy demand of transient users during holiday periods; this has put great strain on existing facilities
—a shortage of land suitable for sewage treatment; the popularity of the coastal zone means that land values are high and competition from alternative land use is strong.

Options available for the discharge of secondary treated effluent include outfall into waterways (rivers, lakes and so on), ocean outfall or through absorption in dune areas. Only the last of these options is addressed here.

The ecosystems in the coastal region are very sensitive and have limited ability to accept changes outside those occurring naturally, that is, drought, fire and storm. Seemingly minor alterations to soil moisture, nutrient status and drainage may lead to dieback in canopy species with stimulation of weed and grass species. When this occurs, the result is aesthetically unattractive and ecologically unstable.

Little is known about the effect on beach stability of increased groundwater levels as a result of effluent disposal in back beach areas. Cursory examination of the behaviour of a dune between a lagoon and ocean would suggest that there is minimal adverse effect if changes in the water table are small. However, more research is required before a definitive judgment can be made.

Because of the high absorptive capacity of large quantities of sandy materials, effluent disposal on beach dunes could be an efficient form of disposal.

However, a thorough appraisal of the environmental impact is required before works are implemented. This appraisal must include assessment of the ability of the vegetation to perform in the changed environment, possible effects on dune stability and likely long-term effects of a rising water table in dunal areas.

18.4 Techniques of Dune Rehabilitation

Where it is not possible to preserve the pristine state of a natural dune system for reasons of changing land use, extreme erosion, fire, disease or the like, then planning, rehabilitation and management must reinstate, as a prime requirement, form and function to the dune.

Areas that have been successfully rehabilitated and revegetated require long-term management and maintenance somewhat different from that which would be undertaken for undisturbed areas.

18.4.1 The Process of Dune Stabilisation

If a self-sustaining cover of vegetation is not present, rehabilitation or maintenance programs are introduced to ensure the development and persistence of vegetative protection. These programs often include enclosure of primary and secondary vegetation by fencing and a continuing program of fertiliser top-dressing, plant replacement and pest control. Weed control also has a major influence on the success or failure of a program.

If the dunal barrier is severely degraded by loss of vegetation and subsequent loss of sand, the dunal barrier must first be re-formed. Subsequent to the severe degradation of the stabilising mantle of vegetation, wind-blown dunal sand migrates landwards, resulting in the formation of gaps or blowouts in the dune barrier. These gaps need to be infilled, using either the original sand or sand from the beach berm. This infilling can be achieved by either sand-trapping fences or the use of heavy earthmoving machinery.

18.4.2 Stabilising the Dunal Barrier

(a) Primary Vegetation

Having re-formed the dunal barrier, all bare sand areas are planted to primary stabilising grasses— either marram grass in the winter months or sand spinifex in the summer months.

Very exposed sites, and areas sown to spinifex seed, require temporary surface stabilisation either by brush matting or by spraying a liquid stabiliser on the soil surface, to hold the sand in place until the primary grass cover becomes established.

All planted areas are fertilised and protected by fencing. Formal access to the beach is provided, where necessary, by board and chain walkways. The erection of suitable marked signposts, both to inform the public of the stabilisation program and to direct pedestrian access to the beach, is critical to the success of the work.

(b) Secondary and Tertiary Vegetation

Once a satisfactory cover of marram grass and sand spinifex has been established, secondary shrub and tertiary tree species can be added to the vegetative cover, depending on location. Generally, this type of vegetation can be established no closer to the sea than the crest of the frontal dune. However, secondary shrub species such as coastal wattle, tea-tree and horsetail oak can be established on the landward face of the frontal dune, in the immediate protection of the dune crest.

Taller-growing tertiary tree species such as banksias, eucalypts, casuarinas and tuckeroo can be progressively introduced, over time, in a gradational landwards succession.

18.4.3 Long-term Management and Maintenance

Any dunal system subject to development, recreation and industrial pressures will require management and maintenance in perpetuity. Good management of undisturbed coastal dune vegetation is as important as continued maintenance of rehabilitated dune areas.

The management strategy should include the maintenance of protective fencing surrounding critical vegetated areas, the maintenance of adequate formal beach access, the prompt repair of any vegetative damage from whatever cause and the progressive upgrading of vegetation in the hinterland.

Weed control and the elimination of fire are also important aspects of good coastal dune management. Fires not only expose the dune to risk of wind erosion but also allow the invasion of more fire-prone plants such as bracken fern and blady grass, which increase the risk of further fires.

The most serious weed threat to coastal dune systems on the north coast is the South African plant, bitou bush. This invasive plant eventually overruns and destroys the native dune plants, and reduces the effectiveness of the dune function. The Department of Land and Water Conservation, Department of Agriculture, CSIRO and Dunecare groups are actively investigating solutions to the bitou bush problem, and encourage bitou bush control programs as part of dune management works.

With the exception of some beaches on the more desolate parts of the coast, it is not possible to restore the dunal system as a 'once only' operation and not engage in a follow-up management plan.

BIBLIOGRAPHY

Anon (1986), *Beach Dunes: Their Use and Management*, Public Works Department and Soil Conservation Service, NSW Government Printing Office, Sydney.

Conacher, P.A., Joy, D.W.B., Stanley, R.J. and Treffry, P.T. (eds) (1988), *Coastal Dune Management*, Soil Conservation Service of NSW, Sydney.

Jacobs, S. (undated), *Understanding Beach Sands*, Soil Conservation Service of New South Wales (now New South Wales Department of Land and Water Conservation).

Soils and Revegetation

CHAPTER 19

W.H. Johnston

Effective long-term soil conservation and sustainable catchment management depend on maintaining ground cover. This chapter outlines some of the principles of soil conservation agronomy, including the importance of recognising soil properties, establish-ing perennial ground cover, improving an existing sward and managing it to meet longer-term objectives. Integration of revegetation with other soil conservation practices is stressed.

19.1 Plants and Soil in Perspective

Soil and vegetation enjoy a mutually beneficial relationship. On the one hand soil provides plants with anchorage, water and nutrients; on the other, plants protect soil from the agents of erosion, chiefly wind and water. Plants also modify the soil physically and chemically and maintain it in a biologically active state, and they are responsible for the main terrestrial component of the hydrological cycle. Herbivores and detritus feeders are also vital components of the ecosystem, forming the third apex of a triangle of biological energy and nutrient-flow pathways which links soils, plants and animals.

Relationships between plants, soils and animals are subtle and complex. Plant root systems gather nutrients that are incorporated into tissue. With the help of herbivores and detritus feeders—fungi and bacteria as well as macro-organisms—nutrients are cycled back to the soil and become concentrated in the surface layers. The ongoing process of soil forma-tion, coupled with this biological activity, causes the topmost soil layer to become the most active and nutrient-rich soil horizon, and therefore the most vital to plant growth.

Although most Australian soils are generally infertile, native plants have no difficulty in making the most of the nutrient pool. In stable native plant communities, rates of long-term dry matter production equal rates of decomposition so that net biomass gain is essentially zero. Species diversity of stable plant communities is high but it may change rapidly if circumstances change, particularly if modifying or dominant plants are affected. In contrast, agricultural plant communities are managed for high net productivity and low species diversity. Nutrient cycling rates in sown crops and grazed pastures are high, nutrient export in produce depletes the nutrient pool and because biomass is harvested, there is little accumulation.

19.2 Plants and the Hydrologic Cycle

The water balance is used to describe the fate of rainfall. Water income (rainfall and water moving from other elements of the landscape, such as runon and lateral flow within the soil) must equal the balance of changes in the amount of water stored in soil minus water loss, which may occur via a number of

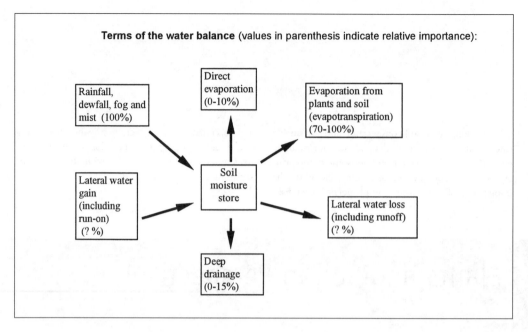

Figure 19.1 Terms of the water balance (values in parentheses indicate relative importance)

pathways including direct evaporation (rainfall that never reaches the ground), surface runoff, evaporation from bare soil and transpiration through plants (often referred to as evapotranspiration), lateral (sideways) movement within the soil, and deep drainage. Rainfall and evaporation components of the water balance are the largest terms (see Figure 19.1).

Changes in any one component of the water balance must appear somewhere else. For instance, removing a forest from the landscape reduces interception storage, which increases the amount of rainfall reaching the ground. The capacity of the soil to take up water is fixed, so if the reduction in direct evaporation is not balanced by increased water use, deep drainage and/or lateral flow must increase.

Direct soil evaporation does not continue at a high rate once the top several centimetres of soil has dried out. Thus the rate of evapotranspiration depends mainly on ground cover by actively transpiring leaf. When ground cover is complete, and water is available to plant roots, transpiration occurs at a rate of about 0.7 times the atmospheric water demand as measured by a standard evaporation pan. As ground cover declines, and soil water becomes less available,

rates of transpiration relative to atmospheric water demand decline also.

Water is not 'pumped' by plants. Evaporation is a passive process and it is impossible to evaporate water at higher rates than the atmosphere will accept. During seasons of low evaporative demand, transpiration is depressed regardless of leaf area and soil water availability. This has important implications for understanding the causes of deep drainage, water logging and dryland salinity, which will be outlined below.

Aside from their role in transpiration, plants also impact on other components of catchment hydrology. They maintain soil infiltration rates by reducing the impact of raindrops, slowing down rates of overland water flow and binding soil particles together, and they arrest the natural tendency for soil particles, and entire soil horizons, to move downhill. The water yield, rates and volumes of runoff from catchments are affected by plant-dependent characteristics of soils such as their water-holding capacity, infiltration capacity, seasonal moisture content and general levels of biological activity. The nature and seasonal moisture demand of the herbaceous ground cover, and the presence or absence of litter, are also important.

19.3 Agricultural Development and Land Degradation

Throughout southern Australia, agricultural development has drastically changed the structure and species composition of the original plant communities. In addition to the wholesale loss of tree cover, the changes, which seem to have been inevitable, can be summarised as a switch in composition from summer- to winter-growing species, from perennials to annuals, and from native to introduced species (Johnston, 1996). In its extreme, the capacity for summer growth has been lost.

Agricultural development has increased rates of nutrient turnover and reversed the annual patterns of moisture use and soil moisture availability. Replacement of plants that actively transpire water in summer with agricultural crops and pastures that use water mainly in winter and spring, leaves soils dry in spring and early summer but wetter in late summer and autumn. This increases rates of deep drainage in winter.

On the other hand, increasing water use in spring creates a more arid environment in summer. In many cases, once-dominant ground plants have disappeared. Complexity has declined in favour of a small group of replacement species that include warm-season grasses such as *Chloris*, *Eragrostis*, *Panicum*, *Sporobolus*, *Aristida* and *Bothriochloa*; the cool season grass species *Stipa*, *Danthonia*, *Hordeum* and *Lolium*; and a range of cool season legumes (mainly species of *Trifolium* and *Medicago*) and broad-leaved weeds.

Many of the changes wrought by agricultural development have been accompanied by increased rates of soil erosion, soil degradation in the form of salinisation, acidification, sedimentation, structural breakdown and other problems, and the movement of soil into streams and rivers.

Destruction of protective plant cover and loss of nutrient-rich topsoil depletes the nutrient capital and results in an inhospitable soil environment for plant establishment and growth. Soil erosion depletes the seed pool while high runoff rates cause most of the seed that falls to be lost. Livestock often prefer to camp on bare ground. If salts are present, they lick the soil and it becomes trampled and dense. Clayey, hardsetting subsoils hold less moisture, they restrict root penetration, and the volume of soil that can be exploited by the roots of plants that manage to establish is greatly reduced. Microsites available for seed lodgement, germination and from which root entry can be gained are also restricted.

Once a focal point for soil erosion has established, erosion usually increases in area unless positive steps are taken to allow or promote the regenerative process. This process depends entirely on plants, and the steps required are aimed to encourage revegetation, which may occur naturally when grazing is discouraged, or may be hastened by practices such as cultivation, deep ripping, sowing seed and applying fertiliser.

19.4 Planning a Revegetation Program

Revegetation recommendations should be based on a sound knowledge of the site, the soil, the species available that are likely to persist, seasonal and other factors that affect germination and establishment, and the range of possible techniques that can be used to overcome or offset the site factors. Background information, such as rainfall data, moisture balance information, temperature data and information regarding seed availability, should be used in planning a program and evaluating risks. Expert advice should be sought if the problem is unusual or complex, and it is useful to discuss, question and criticise proposals to expose weaknesses or practical problems.

Aims and Objectives   19.4.1

Revegetation aims to re-establish ground cover on eroded or disturbed areas. The productive capacity of the ground cover in terms of some form of projected use is a secondary consideration. Plants useful for revegetation may not be favoured for their production. Likewise, highly productive plants may be a nuisance in revegetation situations. Each case must be considered on its merits, and the choice of methods tailored to suit the situation and longer-term landscape goals.

Recommendations for revegetation should consist of a range of best-bet options that take account of costs and practical aspects of implementation.

Because erosion rates increase rapidly at ground cover levels less than 75% (Lang, 1979), revegetation should aim to achieve a vertical projection of ground cover by plant leaves and stems of between 75% and 100% in the first season. If erosion continues, it is unlikely that the newly established ground cover can be maintained.

19.4.2   Revegetation Species

Commercial cultivars of grasses and legumes are mainly used for revegetation because native grasses are not widely available. Commercial pasture plants have been selected mainly for their performance under relatively good soil and management conditions, and under grazing. Apart from climatic suitability, they have been selected using criteria such as dry matter production, digestibility, palatability, lack of toxicity and compatibility with other plants. With the exception of northern New South Wales and the coastal strip, where some warm-season plants are sown in pastures (Zones 5 to 9 in Figure 19.2, and the northern section of Zone 3), most perennial commercially available pasture varieties grow during the cooler months from March to November.

In planning a revegetation program it is important that soils are not disturbed and left bare during periods of high erosion risk. The erosive potential of rainfall is highest in summer (January–March) and least from April to September in northern New South Wales (Figure 19.3), while in the south (Figure 19.4) the erosion risk is highest in February and October and least in the period from March to September. In the north, rainfall is higher and more reliable in summer than winter, consequently ground cover condition in non-cropped situations will tend to offset the high erosion hazard. In southern New South Wales, October is a month of high and reliable rainfall so the risk factors will be low but in February rainfall is both low and unreliable. It is during the late summer and early autumn that the erosion hazard can be expected to be highest because the dominantly cool-season pasture plants, which are actively growing in October, are either dormant, grazed or have set seed and died by February.

The number of species of plants available for sowing in pastures and for soil conservation use in New South Wales is not great. The most important grasses used for soil conservation in northern areas include species of *Cenchrus*, *Panicum*, *Paspalum*, *Chloris*, *Pennisetum*, *Setaria* and *Digitaria*, sown with or without legumes, namely species of *Desmodium*, *Trifolium* and *Medicago*. In central and southern New South Wales and on the northern tablelands, species of *Phalaris*, *Dactylis* and *Lolium* are used, as well as the legume species *Trifolium*, *Medicago* and, in some cases, *Serradella*. Consol lovegrass (*Eragrostis curvula* complex) is the only commercially available warm-season grass that will persist on harsh sites in southern New South Wales.

Plants of value for particular soil conservation situations include: *Thinopyron elongatum* (tall wheatgrass), *Puccinellia ciliata* (marsh grass) and *Trifolium fragiferum* (strawberry clover) for salt area revegetation; *Eragrostis curvula* (Consol lovegrass) and *Cynodon dactylon* (couch) for stabilising acid light soils in lower rainfall areas and roadside batters in higher rainfall zones; *Paspalum dilatatum* (common paspalum) for summer moist areas that receive runoff from, for example, roadsides; and turf grasses, which may be recommended in urban situations. Crop plants such as barley, oats, ryecorn and millet are also used in revegetation to provide immediate or temporary ground cover, or as a cover crop for other plants that are slower to establish. Cover crops should only be used where they are likely to be a real benefit, such as in a temporary revegetation mixture, because they have a high water requirement late in the season and this may reduce the water available to other species and reduce their persistence. Residue from cover crops may also be a fire hazard during summer.

Site Investigations   19.4.3

Before making species recommendations, basic soil tests should be carried out to determine the main factors likely to affect revegetation. These tests should include soil texture, dispersion percentage, shrink-swell potential and pH. If the area is suspected of being salty, electrical conductivity should also be measured. If the pH is below 5.0, a lime requirement test may be in order, and if the soil is cracking, crusting or dispersible and/or has a high shrink-swell potential, a gypsum requirement test will help decide rates of gypsum application necessary to improve structure and stabilise the soil against tunnel erosion and surface crusting. Lime and gypsum requirement tests, although somewhat approximate, provide a guide to application rates that can then be adjusted using local experience. Agricultural lime should be used in preference to gypsum on very acid ($pH_{Ca} <$ 4.5) soil because it may tend to acidify the soil further.

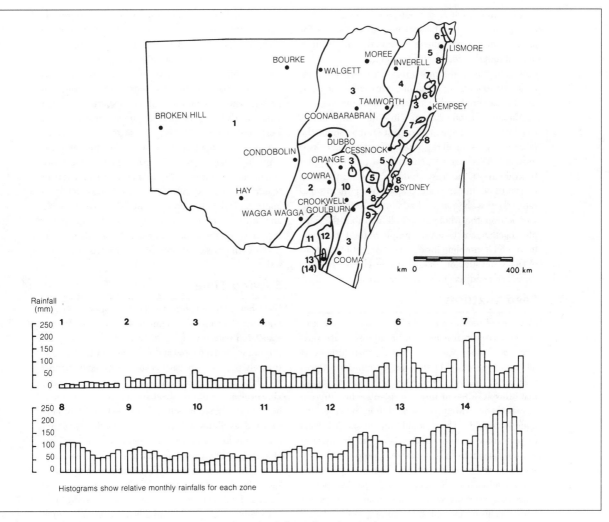

Figure 19.2 Rainfall zones in New South Wales (after Edwards, 1979)

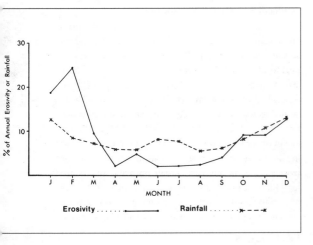

Figure 19.3 Monthly erosivity and rainfall—Inverell

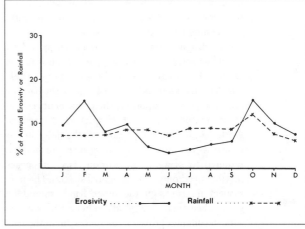

Figure 19.4 Monthly erosivity and rainfall—Wagga Wagga

In addition to these soil tests, it is also wise to drill an auger to a depth of a metre or so and check for hardpans or impermeable soil layers that may restrict normal root development. In general, plant growth is limited by the most limiting site factor, and response to alleviation of, for example, nutrient deficiency, may be masked if other factors are also limiting.

Variation of soil properties in eroded situations is usually high, consequently soil test samples should be taken to represent all the area, not just at one or two locations. This can be achieved by taking samples from the most and least eroded zones and the zone judged to be intermediate, and analysing them separately to give the range of conditions. Alternatively, sample over the whole area, bulk and mix the samples together, and take a subsample of this for testing. It is useful if sampling for chemical or fertility tests to obtain some idea of background levels by sampling topsoil areas adjacent to an eroded area.

19.4.4 Seed Mixtures

Mixtures of seed of a range of plant species are usually preferred in revegetation because sites are variable and some plants do better than others at different locations over the site. Also, mixtures provide better ground cover when they contain plants that grow at different times, or plants with different growth habits. Mixtures should not, however, be unduly complex. They should contain species likely to perform well rather than a 'shotgun' mixture of whatever happens to be available. Competition from species that have little or no chance of persisting beyond initial establishment can jeopardise persistence of those more suited to a site, and only species whose inclusion in a mixture can be justified should be considered.

Recommendations for a major project, such as a large section of highway, should include both a temporary and permanent revegetation mixture. The temporary mixture, consisting of one or two species that are easily established and cheap to buy, should be sown as an ongoing part of the work on all disturbed areas. This should occur immediately after construction and before surfaces set hard or are weathered by exposure. The permanent revegetation mixture is sown at completion of stages of work, or at the completion of the job, and is designed to provide the long-term protection required to prevent soil erosion. The temporary mixture need only contain annual species. The permanent mixture, on the other hand, should be sown when there is a good likelihood of success, using techniques proven to be successful in the particular situation. (In some cases, the temporary mixture may prove sufficiently successful that further sowings are unnecessary.)

Although legumes are essential in agricultural situations, they are less so where the aim is not to create a pasture. The disadvantages of legumes in ungrazed situations are that they smother component grasses, they are highly competitive in the first few seasons then they disappear, they use water more rapidly than grasses, and they raise the nitrogen status of the soil which leads to ingress of nitrogen-responsive weeds. Legumes are only necessary in revegetation where an aim is to match a pasture in which legumes are present. In general, grass-dominated swards are more stable and persistent in ungrazed situations than legume-dominated swards or swards containing an appreciable legume component. The extra trouble and care necessary to inoculate and lime pellet legumes, and the difficulty of establishing them compared with grasses, is often not warranted in revegetation.

Sowing Time 19.4.5

'Best-bet' revegetation requires a good working knowledge of the local climate and the risk factors associated with sowing at different times of the year. It is not essential to sow at the optimum time and this is rarely possible anyway because of other demands, and because revegetation is a year-round activity. However, it is vital to avoid sowing when the likelihood of germination is high but the chance that germinated seedlings will complete their life cycle and set seed is low. Seed can be sown out of season provided it is unlikely to germinate until reliable seasonal conditions are imminent. Temporary seed mixtures should be used during the critical non-sowing period because final revegetation is easier where some seed is already in the ground and a little cover and mulch have been provided. Loss of plants sown in temporary mixtures is not serious compared with losing the results of an intended final sowing.

Timeliness of germination is important for both annual and perennial plants. Annual plants depend on setting seed for their persistence; they need a certain period of growth for this to occur and the later they are sown, the less seed they set. Perennial plants have a number of requirements. They need a sufficiently well-developed root system to withstand summer drought, or they need to have reached reproductive age in order to become dormant and escape stress, or in some cases biomass accumulation is important for frost escape. In any case, small unthrifty plants are less likely to survive stress than plants that have grown for a sufficiently long growing season. Sowing too late is a prominent cause of revegetation failure.

Fertilisers 19.4.6

Using fertilisers is the most common means of altering the productive potential of land. Use of fertiliser alone, however, is not very efficient because plants vary in their response. In order to make the most of the investment, application should be accompanied by sowing plants that give a large response. Fertiliser-responsive plants generally do not do well where fertility is low. This means that in revegetation situations, as fertility declines, species with a fertility requirement will become unthrifty and will disappear. Heavy or prolonged use of fertiliser reduces persistence of many native grasses that are adapted to low fertility, so where a stable native sward is required, fertiliser should be used only sparingly.

Nitrogen and phosphorus are the most commonly applied elements. Nitrogen is more mobile than phosphorus, its availability changes seasonally and rapidly in response to biological activity, and some forms are readily leached. Nitrogen occurs in soil in both mineral and organic forms, and is lost and gained through a number of biological pathways that link together to form the nitrogen cycle. Phosphorus is more likely to be fixed (rendered unavailable to plants) by the soil and is less subject to loss and gain, except in the form of loss by erosion and export as produce, and gain as manure or fertiliser. Grasses are more responsive to nitrogen than phosphorus, and legumes are more responsive to phosphorus than nitrogen. The main contribution of legumes to pasture is fixation of atmospheric nitrogen and its synthesis to protein, which then becomes available to grazing animals. It is eventually returned to the grasses through either decay of legume biomass or excretion when animal intake exceeds requirement.

Although grasses are more responsive to nitrogen than legumes, legumes will use soil nitrogen if it is available. As the nitrogen-fixing ability of legumes is dependent on associated bacteria, uninoculated legumes are not capable of nitrogen synthesis and compete for nitrogen in the same way as grasses.

Nitrogen and phosphorus are the main elements lost through erosion of topsoil. If the soil is naturally deficient in other important elements, such as sulfur, potassium or molybdenum, these are also likely to be lacking following erosion. Soil pH and an assessment of the general thrift of background vegetation can be used as a guide to the nutrient status of a site, but generally the importance of other elements only becomes obvious when the major ones are in good supply and plant growth seems to be constrained. There are a number of diagnostic signs of nutrient deficiency for a range of crop plants given in standard texts which can aid the assessment of nutrient status.

The use of fertilisers in revegetation, particularly in non-pasture situations, is a complex issue. Nitrogen tends to increase leaf area and water use of grasses, and this can cause moisture stress and death of established plants. High rates of water transpiration reduce the time span of water availability and, in dry spring seasons, this can seriously reduce seed set by annual plants. Where leaf area is controlled by grazing, plants persist better. Legumes are also less competitive when they are grazed. The effect of fertiliser phosphorus is less marked in terms of suppression of grasses by legumes in grazed mixed pastures than when legume growth proceeds unchecked.

As a general rule, in ungrazed situations, fertilisers should only be used at modest rates and only to the extent required to achieve ground cover. In grazed and cropped situations, where productivity is important, fertiliser can be applied at higher rates and with greater frequency depending on economic considerations.

Fertiliser should be avoided where its use will promote overgrazing of sown species because of a large contrast in fertility between sown and unsown areas within a paddock or, for example, within a forest or nature reserve. Topsoiling is probably a better alternative in these circumstances than risking overgrazing, unless the whole area can be fertilised and sown to similar species.

Site Preparation 19.4.7

The two basic requirements for successfully establishing pasture plants are:
—the seedbed must provide suitable conditions for seed lodgement, seedling germination, emergence and growth
—competition from other species must be reduced or removed.

In revegetation situations where ground cover is poor or absent, seedbed conditions are also usually poor and likely to limit establishment. Where existing ground cover needs improving, such as in developing a waterway, competition from existing plants may need to managed.

Steps that may be taken to improve establishment of sown or oversown seed include the following.
—Killing all of the existing sward or selected components of the sward using herbicides. This will reduce competition and is advocated in situations where cultivation is not possible or desirable. Broad-spectrum, non-residual contact herbicides (for example, glyphosate) may be used strategically—for instance, to kill annual cool-season legumes and grasses and allow warm-season grasses space to germinate or to promote their early growth. In this case the competing vegetation should be killed in late winter, when there is not much bulk, it is actively growing and there is no risk of soil ero-

sion. Remaining plant residues can be useful to protect the ground until sown plants or winter-dormant species provide cover. Mulch also protects sown seed and conserves soil moisture.

—Conventional cultivation and sowing methods reduce competition and create good seedbed conditions for germination and establishment. Importing and spreading good quality topsoil, organic waste or other suitable material can also be useful on harsh, very eroded sites, but cartage and spreading costs may be high. Wherever possible, topsoil should be stripped and stockpiled for spreading at the completion of work. Even a few centimetres of respread topsoil greatly improves establishment and persistence of sown plants, and improves the rate of development of permanent cover.

—Deep ripping and cultivation using tined implements are recommended where surfaces are hard and compacted. These operations can be used to incorporate fertiliser or ameliorant and should be followed immediately by sowing, so that the surface does not have a chance to set hard again without some seed being buried. Seed will germinate and emerge from hardset soil provided it is buried. If seeds are just sown on the surface without protection, seedlings experience great difficulty in penetrating the hard surface. Theft by ants may then account for most of the seed sown.

If cultivating or deep ripping dispersible or tunnel-eroded areas, care must be taken to break up all the tunnels, otherwise the problem is hidden but not cured. Working on the contour is also important so that water distributes evenly and does not flow along depressions left from the final working.

—Mulching, with or without netting and/or bitumen, use of netting materials alone, bitumen by itself, wood fibre mulches and other products designed to arrest soil movement and enhance seed germination and plant establishment are all useful in specific situations. However, their cost prohibits their everyday use. Specialist advice is available for the use of these products and should be sought when appropriate.

—All soil conservation earthworks should be topsoiled after construction, and most should also be sown using a cheap seed mixture of one or two species. The cooperation of the landholder should be sought, and the need to sow before surfaces set hard cannot be overemphasised. A kilometre of earth bank occupies about one hectare of paddock, which if not revegetated may cost 5 to 10 DSE (dry sheep equivalents) per year in lost production. Topsoiled, sown and harrowed earthworks are the hallmark of a good job; the extra trouble adds little to the overall cost of a job. Revegetated works blend in with the landscape and enhance production. It is difficult to revegetate earthworks once surfaces have settled and weathered. Doing it right the first time is the key to only having to do it once!

Soil Factors Affecting Revegetation

19.4.8

The most common soil factors that may affect the likely success of a revegetation program are:
—surface cracking and crusting
—hardsetting nature of some soils
—salinity
—hardpans and other impediments to root extension
—acid and alkaline soils
—excess aluminium, manganese and heavy metals
—low nutrient availability, chiefly nitrogen and phosphorus; but also potassium, molybdenum and other elements.

Surface Cracking and Crusting

Surface cracking and crusting is mainly a problem in northern New South Wales on black, grey and brown clay soils that are dominated by expansive clay minerals. It is also common in gilgai country on the puffs and around the edges of the hollows, but not in the depressions themselves.

On drying, expansive soils form a surface crust beneath which they crack vertically to varying depths. Cracks may be several centimetres across and metres deep in extreme cases. These soils sometimes also disperse on wetting which, combined with their expansion, seals the surface and reduces water infiltration. The extent of cracking and crusting is determined by the rate of drying after rain and the overall extent of drying. The surface may dry and crust over, even if the soil is still moist at a depth of a few millimetres. Rapid surface drying causes root damage to germinating seedlings and, as roots do not extend into dry soil, seedlings perish if a dry boundary layer occurs under a surface crust. It is also not uncommon to find plants with leaves extended under the crust, but these have no chance of emerging and will perish. Mature plants can be killed by deep cracking, which disrupts root systems and reduces access to soil moisture. Plants with fibrous root systems are more prone to damage than taprooted plants. For further discussion of cracking and crusting problems see Watt (1972, 1974).

Cracking and crusting cannot be overcome easily. The most successful strategies on cracking soils are to use plants with large or awned seeds, which do not fall down cracks easily. Examples are rhodes and buffel grass, medics, seeds with sticky coatings such as paspalum and a number of native species, and cereals. It is important to sow when the chance of prolonged rain is high. Mulching reduces the rate of surface drying and therefore the rate of crusting. Gypsum application at rates up to 4 t/ha has been shown to improve the structural condition of cracking soils and reduce crust strength, but its effectiveness is reduced when the soil surface is not protected by vegetation. When significant subsurface emergence has occurred, the problem may be reduced by waiting until plants are well rooted and lightly harrowing the ground to expose the seedlings, preferably when rain is imminent but while the crust is dry.

Hardsetting Soils

Many soils set hard on drying, forming a massive impenetrable surface structure. This problem is particularly acute for exposed subsoils and pan layers that are naturally more compacted and have greater mechanical strength than organic-rich and better-structured topsoil layers. Once-friable soils can become hardsetting if their structure is broken down by excessive cultivation, traffic or trampling by stock. This often leads to increased runoff and erosion hazard (see also Chapter 16).

Hardsetting soils can be revegetated by deep ripping, incorporation of gypsum, sowing appropriate seed mixtures and, where necessary, mulching. Because these soils tend to have a poor structure, they should not be worked when they are too dry as they may break down into dust, or when too wet as they tend to pug and set hard again on drying out. Their moisture content should be such that they remain cohesive and friable instead of tending to shatter or pug. Surfaces should be left moderately rough, with the final working on the contour, and they should be sown immediately regardless of the time of the year. Once revegetated, hardsetting soils should not be grazed heavily, but should be rested to regain their structure.

Salinity

Dryland salinity and salinisation of irrigated land are increasing problems in New South Wales. Salinity results from gradual concentration of salts at the soil surface because of rising water tables, and/or replacement of deep-rooted summer-active plants with shallow-rooted species with restricted growing seasons. Soils derived from parent material with a high natural salt content are more likely to become saline, but irrigation can cause a rise in saline groundwater which can affect soils regardless of their parent material.

Salinisation is a long-term problem and is difficult to reverse. Salinisation may occur some distance (kilometres) from the recharge zone, so dealing with the problem at its source is often impossible. Salt-tolerant plants include tall wheat grass, marsh grass and strawberry clover, and tree species such as *Eucalyptus camaldulensis* (river red gum), *Casuarina cunninghamiana* (river sheoak), *Atriplex* species (saltbushes), *Melaleuca* species (honey myrtles), *Myoporum insulare* (boobialla) and species of *Tamarix* may grow and persist on salted areas, but they will not alone reverse the trend to increasing salinity.

Revegetation of salted land entails appropriate measures to control water movement and improve drainage; improving the structural stability of affected soil; improving the fertility of the site; controlling grazing; tree planting across suspected zones of underground water movement; sowing an appropriate seed mixture; and mulching to reduce surface evaporation and thus the rate of salt crust formation. There is little likelihood that cropped land will be able to be returned to cropping once salting has occurred, and it is vital that ground cover is maintained on salted land throughout the year even if the problem appears to be successfully overcome (see also Chapter 13).

Hardpans and Impediments to Root Extension

Excessive compaction and cementation of soil reduce root elongation and growth and restrict root activity to channels and zones of weaker soil. Moisture relations are also limited because dense soils hold less water and infiltration is restricted. The total of the water resources available to plants is therefore lessened, and compacted and hardsetting soils are more prone to drought (see also Chapter 16).

Mechanical impedance to root growth is commonly overcome by deep ripping and cultivation. Application of gypsum and organic material is also useful on some soils.

Acid and Alkaline Soils and Nutritional Problems

The pH of soil is a measure of its acidity (or alkalinity), on a logarithmic scale from 0 to 14. The pH of agricultural soils can be measured using one part soil to five parts distilled water. The neutral point on the pH scale is pH 7. Acidity increases as pH falls below 7, and soils become more alkaline as pH increases above 7. Because the scale is logarithmic, each pH unit represents a tenfold change from adjoining units, a 100-fold change from the second next unit, and so on. This means that pH 8 implies a soil condition ten times more acid than pH 9 and 1000 times more alkaline than pH 5. The pH range of most soils is between 4 and 9.

The pH of soils is important because of its effect on biological processes and the availability of most plant nutrients, and the release of elements in quantities toxic to plant growth. Survival of legume nodule bacteria is also affected by pH and this has a direct influence on the thrift and effectiveness of legumes in pasture. Plants also vary in their tolerance of extreme pH and the soil conditions associated with extreme values. It must be noted that extreme pH by itself may not be damaging; more usually plants respond to toxicities and deficiencies that are induced by the pH regime. Also, not all soils respond the same way, or to the same extent, so pH is only an indicator of the likelihood that problems may occur.

The pH of soil can be measured in water (pH_w) or calcium chloride solution (pH_{Ca}) (see Chapter 13). In general, when pH is measured in calcium chloride, values in the range 5 to 7.5 should present no problems. As pH falls below 5 ($CaCl_2$) legume nodulation problems may occur; aluminium and manganese toxicity may occur below a pH_{Ca} of 4.5 and some nutrients may become unavailable, especially calcium, magnesium, molybdenum and phosphorus. Root growth can also be depressed which increases drought susceptibility. If acid soils are encountered in revegetation work, or if sward degeneration is suspected to be due to soil acidity, specialist advice should be sought and further diagnostic soil tests may be required.

Except in soils that contain free calcium carbonate such as black earths, other cracking clay soils and calcareous red earths, pH values above 8 are rarely encountered. The pH of calcium carbonate in equilibrium with carbon dioxide is 8.4; a pH_w greater than this indicates high sodium carbonate content. A pH greater than 9 indicates that salinity and sodicity are likely. However, not all sodic and saline soils are alkaline.

High pH may indicate deficiencies of copper, zinc, boron, manganese and iron. Although high pH is less detrimental to plants than low pH, plants that tolerate acid conditions generally do not grow well in alkaline soils. Lime-induced chlorosis is common in susceptible plant varieties and this is thought to be due to interference with iron metabolism.

Pasture varieties that prefer alkaline conditions include lucerne, medics, red clover, phalaris and tall wheat grass. Subterranean, rose and white clovers, woolly pod vetch, rye grasses and a range of native grasses prefer neutral to slightly acid conditions; tall fescue, cocksfoot, lovegrass, a range of warm-season grasses and *Serradella* are tolerant of acid conditions. In crop plants, rice, oats and ryecorn are more tolerant of acid conditions, followed by wheat, then barley. Barley tolerates high pH and is also useful in sodic soil reclamation.

Calcium compounds are recommended for soil structure amelioration in many instances. Calcium carbonate (agricultural lime) should not be used where soil pH exceeds about 6 because lime is only slightly soluble at these pH levels. Although gypsum is effective over the pH range found in soils, if the pH is less than around 5.5, it is preferable to use lime as this will also improve the pH status of the soil.

19.5 Sward Improvement and Management

19.5.1 Strategic Grazing

Many soil conservation situations require improvement of an existing weak sward instead of its complete replacement. Often this may require strengthening the perennial components or changing the composition of a weedy pasture. Complete replacement involving cultivation incurs a risk that soil erosion will occur before the new pasture is sufficiently well developed to provide protection. This is often not acceptable, particularly if the area is to be used for water disposal, or where water runon is likely.

The easiest way to improve a weak sward is to exclude grazing and allow herbage to accumulate. Fencing and adequate stock management are therefore recommended. Strategic grazing is a more satisfactory and dynamic way of achieving sward improvement than set stocking which should be avoided altogether. Grazing management should aim to maintain more than 75% ground cover. Herbage should not accumulate from season to season, so grazing should be used to maintain green herbage mass in the range between 500 and 1500 kg dry matter per hectare.

Strategic grazing has four objectives:

—maintaining ground cover

—encouraging desirable species to set seed

—opening up the sward to allow desirable species to establish and thus replace moribund older plants

—weed control.

Strategic grazing can also be used in conjunction with other practices designed to improve an existing sward, including oversowing, burning and using selective herbicides to remove competing or unwanted components.

Strategic grazing depends on using high stock numbers (for example, about ten times the average stock-carrying capacity) for short periods of time, and critically judging when stock need to be removed. Stocking rates should be calculated to reduce green herbage mass to about 500 kg/ha within a period of 35 days (five weeks) from when animals are allowed access.

The most important variable in achieving a satisfactory result from grazing is its timeliness in relation to seed-setting of the species in the sward, and germination of seed from previous seasons. In southern New South Wales, heavy grazing in late spring and early summer may reduce competition from annual species which are completing their life cycle, and favour summer-active perennial grasses such as red grass.

Use of high stock numbers achieves uniform grazing by reducing selective grazing (such as preference for leaf over stem, giving a very stemmy sward after grazing); stock tracking is reduced and dung and urine are distributed more evenly. This encourages nitrogen and phosphorus cycling which generally invigorates the sward.

Grazing can be used to control undesirable plants by grazing after stem elongation but before seeding. Several grazings may be required and sometimes the sward should be slashed as a clean-up measure. Grazing heavily prior to expected times of germination is also a useful way of rejuvenating a pasture dominated by mature herbage and, at the same time, encouraging germination of new plants. Heavy grazing when the ground is soft (but not soggy) can be a useful

strategy after oversowing, as this will assist seed burial. Grazing is recommended before using herbicides so that target plants are well exposed. Seedlings of broad-leaved weeds are difficult to kill if they are protected from herbicides by a standing mass of herbage.

Areas such as waterways should not become harbours for weeds and rank vegetation. Not only are they unproductive and unattractive in this state but also their capacity for water disposal will decrease. Strategic grazing of waterways is an important aspect of their use in soil conservation.

Oversowing 19.5.2

Introducing more suitable species into an existing sward is another option for improving existing ground cover which does not involve wholesale cultivation and the consequent risk of soil erosion. Several approaches may be taken, all of which aim to reduce competition from resident vegetation as well as ensuring that surface-sown seed has a good chance of coming into contact with the soil surface. Strategic grazing has already been discussed. There is also the conservation farming technology of removing resident herbage using herbicides before oversowing. Assuming the sward is reasonably open and seed is sown at the optimum time of the year, this can be successful. Small areas can be heavily grazed after sowing to 'hoof bury' sown seed and thus reduce ant theft and other losses. Seed can also be sown using scratch-seeding and minimum tillage techniques. Heavy diamond harrows, and even heavy chains and pieces of timber or railway iron dragged across the ground, can be used to bury oversown seed.

Because surface-sown seed is not protected from loss or environmental stress, surface sowing is less efficient than seed burial, and it is more important that seed is sown when environmental stress and risk factors are low. Also, species differ in their abilities to establish from surface sowing. Small and awned seeds are relatively efficient, and grasses are better adapted than small-seeded legumes; of the grasses, ryegrass is better than cocksfoot which is better than phalaris.

19.6 Sustainable Grazing Management

'Good', 'wise', 'careful' or 'sustainable' grazing is often promoted—but what is it?

Sustainable grazing management aims to:

—maintain ground cover so it is always greater than 75%

—maintain species diversity, productivity and forage quality

—control the tendency for unpalatable species to become more dominant in pasture

—reduce opportunities for weed encroachment

—maximise productivity by minimising waste, including forage that is spoiled and trampled and which goes to waste; water that is not used by the pasture and which contributes to groundwater recharge; and minimising nutrient loss in runoff.

Sustainable management takes control of the pasture away from the animal; it requires a proactive, hands-on approach with the manager deciding to what extent, when and for how long pastures are grazed.

Control of the feed demand/supply situation using PROGRAZE principles (see Simpson and Langford, 1996) is an essential part of sustainable grazing management. Understanding how pastures respond to being grazed and relationships between where different pastures occur and the 'whole farm' landscape are important ingredients of sustainable grazing. Some points to consider are:

—Hilly landscape classes are less likely to support improved pastures than mid-slope or low-slope land. It is not possible to manage different landscapes profitably and sustainably if they are fenced as one unit. A farm plan that integrates land capability and use is an essential tool in farm development.

—All pastures consist of mixtures of species that grow and reach maturity at different times. Set stocking is not a sustainable form of grazing management—it will invariably lead to a decline in palatable species, the translocation of fertility to camp areas, patch grazing and weed invasion. These problems mainly emerge when feed demand lags behind pasture growth, that is, when the pasture is allowed to 'get away'.

— Pastures are only grazed uniformly when animals feel compelled to graze without choice and when demand exceeds supply. Using high stocking rates imposes a social pressure on animals to graze with less choice; stocking rates high enough so that aggregate feed demand exceeds supply ensures high levels of utilisation per grazing.

—Soil erosion rates increase rapidly on sloping land if ground cover declines below about 75%; evapotranspiration depends on having some green leaf present in the pasture, as pastures with less than about 300 kg/ha dry matter are unable to recover green leaf and maintain water use readily. Pastures should therefore not be grazed below a minimum ground cover threshold of 75%, or a dry matter threshold of about 300 kg/ha.

—Because commercial cultivars of perennial grasses lack persistence on hill lands and the productive potential of hilly landscapes is less than low-slope lands, it is generally not economic to sow them to improved pastures. Hill lands should therefore be protected during times of drought by withdrawing grazing to land that can be economically resown if pastures are ruined. Overgrazing native pastures leads to undesirable changes in botanical composition, including weed invasion, which is difficult and costly to reverse. On the other hand, well-managed native pastures can be the first to come away after drought. Money not spent on herbicides is money that may be available to improve facilities, such as fencing and watering, or to improve the productive base of sown pastures on more productive landscapes.

—Hill lands 'go off' earliest, therefore they should be grazed earlier in the grazing cycle than lower-slope lands. Save low-slope pastures for grazing into the summer. As well as making best use of feed when it is at the peak of its quality, grazing out hill pastures in spring will provide competitive relief to summer-active native grasses that commence growth in late spring and early summer.

—Cattle will not graze as selectively as sheep; on the other hand they will also not graze pasture down to the same extent. Grazing the top off pasture with cattle, followed by sheep which will graze off the bulk is more efficient than grazing entirely with the same class of animal.

—Undesirable changes in botanical composition can also occur in response to good years. Good seasons present an opportunity to conserve forage as hay or silage, to allow native pastures to set seed, and to burn off invading woody weeds. In a whole farm context it is important to maintain a grazing strategy that continues to optimise feed utilisation on grazed paddocks. This can be achieved by shutting some paddocks down. Trying to graze too much feed with too few animals is equivalent to low-intensity set stocking—it favours unpalatable species and more mature herbage of palatable plants, and leads to overgrazing, either of patches, or of species that are more palatable.

So, if overgrazing or undergrazing is unavoidable, sacrifice the safest country first.

BIBLIOGRAPHY

Edwards, K. (1979), *Rainfall in New South Wales with special reference to soil conservation*, Technical Handbook No. 3, Soil Conservation Service of NSW, Sydney.

Johnston, W.H. (1996), 'The place of C_4 grasses in temperate pastures in Australia', *New Zealand Journal of Agricultural Research* 39, 527–40.

Lang, R.D. (1979), 'The effect of ground cover on surface runoff from experimental plots', *Journal of Soil Conservation Service of NSW* 35, 108–14.

Simpson, P. and Langford, C. (1996), 'Managing high rainfall native pastures on a whole farm basis', NSW Agriculture, Sydney.

Watt, L.A. (1972), 'Observations on the rate of drying and crust formation on a black clay soil in northwest New South Wales', *Journal of Soil Conservation Service of NSW* 28, 41–50.

Watt, L.A. (1974), 'The effect of water potential on the germination behaviour of several warm season grass species, with special reference to cracking black clay soils', *Journal of Soil Conservation Service of NSW* 30, 28–41.

Soils and Their Use for Earthworks

R.J. Crouch, K.C. Reynolds, R.W. Hicks and D.A. Greentree

Soil conservation earthworks are designed and constructed to minimise soil erosion by controlling runoff for sustainable on-site usage and conveying it to the main stream along an erosion-resistant route. The main earthworks are dams for water retention, bank/channel combinations for controlling water flow across the landscape, and gully control structures for restricting erosion in gullies.

Dams must hold water inside the excavation and against the wall without failing or leaking, and must have an erosion-resistant spillway. Banks are designed to convey or hold water without failing or becoming eroded. For these structures, soil material forms the basis of the excavation, wall and spillway. Soil type also largely determines the runoff rate from the catchment and the subsequent discharge into a structure. Most structures are built to soil conservation specifications that have been found to work on specific soils or in a particular region. However, the variability in soil material for earthwork construction needs to be appreciated, and soil property deviations past critical limits recognised, so that design and/or construction techniques can be appropriately modified.

A most useful general reference text for soil and water conservation engineering works is that of Schwab et al. (1981).

20.1 Design of Earthworks

The design of soil conservation earthworks is primarily concerned with determining structure size, location, capacity and any special characteristics such as channel grade or shape. Four main equations are used:

Rational Formula, $Q = cia/360$ 20.1
Manning's Formula, $V = n^{-1}(A/P)^{.67}S^{.5}$ 20.2
Weir flow equation, $Q = CLH^{1.5}$ 20.3
Flow relationship, $Q = VA$ 20.4

where

A	=	cross-sectional area (m²)	L =	crest length (m)
a	=	catchment area (ha)	n =	Manning's roughness coefficient
C	=	weir coefficient	P =	wetted perimeter (m)
c	=	runoff coefficient	Q =	peak discharge (m³/s)
H	=	hydraulic head (m)	S =	channel slope (m/m)
i	=	rainfall intensity (mm/h)	V =	mean velocity of flow (m/s)

Equation 20.1 relates catchment runoff to catchment size and rainfall intensity. Equation 20.2 relates flow velocity in a channel to channel slope, cross-sectional dimensions and roughness. Equation 20.3 relates discharge over a weir to the hydraulic head of the flow and the weir length. Equation 20.4 relates discharge through a channel to the flow velocity and the channel cross-sectional area.

Known or estimated variables are substituted in these equations and the equations solved to find the unknown. This is covered in more detail in Aveyard (1987). Soil influences design of soil conservation structures by its effect on catchment runoff through the runoff coefficient, erodibility of channel banks and floors, wall permeability, wall stability and optimum spacing of banks.

20.1.1 Runoff

Structure capacity depends on the volume of runoff water that is expected to be stored in, or conveyed through, the constructed earthworks. Land attributes (particularly soils) of a catchment, whether it be hundreds of hectares shedding water into a dam or a few hectares draining into a bank, determine the infiltration capacity and combine with the catchment area, and rainfall intensity and volume, to determine the rate and volume of runoff into a structure. Infiltration and factors that affect it are discussed in Chapter 10. There are a number of ways to estimate runoff, both in terms of peak discharge with a specific probability of occurrence and in terms of the expected minimum runoff volume for certain periods. Some of these involve soil factors.

Peak Runoff

The rational formula requires the estimation of a runoff coefficient 'c', the proportion of the peak rainfall that will run off a catchment. Of the two methods recommended, one method predicts 'c' from statistical catchment data (Pilgrim and McDermott, 1982), another from soil and catchment characteristics (Table 20.1). The latter method is based on the fundamental assumption that the lower the soil's infiltration capacity, the higher the runoff coefficient.

Total Runoff

To estimate catchment yield for small farm dams, Burton (1965) recommends the use of Table 20.2. This applies to small farm dams only and not to major irrigation structures. The percentage figure selected from Table 20.2 is multiplied by the average annual rainfall and the catchment area to determine the runoff volume; for example, for a gradational soil with a 550 mm annual rainfall and a 20 ha catchment area, runoff volume for a year is estimated in the following way:

$$\frac{550 \times 4 \times 20}{10} = 4400 \text{m}^3$$

Channel Erodibility   20.1.2

Flowing water exerts a shear force at the soil/water interface that detaches and transports soil particles. The ease with which particles are moved depends on the erodibility of the unprotected soil in contact with the water. Factors affecting erodibility are considered in Chapter 12. In terms of earthwork design, soil erosion is primarily influenced by the velocity of the flowing water, ground cover and soil characteristics such as slaking and dispersion.

Earthwork channels (bank, waterway, flume or spillway) must be designed to convey flowing water without significant erosion or sedimentation. That is,

Table 20.1 Estimation of runoff coefficient (after Aveyard, 1987)

Catchment Characteristics	Runoff-producing Characteristics							
	Extreme		**High**		**Moderate**		**Low**	
Land Use	Continuous arable cultivation	0.20	Arable land with regular rotations	0.15	Hard grazing	0.05	Light grazing retired land, forest	0.00
Relief	Average catchment slope greater than 20%	0.10	Average catchment slope 11%–20%	0.05	Average catchment slope 5%–10%	0.00	Average catchment slope 0%–5%	0.00
Depression Storage	Steep watercourse Negligible catchment storage Predominantly channelised flow	0.10	Some overland flow Length of natural defined channel flow greater than overland flow	0.05	Some catchment storage in banks and furrows Length of channel flow similar to length of overland flow	0.00	Significant catchment storage in banks and furrows Overland flow lengths greater than defined channel length	0.00
	Use modified Bransby Williams method				Use complex catchment method			
Infiltration and Soil Factors	No effective soil cover Either solid rock or shallow lithosol soils	0.25	Duplex soils with hardsetting surfaces Uf soils	0.15	Gradational soils, duplex soils with non-hardsetting surfaces, Um and Ug soils	0.10	Deep sands and gravel deposits, and Uc soils	0.05
Annual Exceedance Probability	1%	0.40	2%	0.30	5%	0.15	10%	0.10

Plate 20.1 A graded soil conservation bank being constructed in an arable paddock in northern New South Wales

Table 20.2 Minimum dependable yield (80% reliability) as a percentage of average annual rainfall (after Burton, 1965)

Soil Type	Average Annual Rainfall (mm)			
	< 400	400–650	650–1000	> 1000
Uniform sands and loams	1	3	5	10
Gradational soils, non-hardsetting duplex soils and cracking clays	1.5	4	7	20
Hardsetting duplex soils and non-cracking clays	2	5	10	30

flows must have sufficient velocity to transport sediment delivered to them, but not sufficient to detach soil from the channel margins. This non-scouring/non-silting velocity is called the 'maximum permissible velocity', and varies with the soil properties of texture, structure and dispersion, and with vegetation type and cover. It has been determined largely by trial and error, initially for irrigation canals in 1895 (Chow, 1983), and adapted for soil conservation structures. The values shown in Tables 20.3 and 20.4 have been found to apply satisfactorily to New South Wales conditions.

For channel design, the maximum permissible velocity is selected from the appropriate table and substituted with the design discharge in Equation 20.4 to determine the channel cross-sectional area, and Equation 20.2 (Manning's Formula) to determine the required channel slope for graded banks or waterways.

20.1.3 Soil Permeability

The effect of soil permeability on earthwork design depends on the proposed use of the earthworks. For dams intended to store water, final permeability of the wall and the excavation is critical. For flood detention structures or banks, minor 'leaking' does not matter, provided it does not affect wall stability. Permeability is influenced mainly by soil particle size grading, dispersibility, shrink-swell potential and the degree of compaction achieved during wall construction.

To achieve an acceptably low level of permeability, soil must be graded so that there is sufficient fine particles to seal the spaces between the coarser fractions. A well-graded soil has at least 30% clay, 10–20% silt and an even distribution of fine and coarse sand. Soils with as low as 12% clay can form acceptable seals, provided the remainder of the soil is adequately graded. The higher the clay content the lower is the importance of the grading of the coarse fraction.

To be effective, clay must be moist to disperse sufficiently to move and fill voids to form a seal. In some very well aggregated earthy soils, the clay particles are very strongly bonded into fine aggregates that behave as silt or sand-sized particles. This can be easily recognised in dispersion tests such as the Emerson Aggregate Test or dispersion percentage measurement (see Chapter 10).

Soil shrink-swell potential and degree of compaction affect permeability by affecting the amount of space between soil particles. Dam walls are usually built out of relatively dry soil. As the soil wets up it swells, and, when confined, this increases the pressure between soil particles, thus reducing the void sizes. Similarly, the more a soil is compacted, the lower the proportion of voids in the material and the lower its permeability.

20.1.4 Bank Stability

Dam and contour banks constructed from soil become unstable when water flowing over, through or under them develops sufficient volume and velocity to detach and transport soil particles. These situations will be dealt with under the headings of overtopping, throughflow and undermining.

Overtopping

Water will flow over a structure when water depth exceeds the provided freeboard, when wall height has been reduced by settlement, when erosion or slumping has occurred or when channel or spillway capacity is insufficient.

Freeboard, the vertical distance from the floor of the spillway inlet to the wall crest, is made up of allowances for surcharge and settlement. Surcharge calculation, based on peak discharge and spillway width, is outlined in Aveyard (1987). For most small farm dams, a surcharge allowance of 30 to 60 cm is likely to be adequate.

Settlement occurs as soil placed in a wall becomes more compact, as shrinkage and swelling move soil particles into more dense packing patterns, and as slaking and dispersion of aggregates enable particles to pack closer together. Soil placed wet of optimum will be subject to dry settlement; soil placed dry will be subject to saturation settlement. The actual amounts vary considerably, but for small earth dams the allowances in Table 20.5 have been found to be satisfactory.

Erosive forces of raindrop impact and flowing water act on an embankment, just as they do on any steep slope, to transport soil particles from the crest, reducing its effective height. Topsoiling and vegetation establishment reduce the effects of the erosive forces on the soil and reduce erosion if the bank is overtopped. However, over a number of years bank height will inevitably be reduced and will need to be 'topped up' to an acceptable level.

Slumping occurs in soils with a low wet strength due to clay dispersion or high silt and fine sand percentages. When placed on steep slopes, such soils have a tendency to slump or slide, reducing the effective embankment width. Recognition of susceptible soils is particularly important for flood retardation structures where the water level will drop rapidly, leaving a saturated soil exposed on a steep bank (Figure 20.1).

Table 20.3 Maximum permissible velocities for bare soil channels

Erodibility Assessment	Maximum Permissible Velocity (m/s)
Extreme	0.3
Very high	0.4
High	0.5
Moderate	0.6
Low	0.7

Table 20.4   Maximum permissible velocities for vegetated channels

| Cover Type* | Channel Slope (%) | Erodibility Assessment | | | | |
		Low	Moderate	High	V. High	Extreme
Kikuyu and other dense,	0–5	2.6	2.4	2.3	2.2	2.0
high-growing, prostrate	5–10	2.5	2.3	2.2	2.1	1.9
perennials	>10	2.4	2.2	2.1	2.0	1.8
Couch and other	0–5	2.1	2.0	1.9	1.7	1.5
low-growing, prostrate	5–10	2.0	1.9	1.8	1.6	1.4
perennials	>10	1.9	1.8	1.7	1.5	1.3
Perennial improved	0–5	1.7	1.6	1.4	1.2	1.0
pastures	5–10	1.6	1.5	1.3	1.1	0.9
	>10	1.5	1.4	1.2	1.0	0.8
Native tussocky grasses,	0–5	1.4	1.2	1.0	0.8	0.6
sparse, high-growing	5–10	1.3	1.1	0.9	0.7	0.5
legumes (e.g. lucerne)						
and self-regenerating						
annuals**						

* The velocities shown for each cover description assume good (i.e. > 80 per cent) cover conditions.
** Tussocky grassed slopes of >10 per cent gradient are not recommended for vegetated channels because of the channelling effect such vegetation has on flow conditions.

Table 20.5   Expected gross settlement of small earth embankments

Placement Method	Expected Settlement (% height)
Dozer with rolling	5
Scraper without rolling	8
Dozer without rolling	10

Throughflow

Water may erode a subsurface channel through a wall due to soil cracking, tunnelling or piping. Soil cracking can occur following the use of a soil with a high shrink-swell potential (linear shrinkage > 17%) which will form large cracks on drying. If a bank dries for a sufficiently long period, cracks may form that extend right through the wall. If the structure fills rapidly, water will flow through the crack before the soil has time to swell and close it, cutting a channel through the wall. The chance of a crack extending right through a wall can be reduced by building flatter batters. For example, for a dam wall batter, grades could be decreased from the normal 1:3 to 1:4. This increases the mass of soil that must dry out before a crack extends through a wall. Another common cause of vertical cracking is poor construction where soils of different shrink-swell attributes are not properly mixed and layered.

Tunnelling or piping may occur following use of a clay that breaks down into its individual particles when saturated (disperses) and is readily transported through soil pores and cracks. If the clay is sufficiently

dispersible, and a wall is sufficiently permeable to permit transport of clay particles in water moving through the wall, soil will be eroded to form a continuous cavity—a tunnel or pipe (see Plate 20.2).

Wall permeability is often aided by a low clay percentage (< 25%), poor compaction or the formation of settlement cracks. One of the most common causes of tunnelling in small earth dams is inadequate soil moisture during construction and the development of a settlement crack due to saturation settlement. When a small dam first half-fills and water soaks into the wall wetting the lower half of the bank, this soil settles away from the dry top half. If the dam then fills rapidly to above this level, water flows through the cavity eroding the dispersible soil and producing a tunnel. Dispersible soils, inadequately compacted, are most likely to be subject to this form of failure. The volume expansion test is a useful indicator.

Tunnelling due to high wall permeability can occur within hours of the first fill or years after a dam is constructed. It depends on how long it takes percolating water to remove sufficient soil to form a continuous cavity, large enough to permit rapid flow.

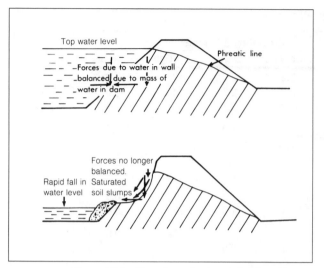

Figure 20.1 Slumping of saturated soil on the upstream side of a dam wall following rapid drawdown

Soil dispersion in banks is usually caused by a combination of a high proportion of sodium on the clay exchange complex (ESP > 6%), and a low concentration of cations in the soil solution within the wall. The addition of calcium ions, such as gypsum ($CaSO_4$) or, in neutral or acid soils, hydrated lime ($Ca(OH)_2$), will reduce both these problems.

The formation of settlement cracks can be reduced by ensuring there are no sharp changes in slope in the foundations and a high level of compaction is achieved throughout the wall, not just in the upper layers. Watering to increase the level of compaction is difficult in dispersible soils and construction may need to be deferred until soil moisture returns to a more suitable level. For large structures built of dispersible soil, sand filters, combined with gypsum-stabilised zones, have been successful in reducing tunnelling failure (Sherard and Decker, 1977).

Undermining

Banks, particularly contour banks built on dispersible or sandy soil, may develop a tunnel or pipe in the foundations. Cracks, cavities or more permeable soil layers not detected during construction may, under the additional head created by the impounded water, permit the passage of water with sufficient velocity to erode soil. This creates a passage under the wall and promotes structural failure.

The chance of this happening with contour banks can be substantially reduced by ripping under the bank, in susceptible areas, prior to construction. Dams should be inspected for possible susceptible layers that should be removed to at least 30 cm and replaced with relatively impermeable clay.

Bank Spacing

20.1.5

Bank spacings are derived to match the quantity and velocity of runoff (volume and erosivity), the soil erodibility and the practical bank size desired. They are then modified to suit local limitations such as outlet location, obstructions (trees to be retained, rock outcrops) and workability of interbank areas with farm machinery.

Plate 20.2 A tunnelling failure in a large farm dam

Soil effects on bank spacing are different for cropping and grazing land. For grazing land, where sheet and rill erosion are not major problems, bank spacing is determined by runoff yield and bank capacity. On cropping land, which is cultivated regularly and where soil is not always protected by vegetation, bank spacing aims to reduce rilling to an acceptable level and depends on soil erodibility, rainfall erosivity and slope. Spacing, therefore, varies for different geographical locations and soil types (see Stewart, 1955).

In general, bank spacing is calculated from:

$$HI = K/S^{0.5} \dots\dots\dots\dots\dots\dots\dots\dots\dots\dots 20.5$$
$$VI = KS^{0.5}/100 \dots\dots\dots\dots\dots\dots\dots\dots\dots 20.6$$

where

HI	=	horizontal interval (m)
VI	=	vertical interval (m)
K	=	bank spacing factor
S	=	slope (%)

The bank spacing factor has been determined by experience in New South Wales and is influenced by rainfall erosivity and soil erodibility. It can be estimated by selecting the appropriate 'K' value from Table 20.6 and multiplying by the appropriate factor from Table 20.7 for soil erodibility.

Table 20.6 Bank spacing factors (K) for specific localities (after Logan, 1968)

Location	Bank Spacing Factor
Inverell	110
Tamworth	140
Gunnedah	150
Wellington	160
Cowra	180
Wagga Wagga	200

Table 20.7 Adjustment to bank spacing factor (K) for soil erodibility

Erodibility	Adjustment Factor
Extreme	0.8
Very high	0.9
High	1.0
Moderate	1.15
Low	1.3

20.2 Construction of Earthworks

Soil conservation earthworks are often the most efficient and economical means to convey runoff water across the landscape while minimising soil loss. Most soil conservation earthworks are constructed using a dozer to build banks, excavate channels and waterponding areas, and cut flumes to suitable grades. In most situations, construction techniques and design based on standard procedures for the most economical means of construction are suitable. However, situations where soil properties deviate past critical limits must be recognised, so that design and construction techniques can be appropriately modified.

There are a number of laboratory tests to assess soil suitability for earthwork construction. However, the relevance of the tests to a particular project depends on how representative the soil tested is of the soil used in the earthwork. Soil in some situations is very variable, and this variation must be assessed if accurate field performance is to be predicted. Fortunately, most of the laboratory analyses can be approximated by simple field tests to assist in the selection of samples for more precise analysis and deciding whether laboratory testing is required.

Soil Sampling for Earthworks   20.2.1

If possible, all earthwork construction should be preceded by soil sampling and testing. Where a gully is present, the profile exposed can be used to examine soil variability in area and with depth. Soil samples should not, however, be taken from the exposed surfaces of the gully. A hole should be dug into the face of the gully to take samples at least 20 cm in from the exposed surface.

In normal situations, the minimum number of test holes recommended is one hole in the middle of the proposed embankment site (using a gully, if present, to gain extra depth) and one at either end of the proposed wall. At least the central test hole should go behind the full depth of the proposed structure. Other test holes should go to the full depth of the proposed structure. For large structures, or for variable soils, further test holes should be bored upstream of the proposed embankment site. Soil samples should be taken that represent the range of soil horizons found in the test holes.

When sampling areas of failed banks and gully control structures, an examination should be made of the bank or wall adjacent to the failed area, from the channel and uphill batter above the failure, and from the borrow area from which repair material is to be taken. In all cases, at least one sample should be of undisturbed soil.

When sampling for leaking gully control structures, it is very important to determine as precisely as possible where the leak is. If the structure drains only to a certain level, the leak is probably at this level. However, the reduction in hydraulic head as the water level decreases can result in a substantial decrease in the rate of water loss. The place of exit of the water downstream of the structure will also give a clue to the location of the leak.

The inside of the structure should be carefully examined for holes, soft spots, rock, gravel or sand seams, layers of flocculated soils or layers of calcium carbonate. If it is a general leak, then test holes should be made into the upstream and downstream batters as well as the bottom. The test holes in the bottom of the structure should go down at least 1 m to determine whether gravel seams are present.

Samples should be taken from any unusual material. Samples from the upstream batter and the bottom should be from undisturbed material. More detailed information on leaking structures will be found later in this chapter.

20.2.2 Field Testing

Prior to the construction of any soil conservation earthworks, the soil must be assessed for its suitability. At most sites this is done by observation, feel and, if appropriate, the reaction of soil to water. Using these simple methods, it is possible to estimate soil texture (particle size), dispersion and shrink-swell characteristics, and from these, inner permeability, tunnelling and cracking susceptibility, and site variability.

Soil Texture

The proportion of different-sized particles in the soil is easily assessed by observation and feel of wet soil to differentiate between sands, sandy loams, loams and clays. For description of the field method for assessing soil texture, see Chapter 10.

To recognise the extremes of soil texture, it is useful to remember that sand can be felt and heard, silt feels extremely smooth and clay is smooth and sticky with resistance to shearing. Organic soil is very spongy with an organic smell, like peat moss. It is unsuitable for earthwork construction. Sand, sandy loams and loams will have a low wet strength and are generally unsuitable for water-retaining structures. Marginal soils require laboratory testing for confirmation.

Soil Dispersion

Dispersible soils can be detected during the texture test and directly observed by conducting a simple Emerson Aggregate Test by dropping a soil aggregate into a glass of rainwater. Highly dispersible soils break down into clay particles, some as soon as they wet up, others after a period of soaking. A cloud of clay in the water indicates dispersion. Laboratory testing is then advisable to determine the degree of dispersion and appropriate modifications to construction technique and design, and/or the need for soil ameliorants. For a full description of the test, see Chapter 10.

Shrink-Swell Characteristics

These are apparent from observation of the soil and a field texture test. Problems are indicated by large cracks when the soil is dry, or cracks with shiny sides that indicate a lot of soil movement. The occurrence of gilgai is also an indicator. In these situations, the soil is probably a heavy clay and warrants testing for confirmation. When in doubt, any soil material assessed as heavier in texture than a clay loam should be tested for shrink-swell potential in the laboratory (see also Chapter 11).

Laboratory Testing   20.2.3

Laboratories normally carry out the following standard tests to determine the suitability of soil for earthwork construction (about 2 kg of soil is required):
(a)   Particle Size Analysis
(b)   Unified Soil Classification System (USCS)
(c)   Dispersion Percentage (DP%)
(d)   Linear Shrinkage (LS) or Volume Expansion (VE)
(e)   Emerson Aggregate Test (EAT)
(f)   Soil pH and Conductivity (EC)
(g)   Hydraulic Conductivity.

In addition to these, tests such as various nutrient, ion and heavy metal levels are assessed where they are applicable and may affect earthwork performance or the local potential for revegetation of earthworks. The sample collected should exclude coarse rock material, which must, however, be accounted for in the soil assessment.

Particle Size Analysis

The method used is the Australian Standard AS 1289.3.6.1, 1995; AS 1289.3.6.2, 1995; and AS 1289.3.6.3, 1994 or adaptations of these methods (see Chapter 10). The concentration of fine fraction is measured using a hydrometer and the coarse fraction is determined by wet sieving. The content of fine sand is determined by difference and therefore the sum of the percentage content of the various fractions is always 100.

The results of the hydrometer and sieve analysis are plotted to give a particle size distribution curve (see Figure 10.1), and the amount of the following arbitrary groups of particle size is interpolated from the curve:

less than 0.002 mm	— clay
0.002–0.02 mm	— silt
0.02–0.2 mm	— fine sand
0.2–2.0 mm	— coarse sand
2.0–6.0 mm	— gravel
greater than 6.0 mm	— stones

Soil behaviour can be characterised in terms of these groupings provided it is understood that:
—there is no sharp change corresponding to the size limit for each group
—a relatively small content of one group may give a soil one particular trait of that group and yet have little effect on other characteristics.

A balance of particle sizes is necessary to form a relatively impermeable seal in water storage structures. For further, detailed information on particle size analysis, see Chapter 10.

Unified Soil Classification System (USCS)

The USCS classifies soils in terms of practical engineering behaviour relative to soil grading (particle size distribution). There are two methods commonly used, details of which are found in Chapter 11.

(i) Rapid Qualitative Assessment (Field Method)

Involves the hand assessment of soil passing an ASTM No. 40 sieve (approximately 0.4 mm). The attributes assessed are:

dilatancy—reaction to shaking moist soil; very fine loam sand is highly dilatant, inorganic silt moderately dilatant and plastic clay non-reactive
dry strength—the resistance to crushing of soil that has been wet-moulded by hand and then dried; clays have high dry strength, silts and silty fine sands low dry strength
toughness—soil 'stiffness' or consistency near the Plastic Limit (that is, that point at which the moisture content causes the soil to change from the plastic (ductile) state to the solid state); the 'stiffer' the soil is near the plastic limit the stronger the clay fraction of the soil.

(ii) Quantitative Assessment Using Atterberg Limits

Two Atterberg limits and one index are determined on the fine-grained fraction of the soil in this test.

The liquid limit is the water content at which the soil passes from the liquid state into the plastic state and is determined by AS 1289.3.1.1, 1995 and AS 1289.3.1.2, 1995.

The plastic limit is the water content at which the soil passes from the plastic state into the solid state and is determined by AS 1289.3.3.1, 1995 (see Chapter 11).

The numerical difference between the liquid limit and the plastic limit is the plasticity index and corresponds to the range of water contents within which the soil is plastic. Soil plasticity is an important property of fine-grained soils. Highly plastic soils have a high value for plasticity index whereas in non-plastic soils, the plastic limit and the liquid limit are the same and the plasticity index is zero.

These limits of consistency are used in the USCS as the basis for laboratory differentiation between fine-grained materials of appreciable plasticity (clays) and slightly plastic or non-plastic materials (silts).

Dispersion Percentage

This test was developed to assess soil susceptibility to failure by tunnelling. It is a modification of Middleton's (1930) dispersion ratio, calibrated to tunnelling of small earth dams in New South Wales by Ritchie (1963) and Rosewell (1970). The measure is the ratio between the amount of soil less than 5 microns in diameter in a soil sample after ten minutes' shaking in distilled water, and the amount determined by particle size analysis on the same sample, expressed as a percentage.

There are a number of factors that affect soil dispersion and subsequent movement in dam walls, including the relative concentration of exchangeable ions, total ionic concentration in the soil solution and soil permeability. The dispersion percentage, measuring the degree of dispersion in distilled water, is only an estimate or a relative rating of the expected level of dispersion in the earthwork. This test can also be used to assess the degree of aggregation of clay and silt into fine sand-sized particles, and from this infer the possibility of a dam leaking. Should the structure be required to hold effluent or other saline waters, then field performance should be further assessed by conducting a second dispersion percentage using actual or simulated storage water.

Linear Shrinkage or Volume Expansion

Linear shrinkage determines the percentage decrease in length of a specimen of disturbed soil on drying from the liquid limit to oven dry. The method used is Australian Standard AS 1289.3.3.1, 1995 (see Chapter 11).

Soil at its liquid limit is placed in a shallow trough 25 cm long by 2.5 cm diameter and dried until shrinkage ceases. The decrease in length is measured and expressed as a percentage of the original length. For earthwork construction, linear shrinkage is used to predict instability due to crack formation through walls or potential leaking problems caused by low clay activity. Very high or very low values can lead to problems.

Volume expansion is a free swell test on disturbed soil when wetted from air dry to saturation. The method is that of Keen and Raczowski (1921) with a modified computational procedure (Wickham and Tregenza, 1973). As the test relies on the soil sample achieving saturation, it can give erroneous results with some slow-wetting dispersible soils (Mills et al., 1980). Some soils may shrink in volume in this test, usually due to their dispersibility, and this may therefore be used as an indicator of likely tunnelling susceptibility.

The test probably has no quantitative value, however it does provide a simple identification of potentially expansive soils. A large volume expansion indicates a large potential expansion on wetting and subsequent shrinkage on drying. Earthworks constructed with such material are susceptible to failure when water enters the shrinkage cracks developed following prolonged dry periods.

A good correlation between linear shrinkage and volume expansion has been found for non-dispersible soils (Mills et al., 1980). The linear shrinkage test is generally favoured, particularly in relation to engineering applications.

The Emerson Aggregate Test (EAT)

This test, developed by Emerson (1967), classifies soil aggregates on the basis of their coherence in water. The interaction with water of clay-sized particles in aggregates may largely determine the structural stability of a soil. The Emerson Aggregate Test is a simple physical test for dividing aggregates into eight main classes, and is fully described in Chapter 10. The expanded test, as described, is used to assess potential soil dispersion, in relation to both soil crusting and soil erosion problems in the field, and is a soil conservation modification of AS 1289.3.8.1, 1995.

Soil pH and Conductivity (EC)

Field pH assesses soil acidity or alkalinity and should be checked in the laboratory. It does not generally affect the structural integrity of earthworks, but could influence the use of the site for specific-purpose water retention.

Conductivity of a soil will affect dispersion and reflect salinity, which may place constraints on the long-term use of the earthworks.

Hydraulic Conductivity

A hydraulic conductivity test assesses water movement through the soil. A falling head hydraulic conductivity is the recommended method for most earthworks because of the low rates of water movement required to be measured. The method commonly used in soil conservation laboratories follows the method used by Klute (1986).

The generally accepted threshold for farm storage is a maximum conductivity of 10^{-7} m/s which is about 0.4 mm/h, or about 3.5 metres per year.

Any pondage that has the potential to contaminate groundwater must be tested to determine if the soil is acceptable and if any special measures such as imported clay, impervious membranes or chemical amelioration are necessary. The current requirement for pollution control is that the hydraulic conductivity of structures should be less than 10^{-9} m/s which is about 0.004 mm/h, which is about 35 millimetres per year down through the soil profile.

Recommendations from Soil Test Data 20.2.4

Recommendations are based on the best available knowledge and experience, with special emphasis on the use of soils in conservation earthworks constructed at what can be considered normal conditions.

The recommendations given in Table 20.8 should always be read in conjunction with the following general remarks and the accompanying notes.

The assessments used in the table are the Unified Soil Classification, particle size analysis, dispersion and shrink-swell potential. Within each of these, criteria are used as follows:

Unified Soil Classification

Reference should be made to Chapter 11 for a full explanation of this system and a qualitative comparison of engineering properties and suitability of the various groups. In general, the field method of characterisation described therein may be used for small earth dams.

GC and SC materials are stable and are generally suitable for water-holding structures if well compacted at the optimum moisture content. If compacted dry, they will be pervious and tunnelling-susceptible. GC and SC materials should never be used for earthworks when very dry.

GM and SM materials should not be used for the core of zoned embankments. They should be protected from erosion if used on exposed batters.

GW, GP, SW and SP materials are not suitable for inclusion in soil conservation earthworks designed to retain water. They may be used in the shell of zoned embankments.

CL and CH materials will retain water. They may be susceptible to tunnelling or cracking so that other test results need to be considered. CH materials with a high content of expansive clay will crack extensively on drying out and also have extremely low wet strengths.

ML and MH materials are semi-pervious if not well compacted. They are also subject to settlement on wetting and are generally erodible.

OL and OH materials are generally not suitable for soil conservation earthworks.

Particle Size Analysis

This test is used for classification of soils by the Unified Soil Classification System, and provides some specific limits on the suitability of soils for earthworks. The following six categories apply in addition to the Unified Soil Classification group.

(i)   Clay percentage less than 10%, silt + clay less than 20%: These coarse-grained soils are considered to be pervious and are not recommended for general use in homogeneous embankments or the impervious zones of water-storage structures. Well-graded soils with a Unified Classification of GC or SC may be suitable for water storage when well compacted at the optimum moisture content. A filter zone may be required.

(ii) Clay percentage 10–25%, silt + clay 20–40%: These are coarse-grained soils that are generally suitable for use in earthworks. Such soils with a classification of GM or SM are not suitable for homogeneous embankments or the impervious zones of water-storage structures. Particular care must be taken with soils of this category that are tunnelling-susceptible.

(iii) Clay percentage 10–25%, silt + clay greater than 45%: These are fine-grained silty soils that have variable permeability and are likely to be erodible. Generally only suitable for upstream and downstream zones of a zoned embankment or in a modified homogeneous embankment with filter zone.

(iv) Clay percentage 25–40%: These are generally fine-grained soils and are suitable for most soil conservation earthworks. Attention should be paid to tunnelling and cracking susceptibility.

(v)   Clay percentage greater than 40%: These fine-grained soils may leak if well aggregated, or crack on drying if they have high volume expansion or linear shrinkage values.

(vi) Clay percentage less than 25%, silt + fine sand greater than 50%: These are generally fine-grained soils with a very low wet strength and high erodibility. Attention should be paid to achieving good compaction, decreasing the batter grades to less than 1:3 and reducing the head of water against the wall.

Dispersion

As discussed under permeability, soil material used in the walls of water-storage dams requires a balance of dispersed clay. There must be enough dispersed clay to seal, but not enough to permit tunnelling failure.

(i) Tunnelling Susceptibility

There are a number of aspects that must be considered in terms of earthwork failure by tunnelling (see Plate 20.2). Earthwork tunnelling is a complex process depending on variables such as the density and moisture content of the soil, soluble salt content of the stored water, rate of filling and soil properties (Rosewell, 1970). When dispersion limits are set, there is a range of uncertainty in the prediction of earthwork performance. Such limits are based on average conditions of construction and filling. In addition, the majority of soils that have been tested and on which experience has been based have been clays of low plasticity, with clay contents in the range 15–50%.

From an examination of the soils and case histories of some 60 earthworks throughout New South Wales, Ritchie (1963) set an arbitrary limit equivalent to 30% dispersion of the < 5 micron soil particles. Soils with a value greater than 30% are susceptible to tunnelling failure. Subsequent experience with many hundreds of soils and case histories shows no reason to change this limit. Studies by Elliott (1976) have indicated additional support.

Recommended limits for the < 5 micron dispersion percentage are:

0–30%	slight risk
30–50%	moderate risk
50–65%	high risk
65–100%	very high risk

Table 10.3 allows interpretation of dispersion percentage in relation to the < 5 micron fraction of

Table 20.8 Recommendations for small farm dams from soil test data

						Linear Shrinkage (%)		
						0–12	12–17	> 17
USCS	Clay %	Clay + Silt %	Silt + FS %	EAT	Dispersion % < 5um	Volume Expansion (%)		
						0–20	20–30	> 30
	< 10	< 20				I*	K	K
CH	> 40			1	65–100	D	F	H & F
				1 or 2	50–65	B	D	G & F
				2 or 3	30–50	B	D	F
				3	10–30	A	G	H
				4–6	0–10	C	C	H
	< 40			1	65–100	D	E	F
				1 or 2	50–65	D	D	F
				2 or 3	30–50	B	D	E
				3	10–30	A	G	H
				4–6	0–10	C	C	H
MH				1	65–100	E	E	F
				1 or 2	50–65	D	D	F
				2 or 3	30–50	B	D	E
				3	10–30	A	B	H
				4–6	0–10	C	C	H
CL	> 25			1	65–100	D	E	F
				1 or 2	50–65	D	D	E
				2 or 3	30–50	B	G	H
				3	10–30	A	G	H
				4–6	0–10	C	G	H
	< 25			1	65–100	D	D	K
				1 or 2	50–65	D	D	K
				2 or 3	30–50	·B	B	K
				3	10–30	A	A	K
				4–6	0–10	C	A	K
	> 50			1 or 2	30–100	D	E	K
				3	10–30	B	B	B
				4–6	0–10	C	C	K
ML				1	65–100	E	E	K
				1 or 2	50–65	D	D	K
				2 or 3	30–50	B	B	K
				3	10–30	A	A	K
				4–6	0–10	C	A	K
SC				1	65–100	D	E	K
				1 or 2	50–65	D	D	K
				2 or 3	30–50	B	B	G
				3	10–30	B	A	K
				4–6	0–10	C	A	K
GC				1	65–100	E	J	K
				1 or 2	50–65	D	D	K
				2 or 3	30–50	B	B	K
				3	10–30	B	B	K
				4–6	0–10	C	A	K
SM & GM				1 or 2	30–100	J	K	K
				3	10–30	B & I	K	K
				4–6	0–10	I	K	K
SW, SP, GW & GP					30–100	J	K	K
					0–30	I	K	K
OL, OH & Pt				All	0–100	J	J	J

* For interpretation see facing page

Notes on Table 20.8

These recommendations are based on experience with farm dams in central and eastern New South Wales. These dams are usually less than 10 000 m³ in capacity and have a top water level of less than 3–4 m above the original ground surface at the upstream side of the wall.

Soil properties are continuously variable; the limits set are not fixed divisions between groups and interpretation will often be a compromise between two recommendations. Soil will vary within a structure site. The relative proportions of the different soil materials need to be considered and construction technique planned accordingly.

The recommendations are based on a well-graded soil. Soil material with 'grading gaps' requires special consideration.

The value limits used in these recommendations have been set for typical farm situations where the consequence of failure only involves loss of stored water. Where the consequence of failure is higher (such as damage to roads or buildings), specifications for batter grades, use of ameliorants and compaction levels should be appropriately modified.

Recommendations for use with Table 20.8

A   Soil is suitable for normal use. Take care to achieve good compaction, preferably with moist soil. If the soil is dry (cannot be moulded without breaking), reduce layer thickness to < 15 cm. Minimum batter grades 1:2.5 upstream, 1:2 downstream, except for CH and MH classifications when they should be decreased to 1:3 and 1:2.5 respectively.

B   This material is similar to A but with high compaction requirements—to at least 85% of maximum dry density. To achieve this, the soil should be close to the optimum moisture content for the compaction plant, be placed in layers < 15 cm thick and compacted with four complete passes of a crawler tractor or roller (or equivalent thereof). As a general guide, the soil should be sufficiently moist to be made into a thread 10 mm thick, but not moist enough to be rolled thinner than 3 mm without breaking. Minimum batter grades 1:3 upstream, 1:2.5 downstream.

C   Aggregated material which may not hold water. Compact to 95% of maximum with at least four passes of a sheepsfoot roller when the soil is slightly wet of optimum (can be rolled into a 3 mm diameter thread). Use of a vibrating roller for dry soils. An ameliorant—STPP or sodium carbonate—could be required. If EAT is Class 6 or dispersion percentage is less than 10, then

the dam is likely to leak unless sealed with better clay or treated with an ameliorant to induce dispersion. A permeameter test may be required.

D   This soil is highly susceptible to tunnelling or piping failure. It must be well compacted *throughout* to reduce permeability and saturation settlement. The soil should be compacted to at least 90% maximum dry density by ensuring adequate moisture content. If drier than optimum, gypsum or hydrated lime should be incorporated into the soil at rates based on laboratory testing. Method to be determined by site and equipment constraints. For additional stability, the structure should be designed to hold no more than 1 m of water against the wall and batter grades should be decreased to 1:3.5 upstream and 1:3 downstream.

E   This soil is very susceptible to tunnelling or piping failure. In addition to Recommendation D, the structure must hold no more than 1 m depth above the original ground surface at the upstream side of the wall, and not be subject to more than 0.3 m/day drawdown (trickle pipes must not be more than 0.3 m below top water level). Gypsum or hydrated lime should be incorporated in the upstream side of the wall. The upstream batter grades should be decreased to 1:4, following compaction to 95% of maximum dry density.

F   This soil is very susceptible to tunnelling or piping failure. Because of the high shrink-swell potential, batter grades must be decreased. In addition to Recommendation D, freeboard must be increased to at least 1 m above surcharge level and hydrated lime or gypsum should be applied at rates determined in the laboratory. Batter grades should be decreased to 1:4 upstream and 1:3 downstream, where compaction at 95% maximum dry density cannot be assured.

G   The high shrink-swell potential of this soil can result in cracks extending through the wall below top water level. To reduce this possibility, a compact central core (at least 95% maximum compaction) must be obtained by constructing when the soil is at near optimal soil moisture. The freeboard must be increased to at least 1 m above surcharge to prevent surface cracks extending below the waterline. Recommended batter grades are 1:3.5 upstream and 1:3 downstream. The structure must be designed to retain sufficient water to keep the wall moist and minimise crack development.

Notes on Table 20.8 (continued)

H   Highly susceptible to failure. Recommendations in G are more critical to prevent failure. Compaction should be to 95% of maximum dry density at least, and batter grades further reduced.

I   Pervious. Not recommended for general use, but may be used in a zoned embankment or mixed with other materials. Recommended batter grades are 1:3 upstream and 1:3 downstream.

J   Not recommended.

K   Usually unsuitable for construction.

NB: Soils C to J need to be assessed in terms of (i) zonal use, (ii) mixing with other on-site soil materials as well as the use of ameliorants. An additional option is the importation of lining materials. Problem soils can usually be corrected by lining with a stable, impervious membrane such as a suitable local clay or incorporating bentonite or a synthetic liner.

the soil. When dispersion percentage approaches the critical levels shown therein, there is a high risk of tunnelling failure.

When interpreting the Emerson Aggregate Test (EAT) assessment, Class 1 aggregates most certainly indicate high tunnelling susceptibility. Detection is difficult in acid soils where there are usually local variations in the percentage composition of the exchangeable cations.

Class 2 aggregates indicate some degree of tunnelling susceptibility, the actual amount depending on other factors such as permeability and resistance to cracking and slumping. Subclass 2(1) material (showing slight dispersion) is desirable for water-storage structures. Subclass 2(2) and 2(3) aggregates should be treated as being unstable, unless indicated otherwise by the < 5 micron dispersion value.

Class 3 aggregates are generally stable and indicate a more desirable material for soil conservation earthworks. It is unlikely that structures built from Class 3 material with sufficient clay will leak but, if they do, they can be readily sealed after construction by compacting when wet with a roller or stock.

Subclasses 3(1) and 3(2) are the most satisfactory materials within this class. Subclasses 3(3) and 3(4) are more subject to dispersion failure following construction and/or working at high moisture content and should, therefore, be used with care.

Class 4, 5, 6, 7 and 8 aggregates are not susceptible to tunnel erosion.

It is not unusual for the dispersion percentage test and Emerson Aggregate Test to give the same soil different ratings. They measure different factors. Where this occurs, careful assessment of all soil attributes and local experience are often the only reliable guides, and erring on the safe side is a wise precaution.

(ii) Permeability

At low levels of dispersion, there are insufficient clay-sized particles in the soil to form a relatively impermeable seal. The structure is therefore likely to leak. Experience has shown that soils with a dispersion percentage less than 10 are likely to fall into this category.

With the Emerson Aggregate Test, soils falling into classes 4 to 8 can have permeability problems but particularly those of Class 6, which may leak even after compaction to 90–95% maximum dry density.

Shrink-Swell Potential

This relates to the capacity of clayey soil material to change in volume with changes in moisture content. It is usually measured by a laboratory assessment of the soil's linear shrinkage. On the basis of experience with soil conservation earthworks, four categories are suggested:

(i)   Soils of *low* shrink-swell potential have linear shrinkage values of 0–12% (volume expansion 0–20%) and are generally suitable for soil conservation earthworks. Soils of very low shrink-swell may not seal and if dispersible can be susceptible to tunnelling failure.

(ii)   Soils of *medium* shrink-swell potential have linear shrinkage values of 12–17% (volume expansion 20–30%) and are susceptible to minor shrinkage on drying and swelling on wetting.

(iii)   Soils of *high* shrink-swell potential have linear shrinkage values of 17–22% (volume expansion 30–40%) and are susceptible to moderate shrinkage and swelling. They are particularly subject to tunnelling if dispersible. Swelling of otherwise stable soils will decrease permeability.

(iv)   Soils of *very high* shrink-swell potential have linear shrinkage values in excess of 22% (volume expansion > 40%) and are susceptible to extensive shrinkage and cracking on drying and swell considerably when wetted.

A soil containing greater than 40% clay and having a very high shrink-swell potential will be susceptible to extensive cracking, to the extent that it is not suitable for use in the critical parts of an earthwork.

Soils with a medium to high shrink-swell potential are prone to cracking, but are amenable to suitable control measures.

Soils with a low shrink-swell and moderate to low level of dispersion are generally considered stable for use in earthworks.

Soils with high soil dispersion, with high or very high levels of shrink-swell potential, are highly tunnelling-susceptible, as the cracking or drying may provide suitable pathways for tunnel formation when the soil is rewetted.

High soil dispersion combined with low shrink-swell potential may result in settlement on saturation when the structure fills. The gap remaining between the settled saturated zone and the unsettled unsaturated zone may induce tunnelling failure, particularly in high tunnelling-susceptible soils. The settlement may also induce cracking.

20.2.5 Practical Considerations

At any proposed earthwork site, it is possible to get varying soil materials, both with depth and in area. Very often these different soils have vastly different suitabilities as construction materials.

The major considerations are for the compacted earthwork to have sufficient strength and for permeability to be sufficiently low to retain the stored water under all conditions of operation. All soil materials are pervious to water and the through-seepage flow has to be considered in the ultimate stability of the embankment.

The through-seepage in small structures can represent a major loss from the stored water. Seepage losses in farm dams constructed to standard specification can range from about 3 m per year down to about 3 cm per year. Through-seepage results in saturation of part of the embankment. This saturation has two undesirable effects on the stability of the downstream portion of the embankment. First, the total weight of the soil is greater when saturated than when unsaturated. Second, the water pressure in the soil pores reduces the shear strength of the soil. If the downstream batter is too steep, a failure, in the form of a slide or slip, may occur. This is more likely the higher the permeability of the soil because the steady state seepage, or phreatic line, reaches the downstream batter at a higher level. This problem may also be greater in soils of high clay content owing to their lower shear strength when saturated.

Similarly, a slip failure of the upstream batter may occur when the water level is reduced quickly after being held at a high level for a long period. This 'drawdown' type of failure is the result of the combined effect of the removal of the water load on the upstream face together with the residual water pressure remaining in the pores of the 'impervious' fill (see Figure 20.1).

In a homogeneous embankment constructed from uniform soil material, the usual protective measure is to have a relatively flat downstream batter and to cover it with a layer of fertile, organic topsoil, and to maintain a dense vegetative cover. The topsoil is normally pervious and allows drainage of the seepage water down to the toe of the embankment and, in addition, the root system of the vegetation helps to resist piping and uses some of the seepage water for transpiration.

Where different soil materials are present at the site, then it is advantageous to design and construct an embankment with a zoned cross-section. A zoned cross-section makes best use of the available material and enables the construction of a dam where there is insufficient suitable material for a homogeneous wall. Examples of zoned embankments are shown in Figure 20.2.

In all cases, the best (least pervious) material is placed in the impervious core and this is supported by well-compacted, more pervious material. The pervious zone must be equally well compacted as the impervious zone, as settlement of the former will cause cracking of the latter. Ideally, the material should grade from one zone to another.

The type of zoning used will depend on the amount and location of the soil material. The central core is the most versatile of all embankment cross-sections. The low-strength impervious core material is supported by the stronger shells, which, in addition, provide drainage on the downstream side and resist the effects of sudden drawdown of the stored water on the upstream side.

The upstream blanket zoned cross-section shown can be used on embankments where the quantity of impervious material is limited. This impervious layer can be added later as a repair measure when through-seepage of a homogeneous cross-section is excessive.

Soil Compaction   20.2.6

The single most important task in many engineering works is to achieve good soil compaction. Compaction substantially influences the future behaviour of any earth structure or foundation, as poor compaction results in low strength, high permeability, susceptibility to tunnelling in dispersible clays, high erosion risk and large settlements. The dangers of slip failure or collapse are enhanced and, for expansive clays, the potential for shrink and swell is more readily realised. Compaction is defined as the process by which an immediate increase in density in an earth mass can be effected by the displacement of air. It should not be confused with consolidation, which is a

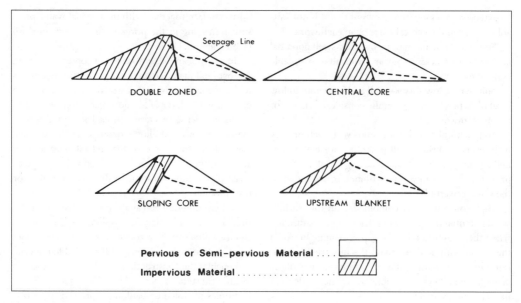

Figure 20.2 Zoned cross-section embankments

slow increase in density due to the gradual rearrangement of soil particles over time, often associated with the slow removal of water from the soil.

For most soil conservation earthworks, good compaction is essential to quickly achieve a stable, water-holding or functional structure. This is particularly so in the case of sodic or dispersible soils.

Soil Compaction Characteristics

(i) Compaction and Moisture Content

The amount of compaction that can be achieved in a soil depends on the moisture content. Hence, the moisture content at which the soil is compacted determines the effectiveness of the contact pressure applied.

Just how this happens is shown in Figure 20.3. As can be seen, there is an optimum moisture content at which a maximum dry density can be achieved. Wetter than this, there is insufficient confining force to compact the soil, which simply flows away from the compaction zone due to loss of shear strength in the soil. Drier than the optimum, soil strength tends to be greater than the compactive effort and the soil resists compaction. Optimum moisture content and maximum dry density can be determined in the laboratory for a soil using the maximum compaction test.

AS 1289.5.1.1, 1993 (previously referred to as the Proctor test) sets out a method for the determination of the relationship between the moisture content and the dry density of a soil, when compacted, using standard compactive effort (596 kJ/m³). Compaction is conducted over a range of moisture con-

tents so as to establish the maximum mass of dry soil per unit volume achievable for this compactive effort and its corresponding optimum moisture content. The procedure is applicable to that portion of a soil which passes the 37.5 mm sieve. Soil which all passes the 19.0 mm sieve is compacted in a 105 mm mould. Soil that contains more than 20% of material retained on the 19.0 mm sieve is compacted in a 152 mm diameter mould. Corrections for oversize material are not made in this method but may be made using method AS 1289.5.4.1 when required for compaction control.

This is a standard test and is used as a yardstick to assessing the degree of compaction achieved, usually

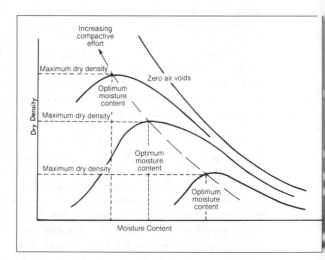

Figure 20.3 Soil compaction characteristics

expressed as a percentage of the maximum dry density attained during the test. It also provides a guide for the optimum moisture content for compaction. For example, for compaction with a dozer, the soil should be slightly wet of the optimum indicated by the test; for compaction with a sheepsfoot roller, it should be slightly drier.

In Figure 20.3, the line known as the 'zero voids line' represents the densities that might theoretically be achieved if no air voids remained in the soil. In practice, there is usually a range of moisture contents over which adequate compaction can be achieved for small earth walls (< 5 m high) built of stable non-dispersible materials. The dry end of the range should be avoided, particularly in dispersible materials. Sometimes aggregated soils are worked wet of the optimum moisture content to facilitate clay dispersion.

Both optimum moisture content and maximum dry density can be estimated from the Unified Soil Classification System (see Table 11.6).

(ii) Measuring Compaction and Moisture Content in the Field

The monitoring of soil density, and hence compaction, can be done in the field, as can moisture content. Sand displacement methods (AS 1289.5.3.1, 1993 or AS 1289.5.3.2, 1993) are usually used, but specialised equipment such as a nuclear surface moisture-density gauge (AS 1289.5.8.1, 1995) are now used by some field workers. Older field methods such as AS 1289 E3.4, 1977 (balloon) and AS 1289 E3.5, 1977 (water displacement) are no longer in common use.

Techniques for assessing field bulk density (similar to those used in agricultural soils) such as AS 1289 E3.3, 1977 (core cutter) or the simple 'clod' displacement test (Department of Land and Water Conservation soil physical test number P141BD) are quickly and easily carried out and can provide useful complementary assessment of field density.

Moisture content has been assessed using specialised equipment or by burning the soil with methylated spirits (AS 1289 B1.3, 1977), but is best measured under more controlled conditions such as AS 1289.2.1.1, 1992 (oven drying) or for more rapid results, AS 1289.2.1.4, 1992 (microwave drying).

Results for compaction are often reported as a percentage of maximum compaction. A percentage of 85 is usually considered minimal for small earth walls, representing 0.85 of the maximum compaction achievable by the compaction test.

Achieving Compaction in the Field

A number of factors are relevant when attempting to achieve compaction in the field.

(i) Soil Particle Packing and Compaction

The effect of particle crushing due to rolling may be included as part of a more general principle of compaction, that highest densities are achieved by mixtures of different particle sizes.

Compaction is achieved by filling the voids between the larger particles with particles of smaller size. Overworking and reuse of soil, however, can create excessive fines, drop the shear strength and occasionally destroy natural cohesion in a soil, such that loss of density occurs by excessive rolling.

Soil particles with a flat and flaky form normally pack much more densely than those with rounder form (provided they are greater than 20 microns in size). The minute flat sheets of clay particles and micas possess considerable surface repulsion forces because of their small size, and thus resist compaction so effectively that they are among the most difficult earthen materials to compact.

Particles with a smoother surface will pack more easily than rough and angular particles owing to their lower interparticle friction, but they are also displaced more easily under shear forces.

However, it should be remembered that packing is finally dependent on the type of equipment used and moisture content.

(ii) Maximum Compaction

If maximum compaction is to be achieved, it is clear that the soil must be confined, that is, a constraining layer around and below the soil being compacted is required. The confining mass may be the soil itself, provided the applied compressive forces dissipate within an acceptable distance from the point of application. Where a layer of saturated soil or other soil of low bearing capacity underlies the layer to be compacted, no real compaction can be achieved as the soft layer will deform so as to negate much of the applied compactive effort. In such circumstances (and indeed in all good compaction practice), stage compaction is adopted, that is, the soil is first rolled with light rollers to develop its own strength, then heavier rollers are applied to complete the compaction. This change in bearing pressure can be achieved by selection of different rollers or by adjusting the ballasting of a roller, or for tyred rollers, adjusting the inflation pressure, ballasting or both.

If a compactive force is applied in one direction, it will produce a compacted state with preferred orientation of the voids. Second, a static force is usually less effective than a dynamic force of the same magnitude. Shear forces provide more rapid compaction than compressive forces, and vibratory forces often provide the most rapid compaction of all. To achieve compaction, the use of various types of rollers is widespread, and these are listed in Table 20.9.

The most important features of roller compaction are as follows:

—Passes of the same roller have less compaction effect at the same moisture content. This behaviour results since the increase in density leads to increases of both internal friction and cohesion, and hence resistance to further compaction becomes progressively greater for a given compactive effort.

—The optimum moisture content for the first roller pass may well be slightly higher than for the second pass, and so on; therefore, some drying time can be advantageous between successive roller passes.

—The speed of the roller has little effect on the compaction achieved.

—Different rollers are better suited to different materials. Plain and vibrating drum rollers are best for sandy-type soils at low moisture contents; sheepsfoot and drum rollers for loam and clay soils at low moisture contents; the pneumatic-type roller for clays at high moisture contents.

—No roller adequately compacts the soil if the layer is greater than 45 cm thick in a loose state.

—Use of the heaviest roller that does not cause subgrade failure is always recommended.

—A sheepsfoot roller requires more passes than other rollers because of its limited bearing area and hence poor coverage. The loading intensity under a small weighted sheepsfoot roller may be as high as 700 kPa, whereas the loading intensity under the track of a typical dozer-type machine is about 50 kPa (D5). Despite this disparity in compacting pressures, satisfactory compaction can be achieved with a track-type tractor under certain conditions.

A tracked vehicle can only attain good compaction at a moisture content slightly higher than the optimum determined in maximum compaction testing.

Work in England (Transport and Road Research Laboratory, 1982) suggests that eight passes of a smooth wheel roller or pneumatic tyre roller generally approaches the maximum compaction achievable by the roller at optimal soil moisture. However, compaction by a sheepsfoot roller, because it only acts on a portion of the soil, generally requires 24 passes of the machine. In the long run, however, the sheepsfoot roller does give higher compaction. The vibrating rollers appear to be most effective despite their lower weight. A review of compaction equipment is given by the Transport and Road Research Laboratory (1982).

(iii) Compaction Control in the Field

It is not always possible to measure the most critical parameters by simply running field tests, so relationships must be established with more easily measured parameters, and the latter used for field control.

Density and moisture content are by far the best compaction control parameters, since they allow the determination of what air voids remain in the soil;

Table 20.9 Various roller types, their compaction pressures and uses

Roller Type	Usual Weight (t)	Usual Width (m)	Estimated Load per Unit Width (/m)	Likely Pressure (kPa)	Remarks
Dozer				50	Not very efficient
Smooth Wheel	8–12	1.9	4.2–6.3	420–650[1]	Better for granular materials. Will operate successfully above optimum moisture content (if not too wet)
Vibratory	4	1.8	2.2	220[1]	Better for granular materials than smooth wheel rollers since only half the weight gives equal compaction. Frequency about 2000 Hz (cycles/min)
Sheepsfoot	3–5 unballasted	1.7	Not applicable	480–1725 on the feet	Best suited for clays, especially in semi-arid zones
	5–7 ballasted				The foot pressure may be too high for saturated high moisture clays and too low for very dry clays
Pneumatic	8–12	2.3	4.3–6.5[2]	430–650[2] depending on tyre inflation (may be as low as 250)	4–11 wheels with 0.9–2.2 t per tyre. Best for cohesionless and low cohesion soils, and for surface finishing

1 Estimated assuming 10 cm × roller width is contact area of roller with the ground.
2 Estimated as above, but also assuming that tyre inflation is such as to give 80 per cent coverage of ground.

and achieving about 85% maximum density should be the general aim on most soil conservation works. A compaction of 85% of maximum dry density is a minimum requirement for water retention structures when working with the better class of construction soils with adequate soil moisture. Suspect soils would normally be expected to require compaction of 90–95% of maximum dry density with soil moisture being critical.

A series of curves (Figure 20.4) relating optimum moisture content and dry density to the Atterberg limits is useful as a means of arriving at a first estimate of compaction optima.

20.2.7   Soil Amelioration

Amelioration techniques can be used when soil properties are inadequate for proposed earthworks and the cost is warranted. The most common techniques used for soil conservation earthworks involve the use of hydrated lime or gypsum for stabilising dispersible clays, and sodium tripolyphosphate and sodium carbonate for dispersing and sealing highly aggregated clays. A range of measures for sealing leaking structures is discussed later in this chapter.

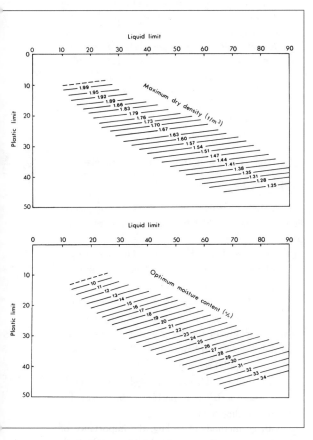

Figure 20.4 Compaction characteristics and Atterberg limits

Stabilising Dispersible Clays

Dispersible clays are difficult soils for building earthworks as they are susceptible to tunnelling failure. Therefore, amelioration is recommended where such soils have to be used for construction purposes.

As most of the dispersible nature of the clays is a result of high sodium levels, the most common method of amelioration is to add a source of calcium ions, usually as gypsum ($CaSO_4$) or hydrated lime ($Ca(OH)_2$). As well as replacing the sodium ions, often the addition of gypsum or hydrated lime increases the ionic strength of the soil solution, which also increases clay stability (see Figure 10.4). Lime may be converted to calcium carbonate ($CaCO_3$) in alkaline conditions and rendered insoluble. Therefore, lime should only be used on acid or neutral soils.

For earthworks, rates of about 1 t/1000 m³ of earth are generally recommended for both gypsum and hydrated lime. Specific requirements can be assessed by laboratory testing. In agricultural situations, about 2 to 5 t/ha are typically recommended for improving the aggregation of field soils.

It should be noted that these rates of hydrated lime for earthworks represent only about 0.1%, which is very much lower than the rates used for lime stabilisation of soils (usually about 7%) for engineering purposes.

When using gypsum, it is advisable to check its solubility or flocculating potential, as gypsum from different sources can vary considerably in its solubility and hence its ability to flocculate dispersible clays.

Aggregated Clays and Sodium Tripolyphosphate (STPP)

The details of the use of STPP to seal dams built from aggregated clays are discussed later in this chapter. It is a soil additive that has use in some soil conservation structures, where a measure of dispersion is required to assist in dealing with highly aggregated soils.

Soil Stabilisation for Engineering Purposes

The stabilisation of soils for engineering purposes generally requires the addition of large quantities of lime, cement or bitumen. Quantities of at least 10 t/1000 m³ of soil, and frequently much higher, are required for lime stabilisation. Similar quantities are required for cement stabilisation. These quantities and techniques are largely used for highway construction and other such projects, and are not usually used for everyday soil conservation work. The techniques may have potential for flume construction or batter stabilisation. Any reader requiring further information is referred to the publication by Ingles and Metcalf (1972), which details these techniques.

Specialised Soil Stabilisation Techniques Against Erosion

In critical areas, and at critical points in a soil conservation program, erosion control may be best achieved by specialised soil stabilisation techniques. Included is the use of meshes made of jute or polythene, bitumen, sandbags, straw or paper mulches or concrete structures. Lime and cement stabilisation, as mentioned previously, are possible methods, but they have not been widely used.

For more information on these techniques, the reader is referred to the following publications: Ingles and Metcalf (1972), Reynolds (1976, 1977) and Quilty et al. (1978).

20.3 Sealing Leaking Structures

Seepage in earth dams and gully control structures is a result of high permeability, due to either inadequate clay content, inadequate compaction/consolidation, unsuitable clay type, the presence of sand/gravel seams or major inclusions such as tree stumps and other biodegradable matter. These problems can usually be avoided by adequate site inspection, laboratory or field soil testing and careful construction.

It is largely the clay fraction of the soil that determines its permeability. Under certain conditions, dams can be built successfully with soils containing as little as 10% clay, but the clay must be of the right type and the soil must contain a similar proportion of silt and fine sand. Correct compaction at a moisture content close to the optimum is also necessary.

Topsoils and/or soils rich in organic material should never be incorporated into the wall of a water-retention structure. Soils with a high organic content make an excellent surface mulch and provide an optimal environment for plant establishment, and should be stockpiled away from the excavation and used to 'topsoil' the completed structure.

20.3.1 Finding the Leak

Locating the site of a leak in a structure is well worthwhile as the leak may be in an isolated spot and readily treated with a minimum of expense. First, determine if the structure is leaking by marking the waterline with a peg each week. Stock should be excluded and allowance made for evaporative losses by placing an open vertical-sided container at the edge of the dam and regularly maintaining a constant level therein. The depth of water added will provide a reasonable estimate of evaporative loss. Alternatively, the nearest DLWC Research Centre will provide an estimate of expected evaporative loss.

A slow leak of 15 cm loss of depth a week may indicate either a general leak due to lack of compaction or a clay that does not seal well; or it may be a small porous seam. A loss of 15 to 30 cm a week could be due to a large sand or gravel seam. Anything over 30 cm a week could indicate that a 'pipe' or 'tunnel' has formed.

If the rate of leaking decreases sharply at a certain depth, a careful search around the waterline with a crowbar or soil auger will often reveal a porous seam. At least it will show how much or what areas of the dam have to be sealed.

Sometimes wet patches downstream of the dam can give some clue to the location of the leak.

20.3.2 Should the Leaking Structure Be Repaired?

If a suitable alternative site is available, it is frequently less costly to rebuild rather than repair an existing structure. Moreover:

—structures sealed with bentonite, bitumen or flexible membranes must be fenced to prevent damage by stock
—soils sealed by bitumen or chemical amelioration need subsequent maintenance treatments
—the durability of flexible membranes (especially the cheaper membranes normally used for rural water supply) is uncertain
—structures located in steep or rough terrain usually increase installation problems and costs for most sealing techniques.

With all methods of sealing, the importance of adequate compaction cannot be overemphasised. Many failures can be attributed to poor compaction.

Compaction is the cheapest part of any sealing method.

Sealing Structures Without Drainage

Puddling

A very low-cost and frequently effective technique. A temporary fence is erected around the dam and stock (preferably cattle) concentrated on the site. Two tractors towing a sheepsfoot roller backwards and forwards through the dam can produce a similar effect.

Should puddling prove ineffective or only partially effective, the addition of a suitable clay (should it be readily available) and 'repuddling' the structure can often produce an economical seal.

Bitumen Emulsion

A moderate- to high-cost method. Bitumen can effectively seal a large range of soils of varying porosities. Normally anionic bitumen emulsion would be suited to most applications, but cationic emulsions will probably produce a more effective seal on acid soils, especially those with a high silica content.

Normal application rate is 3 L/m^2 of wetted soil surface, which should reduce seepage by 70–90% within two days of application. Should a more effective seal be required, a rate of 5 L/m^2 can be used. The area must be permanently excluded from stock, but water is usually fit for consumption after 48 hours.

No information is available as to the durability of the treatment in reservoirs, but irrigation canals require additional maintenance applications at, approximately, three-year intervals.

Bentonite

A high-cost method. The application of bentonite to storage water is recommended only for an emergency seal. Information on bentonite and the normal methods of application is to be found below.

Specialised Liquid Sealants

A high-cost method. A range of commercial sealants has been produced. None is currently recommended for soil conservation structures.

Sealing Methods Requiring Drainage

Sealing by Compaction Alone

A low-cost method. This method is effective with well-graded soils containing sufficient clay (10% or more) to effect a seal. If the clay is highly flocculated, the technique will only be effective if compacted wetter than optimum using compaction equipment with high contact pressure. Flocculated clays usually require the use of chemical additives to effect a seal. Dispersible soil material that is tunnelling-susceptible must be treated with a calcium ameliorant (gypsum and/or hydrated lime are recommended) in conjunction with compaction to provide adequate soil stability.

Water should be added to bring the soil to the optimum moisture content for compaction. The water is best added in multiple light sprinklings, with a minimum of surface slaking, to a soil scarified to a depth of 20 to 25 cm so that maximum penetration is achieved. Should water exude from the soil upon compression, moisture content is in excess of the optimum.

Seals thicker than 15 cm should be compacted in two or more layers, each no more than 15 cm thick (that is, the top layer has to be removed and stockpiled).

The necessary compaction can often be achieved by 'puddling' after rain, and particularly if the stock have access to the dam while the stored water is rising or falling.

Chemical Compacting Agents

Moderate cost in addition to compaction. None of the commercially available products is currently recommended for use in soil conservation structures.

Sealing with Blankets of Earth

(i) Local Clay

Moderate/low cost in addition to compaction. Seepage can be sealed by an earth blanket if there is a suitable borrow area close enough to permit hauling at reasonable cost. The blanket material should be well graded with at least 20% clay. The suitability of the material for sealing and the required thickness should be based on laboratory tests or on local experience.

(ii) Bentonite

A high-cost method, to be used in conjunction with compaction. Bentonite is suitable for use on soils having a high proportion of coarse-grained particles and insufficient clay. It can be used on silty, sandy or gravelly soils and for sealing porous patches and gravel seams. A lump of dry soil, which is suitable for bentonite sealing, will crumble readily when crushed between the fingers. If the moistened soil can be rolled out to form a thin pencil, it contains too much clay for the use of bentonite.

The normal rate of application is 5 to 7 kg/m^2 (Elliott, 1980). A laboratory analysis on the soil to determine the rate of application is recommended. A discrete layer of bentonite applied to the surface provides the most efficient treatment, but is not recommended in structures with large fluctuations in water level, on steep batters and where stock are not excluded. A protective cover of soil over the bentonite is beneficial, but not considered sufficiently reliable in these situations.

Bentonite should be uniformly mixed with the soil material and, wherever economically practicable, covered with a further protective layer of soil.

Chemical Sealants

(i) Sodium Phosphates

Moderate/low cost to be used in conjunction with compaction. Sodium tripolyphosphate (STPP) is the most suitable form currently available in New South Wales (Wickham and Harris, 1976). The treatment is used on flocculated soils with 50% of material finer than silt size, and at least 15% clay. STPP is relatively ineffective on soils with a high soluble salt content and on those containing more than 20% free calcium carbonate. In order to assess suitability, a wet ball of soil is formed in the hand and a pinch of STPP added to it and the whole worked again. If the soil appears softer and more moist, it is suitable.

The normal rate of application is 0.25 kg/m^2, but rates up to 0.5 kg/m^2 may sometimes be required. A laboratory test is desirable.

Caution is necessary when sealing a dam where the total salt content of the stored water is greater than 400 ppm. In the absence of a chemical analysis, if an STPP-treated soil disperses, and stays dispersed for a day or longer in the water which is to be stored, then the salt content of the water is not too high.

Where water is added to bring the soil to optimal moisture content, the STPP can be applied by dissolving it in the water and sprinkling it on at the appropriate rate.

For one or more years after treatment, the water in the dam will be turbid, indicating satisfactory dispersion of the soil particles and optimum conditions for effective sealing.

(ii) Sodium Carbonate

A moderate/low cost method. Soils that will respond to treatment by sodium phosphates should generally respond equally well to treatment by sodium carbonate. However, sodium carbonate may be superior to sodium phosphates in sealing *uncompacted* calcium-montmorillonite clays and clay soils containing free calcium carbonate (soils with pH levels higher than 8).

A few years after treatment, the seepage rate may increase as sodium is lost (treatment does not usually completely stop seepage) and calcium and magnesium in the dam water will eventually replace the sodium applied in the treatment. The treatment may then have to be repeated.

Sealing with Flexible Membranes

These methods involve high cost and have critical installation requirements. There is a wide variety of flexible membrane material suitable for sealing dams, such as polyethylene (PE, generally known as polythene), polypropylene (PP), reinforced polythene fabrics, polyvinyl chloride (PVC, generally known as vinyl), reinforced vinyl fabric, bitumen-impregnated synthetic fabrics and synthetic rubber.

The use of synthetic fabrics should only be considered where an impervious seal is essential and lower-cost alternatives are considered to be inadequate. The membrane must be protected by the careful preparation of soil under it and, with the exception of the most expensive materials, a layer of soil over it to ensure that the sheet remains in position and that it is not damaged during the normal usage of the dam. Batter grades no steeper than 1:3 are recommended, as the material becomes more difficult to install on steep slopes and is more prone to damage.

Correct site preparation is essential to ensure that damage does not occur from sharp projections in the structure, subsidence, plant growth, yabbies, rodents and entrapped methane gas from rotting organic matter. All large clods, brush, roots, rock, sod or foreign material that might puncture the lining material should be eliminated from the embankment area. Harrowing, scarifying, rolling or surface compaction is encouraged to provide an extra measure of safety and to enable good bonding, and to reduce the possibility of puncture by piercing of the membrane. The polypropylene and polyethylene sheets, in particular, are subject to damage from wind and abrasion during installation.

Liners should be approximately 10% oversize to allow for shrinkage due to ageing and minor settling of the structure. Allowance should also be made for the cost of cutting back the batters above top water level to locate the flap and protect it with soil. Liner membranes should preferably be polyethylene sheeting manufactured from good quality reprocessed resin with a minimum thickness of 0.2 mm and minimum strength equivalent to 'Medium Impact Resistance' as specified in the appropriate Australian Standards (AS 1326, 1972 and AS 2870, 1986). Alternative materials may be used provided the supplier provides certification that the material meets the strength, thickness, durability and impermeability criteria of these standards.

Membrane lining should be supplied in sections as large as practical. However, availability of equipment to handle large rolls or packages may limit the section size. Suitable material is generally available in 4 m-width rolls of 50 m length.

The sequence for placement of the membrane will be determined by the characteristics of the job and site. Generally the lining material should be laid horizontally, starting from the existing ground level, or the point having zero head of water storage in relation to the downstream toe of the embankment, to a minimum level of 0.5 m above the spillway. The material should not be stretched during installation but installed in a relaxed state with slack allowed in both directions. After laying the first strip, soil cover is pushed over the membrane to hold it in place while the next strip is laid.

The membrane strips are progressively joined using 50 mm ducting tape or similar, with a minimum strip overlap of 0.3 m. All joins should be in accordance with manufacturers' recommendations using material specified for the purpose. The completed joints must be watertight and maintain their integrity through the expected life of the lining. Sections of lining to be joined must be laid flat against one another. That is, one section should not be drawn taut over an adjoining section that is wrinkled. Strips are progressively covered with soil as the laying and joining proceeds.

When total lining of a structure below top water level is required, or where repairs to an existing structure are needed, consideration may be given to placement of the lining up the batter slope rather than across it. This reduces stress on the membrane joints and may fit in better with construction technique and machinery. Pipes, drains and spillway structures that penetrate the membrane lining must be suitably flashed and sealed to prevent leakage.

After laying, the membrane is covered completely with soil to a minimum depth of 0.3 m. Cover material, preferably clay, is required to protect the membrane from mechanical damage as well as reducing the seepage loss through any holes in the lining. Cover material must be properly compacted and placed in such a manner that the lining will not be unduly displaced or damaged by equipment or overburden. It is desirable that the whole operation be carried out in moderate temperature conditions, that is, between 0 and 38°C.

Further information on the various methods of sealing leaking structures, and the relative costs involved, are to be found in Reynolds and Rosewell (1975) and Gardiner (1990).

20.4 Soil Drainage

Rain will infiltrate into soil at a steady rate until it reaches a zone of different permeability—a highly permeable sand or gravel seam, or it may be a less permeable clay or rock. In the case of impermeable or semi-permeable layers, the water will move downslope along the top of the layer. If this impermeable layer approaches the surface or outcrops, the water flowing along it will emerge as a spring.

Terrain variations also influence the movement and accumulation of subsurface water. Where soil properties permit, groundwater moves under the influence of gravity, normal to the contour, from high to lower elevations in the landscape—footslopes, drainage plains and flood plains. Drainage from these low-lying positions is often slow because of gentle slope gradients and low hydraulic gradients, thus allowing groundwater to accumulate.

Useful soil drainage classes are defined in the following section. The imperfectly and poorly drained classes are those most likely to cause problems.

Soil Drainage and Soil Type   20.4.1

Soils having imperfect or poor drainage (seasonally wet soils) generally have one or more of the following features:
—bleached A_2 horizons
—small reddish or orange flecks within the A_1 and A_2 horizons
—manganese and iron nodules (these may be relict features)
—a mottled subsoil, usually with red, orange or yellow mottles on a grey or dull yellowish brown background
—a neutral or alkaline soil reaction trend.

Types of soils include a large proportion of the mottled yellow duplex soils (Dy3.4, Dy3.8), some black duplex soils (Dd), some yellow earths (Gn2.6, Gn2.9) and those black cracking clays with deep B horizons of mottled grey or yellow (Ug5.16). Great soil groups include the yellow podzolic soils, yellow solodic/soloth/solodised solonetz, yellow earths and the wiesenbodens.

Drainage is a useful term to summarise local soil wetness conditions, that is, it provides a statement about soil and site drainage likely to occur in most years. It is affected by a number of attributes, both internal and external, which may act separately or together. Internal attributes include soil structure, texture, porosity, hydraulic conductivity and water-holding capacity, while external attributes are source and quality of water, evapotranspiration, gradient and length of slope, and position in the landscape. Specific drainage classes adopted for Australian conditions are as follows (McDonald et al., 1984).

1 *Very poorly drained*

Water is removed from the soil so slowly that the water table remains at or near the surface for most of the year. Surface flow, groundwater and subsurface flow are major sources of water, although precipitation may be important where there is a perched water table and precipitation exceeds evapotranspiration. Soils have a wide range in texture and depth, and often occur in depressed sites. Strong gleying and accumulation of surface organic matter are usually features of most soils.

2 *Poorly drained*

Water is removed very slowly in relation to supply. Subsurface and/or groundwater flow, as well as precipitation, may be significant water sources. Seasonal ponding resulting from runon and insufficient outfall also occurs. A perched water table may be present. Soils have a wide range in texture and depth; many have horizons that are gleyed, mottled, or possess orange or rusty linings of root channels. All horizons remain wet for periods of several months.

3 *Imperfectly drained*

Water is removed only slowly in relation to supply. Precipitation is the main source if available water-storage capacity is high, but subsurface flow and/or groundwater contribute as available water-storage capacity decreases. Soils have a wide range in texture and depth. Some horizons may be mottled and/or have orange or rusty linings of root channels, and are wet for periods of several weeks.

4 *Moderately well drained*

Water is removed from the soil somewhat slowly in relation to supply, owing to low permeability, shallow water table, lack of gradient or some combination of these. Soils are usually medium to fine in texture. Significant additions of water by subsurface flow are necessary

in coarse-textured soils. Some horizons may remain wet for as long as one week after water addition.

5 *Well drained*

Water is removed from the soil readily but not rapidly. Excess water flows downward readily into underlying moderately permeable material or laterally as subsurface flow. The soils are often medium in texture. Some horizons may remain wet for several days after water addition.

6 *Rapidly drained*

Water is removed from the soil rapidly in relation to supply. Excess water flows downward rapidly if underlying material is highly permeable. There may be rapid subsurface lateral flow during heavy rainfall provided there is a steep gradient. Soils are usually coarse textured, or shallow, or both. Normally no horizon remains wet several days after water addition.

Those soils having very poor drainage (wet for the majority of the year) generally have one or more of the following features:

—fibrous or peaty organic horizons at the surface

—bleached A_2 horizons (however, A_2 horizons tend to form only when there is a significant through-flow of water, which may not happen in many of these soils)

—grey or bluish grey subsoils that are frequently mottled with reds and yellows

—deeper subsoils that are often very light grey to bluish grey with bluish and greenish mottles.

Types of soils included here are mainly the gley duplex soils (Dg), but may also include some uniform clay soils (Uf6.6, Uf6.4). These are mainly the humic gley and gleyed podzolic great soil groups.

Problems Associated with Poor Soil Drainage 20.4.2

In areas of imperfect or poor drainage when spring water or throughflow accumulate, the following types of problems may occur:

—salinity

—increased surface runoff

—soil dispersion

—mass movement

—rank and undesirable plant growth

—access difficulties for construction equipment

—differential settlement of buildings

—undermining of pavements, particularly of roads

—flooding and dampness of basements

—salt damp in buildings located on saline soils

—waterlogging of recreation areas.

Many of these problems may be severe and require subsurface drainage to allow agricultural and urban development.

Factors Contributing to Poor Soil Drainage

Climate

Where average precipitation exceeds average evaporation for several successive months in the year, the potential for drainage problems is high. This is particularly so in lower slope positions such as on footslopes and in drainage depressions.

Underlying Rock/Geology

Where impervious rock comes close to the surface, even on a sideslope, a spring can occur as the groundwater follows the line of the impervious rock. If the soil is relatively permeable, allowing most rainfall to penetrate the soil rather than flow away as runoff, then the spring phenomenon is exaggerated, particularly if the country rock is impermeable. Such is the case with many granite and sandstone landscapes where springs tend to be common.

Terrain

Because of gravity, water tends to accumulate in the lowest parts of the landscape and this generally makes areas such as footslopes and drainage depressions subject to drainage problems. Thus, a sequence of soil water regimes down a slope is set up, with the higher areas such as crests and sideslopes not as prone to drainage problems.

Soils

This effect is considerably enhanced if highly permeable subsoils occur on crests and slopes, greatly increasing the amount of groundwater available to accumulate lower down the slope.

If B horizons have low permeability, as is the case with sodic B horizons, the surface soils become more readily saturated than otherwise would be the case. Perched water tables are formed. To this extent, those soils with sodic B horizons (the solodic group of soils) are more prone to soil drainage problems.

Possibly the worst combination is to have highly permeable sandy soils on sideslopes and crests with sodic soils on the lower slopes and in depressions. Such is the case for many granite landscapes throughout the tablelands of New South Wales.

Vegetation

Vegetation can be an indicator of drainage and is particularly useful when visiting an area during summer or in seasonally dry conditions. The growth of sedges and rushes is often a good indication of poor drainage. Other indicator species include Yorkshire fog (*Holcus Ianatus*) and dock species (*Rumex* species).

Areas that show green plant growth for longer periods than surrounding areas can also be considered areas of poor drainage. Saline patches, indicated by poor vegetative growth or salt efflorescences, can also be considered problem areas.

Treatment of Drainage Problems

Before a comprehensive subsoil drainage program is undertaken, the benefits of draining the soil should be assessed to justify the cost of installing the drainage. The soil must have sufficient depth and permeability to permit effective drainage, and a stable and adequate water disposal outlet must be available.

Objectives of Subsoil Drainage

A subsoil drainage program may be implemented to achieve one or more of the following objectives:
—to improve the soil environment and vegetative growth by regulating the water table and groundwater flow
—to intercept and prevent groundwater movement into a wet area, so improving stability, moisture status and general surface condition of that area
—to carry base flow in table drains, grassed waterways, or other grassed drainage depressions (often in conjunction with a low-flow pipe or channel)
—to remove surface water from depressions or low-lying areas
—to provide internal drainage of slopes, batters and filled areas to improve their stability and reduce soil erosion
—to improve the drainage or to hasten drying of intensively used agricultural and recreational areas.

Design and Installation

Subsoil drains provide a channel for free water movement through the soil. There are several common types:

(i) Mole Drain

Made by dragging a torpedo-shaped implement through the soil, producing an unlined underground channel.

(ii) Rubble Drain

Obtained by placing rubble in an excavated trench, preferably lined with geotextile, and refilling with soil.

(iii) Perforated Pipe, Pervious Pipe or Tile

Several forms of drainage pipe that are installed in the bottom of an excavated trench, and re-covered with soil.

(iv) Open Drain

An excavated trench or channel.

Pipes are the more permanent of these structures. The basic criteria to consider when selecting these or any other form of subsoil drain are capacity, gradient and cost benefit. Capacity or pipe size can be estimated by measurement of discharge on site, or by comparison with similar sites where subsoil drain yields have been measured previously. To ensure drains will be self-cleansing, their gradient must be sufficient to provide a flow velocity greater than 0.4 m/s. Also, there should be no local depressions within a drain, as these will result in gradient reversal, promoting sediment deposition and subsequent drain blockage.

When pipes are used, they should, where possible, be placed at a minimum depth of 60 cm in an excavated trench. Their spacing will be determined by the depth of installation, the permeability of the soil and the degree of drainage required.

Bedding of pipes can be improved by laying them in gravel, or in a sand and gravel envelope. The envelope should extend for a minimum depth of 8 cm above and below the pipe.

This envelope is also appropriate as a filter in fine soils, or in situations where low flow velocities in drains allow the accumulation of sediment in the conduits. The filtration effect can be improved by covering the surface of the envelope with a semi-permeable membrane before backfilling above it with soil.

Where the flow of groundwater pipes is impeded as a result of low soil permeability, it can be improved by increasing the amount of envelope material. This is achieved by extending the material to the level of the uppermost seepage strata: to the surface, if there are no seepage strata; or behind retaining walls, to within 30 cm of the top of the structure.

A continuous section of pipe without open joints or perforations should be used over the final 3 m to the outlet of a subsoil drainage system.

Soil erosion may occur around drainage pipes on steep gradients. This can result in subsidence of backfill in the trench above the pipes and failure of the system. The problem can be overcome by wrapping open joints with filtration geotextile, enclosing continuous perforated pipe with filter materials as previously described, revegetating areas disturbed during pipe installation and protecting the drain outlet against scour.

BIBLIOGRAPHY

Australian Standards, AS 1289 (Testing soils for engineering purposes)
AS 1326-1972 (Polythene film for packaging and allied purposes)
AS 2870-1986 (Residential slabs and footings)
Aveyard, J.M. (ed.) (1987), *Design Manual for Soil Conservation Works,* Technical Handbook No. 5, Soil Conservation Service of NSW, Sydney.
Burton, J. (1965), *Water Storage on the Farm I,* Bulletin No. 9, Water Research Foundation of Australia, Sydney.
Chow, V.T. (1983), *Open Channel Hydraulics*, McGraw-Hill, New York.
Crouch, R.J. (1979), 'The causes and processes of tunnel erosion in the Riverina', M.Sc. (Agr.) Thesis, UNE, Armidale.
Edwards, J.A. (1964), *Storing Water in Farm Dams,* Bulletin 467, Dept. of Agric., South Australia.
Elliott, G.L. (1976), *Some Properties of Tunnelling Soils in the Hunter Valley*, Tech. Bull. No. 15, Scone Research Centre, Soil Conservation Service of NSW, Sydney.
Elliott, G.L. (1980), 'Sealing soil conservation earthworks with bentonite', *Journal of the Soil Conservation Service of NSW* 36, 87–9.
Emerson, W.W. (1967), 'A classification of soil aggregates based on their coherence in water', *Australian Journal of Soil Research* 5, 47–57.
Fietz, T.R. (1969), *Water Storage on the Farm II,* Bulletin No. 9, Water Research Foundation of Australia, Sydney.
Gardiner, T. (1990), 'Dam plastic—fantastic!', *Australian Journal of Soil and Water Conservation* 3(1), 19–22.
Ingles, O.G. and Metcalf, J.B. (1972), *Soil Stabilization: Principles and Practice*, Butterworths, Sydney.
Keen, B.A. and Raczowski, H. (1921), 'The relation between clay content and certain physical properties of a soil', *Journal of Agricultural Science* 11, 441–9.
Klute, A. (1986), *Methods of Soil Analysis*, Part 1 (Second Edition), 18, 700–3.
Laing, I.A.F. (1975), *A Report of a Tour of Water Conservation Research Institutes in Some Arid Areas of the USA and Israel (1974)*, Soils Division, Department of Agriculture, Western Australia.
Logan, J.M. (1968), 'The design of earthworks for soil conservation in eastern New South Wales. IV. Banks', *Journal of the Soil Conservation Service of NSW* 24, 185–209.
McDonald, R.C., Isbell, R.F., Speight, J.G., Walker, J. and Hopkins, M.S. (1984), *Australian Soil and Land Survey Field Handbook,* Inkata Press, Melbourne.
Middleton, H.E. (1930), *Properties of Soil Which Influence Soil Erosion*, USDA Tech. Bull. 178, Washington DC.
Mills, J.J., Murphy, B.W. and Wickham, H.G. (1980), 'A study of three simple tests for the prediction of soil shrink-swell 'behaviour', *Journal of the Soil Conservation Service of NSW* 36, 77–82.
Pilgrim, D.F. and McDermott, G.E. (1982), 'Design floods for small rural catchments in eastern NSW', *Journal of the Institute of Engineers, Australia* 24, 226–34.

Quilty, J.A., Hunt, J.S. and Hicks, R.W. (1978), *Urban Erosion and Sediment Control*, Technical Handbook No.2, Soil Conservation Service of NSW, Sydney.

Reynolds, K.C. (1976), 'Synthetic meshes for soil conservation use on black earths', *Journal of the Soil Conservation Service of NSW* 32, 145–60.

Reynolds, K.C. (1977), 'Synthetic meshes for soil conservation stabilization of steep batters', *Journal of the Soil Conservation Service of NSW* 33, 117–27.

Reynolds, K.C. and Rosewell, C.J. (1975), *Sealing Earth Dams and Gully Control Structures,* Technical Handbook, Soil Conservation Service of NSW, Sydney.

Ritchie, J.A. (1963), 'Earthwork tunnelling and the application of soil testing procedure', *Journal of the Soil Conservation Service of NSW* 19, 111–29.

Rosewell, C.J. (1970), 'Investigations into the control of earthwork tunnelling', *Journal of the Soil Conservation Service of NSW* 26, 188–203.

Schwab, G.O., Frevert, R.K., Edminster, T.W. and Barnes, K.K. (1981), *Soil andWater Conservation Engineering* (Third Edition), John Wiley and Sons, New York.

Sherard, J.L. and Decker, R.S. (eds.) (1977), *Dispersive Clays, Related Piping and Erosion in Geotechnical Projects,* ASTM STP 623, Philadelphia, USA.

Stewart, J. (1955), 'Spacing of graded banks', *Journal of the Soil Conservation Service of NSW* 11, 165–71.

T.R.R.L. (1982), *Soil Mechanics for Road Engineers*, Dept. of Environment, Transport and Road Research Laboratory, HMSO, London.

Wickham, H.G. and Harris, G.L. (1976), 'Sodium tripolyphosphate prevents seepage in red earth structures: Coffs Harbour', *Journal of the Soil Conservation Service of NSW* 32, 68–77.

Wickham, H.G. and Tregenza, G.A. (1973), 'Modified computation procedure: Keen-Raczowski volume expansion test', *Journal of the Soil Conservation Service of NSW* 29, 170–6.

Soils and Urban Land Use

R.W. Hicks and C. Hird

Soils can have a major influence on urban land use, particularly in relation to erosion during the construction phase of development and eventually on the performance of building and road foundations. As a response to this, authorities in New South Wales have conducted numerous soils investigations in relation to urban land use, examples of which are Quilty et al. (1975), Manson and Harte (1978), Houghton and Emery (1981), Hunt et al. (1982), and Murphy and Attwood (1985).

Knowledge of soil properties, and how they react in a particular location, is of major benefit to those involved with the land development and building industries, particularly engineers, planners and building inspectors.

In the building and construction fields, the type of soils information generally required can be split into two broad groups:

—*soil properties* such as depth, shrink-swell potential, bearing strength, drainage properties, erodibility, salinity and pH
—*terrain or landscape properties* such as susceptibility to mass movement, water table conditions, subsidence and flooding.

For more detail on the soils information collected for urban capability studies, see Chapter 7.

21.1 Soils and Construction Practice

Major soil limitations to urban and related developments are brought about by the following properties and processes.

21.1.1 Mass Movement

Identification of existing and potential areas of mass movement, including landslips and earthflows, is an important part of urban soil surveys. Identification of these parameters enables planners to determine which areas should be avoided and which are capable of development subject to specialised investigation and controls.

Areas with a risk of failure have particular combinations of gradient, terrain type, site drainage, soil properties, geology and climatic features. In some regions, local indicators such as a particular slope or vegetation type can be used to indicate areas of potential risk. For example, at Castle Hill in the north-west of the Sydney metropolitan area, slope failures are prevalent on deep red podzolic soils with gradients between 7% and 10%. Similarly, in the Gosford area on the New South Wales central coast, south- to south-easterly-facing slope facets in excess of 20% gradient have been shown to present a risk of failure.

21.1.2 Expansive Soils

Soils that shrink and swell with changes in moisture content can result in major long-term stability problems in urban areas, owing to the possibility of unstable foundations for buildings and other infrastructure.

Expansive soils are often derived from basic rocks such as basalt, or from shales, mudstones or alluvia containing materials of volcanic origin. However, any soil with significant amounts of clay (15% or more) can be subject to movement. Therefore, it is advisable to test clay soils wherever they occur in proposed development areas, as set out in Chapter 11.

21.1.3 Soil Erodibility

This property is significant owing to the impact of erosion and sedimentation on the environment, especially streams and other waterbodies, and the costs it engenders in the development of new urban areas. Effective identification of areas most at risk from soil erosion will enable good planning and construction practice to minimise the cost to the community. Major costs are borne by councils and the community when service facilities are adversely affected. For

Plate 21.1 Inadequate drainage and unstable soils led to this mass movement on an urban subdivision

example, stormwater pipe inlets may become blocked, and the capacity of drainage channels reduced by sediment. Individual homeowners often have to bear considerable expense and inconvenience by importing topsoil to establish gardens on building lots denuded of soil.

The erodibility of a particular soil type generally depends on the proportion of fine sand and silt-sized particles, and the dispersibility of the clay fraction. Highly erodible soils usually have a high content of one or more of these fractions and a clay content that is dispersible in water. Any non-cohesive soils such as sands, particularly if they show signs of being water-repellent, and any sodic soils should be regarded as highly erodible. Hardsetting soils should also be treated with caution.

Further, soils with a high proportion of dispersible clay minerals may adversely affect the use of on-site absorption systems for effluent disposal. Localised tunnelling may also occur in backfilled service trenches where unstable clays are present.

Soil Drainage and Depth

21.1.4

Terrain type, slope and underlying soil and rock characteristics influence the subsurface movement of water. Most building foundations will be adversely affected if there is excessive moisture or fluctuations of moist content in their vicinity. Further, control of major problems of slope failure and expansive soils is very much dependent on effective management of subsurface water movement.

Soil surveys can identify areas subject to seasonal waterlogging, permanently high water tables, lateral seepage and inundation. Identification of areas requiring subsurface drainage controls, during the planning stages of a development, will greatly assist in the orderly preparation of building approval conditions and pinpoint critical areas that need to be closely supervised by the building inspector.

In the urban development sphere, soils with less than 50 cm to hard rock impose a constraint to development; in particular to the excavation of service trenches for reticulated services such as water and sewerage, and in cutting and filling for roads and building foundations.

Shallow soils will also lead to a build-up in subsurface moisture levels during wet periods, thus necessitating drainage in critical areas and restricting the opportunity for effluent disposal using absorption techniques. Highly permeable soils composed of sandy or gravelly materials will require frequent watering if used in parks or gardens, and are also usually unsuitable for effluent disposal. Soils with low water transmission properties (heavy clays) are likewise unsuitable for effluent disposal, and may also give rise to waterlogging problems in urban situations.

21.1.5   Soil Salinity

In the urban situation, salinity can also be a serious problem, where it is caused by similar processes to those that result in salinisation in rural areas. Rising water tables can occur owing to a variety of factors, including overclearing of trees for urban development, excessive watering of gardens and lawns, over-irrigation of public parks and recreation areas, disruption of natural drainage lines and overflow of septic tanks and sullage pits.

Indicators similar to those in rural situations are found, such as bare areas, sometimes with encrusted salt on the surface, unhealthy vegetation and greasy or waterlogged ground. Additionally, warning signs may be found on buildings and other infrastructure, including salty crusts on brickwork, rising damp, and salt damage to footings, pipes, railway lines and road surfaces.

Management to control this urban salinisation mainly centres on limiting the amount of water reaching the water table. This can be achieved through strategic clearing prior to development, minimising irrigation and garden watering after development, and the use of plants that have minimal water requirements for gardens and lawns. Drainage from roofs must be conducted away via the stormwater system rather than disposed of over the land surface.

Other Problem Soils   21.1.6

Soils that may cause other problems in urban development include infertile, toxic and low-strength

Plate 21.2   Areas of unstable soils can be a major constraint to the development of new housing subdivisions

soils. Infertile soils are difficult to stabilise with vegetation and are therefore likely to be eroded rapidly. Soils that contain chemical pollutants or show extremes of pH (for example, acid sulfate soils) are also going to be difficult to revegetate and may present a health hazard in an urban environment.

Unconsolidated materials, particularly those with high levels of organic matter, are unsuitable for building foundation purposes, and must be used in situations where slumping or settlement over time will not present a major hazard. Such materials are also likely to be highly erodible.

21.2 Treatment Measures

This section sets out broad guidelines to deal with urban development situations where the problems of mass movement, expansive soils, erodible soils, drainage and soil salinity occur. In addition, Table 21.1 shows how these problems interact with slope and terrain components to determine suitable urban land uses.

21.2.1 Mass Movement

The process of mass movement is discussed in Chapter 2.

It is undesirable to develop land that is subject to mass movement, as the problem is such a serious one and it is often impractical or uneconomic to counter it. If, however, development occurs on land where such hazard is marginal, or the cost of control measures is warranted, a number of techniques can be adopted to reduce the hazard.

(i) Cut and fill operations should be restricted to a minimum, and deep cuts and excessive fill should be avoided to reduce slope loading and avoid reducing shear strength. Location of roads up and down rather than across the slope, and construction of houses raised above ground level, can assist this aim (for example, pole houses).

(ii) Where batters are formed, low angles of cut or fill are desirable. High batters should, if possible, be benched and provided with adequate drainage. A structural facing such as gabions or a crib wall can be used to strengthen batters against failure, subject to engineering design.

(iii) Removal of subsoil moisture assists stabilisation of areas prone to mass movement by reducing the amount of soil water available to trigger failures.

(iv) Efficient surface drainage upslope of slip-prone areas will also reduce the hazard of failure, removing surface runoff before it can enter the rock and soil of the unstable zone. Diversion channels may be installed immediately upslope of batters for this purpose.

(v) Sealed diversion drains formed at intervals down the face of high batters, or located on berms if these have been formed on the batter face, will prevent the accumulation of local runoff.

(vi) An impervious surface course is desirable on pavements to limit infiltration and water movement into the subsoil.

(vii) Surface drainage of road pavements should not direct runoff into fill batters.

(viii) All household and road drainage should be carried in pipes or sealed drains away from unstable areas. The use of vegetated waterways and runoff-retarding measures is not generally applicable to such areas, as the primary aim must be to remove surplus water as quickly as possible.

(ix) Apart from drainage and structural techniques, the stability of slip-prone areas can be improved, though not so rapidly, by extensive planting of silver-leaf poplars or shrub willows. Their strong and extensive root systems bind the soil, and the trees significantly reduce soil moisture. They will regenerate if their roots are broken by soil movement. The roots are, however, a hazard for sewer lines and drainage pipes.

(x) Further engineering techniques such as grouting, electro-osmotic draining and chemical treatment of highly plastic clays are also sometimes used to stabilise slip-prone areas where development must take place on them.

While all these measure may reduce slip hazard, they will not always eliminate it. As a general rule, where development is proposed on suspect areas, prior geotechnical survey should be carried out. If

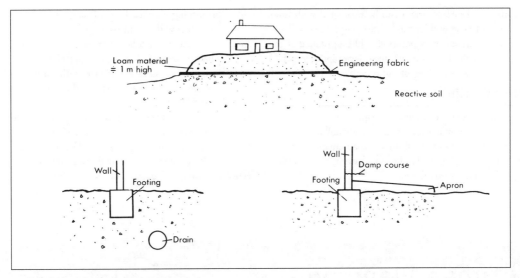

Figure 21.1 Treatment of swelling soils

this survey confirms a hazard of mass movement, advice should be obtained from professionals experienced in the field of slope stabilisation for the design and installation of all cuts, fills, foundations and drainage.

21.2.2 Expansive Soils

Expansive soils are dealt with in Chapter 11.

Where urban development is planned for areas with expansive soils, the following measures should be considered:

(i) stabilisation of foundation soils with cement or lime (see Chapter 20)

(ii) preparing a mound of non-reactive soil materials on which to site house (see Figure 21.1)

(iii) designing houses to cope with movement (articulated design), that is, timber panelling above doors, above and below windows, flexible service pipe connections and so on

(iv) use of specialised footings and slabs (requiring engineering design) or pole foundation designs

(v) minimising variation in soil moisture content near house footings, to reduce the extent of soil movement, by installing subsoil drains below the level of the footings on the outside (see Figure 21.1)

(vi) installing a 1 m-wide concrete apron around the house with fall away from house (see Figure 21.1)

(vii) installing lot drainage to direct surface water away from house

(viii) avoiding placement of garden beds against or very close to house

(ix) promptly repairing leaking taps, water pipes and drainage pipes

(x) avoidance of planting trees, or having them within half their mature height from the house.

Soil Erodibility 21.2.3

Soil erodibility is generally discussed in Chapter 12.

Erodible or dispersible soils in urban developments can be dealt with by the following measures:

(i) control of soil dispersion by incorporation of hydrated lime in critical areas, for example culvert aprons, stormwater channels and so on

(ii) use of erosion and sediment control techniques in construction areas, for example catch drains, perimeter banks, sediment traps and so on (refer Quilty et al., 1978)

(iii) use of adequate topsoil and revegetation techniques on disturbed areas to ensure good grass establishment so as to minimise soil erosion risk

(iv) design of subdivision and lot layout to avoid concentrating runoff in critical risk areas.

Soil Drainage 21.2.4

Soil drainage is dealt with in Chapter 20.

Drainage problems in urban development areas can be reduced if the following points are borne in mind:

(i) soil drainage is critical in reducing problems associated with slope stability and expansive soils

(ii) using cutoff or interceptor drains is a means of protecting critical areas

(iii) grassed surface dish drains can convey runoff away from houses and other buildings to safe disposal points

(iv) subsurface drains can be protected with filter fabrics and/or sand/gravel envelopes

(v) surface entry points should be installed at the highest point in the subsurface pipe network to enable flushing of the system.

21.2.5 Soil Salinity

Soil salinity is dealt with in Chapter 13.

The following points are important in reducing the effects of saline soils in an urban environment:

(i) reducing the level of the water table by surface and subsoil drainage

(ii) promoting the leaching of salts to a lower level, below the plant root zone

(iii) establishing salt-tolerant trees in affected and adjacent areas, to lower the water table

(iv) installing subsurface drainage with outlets to safe points, such as a trickle flow pipe in a stormwater disposal system

(v) sowing salt-tolerant grasses using specialised techniques (refer Quilty et al., 1978)

(vi) minimising the watering of parks and gardens

(vii) limiting lawn areas and high water-use gardens

(viii) providing adequate stormwater disposal systems.

21.3 Urban Planning

21.3.1 Change from Rural to Urban Use

Conversion of land from rural to urban use considerably modifies the natural environment, particularly with respect to the land surface and its hydrological characteristics.

These changes can lead to:

—sheet and rill erosion on disturbed sites

—erosion of natural drainage lines

—localised flooding

—sedimentation of stormwater pipes and waterways.

To minimise their impact, it is essential to plan land use carefully, to schedule development and install soil erosion control measures in conjunction with stormwater disposal systems.

While these types of problems are likely to occur on small-scale, intensive development sites, they generally can be avoided if considered when major strategy plans and development control plans are being prepared.

A wide range of possible land uses can be adopted on various types of terrain, but their suitability also depends on slope gradient, soil type and drainage characteristics. Some common terrain components and their impact on urban land use are shown in Table 21.1. Other types of terrain that may influence the capability for urban land use are coast features, such as frontal dunes and hind dunes, and formerly disturbed areas such as waste disposal sites or extraction sites for soil, gravel and minerals.

21.3.2 Urban Capability Classification

The classification system developed and used by the former Soil Conservation Service incorporates five primary classes (Hannam and Hicks, 1980). These are as follows:

Class A

Areas with little or no physical limitations to urban development.

Class B

Areas with minor to moderate physical limitations to urban development. These limitations may influence design and impose certain management requirements on development to ensure that a stable land surface is maintained both during and after development.

Class C

Areas with moderate physical limitations to urban development. These limitations can be overcome by careful design and by adoption of site management techniques to ensure the maintenance of a stable land surface.

Class D

Areas with severe physical limitations to urban development that will be difficult to overcome, requiring detailed site investigation and engineering design.

Table 21.1 Physical criteria and urban land use

Slope Class	Terrain Component	Potential Hazards Related to Topographic Location and Slope and Which Will Affect Urban Land Use	Suitable Urban Land Use
0–5%	Drainage plain	Flooding, seasonally high water tables, high shrink-swell soils, high erosion hazard	Drainage reserves/stormwater disposal
	Floodplain	Flooding, seasonally high water tables, high shrink-swell soils, saline soils, gravelly soils	Open space areas, playing fields
	Hillcrests	Shallow soils, stony/gravelly soils	
	Sideslopes	Overland flow, poor surface drainage and profile drainage	Residential; all types of recreation; large-scale industrial, commercial and institutional development
	Footslopes	Drainage impedance in lower terrain positions, deep soils	
		Others — swelling soils, erodible soils, dispersible soils	
5–10%	Hillcrests	Shallow soils	Residential subdivisions, detached housing, medium-density housing/unit complexes, modular industrial, active recreational pursuits
	Sideslopes	Overland flow	
	Footslopes	Deep soils, poor drainage	
		Others — swelling soils, erodible soils, dispersible soils	
10–15%	Sideslopes	Overland flow Geologic constraints — possibility of mass movement Swelling soils Erodible soils	Residential subdivisions, detached housing, medium-density housing/unit complexes, modular industrial, passive recreational
15–20%	Sideslopes	Overland flow Geologic constraints — possibility of mass movement Swelling soils Erodible soils	Residential subdivisions, detached housing, medium-density housing/unit complexes, modular industrial, passive recreational
20–25%	Sideslopes	Geologic constraints Possible mass movement High to very high erosion hazard	Residential subdivision, passive recreational
25–30%	Sideslopes	Geologic constraints Possible mass movement High to very high erosion hazard	Upper limit for selective residential use, that is, low-density housing on lots greater than 1 ha, passive recreational
Greater than 30%	Sideslopes	Geologic constraints Mass movement Severe erosion hazard	Recommend against any disturbance for urban development

Class E

Areas where no form of urban development is recommended because of very severe physical limitations to such development that would be very difficult to overcome.

Within these primary classes, a number of subclasses can be defined on the basis of the dominant physical limitations that will restrict development potential. The combination of several subscripts indicates a number of limitations that interact to place a constraint on development. The dominant limitation is indicated first. Where a subscript is in brackets, it is considered to be of lesser importance among the limitations listed for the subclass. Letter subscripts have been used to define these physical limitations as follows:

o no significant limitations

Soil Limitations

c very high permeability
d shallow soil
e erodibility
g low wet strength
l salinity
p low permeability
v shrink-swell potential
z gilgai

Other Limitations

f flooding
m mass movement
r rock outcrop
s slope
t topographic feature (rock fall, runon, and so on)
w seasonal waterlogging
x extraction/ disposal site
y swamp (that is, permanently high water table)

The capability indicated for each subclass refers to the most intensive urban use that areas within that subclass will tolerate without the occurrence of serious soil erosion and sedimentation in the short term, and possible soil instability and drainage problems in the long term. In assessing this capability, no account is taken of development costs, social implications, aesthetics or other factors relating to ecology and the environment. Using urban capability for planning at the conceptual level will, however, take account of soil and landform limitations, while providing for development that is generally consistent with preservation of an aesthetically pleasing landscape and minimisation of long-term repair and maintenance costs.

Categories of urban development provided for in this form of planning include extensive building complexes, residential, low-density residential, strategic development, reserves, drainage reserves and no development.

Assessment of urban capability is objectively based on physical criteria alone. Thus, the classification of various areas as capable of accepting certain forms of urban development is an assessment of the capacity of those areas to sustain the particular level of disturbance entailed. It is not a recommendation that such a form of development be adopted. Rather, it forms the basis onto which other town planning considerations may be imposed to derive a development plan.

The urban capability assessment is not a substitute for specific engineering and design investigations, which may be required to more accurately define constraints applying to the location of roads, individual buildings or recreational facilities. Nor does it imply the capacity of a site to support multistorey units or other major structures. Before structural works of such magnitude are undertaken, a detailed analysis of engineering characteristics of the soil, such as bearing capacity and shear strength, may be necessary on a specific development site.

21.4 Waste Application to Soils

In recent years the community has recognised that many 'waste' products can benefit plant and animal enterprises, reduce costs associated with sewage treatment processes and significantly reduce the need to discharge to waterways or dispose of in landfills.

Typical applications of wastes to soils include:
—on-site waste management systems for domestic households
—incorporation of biosolids (sewage sludge) onto cropping, pasture or forest lands
—irrigation of effluent from sewage treatment plants
—land application of manures and waste waters from intensive animal industries such as dairies and feed lots
—land application of other waste products such as out-of-date food products (for example, beer, soft drinks)
—land application of by-products from industry (for example, fly ash from burning coal, grease-trap wastes from food industries)
—irrigation of effluents from mining and industry (for example, wash-downs from coal overburden dumps).

Liquid waste products are irrigated onto soils using a variety of equipment including sprays, flood and underground systems. Solid wastes can be spread and incorporated into soil using conventional cultivation machinery; semi-solid sludges are usually injected under the soil surface using liquid injection equipment attached to an agro-plough. A review of waste application to soil is presented in Cameron et al. (1997).

Potential Impacts on Soils   21.4.1

Beneficial effects can include one or both of the following:
—the addition of nutrients and/or water to the soil to directly enhance plant and/or animal productivity
—the addition of organic matter or other materials (for example, calcium compounds) that improve soil physical properties.

Potential adverse impacts include:
—waste product constituents (for example, salt, sodium, nutrients, heavy metals, sugars, fats, and so on) downgrading the potential use of

the soil and/or nearby ground or surface water resources

—waste product constituents directly affecting the productivity or quality of plant and animal products.

21.4.2 Properties of Waste Products

The general physical and chemical properties of common waste products are shown in Table 21.2.

Liquid products applied to the soil by irrigation can improve plant or animal yield. However, over-application of liquids can waterlog certain soil types as well as providing the means to transport any contaminants to surface or groundwater resources.

Organic matter (OM) in waste products is almost always beneficial to soil structure and water- and nutrient-holding capacity. Fine sand and silt-size mineral particles are found in waste products such as fly ash and some intensive animal by-products. These may also benefit soil structure and water- and nutrient-holding capacity, however further investigation of the potential beneficial nature of the mineral material is required.

Nutrients such as nitrogen, phosphorus, potassium, sulfur and calcium are also beneficial to plant and animal growth. However, unless composted, the range of nutrients supplied in waste products is not usually supplied in the ratio required by plants.

Fat is found in grease-trap wastes and by-products of some food processing industries. Fats are slow to break down in soil and can upset the soil microbial population. Furthermore fats can clog up soil pores, thereby reducing soil aeration.

Sugars are found in soft drinks, beer, jams and honeys. Sugar on pastures can improve palatability. However, sugars raise the carbon content of the soil at the expense of nitrogen, upsetting microbial activity. Applications of nitrogen fertiliser are usually required to overcome this constraint.

Salt applied to soil, particularly as an irrigated effluent, can raise soil salinity. The concentration of salt (measured as electrical conductivity or total dissolved solids, TDS) is important. Where TDS is less than 500 mg/L, salt is not considered a concern and in fact may improve salinity in already saline soils. Slightly salty water can also improve permeability on sodic soils. The salinity of solid waste products is generally not a concern unless the solid is applied at a very high application rate.

The sodium adsorption ratio (SAR) represents the relative amount of sodium in a product compared to calcium and magnesium. High SAR (that is, greater than 6) in wastes can increase soil sodicity. Conversely, when SAR is very low (< 2), improvements in soil sodicity can occur.

Chemical contaminants include organochlorines, PCBs, heavy metals and other compounds known to have a deleterious effect on plant or animal health. The level of these contaminants should always be measured in waste product before it is applied to soil.

Pathogens such as bacteria, viruses, protozoans and helminths can occur in human and animal waste products. These are of special concern where plants are irrigated for human and animal consumption.

Table 21.2 Properties of some common waste products that can be beneficially applied to soils

Product	Form	OM	Minerals	Nutrient	Fat	Sugar	Salt	SAR	Contaminants	Paths
Domestic waste	S	+++	−	++	−	−	++	5		
Biosolids	L/S	+++	−	++	−				++	++
STP* and domestic on-site effluent	L	−	−	+	−	−	+++	5	+	+
Intensive animal industry effluent	L	+	+	++			+++	3.15	+	+
							++ −			
Animal manures and sludges	S	+++	−	++	−	−	+		−	+
Drinks	L	−	−	+	−	++	++	< 2	−	−
Grease-trap waste	S	+++	−	++	++	−	+	< 2	−	−
Fly ash	S	−	+++	+	−	−	+	< 1	+	−
Coal wash	L	−	−	+	−	−	++	variable	−	−

L liquid, S solid, − not significant, + low, ++ moderate, +++ high

* Sewage treatment plant

21.4.3 Conditions Where Wastes Should Not Be Applied

Wastes should not be incorporated or irrigated into the soil in the following locations:
—where slopes exceed 20%
—within drainage plains, swamps or wetlands or within 5 m of streams
—within 50 m of an environmentally sensitive surface water resource
—in ecologically significant areas (such as national parks, nature reserves, and so on)
—in rocky areas or areas where soil depth is generally less than 50 cm to bedrock or hardpan
—where any groundwater table occurs in the top 50 cm of soil
—where an existing or potential groundwater resource occurs in the top 1 m of soil.

The following locations are generally unsuitable for most waste products:
—saline areas
—sodic soils (unless the waste product has an SAR of less than 2)
—impermeable soils and soils subject to waterlogging (liquid waste products)
—where a sensitive groundwater table lies in the top 3 m of soil (unless it can be demonstrated that the overlying soil mantle will effectively immobilise any groundwater contaminants)
—acid soils (biosolids containing toxic metals such as cadmium and zinc).

21.4.4 Acceptable Application Rates

The allowable application rate of any waste to land is the sum of the amount removed by the crop grown, plus the amount that can be stored in the soil, plus an allowance for any gaseous losses, and/or any *acceptable* leakages from the system to groundwater and surface water. These rates are determined by constructing water and nutrient budgets.

The amount of waste water applied is determined by solving the following equation:

irrigated waste water + rainfall =
     evapotranspiration + percolation + runoff

Irrigated waste water should not be allowed to percolate or run off, however these phenomena will occur as a result of natural rainfall events and need to be accounted for in the budget. The land area required and wet weather storage for any waste water scheme are determined by solving the above equation over a number of years, preferably on a daily basis.

Most waste products contain varying amounts of nitrogen, phosphorus, potassium and, to a lesser extent, calcium, which are important plant nutrients. However, as the ratio of these nutrients in wastes is rarely the same as the plant uptake requirement, waste products must be applied at the rate of the most limiting nutrient. Conventional fertilisers are then applied to make up plant requirements for other nutrients. For example, the ratio of phosphorus to nitrogen is usually much higher in waste products than the ratio of these nutrients required by plants. Hence the application rate could be determined by the plant requirement for phosphorus, with nitrogen fertiliser being applied to make up plant requirements. However, as many Australian soils are capable of absorbing much of this phosphorus (thereby rendering the phosphorus unavailable to plants), the soil's potential to absorb phosphorus should be taken into consideration when determining the relevant application rates.

When assessing plant nutrient uptake requirements, the form in which the nutrient occurs is also important. Plants generally take up nitrogen as nitrate. Soil microbes convert organic nitrogen compounds to nitrate as well as other forms such as nitrogen and ammonia, which are lost to the atmosphere.

Nutrient budgets should take into account:
—plant uptake requirements
—expected plant yield
—any recycling of nutrients from decaying plant material or animal faeces and urine
—the initial nutrient compounds and the expected rate of breakdown
—any potential for losses to the atmosphere (N) or sorption to the soil (P).

Nutrient budgets should be simulated for a minimum of five years to ensure that inputs from nutrient compounds that break down slowly are fully accounted for in the budget.

21.4.5 Salinity and Sodicity of Waste Products

Plants and animals do not necessarily make use of all constituents in waste products. Nutrients and carbon (C) compounds are recycled by soil micro-organisms with some N and C ultimately lost to the atmosphere. However other compounds (salts) will build up in soils. Hence it is important that allowance is made for leaching of these salts to the groundwater table. This process can degrade the groundwater resource unless carefully managed. Models can be used to predict the movement of salt through the soil to the groundwater table. The characteristics of the groundwater table must then be assessed to determine its capacity to absorb potential contaminants in waste products.

Applications of low-strength effluents (TDS < 500) and solid wastes are usually not a concern unless the soils are already saline, or the subsoil materials have a low permeability. In the latter case salt will build up in the impermeable layer.

For moderate- to high-strength effluents (for example, those from intensive animal industries), the location, size and quality of the groundwater resource should be assessed to determine the level of salt that can be tolerated by the resource. Moderately saline and impermeable soils are not suitable for applications of high-strength effluent.

Where wastes with a relatively high SAR are applied to land, care must be taken that calcium ions held by the soil are not replaced by sodium ions. When this occurs, structural degradation of the soil results with a loss of soil permeability and aeration. Gypsum or lime should be applied where this occurs.

21.4.6 Soil and Land Management

Information is required to assess the ability of the soil to grow plants, as well as the soil's ability to immobilise any potential contaminants in waste products. The following soils information should be collected:
—pH
—water-holding capacity
—nutrient-holding capacity
—soil permeability (saturated hydraulic conductivity)
—soil depth to an impermeable layer or seasonally saturated layer
—salinity
—concentration of specific chemicals known to adversely affect plant and/or animal growth
—erodibility
—phosphorus sorption capacity
—soil microbial activity.

Soil management practices that will reduce potential impacts include:
—ensuring a sustainable application rate (that is, undertaking a water and nutrient budget)

—ensuring plant growth is optimised through managing soil structure and applying additional fertilisers where necessary
—monitoring soil properties so that any nutrient or chemical imbalances can be corrected.
Special practices include:
—application of lime to overcome acidity in soils
—application of gypsum to overcome sodicity in topsoils.

BIBLIOGRAPHY

Anzecc, NH, Armcanz, MRC (1998), *Draft Guidelines for Sewerage Systems—Reclaimed Water*.

Cameron, K.C., Di, H.J., and McLaren, R.G. (1997), 'Is soil an appropriate dumping ground for our wastes?', *Australian Journal of Soil Research* 35, 995–1035.

EPA (1997), *Environmental Management Guidelines for the Use and Disposal of Biosolids Products*.

EPS (1977), NSW Health and DLWC, *On-site Waste-water Management Systems for Domestic Households*.

Hannam, I.D. and Hicks, R.W. (1980), 'Soil conservation and urban land use planning', *Journal of the Soil Conservation Service of NSW* 36, 134–45.

Houghton, P.D. and Emery, K.A. (eds) (1981), *Land Resources Study of the Bathurst-Orange Region*, Soil Conservation Service of NSW, Sydney.

Hunt, J.S. (1992) (ed.), *Urban Erosion and Sediment Control* (Revised), Department of Conservation and Land Management, Sydney.

Hunt, J.S., Elliott, G.L., Hird, C. and Thomas, D.K. (1982), *Land Resources Study of the City of Greater Cessnock*, Soil Conservation Service of NSW, Sydney.

Manson, D.A. and Harte, A.J. (1978), *Urban and Rural Capability Study: Casino Municipal Region*, Soil Conservation Service of NSW, Sydney.

Murphy, B.W. and Attwood, R.D. (1985), *Reconnaissance Soil Survey: Forbes Township*, Soil Conservation Service of NSW, Sydney

Quilty, J.A., Hunt, J.S. and Hicks, R.W. (1978), *Urban Erosion and Sediment Control*, Tech. Handbook No. 2, Soil Conservation Service of NSW, Sydney.

Quilty, J.A., Wickham, H.G. and Houghton, P.D. (1975), *Urban Capability Study: Hawthorne Estate*, South Grafton, Soil Conservation Service of NSW, Sydney.

Soils and Extractive Industries

CHAPTER 22

G.L. Elliott and K.C. Reynolds

As a form of land use, extractive industries have substantial national and social values. Their products contribute to manufacturing viability, earn national income and provide employment. Returns to extractive industries generally greatly exceed alternative returns to agriculture for the areas of land involved.

In the past, several factors have contributed to low levels of rehabilitation effort after mining activities. These factors included local community dependence on mining which excluded other considerations, the relatively small scale of extractive operations, lack of planning for the preservation of the soil overburden, the relatively small opportunity cost of not rehabilitating mined areas and the general lack of environmental awareness in the community.

These factors have changed. Extractive operations have benefited from economies of scale and, both individually and in total, large areas are disturbed by mining; suitable equipment for selective handling of overburden is commonly available; increasing pressure on land has created multiple land use concepts; rehabilitation is accepted as a cost of production by both producers and consumers; and public education has led to environmental awareness, concern and appreciation of the benefits of rehabilitation. This public recognition has been reflected in government legislation, which, supported by industry, has established statutory requirements for rehabilitation.

22.1 Importance of Natural Topsoil

Subsequent to large-scale disturbance (such as open-cut mining), restoration of the original environment is frequently impractical and in some instances not the most suitable option (for example, many mining sites are on degraded lands). Mined lands are therefore generally rehabilitated to what is considered the most suitable land use option rather than restored to original condition.

It is generally accepted that for successful rehabilitation there is no substitute for natural topsoil. Research results generally confirm that equivalent growth may be obtained from natural soil and a similar soil that has been disturbed by stripping, storage and respreading. There is, therefore, everything to be gained by preserving the topsoil resource. For example, the use of topsoil generally promotes high growth rates and absolute levels of plant production, species diversity, microflora populations, the attainment of sustained, relatively high levels of both infiltration and porosity, the attainment of relatively low levels of bulk density in the plant root zone, and the provision of a balanced pool of nutrients in equilib-rium with a relatively high level of 'stored' soil organic matter. These effects of naturally occurring A_1 horizon soil material can only be duplicated at great cost.

An inventory of the soil resources should, therefore, form part of the mine planning phase. There are two reasons for this. First, the inventory establishes the minimum and maximum standards of rehabilitation by contributing to a general understanding of soil fertility and soil variability. Second, it defines the soil material with which this rehabilitation may be attempted. No matter how detailed the information collected during the soil mapping and testing phases of this inventory, the end results must provide sufficient information to determine soil-related limitations to land capability and locations and volumes of soil which should be recovered in order to achieve satisfactory rehabilitation.

The limitations of fertility to land use generally are expanded in Chapters 13, 14 and 15. Local variation in factors such as stoniness, soil moisture storage (soil thickness and/or texture effects) and existing erosion will contribute to establishing constraints on land use.

Open-cut operations generally result in a 20–30% volumetric increase over existing topography. Volumes of recoverable topsoil to meet rehabilitation requirements are therefore always less than those required to provide a depth of topsoil similar to that of the pre-mined soil. In many instances the volume of topsoil may be adequate particularly when the original soils have very thin surface horizons, or have been eroded or degraded. Soil losses during stockpiling and erosion losses during establishment of stabilising vegetative cover can further reduce the final amount of topsoil on the rehabilitated site. Sometimes the surface soils are unsuitable for resoiling (for example, gravels/coarse sands, highly saline and chemically contaminated soils). In these circumstances, substitute topsoil material for surface reconstruction may be selected from other materials characterised during the soil inventory. The objective criteria for selection are presented schematically in Figure 22.1.

When topdressing material is first respread over reshaped areas preparatory to seeding, the erosion hazard is very high. Soil material is loose and erosion is limited only by the transport capacity of runoff. Vehicle tracks intercept and concentrate runoff, and serious rill erosion may occur. Minimising exposure to erosion is critical to success. Selection of topdressing material should be based on properties that indicate resistance to erosion (see Chapter 12). For example, crusting and the re-formation of dense macro-aggregates that restrict plant root growth and water movement are undesirable attributes of soil used in surface reconstruction.

22.2 Alternative Topdressing Materials

The availability of alternative topdressing material to assist the achievement of high levels of surface vegetative cover increases the options available to rehabilitation managers. With a greater amount of suitable growth media, critical areas can be treated unstintingly. Critical areas may include final discharge areas, waterways and planned arable sections of old lease areas. On the other hand, where the original topsoils are not available in sufficient quantity or are found to be unsuitable for rehabilitation, selective use of overburden may provide an acceptable alternative. A number of sedimentary overburden rock materials with the appropriate treatment have been found to be suitable.

22.2.1 Soil Attributes for Topdressing

Figure 22.1 should be considered an aid to the selection of suitable material in situations where insufficient topsoil exists. It was developed for field application in the Hunter Valley area of New South Wales and is subject to modification. For example, soils that are permeable but which contain mottles owing to groundwater presence, and not owing to impeded moisture movement, would not necessarily be unsuitable for topdressing. Features additional to those that appear in the figure and which are useful in marginal decision cases are also discussed.

(a) Soil Structure

Water entry must be considered in the selection of topdressing material because water is essential for the germination of plants and available water is needed for plant establishment. Infiltration generally varies with structure grade, and is known to depend on the proportion of coarse peds in the soil surface.

(b) Soil Coherence

Russell (1974) considered that surface soil should be crumbly and that the crumbs should be large enough not to blow away, but be small enough to allow good germination of seed and be sufficiently non-sticky when moist to maintain their individuality when disturbed.

Soil material with the higher grades of structure (Butler, 1955), or the more pedal soil, is considered suitable for use in topdressing. Coherence in the less pedal soils is used as a measure of the ability of soil to maintain its structure grade. Soils that are structureless, or in which structure is likely to be destroyed by mechanical work associated with the extraction, transportation and spreading of topdressing material, are not considered suitable for revegetation. Surface sealing and reduced infiltration of water will restrict germination and establishment in these soils.

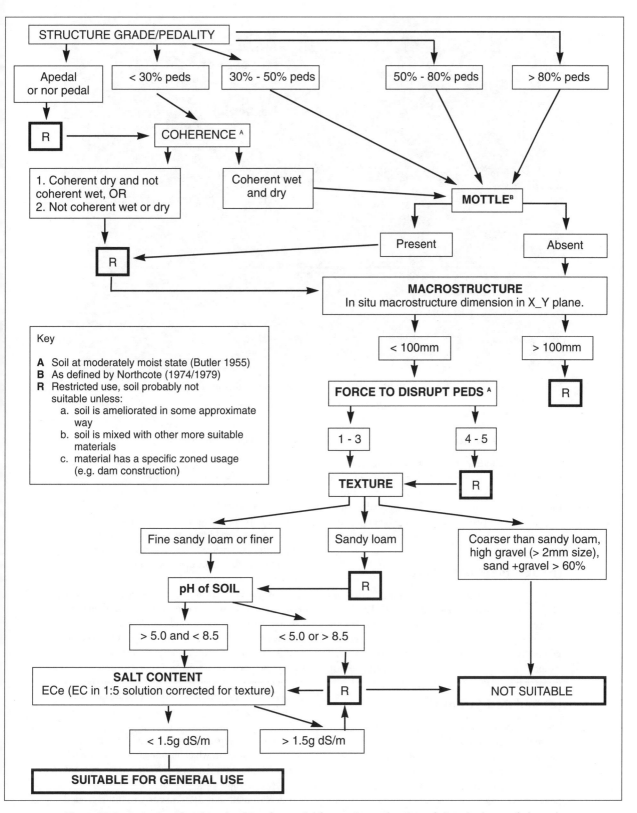

Figure 22.1 Procedure for the selection of material for use in topdressing of disturbed areas (adapted from Elliott and Veness, 1981)

Plate 22.1 An open-cut coal mine in the Hunter Valley showing substantial areas of land rehabilitated after mining

(c) Soil Mottling

Mottling is an indicator of poor drainage conditions (Bouma, 1977). Poor moisture transmission rates will unfavourably affect infiltration, available moisture and air porosity. Mottles, especially value-chroma Class Two mottles (Northcote, 1979), indicate material that is generally not suitable for use in revegetation.

(d) Soil Macrostructure

Macrostructure refers to the form and orderliness of the arrangement of peds in the soil. The size of macrostructure units is an indication of the tendency of a soil horizon to form a massive structure when it is wet. The void space is reduced in such situations and thus the tendency is undesirable in top-dressing materials.

(e) Soil Ped Strength

The force to disrupt peds, assessed on soil at about the moderately moist state, is taken as an indicator of the method of ped formation. Deflocculated soil will dry to a hard ped and rewet to a dispersed state. Flocculated soil dries to crumbly peds and retains these peds on wetting. Soils whose aggregates require a strong force to break them are considered less suitable for revegetation work.

(f) Soil Texture

Soil texture is a primary determinant of suitability. Generally materials coarser than a sandy loam or of very high sand content can be extremely erodible, and it is also considered that sandy soils with low available water are poorly suited to plant growth. Schafer (1979) considered that materials with textures coarser than sandy loam were less desirable for use in strip mine reclamation than medium-textured soils. Favourable topographic situation or reliable climate may obviate the adverse effects of using sandy material, and good moisture relations in the underlying layers may permit the use of sandy topdressing materials.

It is possible that topdressing material that retains a large amount of available water, especially in relation to underlying strata, may not be suitable for revegetation in the long term. This is because roots of established plants can concentrate in this layer with the following effects:

—inhibition of natural regeneration by competition with germinating seedlings and by desiccation of the zone in which germination is occurring
—poor drought tolerance because of poor root distribution
—establishment of isolated plants with little ground cover between plants;

leading to:

—increased erosion

—weed invasion in periods favourable for growth.

The finer-textured soils that have relatively high available water-holding capacities have generally given satisfactory short-term establishment.

(g) Gravel and Sand Content

If the combined amount of gravel and sand exceeds 60% of the soil, plant growth may be retarded. Too much coarse-grained material can result in weakened soil structure and reduced moisture-holding capacity.

(h) Soil Acidity and Salt Content

Before the final assessment of the suitability of material is made, both pH and salt content should be measured.

Adverse affects of high salt content and both high and low pH values are well known. Values of pH should be between the limits 4.5 and 8.5, or preferably between 5.5 and 7.5 (measured in a 1:5 soil:water suspension). Values of electrical conductivity, considered as an index of soluble salt levels, should be less than 1.5 dS/m measured in a 1:5 soil:water suspension (Blake, 1965).

(i) Soil Colour

Colour is an important indicator of soil types and attributes. Colour is indicative of the soil's fertility, permeability and stability. The darker the soil, usually the higher the organic content and the better the fertility. Brown and red soils are usually more pervious than yellow soils. Gleyed/grey soils and mottled soil are signs of poor soil drainage. Where red and yellow soils occur in toposequence, as a general rule the red soils will be the more stable with respect to soil conservation structures, cultivation and erosion. Stapledon and Casinader (1976) found that 45% of yellow-brown B/C horizons they examined were dispersible, compared with less than 25% of red-brown B horizons. Most of the instability in red-brown soils was caused by slaking along fracture planes. Murphy (1980) presented data on aggregate dispersion in duplex red and yellow soils, showing more stable aggregation in the red soils. Soils as red as, or redder than, 7.5 YR, and with value-chroma Class One or Five, or with value-chroma Class Four and colour value less than or equal to 6, are generally suitable with respect to aggregate stability and to erodibility. Soils as yellow as, or yellower than, 7.5 YR, and with value-chroma Class Four and colour value greater than or equal to 7, are often not suitable with respect to stability and erodibility.

(j) Soil Cutans

The presence of cutans in the soil has been found to be a useful indicator of permeability, especially when the distribution of cutans is considered. A uniform distribution of cutans indicates uniform and deep wetting. A discontinuous distribution indicates restrictions to permeability or a limitation to the expansion of a wetting front. Following Brewer's (1964) terminology, grain cutans, tertiary ped cutans, and channel and plane cutans are useful indicators of satisfactory permeability. Void cutans and cutans restricted to primary peds may indicate restricted permeability.

(k) Other Factors

In addition to the preceding factors, particular studies should be made of attributes that are important in individual circumstances. For example, chromium and nickel toxicities may occur in some areas associated with antimony mining, and sulfide-rich overburdens may occur in association with coal deposits.

(l) Soil Mixtures and Zonal Use

Having decided the suitability of each layer in the soil profile, it is possible that some materials that are suitable for use occur alternately with materials that are not. Physically unsuitable materials may be included in the topdressing mixture as soil horizons if they are less than 20 cm thick and comprise less than 30% of the total material to be used. If the layer of unsuitable material is more than 20 cm thick, it is probably practical to selectively remove and discard it.

Alternatively, soil materials that could be normally considered unsuitable or of restricted suitability could be mixed with other soil materials to provide a suitable topdressing or be used zonally (for example, in earthworks).

(m) Amelioration

Soils can often be ameliorated for physical attributes that might otherwise preclude their general usage by:

—mixing with other complementary soil materials (for example, coarse soil materials with fine soil materials and acid soil materials with alkaline soil materials)

—the use of gypsum (or lime/dolomite) to overcome dispersion problems and improve permeability

—the use of lime/dolomite to overcome acidity problems

—the use of acidifying fertilisers to reduce alkalinity (high pH)

—leaching saline soils.

Handling Topdressing Material `22.2.2`

Having identified materials suitable for topdressing, recovery, storage and respreading should be performed in ways that preserve, as far as possible, desirable attributes. This is more critical for less well-structured soils. As a generalisation, soil should be mechanically handled at moisture contents between the shrinkage limit (determined at the break of slope of the normal section of the shrinkage curve) and the plastic limit. This minimises both brittle failure that may occur in a dry soil and compressive failure that may occur in soils that are too wet. For example, measurements of structural attributes of stockpiled soils and adjacent undisturbed soils indicated that mean weight diameter values decline by 25–43% for topsoil of a yellow podzolic soil handled in too dry a condition. In contrast, handling at a preferred moisture content actually resulted in an increase in mean weight diameter for topsoil of a very similar soil. This result is consistent with desirable changes in aggregate size distribution. (In context, it should be recognised that these effects on soil represent no more severe disruption than those induced by normal agricultural practices.)

Mine machinery (and equipment) is designed to move large volumes of materials at low unit cost. The machinery is inappropriate in many instances for the removal and emplacement of topsoiling materials because of the high potential for structural damage, compaction and inadequate precision/high wastage in the exclusion of non-desirable soil materials. A number of operational strategies to minimise damage to topsoiling materials have been developed by the industry:

—the restriction of heavy mining equipment/ machinery to designated road systems
—the off-road use of equipment/machinery specifically suited to earthmoving with low ground pressure loadings for soil removal and emplacement
—regular monitoring of soil suitability for removal and emplacement
—elevating soils for stockpiling in a single lift to eliminate compaction problems from dumping.

Grading or pushing soil into windrows with light graders or track-driven bulldozers for later collection by elevating scrapers, or for loading into rear dump trucks by front-end loaders, are examples of less aggressive soil-handling systems. They minimise compression effects of the heavy equipment which is often necessary for most economical transport of soil material.

Soil transported by dump trucks may be placed directly in storage for the first stockpile layer without tracking over stored material. Soil transported by bottom-dumping scrapers should be pushed into place by other equipment to avoid tracking over previously laid soil by the scraper. An elevator can readily raise dumped materials to 3 m storage with minimal compaction and soil damage. The surface of stored soil dumps should be left in as rough a condition as possible in order to promote infiltration and minimise both wind and water erosion until vegetation is established.

The placement of soil in temporary storage will depend on available space and projected storage time, but research has shown that large shallow dumps provide opportunities for the preservation of desirable soil attributes. Microflora and macroflora populations and soil structure are more likely to be retained and conditions are better for the regeneration of attributes affected by plant growth (such as the size of aggregates and aggregate stability). In shallow storages, the net survival of pre-existing flora is increased relative to that in thicker storages. Maximum depths of up to 60 cm appear to be appropriate for the regeneration and preservation of soil attributes influenced by plant growth. If soil profile reconstruction is to be attempted, subsoil storages may be much thicker than the recommended 60 cm.

Dumps of topdressing material should be revegetated sequentially as soon as possible after emplacement to minimise any risk of erosion/soil degradation, and reduce weed competition and establishment costs. This will generally mean that seasonal conditions will frequently be unsuitable for desired permanent pasture species and may probably necessitate the use of cover crops such as ryecorn and millet. In many instances the use of self-sterile hybrids is desirable to subsequently minimise plant competition when more permanent cover is being established.

A vegetative cover minimises erosion, improves soil attributes, increases the chances of successful final, stable rehabilitation, and reduces the invasion and establishment of weeds. Sowing of all species and spreading of fertiliser should be carried out to minimise any compaction of the stockpile surface and ensure that any wheel tracks are removed (preferably using a tyred implement). Any cultivation/harvesting must be carried out to minimise any concentration of runoff that could result in rill erosion.

BIBLIOGRAPHY

Blake, C.D. (ed.) (1965), *Fundamentals of Modern Agriculture*, Sydney University Press, Sydney.

Bouma, J. (1977), *Soil Survey and the Study of Water in Unsaturated Soil: Simplified Theory and Some Case Studies*, Soil Survey Paper 13, Soil Survey Institute, Wageningen, The Netherlands.

Brewer, R. (1964), *The Fabric and Mineral Analysis of Soil*, CSIRO Division of Soils, Melbourne.

Butler, B.E. (1955), 'A system for the description of soil structure and consistence in the field', *Journal of the Australian Institute of Agricultural Science* 21, 239–49.

Elliott, G.L. and Veness, R.A. (1981), 'Selection of topdressing material for rehabilitation of disturbed areas in the Hunter Valley', *Journal of the Soil Conservation Service of NSW* 37, 37–40.

Murphy, B.W. (1980), 'Variations within some taxonomic units of the Northcote Factual Key for the recognition of Australian soils', *Journal of the Soil Conservation Service of NSW* 36, 109–15.

Northcote, K.H. (1979), *A Factual Key for the Recognition of Australian Soils*, Rellim Technical Publications, Glenside, South Australia.

Paton, T.R. (1978), *The Formation of Soil Material*, George Allen and Unwin, London.

Russell, E.W. (1974), *Soil Conditions and Plant Growth*, Longman Green and Co. Ltd., London.

Schafer, W.M. (1979), 'Guides for estimating cover soil quality and mine soil capability for use in coal strip-mine reclamation in the western United States', *Reclamation Review* 2, 67–74.

Stapledon, D.M. and Casinader, R.J. (1976), 'Dispersive soils at Sugarloaf Dam site, near Melbourne, Australia', in J.L. Sherard and R.S. Decker (eds), *Dispersive Clays, Related Piping and Erosion in Geotechnical Projects*, American Society for Testing and Materials STP 623, 432–66.

Thomasson, A.J. (1978), 'Towards an objective classification of soil structure', *Soil Science* 29, 38–460.

Soils and Sustainable Development — a Concluding Perspective

CHAPTER 23

P.E.V. Charman

Many Australian soils are fragile, owing to low levels of organic matter and poor aggregation, and careful management is essential if they are to remain productive into the twenty-first century. It was the introduction of European-style agriculture in the late eighteenth century that started a process which is now threatening the productivity of our soils. Agricultural practices expose the soil to a variety of climatic and human-made forces that fuel the process of soil degradation. This degradation leads not only to lower crop yields but also to less efficient agricultural production on a broad scale.

The last two decades have seen a widening of perspective on soil degradation, so that it has increasingly been recognised, not just as an agricultural problem, but as part of the broader problem of land degradation or, even more broadly, ecosystem degradation. Land includes soils, vegetation and water resources occurring in and on the land, and their degradation occurs irrespective of property boundaries. The problems cannot be dealt with just on each individual farm but must be tackled across whole catchments or even across river basins that are made up of a number of catchments. The work of the

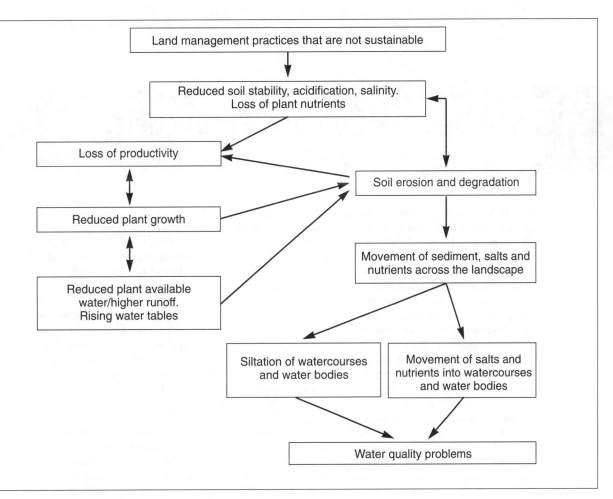

Figure 23.1   The interaction of soil, plant, land and water processes following poor catchment management.

Murray-Darling Basin Commission is an example of this approach.

Community involvement in dealing with land degradation has also increased markedly, along with the greater public consciousness of the environment and its problems. The landcare movement of the eighties and nineties has been a response to these concerns, and it seems to be the community's view that working with the government and research institutions is the best way to approach what is, after all, a national set of land-related issues. People in different parts of catchments have recognised that their environmental concerns are related, and have been prepared to join landcare groups and catchment committees to work together for the benefit of not only their own properties or localities but the catchments overall. The New South Wales program of Total Catchment Management embraces these concepts with a view to addressing the statewide problems of soil, water, vegetation and land degradation on a catchment basis. Figure 23.1 shows how such degradation affects the functioning of ecosystems.

To sustain development on this broader front we must think in terms of ecosystems. Soils are a vital component of the state's environmental resources and an essential base for the productive capacity of our land. The health and management of our soils has a large impact on the health, diversity and productivity of the state's ecosystems. These include catchments, forests, rangelands, pastures, cropping paddocks and waterways. Degradation of soil resources damages all these ecosystems, and therefore the prevention of soil degradation is an essential objective of ecologically sustainable development.

Ecologically sustainable development should be a community goal. It requires:
—the effective integration of economic and environmental considerations in decision-making processes
—the maintenance of the health, diversity and productivity of the environment for the benefit of future generations
—conservation of biological diversity and ecological integrity
—improved valuation and pricing of environmental resources.

23.1 Soil Degradation and Productivity

The effects of soil degradation on productivity result in a direct loss of crop yields, owing to a reduction in soil volume, a decline in soil quality or a combination of both. Reduction in total soil volume, beyond a critical point, means that the soil cannot provide sufficient moisture and plant nutrients to realise crop yield potential for the given climate. Reduction in soil quality may involve removal of part of the soil, the degradation of its physical, chemical and biological attributes, or both.

Erosion also affects the water-holding capacity of the soil by changing such characteristics as the level of organic matter or the structure profile. Coarser sands, with limited moisture-storage ability, will remain after erosion at the expense of finer components. Removal of surface soils by erosion increases the likelihood of denser subsoil zones being required to support root systems. Soil mechanical impedance may then restrict root development and this is also likely to reduce plant yields.

Soil erosion is generally reported in terms of the weight of soil loss per unit area. Often this is misleading in that it does not indicate the quality of the soil removed, the textural fractions of soil removed or the quantity of soil degraded. Soil degradation predisposes the soil to erosion while erosion accelerates the rate of soil degradation. The relationship between soil erosion and production is conceptualised in Figure 23.1.

Loss of soil quality can be partially or wholly reversed by the addition of extra fertiliser, the introduction of pasture ley phases or green manure crops into a rotation, or by the amelioration of subsoil material. This latter process is enhanced by the mixing action of soil tillage. Thus, it is possible, particularly in deeper soils, for the full effects of soil loss to be concealed, perhaps for many years. Cereal production, which is the most important type of crop production, is mainly carried out under dryland conditions in New South Wales. In the main wheat-growing belt, the average annual rainfall usually varies between 400 and 900 mm. In recent years, the adoption of conservation farming practices in the cereal-growing belt has partly addressed the widespread occurrence of soil erosion resulting from traditional cropping practices. Nevertheless, arable land use still poses the greatest erosion hazard associated with the agricultural use of land in New South Wales.

Some details of actual soil loss rates measured in New South Wales over varying periods are given in Chapter 3.

Aveyard (1983) recognised that, particularly for Australian conditions, the effect of soil erosion on productivity was irregular. This is because erosion patterns reflect the spasmodic nature of Australia's erosive rainfall events. Major soil loss events occur on a very irregular basis, and this means that a particular management system may not necessarily be a safe one even though soil erosion has not occurred for a number of years.

There is a high risk of erosion occurring over a large proportion of the New South Wales wheat-belt, either because the soils are highly erodible or the land cultivated is steep, or both.

Recent field evidence suggests that not only is the area of land affected by erosion increasing, but also that soil loss may be occurring in very much larger amounts than indicated by plot and catchment data. On the outskirts of Wagga Wagga, for example, field measurements have estimated a soil loss of 21.7 t/ha/yr over a 50-year period from a 90 ha cropland area (Ring, 1982). In the Inverell district, soil losses of 150 t/ha and 450 t/ha have been reported from arable slopes of up to 1.5% after a single rainfall event (Enright, 1983).

The study of the effect of soil loss on productivity remains the most neglected area of soil erosion research. Of the work completed on a global basis, Stocking (1984) acknowledged that 87% of studies have been undertaken in the USA, and 9% in Australia. Investigations usually concern artificial removal of topsoil (erosion simulation), small catchment studies, determination of soil loss tolerance and values, and model building. A review of erosion/productivity research in the USA is presented in Follett and Stewart (1985).

Stocking (1980), in an FAO review on erosion and productivity, recognised that the degree to which a unit quantity of erosion reduces productivity is dependent on a range of soil, crop and environmental characteristics. For example, productivity restoration after erosion of duplex soils may be very difficult. These soils are characterised by low to moderate nutrient levels that vary widely in their suitability for root growth. Examples of these soils in New South Wales are the red-brown earths of the wheat-belt and the podzolic and solodic soils of the tablelands. Alternatively, productivity restoration on soils that have moderate to high fertility and favourable physical properties may be accomplished more easily by suitable management. Examples of this group of soils in New South Wales are black earths, krasnozems and euchrozems.

Stocking also identified a definite exponential decay in yield as erosion progresses, with the drop being most dramatic in the early stages of erosion. It is, therefore, preferable to concentrate soil conservation resources on the lands that are already better managed and more productive than on those most severely degraded. This author further pointed out the interdependence between productivity and erosion, adding that there are other factors responsible for yield declines that should be distinguished from erosion. Climate and soil structural degradation are but two of these factors. The latter of these is an important precursor to erosion in many situations, but on its own is regarded widely in Australia as having a more significant effect on productivity losses.

Unfortunately, there is very little quantitative information in Australia on the direct effect of soil erosion on crop yield. Investigation of the relationship is a most difficult experimental problem because of the irregular occurrence of major erosion events, as discussed previously, and the fact that soil erosion is just one of the factors influencing crop yield.

A Soil Conservation Standing Committee Working Group examined the available data on erosion and production from all over Australia (Aveyard, 1984). Among the conclusions reached, the following were pertinent to the New South Wales situation.

—Soil erosion and degradation are important factors in crop production in Australia, in both the short and long term.

—In most cases, soil erosion and other forms of degradation reduce crop yield. In some instances, notably in southern New South Wales, the reported on-site effects are so large as to suggest that it must be economical for individual farmers to adopt conservative practices even in the short term. It was also noted that the reported magnitude of reduced yields, particularly in south-east Australia, is sufficiently large to make soil degradation a contributing factor to the plateau in wheat yields that is presently being observed.

—The magnitude of crop yield reduction depends on soil type and season as well as on soil and crop management. Duplex soils such as the red-brown earths, podzolic and solodic soils, found in the southern wheat-belt, and soils that show a sudden reduction at shallow depth in the concentration of a limiting plant nutrient, appear to suffer a larger loss of production per unit of soil erosion than do some other soils.

23.2 Soils and Catchment Health

The catchments of New South Wales are basically formed by the Great Dividing Range, which along its length sheds water in an easterly direction to the coast and westerly to the inland. With the range running roughly parallel to the coastline and about 150 km inland, the coastal rivers are generally fairly short. They have steep, well-watered catchments that drain quickly to the sea without extensive areas of low-slope land. In contrast the western catchments are large, with their rivers draining wide tableland areas and extensive slopes and plains that extend for hundreds of kilometres before flattening out to the floodplains of the many rivers that drain into the Murray-Darling system.

Each catchment, whether it be a small coastal one like the Hastings or a large western one like the Lachlan, has a complex pattern of land and land uses. Its natural resources comprise an array of geology, landforms, soils, slopes, drainage patterns and vegetation. The uses of the land they form include agricultural, urban, industrial, mining, forestry enterprises and national parks, as well as essential community infrastructure such as airports, recreational facilities and sewage treatment works.

To put soils in a catchment context, they have an important role to play in all these uses, but also vital is their interaction with rainfall to give rise to our water supplies. Both quantity and quality are important here, particularly with regard to water for domestic and industrial use, for irrigation, and for aquatic habitats. Land use practices since European occupation have tended to increase water yield and variability, owing to tree clearing (and replacement with crops and pastures that use less water), reduction of ground cover and increased soil compaction. Unfortunately such practices have also tended to decrease water quality, owing mainly to erosion and sedimentation, overuse of fertilisers and chemicals, overgrazing, overclearing and salinisation (Cullen, 1991; Costin, 1991; Thomas, 1991).

Considering their interaction with rainfall, soils may be broadly grouped into three categories in a conceptual model. First, there are those that generate runoff—the hardsetting soils, the shallow soils, typically soils some with some permanent vegetative cover situated on upper slopes. During heavy rainfall these soils tend to shed runoff that is relatively 'clean', that is, with only modest sediment or nutrient content. To maintain or improve this situation, vegetation management is the number one priority for these soils, particularly on steep slopes. This may involve clearing, burning and grazing controls, and fertiliser application.

Second, there are those soils that tend to generate soil loss—poorly structured soils, sodic soils, and medium-textured silty soils, often on intermediate slopes that are cultivated for cropping. Typically this is the part of the landscape prone to rilling and gullying. During heavy rain these soils produce runoff that is 'dirty', that is, with high soil and nutrient content. The priority for these soils is erosion management, which may involve soil conservation works, contour cultivation, gully control, and so on.

Third, there are those soils that tend to absorb rainfall—deep soils, well-structured soils, and fertile soils with good cover, often occurring on flatter slopes. During heavy rainfall these soils shed little runoff, and rainfall infiltrates the soil and percolates to the water table. The priority for these is soil management, which may involve conservation tillage, rotations and best farm practice.

The second of these groups is clearly the one that yields water to the catchment drainage system and which contains most soil (sediment), organic matter and nutrients (either dissolved or particulates). Put another way, the second soil category has the greatest deleterious effect on water quality, and this highlights the important interaction between soil and water resources in a catchment context. The management requirements of the three categories reinforce the fact that the quality of water in a storage or river reflects the catchment's health. Health as used here means the ability of a catchment to yield water of high quality to its streams, rivers and storages.

The main contaminants that have an adverse effect on water quality are as follows:
—sediment: this is a serious problem, which results from soil erosion in the catchment; water turbidity and sedimentation are the result; sediment may also carry nutrients
—salinity: this is particularly serious in irrigation districts; results from substandard irrigation practices, but can also occur in dryland areas owing to overclearing of catchment slopes
—nutrients: nitrogen and phosphorus are the main concerns; results in possible eutrophication of

rivers and water storages and adverse effects on the ecology of streams and wetlands

—organic matter and microbes: these come mainly from animal waste that may be washed directly into waters in runoff, or from sewage treatment works' effluent flowing into streams; plant remains can also be involved

—agricultural chemicals and heavy metals: these come from pesticides, herbicides and fertilisers used on farms, or runoff from industrial sites and roads.

These contaminants find their way into water bodies and streams from both individual sites (point sources) and broad areas (non-point sources), via runoff and percolation through the soil or a combination of both (Thomas, 1991). Nutrient pollution has been of particular concern in recent years, following the blue-green algae blooms of the early nineties, mainly caused by a combination of high nutrient input to rivers and low flows. Nutrients from agricultural land are thought to come largely from gullies and streambanks on grazing land (tablelands, slopes), but mainly from sheet erosion on cropping land (slopes, plains).

Management to improve catchment health and contain these problems focuses primarily on:

—vegetation control—grazing management to maintain cover within the catchment, stabilisation of critical areas including streambanks, buffer strips along streams, strategic fire control, and so on

—chemical control—fertiliser, pesticide and herbicide management, waste handling and disposal, and so on

—soil conservation—strategic works, contour farming, runoff control, gully control, conservation tillage, rotations and strip cropping, sediment ponding where necessary, and so on.

Catchment health requires the application of these principles to all parts of the catchment by landholders and industries in cooperation with the various levels of government. The formation of landcare groups and wider coordination by catchment committees are seen as being the important framework within which the necessary on-ground work can be undertaken, to achieve optimum land use throughout catchment areas and optimum yields of high quality water for human and environmental purposes.

23.3 Conclusion

In conclusion, it must be said that there is an urgent need for land users to adopt land management practices that will reduce the rate of soil erosion and other forms of soil degradation, while sustaining productivity and water quality in the long term. This particularly applies to arable lands where the effects of soil erosion and structural degradation are currently most obvious. However, the effects of other, perhaps more insidious, forms of soil degradation such as soil acidification and salinisation, currently evident on many other lands in New South Wales, must be constantly monitored.

It is, therefore, important that, in order to promote long-term soil stability and productivity as well as water quality, the soil conservationist, agronomist, catchment planner or other practical soils adviser should keep the following points in mind:

—The rate of soil formation under most Australian conditions, for all practical purposes, is close to zero. It is, therefore, important to understand the nature of soil and the way it forms, particularly when erosion rates at current lev-

els are far in excess of the production rate of 'new' soil.

—Soil erosion occurs very irregularly. Knowledge of the processes that lead to the various forms of erosion is thus vital in the development of land management practices that will conserve the soil resource.

—The various forms of soil degradation predispose soils to erosion. An understanding of all forms of soil degradation is important in the context of an integrated and comprehensive approach to soil conservation and catchment health.

—Soil erosion results in higher production costs and lower yields. It is thus imperative that soil conservationists have a good knowledge of soil properties that are responsible for good crop and water yields, that are adversely affected when erosion takes place and/or can be modified to improve soil stability.

—The effects of soil degradation are greater for some soils than for others. This implies a need to understand soil profiles as a basis for classification of different soils, and their mapping and occurrence in relation to landscapes.

—The effects of soil degradation will also vary according to seasonal conditions and management practices. A knowledge of soils in relation to climate, land use and farming practices thus becomes essential to the soils adviser, whether advice is required in a cropping, grazing, forestry or catchment context.

—The effects of soil erosion on farm production and catchment health can be very large. Vegetation and structural measures are key elements in the confinement of erosion to manageable levels, and an understanding of their dependence on soils is essential to the choice of either in a soil conservation or catchment management plan.

—The rate of soil recovery after erosion is variable, but it can be very slow. Lost soil, however, is virtually irreplaceable by natural means and a good knowledge of soils must form the basis for a preventative approach to soil erosion control.

—The costs of soil erosion are frequently manifest in off-site effects, such as sedimentation of roads or reservoirs, adverse effects on water quality, or the encroachment of drifting dune sands onto an urban development. Knowledge of soil materials in relation to high-cost infrastructure such as mines, urban areas or water storages thus becomes a key part of soil conservation advice in an industrialised society.

—Soil degradation is generally the result of poor management of the land, whether on a farm or beach, an urban subdivision, a mining lease or a whole catchment. The aims of soil management should be to ensure that sustainable productive and economic use of the land is possible, consistent with the maintenance of long-term soil stability and prevention of any form of land or soil degradation. On a catchment basis, these aims should also include land management to ensure the sustainable production of sufficient high quality water for community needs.

BIBLIOGRAPHY

Anon. (1980), *Agricultural Sector, Land Use, Artificial Fertilizers and Other Improvement, New South Wales*, Publication 7104.1, Australian Bureau of Statistics, New South Wales Office, Sydney.

Anon. (1981), 'Soil erosion effects on soil productivity', *Journal of Soil and Water Conservation* 36, 82–90.

Aveyard, J.M. (1983), *Soil Erosion: Productivity Research in New South Wales to 1982*, Wagga Wagga Research Centre Tech. Bull. No. 24, Soil Conservation Service of NSW, Sydney.

Aveyard, J.M. (1984), *Report of Working Group on Soil Erosion and Production*, prepared for the Standing Committee on Soil Conservation.

Costin, A.B. (1980), 'Runoff and soil and nutrient loss from an improved pasture at Ginninderra, Southern Tablelands, NSW', *Australian Journal of Agricultural Research* 31, 533–46.

Costin, A.B. (1991), 'Land use and water quality—the importance of soil cover', *Australian Journal of Soil and Water Conservation* 4(3), 12–17.

Cullen, P. (1991), 'Land use and declining water quality', *Australian Journal of Soil and Water Conservation* 4(3), 4–8.

Edwards, K. (1980), 'Runoff and soil loss in the wheat belt of NSW', *Proceedings of the Agricultural Engineering Conference*, Geelong, 94–8.

Enright, N.F. (1983), *Annual Research Report*, Inverell Research Centre, Soil Conservation Service of NSW, Sydney.

Follett, R.F. and Stewart, B.A. (eds) (1985), *Soil Erosion and Crop Productivity*, American Society of Agronomy, Inc., Crop Science Society of America, Inc. and Soil Science Society of America, Inc..

Hamilton, G.J. (1970), 'The effect of sheet erosion on wheat yield and quality', *Journal of the Soil Conservation Service of NSW* 26, 118–23.

Kaleski, L.G. (1945), 'The erosion survey of NSW', *Journal of the Soil Conservation Service of NSW* 1, 12–20.

Lal, R. (1984), 'Mechanized tillage system effects on soil erosion from an alfisol in watersheds cropped to maize', *Soil and Tillage Research* 4, 349–60.

Marsh, B. a'B. (1971), 'Immediate and long-term effects of soil loss', *Proceedings of the Australian Soil Conservation Conference*, Melbourne.

Mbagwu, J.S.C., Lal, R. and Scott, T.W. (1984), 'Effects of desurfacing of alfisols and ultisols in Southern Nigeria', *Soil Science Society of America Journal* 48, 828–38.

Reuss, J.O. and Campbell, R.G. (1961), 'Restoring productivity to levelled land', *Soil Science Society of America Proceedings* 25, 302–4.

Ring, P.J. (1982), 'Soil erosion, rehabilitation and conservation: comparative agricultural engineering economic case study', *Proceedings of the Agricultural Engineering Conference, Armidale, NSW*.

Saffigna, P. (1983), 'Decline in yield of sunflowers growing on artificially eroded soils' in *Sunflower* (May), Australian Sunflower Association.

Stewart, J. (1968), 'Erosion survey of New South Wales Eastern and Central Divisions Re-Assessment 1967', *Journal of the Soil Conservation Service of NSW* 24, 139–54.

Stocking, M. (1980), *Erosion and Productivity: A Review*, Land and Water Development Division, AGLS, FAO, Rome.

Thomas, P. (1991), 'Land uses and some water quality problems', *Australian Journal of Soil and Water Conservation* 4(3), 27–30.

Walker, P.H. (1980), 'Soil morphology, genesis and classification in Australia', *Proceedings of the Australian Society of Soil Science National Conference*, Sydney.

White, P.J. (1986), 'A review of soil erosion and agricultural productivity with particular reference to grain crop production in Queensland', *Journal of the Australian Institute of Agricultural Science* 52, 12–22.

Williams, J.R. and Renard, K.G. (1985), 'Assessments of soil erosion and crop productivity with process models', in R.F. Follett and B.A. Stewart (eds), *Soil Erosion and Crop Productivity*, American Society of Agronomy, Inc., Crop Science Society of America, Inc. and Soil Science Society of America, Inc.

Broad Correlation between Australian Soil Classifications and the US Soil Taxonomy

AUSTRALIAN SOIL TAXONOMY Great Soil Groups	Australian Soil Classification	US SOIL TAXONOMY Order	Suborder
Alluvial soils	Rudosols Tenosols	Entisols	Fluvents
Lithosols	Rudosols Tenosols	Entisols Inceptisols Aridisols	Orthents Ochrepts Orthids Argids
Calcareous Sands	Rudosols	Entisols	Psamments
Siliceous Sands	Rudosols	Entisols	Psamments Orthents
Earthy Sands	Tenosols	Aridisols Entisols	Argids Psamments Orthents
Grey-brown and Red Calcareous Soils	Calcarosols	Aridisols Alfisols	Orthids Argids Xeralfs
Red and Brown Hardpan Soils	Kandosols	Aridisols	Orthids Argids
Rendzina	Calcarosols (Rudosols)	Mollisols Aridisols	Xerolls Ustoll Orthids
Black Earths	Vertisols	Vertisols	Usterts Torrerts
Grey, Brown, and Red Clays	Vertisols	Vertisols	Usterts Xererts Torrerts
Chernozems and Prairie Soils	Dermosols	Mollisols Entisols Alfisols	Ustolls Xerolls Fluvents Ustalf Xeralf

Great Soil Groups	AUSTRALIAN SOIL TAXONOMY Australian Soil Classification	US SOIL TAXONOMY Order	Suborder
Wiesenboden	Vertisols	Vertisols	Usterts Xererts
Solodic Soils (solonetz, solodised solonetz, solodic soloth)	Sodosols	Alifsols Aridisols	Xeralf Ustalf Argids
Solonised Brown Soils	Calcarosols	Aridisols Mollisols Alfisols	Orthids Argids Xeroll Xeralf
Desert Loams	Sodosols	Aridisols Alfisols	Argids Xeralf
Red-Brown Earths (sodic)	Sodosols	Alfisols	Xeralf Ustalf
Red-Brown Earths (non-sodic)	Chromosols	Alfisols	Xeralf Ustalf
Non-calcic Brown Soils	Chromosols	Alfisols	Ustalf Xeralf
Chocolate Soils	Dermosols Chromosols	Mollisols	Xeroll Ustoll Ustalf
Calcareous Red Earths	Kandosols	Ardisols	Argids Orthids
Red Earths	Kadosols	Alfisols Oxisols Aridisols Ultisols	Ustalfs Xeralfs Torrox Argids Ustults Udults
Yellow Earths	Kandosols	Alfisols Ultisols Aridisols	Ustalfs Xeralfs Ustults Argids
Terra Rossa Soils	Dermosols	Alfisols Inceptisols Entisols	Xeralfs Ustalfs Ochrepts Orthents
Euchrozems	Dermosols/Ferrosols	Alfisols Inceptisols	Ustalfs Ochrepts

Great Soil Groups	AUSTRALIAN SOIL TAXONOMY Australian Soil Classification	US SOIL TAXONOMY Order	Suborder
Xanthozems	Dermosols	Oxisols	Orthox Humox
		Ultisols	Humults Ustults
Krasnozems	Ferrosols	Oxisols	Humox Orthox Ustox
		Alfisols	Ustalfs
		Ultisols	Humults
Red Podzolic Soils	Chromosols (Dermosols, Kurosols)	Ultisols	Ustults Udults Xerults
		Alfisols	Ustalfs Xeralfs
Yellow Podzolic Soils	Chromosols (Dermosols, Kurosols)	Ultisols	Ustults
		Alfisols	Ustalfs Xeralfs
Lateritic Podzolic Soils	Kurosols	Ultisols	Ustults Xerults
		Alfisols	Xeralfs Ustalfs Aqualfs
Podzols	Podosols	Spodosols	Orthods
		Entisols	Psamments
Alpine Hummus Soils	Tenosols	Inceptisols	Umbrepts Ochrepts
Humic Gleys and Gleyed Podzolic Soils	Hydrosols	Ultisols	Aqults
		Alfisols	Aqualfs
		Inceptisols	Aquepts

Identification of Soil Groups

UNIFORM SOILS (U)

Great Soil Group	Northcote Codings	Further Diagnostic Features	Notes
Solonchaks	Some Um1 and Uf1 soils Some Uf6 such as Uf6.51 Db0 and Dy1 soils ending in 3	Soils dominated by salt accumulation with one or more of the following features: a. salty encrustations b. surface flaking of clay crusts c. polygonal cracking of surface d. powdery structure e. lack of normal plant growth except for salt-tolerant species	Primary solonchaks show little pedologic organisation, and are uncommon. Secondary solonchaks may show features of other soil groups, and are best described as saline forms of those groups (such as Dy1.43). These soils form when climatic and/or topographic conditions allow the accumulation of free salt in the soil profile. Soil salinity leads to revegetation problems and erosion hazard.
Alluvial Soils	Some Um1 and Uf1 soils Um5.2 soils	Juvenile soils formed by deposition from still or moving water. Little pedologic organisation, beyond some accumulation of organic matter at the surface and sometimes minor changes in colour, consistence, texture and structure throughout the profiles.	Extreme variation. Soils may show differential layering of sedimentary material, but no well-developed pedogenic horizons. (NB The majority of soils are alluvial or colluvial to a certain extent.) Generally no erosion problems.
Lithosols	Some Uc4, Uc6 Um1 and Uc1 soils Some Uf1, Um2, Um4, Um5.41, Um5.51, Um5.21	Shallow stony soils dominated by presence of underlying rock material. Little pedologic organisation.	Characteristic of steeper exposed slopes where natural erosion restricts profile development. Some have minimal A_2 horizons (Um2 and Um4). Erosion hazard due to slope and low fertility.
Sands	Some Uc1 and Uc5 soils Uc2.1, 2.2, Uc3 and Uc4 Uc6 soils	Soils with little or no pedologic organisation, generally deep, characteristic of coastal dune areas and inland deserts	Calcareous, siliceous, ferruginous, dense, leached and earthy variants occur. Leached sands — Uc3 and Uc4, Uc2.1. Structured sands — Uc6. Generally erodible because of poor structure and low fertility.
Calcareous Desert Soils	Um1.3, Um5.1 and Um6.2 soils Gc1.1, Gc2.1 soils Some Um6.4, Um5.6	Shallow soils of arid areas, on calcareous sedimentary rocks, with finely divided $CaCO_3$ throughout the solum.	Grey-brown and red variants occur, also more clayey types (Gc1.1).
Hardpan Soils	Um5.3 soils	Shallow massive and earthy soils overlying an indurated pan.	See also Dr1 soils.
Grey Clays	Ug3.2, Ug5.2 and Ug5.5. Ug6.2 and Ug6.5 soils	Deep, strongly structured soils with high clay content.	Cracking soils with high volume expansion coefficients. Formed on fine-textured alluvium associated with inland river systems, as are brown clays also. Ug5.2 typical.
Brown Clays	Ug5.3 and Ug5.6, Ug6.3 soils Ug5.32–5.35 Brown Ug5.37–5.39 Red	Deep, strongly structured soils with high clay content.	Cracking soils with high volume expansion coefficients. Includes red variants (hue of clay horizon as red or redder than 5YR). Prone to gilgai formation. Ug5.3 typical. Parent material as above and probable aeolian influence.

UNIFORM SOILS (U) (Cont.)

Great Soil Group	Northcote Codings	Further Diagnostic Features	Notes
Black Earths	Ug5.1 and Ug5.4, Ug6.1 soils	Dark-coloured well-structured clay soils with a self-mulching surface. Weisenboden types have a water table present.	Includes Weisenboden (hydromorphic variant). Ug5.16, Ug5.17 prone to gilgai formation and cracking. Often basaltic in origin. High volume expansion coefficients. Ug5.1 typical. Unstable when wet. Erodible under intensive cultivation.
Rendzina	Uf6.11, possibly Uf6.22 Some Uc6.12 (Sandy) Some Um6.21 Some Ug5.11	Calcareous parent materials present. Soils rest directly on fresh rock. Usually well structured.	Similar to shallow friable black earths on calcareous parent material. Uf6.11 typical. More stable than black earths due to less cracking.
Chernozems	Um6.11	Deep, dark loamy soils of alluvial origin.	Occurrences generally isolated, and restricted to alluvial parent materials with basic constituents. Few erosion problems. Similar to coarser textured, friable and porous black earths. Um6.11 typical. Organic soils normally very fertile.
Terra Rossa	Um6.24, Um6.33, Um6.43, Uc6.13	Calcareous parent materials present. Soil rests directly on fresh rock.	Normally shallow, structured, red soils on calcareous parent material. Texture varies with texture of parent material.
Podzols	Uc2.3, Uc3.3 soils Some Uc4.3 soils Um2.32 Some Uc2.2	Acid soils with organic A_1 horizons and accumulation of humus and/or sesquioxides in the B horizon.	Peaty and humic variants occur. There is sometimes minor clay accumulation in the B horizon. These soils occur mainly in cool wet climates and on siliceous parent materials — particularly unconsolidated sands. Uc2.36 is the most typical podzol in the true sense. Low fertility — erodible when exploited.
Alpine Humus Soils	Um7 soils ending in .11, .12	Dark-coloured loamy soils, high in organic matter.	Shallow organic soils which are acid throughout — typical of cold, wet situations of the SE highlands. Erosion likely when native vegetation destroyed.
Humic Gleys	Possibly Uf6.4 Uf6.6	Organic soils with evidence of gleying in subsoil.	Very variable — occur in low-lying areas subject to periodic inundation. May be subject to gullying on lower slopes. See also gleyed podzolic soils.
Peat Soils	O Acid O Neutral	Highly organic soils in swampy situations.	Soils dominated by plant remains for at least the top 30 cm. Few erosion problems due to situation.

GRADATIONAL SOILS (G)

Great Soil Group	Northcote Codings	Further Diagnostic Features	Notes
Prairie Soils	Gn3.2, Gn3.3 Gn3.4, Gn3.9 and Gn3.0 soils, Gn4.3, Gn4.4 Possibly Uf6.22	Thick dark well-structured A horizon. No free carbonates in solum.	Gn3.42 is modal profile code. These are dark soils, often associated with black earths or krasnozems in higher rainfall areas. Form on basic parent materials usually. Some browner soils are included (Gn3.2). Bleached prairie soils — Gn3.3, minimal prairie soils — Gn3.9, Gn3.0. Erodible under intensive usage.
Solonised Brown Soils	Gc1 and Gc2 soils Possibly Gn1.13	Soils dominated by calcareous material throughout the profile.	Mallee soils. These develop from calcareous parent material of transported origin, in semi-arid regions. Gc1.12 typical. Subject to wind erosion when native vegetation removed.
Red Earths	Gn2.1 soils	Dark A_1 horizon normally deep and sandy textured. B horizon is generally massive, but with an earthy porous fabric.	Calcareous red earths (normally Gn2.13) also occur. Lateritic or podzolic variants occur, but less frequently. Gn2.11 is modal profile code. Gn2.17, 2.18, 2.19 — red leached earths, also Gn2.111, 2.112, 2.113. Stable soils, especially under pasture.
Yellow Earths	Gn2.2, Gn2.3, Gn2.4, Gn2.5, Gn2.6, Gn2.7 soils	Massive, highly porous, earthy soils with weak profile differentiation.	Similar to red earths except for colour. Red and yellow earths are typically deep soils developed from siliceous parent materials. Gn2.3, 2.7 — yellow leached earths. Gn2.4 — yellow-brown earths. Gn2.5 — yellow-brown leached earths.
Euchrozems	Gn3.12 and Gn3.13 soils Gn3.15–Gn3.17 Gn4.12, Gn4.13 Possibly Uf6.21, Uf6.31	Types with A_2 horizons (sometimes bleached) are included, but are not typical.	Gn3.12 is the modal profile code. These are the less acid and less friable equivalent of the krasnozems (see below), generally found on basic rocks or colluvium. Erosion likely on cultivated slopes.
Xanthozems	Gn3.5, Gn3.6, Gn3.7 soils Gn3.8 Possibly Uf5.23	Types with A_2 horizons (sometimes bleached) are included, but are not typical.	These soils are essentially yellow krasnozems, which have been formed under less well drained conditions (see below). Gn3.71 typical. Not common in NSW.
Krasnozems	Gn3.10, Gn3.11, Gn3.14, Gn4.11, Gn4.14, Gn4.5, Gn4.6 Possibly Uf5.21	Types with A_2 horizons (sometimes bleached) are included, but are not typical.	Gn4.11 is the modal profile code. Outside the tropics these soils are found mainly on basic parent materials, where annual rainfall is over 1200 mm. Typically they are red, deep, well-structured, acid and porous clay soils. Minimal krasnozems — Gn4.5, Gn4.6. Erosion likely on cultivated slopes.

DUPLEX SOILS (D)

Solonetzic Soils	Dr soils ending: 2.43, 2.83, 3.33, 3.42, 3.43, 3.83, 4.43, 5.32, 5.33, 5.42, 5.43 Dy soils ending: 2.33, 2.43, 2.73, 2.82, 2.83, 3.32, 3.33, 3.42, 3.43, 3.73, 3.82, 3.83, 4.33, 4.43, 4.83, 5.32, 5.33, 5.43, 5.82, 5.83	A_2 horizon: poorly developed, less than 3 cm thick (solonetz); well developed, more than 3 cm thick (solonetzic). A/B boundary abrupt (all soils). B horizon: v. hard/tough consistence (all soils). Prismatic to columnar structure (solonetzic).	Solonetz soils are normally formed on medium-textured colluvium or alluvium associated with sedimentary parent material. Solodised solonetz soils are characterised by the domed columnar structure of the B horizon. The whole solonetzic group is typically poor in both chemical and physical fertility, erodibility

DUPLEX SOILS (D) (cont.)

Great Soil Group	Northcote Codings	Further Diagnostic Features	Notes
	Db and Dd soils ending: 1.33, 1.43, 2.32, 2.33, 2.42, 2.43, 3.33, 3.42, 3.43, 4.43	Columnar or blocky structure with doming (solodised solonetz). Laboratory confirmation of the sodic horizon (ESP > 6) is desirable; otherwise these soils may fall into other non-sodic great soil groups (such as podzolic soil).	is extreme, dispersibility moderate to high and tunnelling common. Soils with an acid reaction trend are excluded. The group intergrades closely with solodic soils (see below). Individual members are named according to the colour of the B horizon (such as red solonetzic soil).
Red-Brown Earths	Dr soils ending: 2.13, 2.22, 2.23, 2.63, 3.13, 3.22, 3.23, 4.33, 4.73 Db soils ending: 1.13, 1.22, 1.23, 1.33, 1.62, 2.13, 2.22, 2.23, 3.22, 3.23 Possibly Dd4.13	A_2 horizon: poorly developed. A/B boundary abrupt to clear. B horizon: hard/firm to friable consistence. Blocky structure. Soils ending .33 or .73 tend to have solonetzic features, and could be regarded as intergrades.	Dr2.23 is the modal profile code. These soils should not be confused with red solodic soils. The important features are the character of the A_2 horizon, structure in the B horizon, and reaction trend. They form on a wide range of parent materials, except for the most acid and most basic igneous rocks. Erosion is widespread on these soils, particularly under continuous cultivation or overgrazing. Structural deterioration of the A_1 horizon can be very rapid.
Solodic Soils	Dr soils ending: 2.32, 2.33, 2.42, 2.53, 2.62, 2.72, 2.73, 2.82, 2.83, 3.32, 3.62, 4.42, 4.72, 4.82, 5.22, 5.23, 5.62 Dy soils ending: 2.13, 2.23, 2.32, 2.42, 3.13, 3.23, 3.53, 3.63, 3.82, 4.13, 4.23, 4.32, 4.42, 5.13, 5.23, 5.42, 5.63, 5.82 Dd soils ending: 1.13, 1.23, 1.32, 1.33, 1.42, 2.12, 2.13, 2.22, 2.32, 3.13, 3.23, 3.32, 4.13, 4.23, 4.32, 4.63	A_2 horizon: well developed. A/B boundary abrupt. B horizon: hard/v. tough consistence. Blocky structure but sometimes massive. Laboratory confirmation of the sodic nature of the B horizon (ESP > 6) is desirable; otherwise these soils may fall into other non-sodic great soil groups (such as red-brown earth).	These soils are similar to, and intergrade with, solodised solonetz soils. The structure of the B horizon tends to be less coarse. Subsoils are usually dispersible, giving rise to tunnelling and high erodibility. They form on a wide range of parent materials except the most basic. Individual members are named according to the colour of the B horizon (such as yellow solodic soil).
Solods (soloths)	Dr soils ending: 2.14, 2.81, 3.31, 3.41, 3.71, 3.81, 4.41, 4.81, 5.41, 5.81 Dy soils ending: 2.41, 2.81, 3.31, 3.41, 3.71, 3.81, 4.41, 4.81, 5.31, 5.41, 5.71, 5.81 Db soils ending: 1.41, 1.81, 2.31, 2.41, 3.40, 3.41 Dd soils ending: 1.31, 1.41, 1.81, 2.31, 2.41, 3.41	A_2 horizon: well developed. A/B boundary abrupt. B horizon: hard/v. tough consistence. Blocky structure but sometimes massive. Laboratory confirmation of the sodic nature of the B horizon (ESP > 6) is desirable; otherwise these soils may fall into other non-sodic great soil groups (such as yellow podzolic soil).	Solods are the acid members of the solodic group, and have somewhat less abrupt A/B boundaries, and more friable consistence in the B horizon. These soils are less dispersible than solodic soils generally.
Desert Loams	Dr1 soils (excepting those below)	No pans. Alkaline.	Dr1.33 is the modal profile code. These soils are found on fine-textured sedimentary rocks and alluvium of the arid inland. Crusting surface can give rise to low infiltration and high runoff rates.
Hardpan Soils	Dr1 soils ending: .14, .15, .16, .54, .55, .56	Duplex soils with a hardpan in, or immediately below, the clayey B horizon.	See also Um5.3. These soils are common in the central part of WA, but not as common in NSW.
Non-calcic Brown Soils	Dr2 or Db1 soils ending in .12, .52 Possibly Db1.22, Dr2.22	No A_2 horizon, solum is carbonate-free.	Dr2.12 is the modal profile code. Differ from red-brown earths in lack of A_2 horizon, and less alkaline reaction trend. (In WA, Dr2.22 is typical.) Erodibility similar to that of red-brown earths.

DUPLEX SOILS (D) (Cont.)

Great Soil Group	Northcote Codings	Further Diagnostic Features	Notes
Chocolate Soils	Db3.1 soils Dr4.1, Dr4.2, Dr4.5 and Dr4.6 soils Also Db3.21	A_2 horizons are generally indistinct. Good surface crumb structure.	Shallow well-structured and friable reddish brown soils on basalt and similar rocks. Some types are more uniform, tending towards Um6 or Uf6. Normal chocolate soils are clay-loam over clay soils with an acid reaction trend. Probably the most stable of the duplex soils.
Grey-Brown Podzolic Soils	Db1 and Db2 soils ending .21, .61 Also Db1.31, Db4.21	A_2 horizon well developed. A/B boundary abrupt to clear. B horizon has: firm/plastic consistence. Blocky structure.	A_2 horizon sometimes bleached. Drab colours predominate in these soils as compared with red and yellow podzolic soils. Are sometimes associated with brown podzolic soils in drier parts of the eastern highlands. They are occasionally dispersible, and occur on a range of parent materials. Gully erosion is fairly common in these soils.
Red Podzolic Soils	Dr2, Dr3, Dr5 soils ending in .11, .21, .31, .61 Also Dr2.51, Dr2.71 (excepting Dr3.31, Dr5.31, Dr3.51)	A_2 horizon well developed. A/B boundary clear to gradual. B horizon has: firm/friable consistence. Fine angular blocky structure.	Dr2.21 is the modal profile code. These soils are typical of coastal and subcoastal regions in NSW often grading into red earths or krasnozems in wetter areas, into red solodic soils in drier areas, or into yellow podzolic soils in lower catenary positions. They are more stable than other podzolic soils, and occur on a wide range of parent materials.
Yellow Podzolic Soils	Dy2, Dy3, Dy4, Dy5 Soils ending in .11, .12, .21, .22, .51, .52, .61, .71, Dy2.31	A_2 horizon well developed. A/B boundary clear to gradual. B horizon has: firm/friable consistence. Fine polyhedral or blocky structure.	Dy2.21 or Dy3.21 is the modal profile code. These soils form in catenary association with red podzolic soils in poorly drained situations, and on a wide range of parent materials. A_2 horizon is less bleached than that of the yellow solodic soil, often more so than in the red podzolic soil. Are sometimes dispersible, especially in coastal areas.
Brown Podzolic Soils	Db soils ending in .11, .51 (except Db3)	A_2 horizon not evident. A/B boundary gradual to diffuse. B horizon has: Friable consistence. Weak to blocky structure.	The A horizon has higher organic matter content than the other podzolic soils, which masks the strong horizonation typical of them. These soils are found on the eastern highlands in cool, moist situations (e.g. Monaro, New England plateau). The brown colouration applies to the whole of the profile. They tend to be more stable than other podzolic soils.
Lateritic Podzolic Soils	Dy2, Dy3, Dy4 and Dy5 Soils ending in .84, .85, .86 Some Dy3.61, 5.41, 5.61	Ironstone layer at base of thick sandy A horizon. Mottled kaolin clay layer at base of B horizon, but this is sometimes absent.	Any podzolic soil with laterite included. Dy3.84 is the modal profile code. These are generally relict soils not common in NSW.
Gleyed Podzolic Soils	Dg soils	Gleying: coarse mottling throughout B horizon. Evidence of high water table (grey or blue colours).	Any podzolic soil showing gleyed features is included. The gleying may be due to prolonged seepage in the soil, or to high water table. See also Humic Gleys. May be gully eroded on lower lands.

Glossary of Soil Science Terms

THE FOLLOWING TERMS are taken from *A Glossary of Terms Used in Soil Conservation* by Houghton and Charman (1986).

A, A₁, A₂ Horizon See **Soil Profile**.

Acid Soil A soil giving an acid reaction throughout most or all of the soil profile (precisely, below a pH of 7.0; practically, below a pH of 6.5). Generally speaking, acid soils become a problem when the pH drops below 5.5. At this level, and particularly below 5.0, the following specific problems may occur — aluminium toxicity, manganese toxicity, calcium deficiency and/or molybdenum deficiency. Such problems adversely affect plant growth and root nodulation, which may result in a decline in plant cover and increase in erosion hazard.

The term is frequently used to describe soils with the acidity problems just described. Correction of the acidity is normally carried out by the application of appropriate amounts of lime to bring the soil pH to a level of 6.0–6.5. See also **pH**.

Acidity The chemical activity of hydrogen ions in soil expressed in terms of pH. See also **Acid Soil**, **pH**.

Adsorption The interaction of ions with the surfaces of soil materials, particularly with organic matter and inorganic colloidal substances comprising part of the clay fraction. In this way, nutrients in solution, made up of ions, become attracted to sites on soil particles with an opposite charge.

Aerobic Describes soil conditions in which free oxygen is plentiful and chemically, oxidising processes prevail. Such conditions are usually found in well-drained soils with good soil structure. Ant. **Anaerobic**.

Alkaline Soil A soil giving an alkaline reaction throughout most or all of the soil profile (precisely, above a pH of 7.0; practically, above a pH of 8.0). Many alkaline soils have a high pH indicated by the presence of calcium carbonate, and are suitable for agriculture. However, others are problem soils because of salinity and/or sodicity. Soils with a pH above 9.5 are generally unsuitable for agriculture. See also **Sodic Soil**, **Saline Soil**, **pH**.

Alkalinity The chemical condition of soil with a pH greater than 7.0. Often associated with saline soils and sodic soils. See also **pH**.

Alluvial Describes material deposited by, or in transit in, flowing water. See also **Alluvium**.

Alluvial Soil A soil developed from recently deposited alluvium, normally characterised by little or no modification of the deposited material by soil-forming processes, particularly with respect to soil horizon development.

Alluvium An extensive stream-laid deposit of unconsolidated material, including gravel, sand, silt and clay. Typically it forms floodplains that develop alluvial soils.

Ameliorant Syn. **Soil Ameliorant**.

Anaerobic Describes soil conditions in which free oxygen is deficient and chemically, reducing processes prevail. Such conditions are usually found in waterlogged or poorly drained soils in which water has replaced soil air. Ant. **Aerobic**.

Antecedent Moisture Content The moisture content of a soil prior to a rainfall event. It has an important influence on the likelihood of runoff occurring as a result of subsequent rainfall, because antecedent moisture restricts the amount of infiltration which can take place.

Apedal Describes a soil in which none of the soil material occurs in the form of peds in the moist state. Such a soil is without apparent structure and is typically massive or single-grained. When disturbed, it separates into primary particles or fragments which may be crushed to primary particles. Ant. **Pedal**.

Aquifer A porous soil or geological formation, often lying between impermeable subsurface strata, which holds water and through which water can percolate slowly over long distances and which yields groundwater to springs and wells. Aquifers may, however, be unconfined and the water level subject to seasonal inflow. An AQUITARD is also a groundwater-bearing formation, but is insufficiently permeable to transmit and yield water in usable quantities.

Association Syn. **Soil Association**.

Atterberg Limits The soil water contents at the solid/plastic and plastic/liquid boundaries. Atterberg limits are based on the concept that a fine-grained soil can exist in any of three states depending on its water content. Thus, on the addition of water, a soil may proceed from the solid state through to the plastic and, finally, liquid states. The water contents at the boundaries between adjacent states are termed the plastic limit and the liquid limit. Water content is expressed as a percentage of the oven-dry weight of soil.

PLASTIC LIMIT (PL)
The plastic limit of a soil is the water content at which the soil passes from the solid to the plastic state. It is arbitrarily defined as the lowest water content at which the soil can be rolled into threads 3 mm in diameter without the threads breaking into pieces.
(Australian Standard AS 1289 C2.1 — 1977)

LIQUID LIMIT (LL)
The liquid limit of a soil is the water content at which the soil passes from the plastic to the liquid state. It is arbitrarily defined as the water content at which two halves of a soil cake will flow together for a distance of 12 mm along the bottom of the groove separating the two halves, when dropped 25 times from a distance of 10 mm at the rate of 2 drops per second, using standard apparatus.
(AS 1289 C1.1 — 1977)

PLASTICITY INDEX (PI)
The plasticity index of a soil is the numerical difference between the plastic limit and the liquid limit. (AS 1289 C3.1 — 1977)

These limits are used to categorise soil materials in terms of their likely engineering behaviour at different moisture contents, such as in building or road foundations. Prior to earthwork construction, their determination assists the specification of required compaction levels, batter slopes and optimum moisture levels for construction. See also **Unified Soil Classification System**, **Plastic**, **Soil Plasticity**, **Linear Shrinkage**.

Available Nutrient The portion of any element or compound in the soil that can be taken up and assimilated by plants to enhance their growth and development.

Available Soil Water	That part of the water in the soil that can be absorbed by plant roots. The amount of water held between the moisture content prevailing at any point in time and the moisture content at which plant growth ceases.
	Plants have difficulty extracting moisture when the available soil water approaches wilting point. Where soil moisture can be controlled, the aim is to maintain it at a level where it can be readily extracted by plants, that is, to prevent moisture depletion below about 50 per cent of the available range. See also **Field Capacity**.
Available Water-holding Capacity	The amount of water in the soil, generally available to plants, that can be held between field capacity and the moisture content at which plant growth ceases. Sometimes also known as the **Plant-available Water Capacity**.
B, B$_1$, B$_2$ Horizon	See **Soil Profile**.
Bedrock (Country Rock)	Solid rock underlying the soil profile or other surface materials. It does not necessarily represent the parent material of the overlying soil.
Bentonite	A clay usually formed by the weathering of volcanic ash, and which is largely composed of montmorillonite type clay minerals. It has great capacity to absorb water and swell accordingly. For this reason, it is used to seal dams and/or earth embankments built of coarse materials or which contain a coarse-textured seam causing them to leak.
Biological Fertility	See **Soil Fertility**.
Bleaching	The near-white colouration of an A$_2$ horizon which has been subject to chemical depletion as a result of soil-forming processes including eluviation. The colour is defined for all hues as having a value of 7 or greater with a chroma of 4 or less, on dry soils. CONSPICUOUS BLEACHING means that 80 per cent or more of the horizon is bleached, whereas SPORADIC BLEACHING means that less than 80 per cent of the horizon is bleached, with affected portions appearing irregularly through the horizon. See also **Soil Colour**.
Bolus	A small handful of soil which has been moistened and kneaded into a soil ball which just fails to stick to the fingers. The behaviour of the bolus, and of the ribbon produced by pressing it between thumb and forefinger, characterises soil texture.
Buffer Capacity	The ability of a soil to resist changes in pH. The buffering action is due mainly to the properties of clay and fine organic matter. Thus, with the same pH level, more lime is required to neutralise a clayey soil than a sandy soil, or a soil rich in organic matter, than one low in organic matter.
Bulk Density	The mass of dry soil per unit bulk volume. The bulk volume is determined before drying to constant weight at 105°C. The unit of measurement is usually grams per cubic centimetre.
	Bulk density is a measure of soil porosity, with low values meaning a highly porous soil and vice versa. It does not, however, give any indication of the number, sizes, shapes, distribution or continuity of soil pores.
	This parameter is also used as an indicator of the structural condition of a soil, with low values indicating a better state of aggregation than high values. The range for soils in natural condition would typically be from 1 to 2 g/cm^3.
Buried Soil (Paleosol)	One or more layers of soil which were formerly at the surface, but which have been covered by a more recent deposition, usually to a depth greater than the thickness of the solum.
C Horizon	See **Soil Profile**.

Calcareous Refers to materials, particularly soils, containing significant amounts of calcium carbonate. Describes rocks composed largely of, or cemented by, calcium carbonate. A calcareous soil is one containing carbonate in sufficient quantity to effervesce visibly when treated with cold dilute (N) hydrochloric acid.

Capillary Fringe The zone above the water table in an unconfined aquifer where the formation is saturated, but the water in the zone is at less than atmospheric pressure.

Capillary Porosity See **Soil Porosity**.

Catena A repetitive sequence of soils generally of similar age and parent material, encountered between hillcrests and the valley floor.

The soils in the sequence occur under similar climatic conditions, but have different characteristics due to variation in relief, drainage and the past history of the land surface. Such variations are normally manifest in differential transport of eroded material and the leaching, translocation and redeposition of mobile chemical constituents.

In soil mapping, the use of this term has been largely replaced by the more general term *toposequence*.

Cation Exchange Capacity The total amount of exchangeable cations that a soil can adsorb, expressed in centimoles of positive charge per kilogram of soil. Cations are positive ions such as calcium, magnesium, potassium, sodium, hydrogen, aluminium and manganese, these being the most important ones found in soils. Cation exchange is the process whereby these ions interchange between the soil solution and the clay or organic matter complexes in the soil. The process is very important as it has a major controlling effect on soil properties and behaviour, stability of soil structure, the nutrients available for plant growth, soil pH and the soil's reaction to fertilisers and other ameliorants added to the soil.

Cemented Describes soil materials having a hard, brittle consistency because the particles are held together by cementing substances such as humus, calcium carbonate or the oxides of silicon, iron and aluminium. The hardness and brittleness persist even when the soil is wet. Cf. **Indurated**.

Chemical Fertility See **Soil Fertility**.

Chroma See **Soil Colour**.

Clay Soil material consisting of mineral particles less than 0.002 mm in equivalent diameter. This generally includes the chemically active mineral part of the soil. Many of the important physical and chemical properties of a soil depend on the type and quantity of clay it contains. Three broad classes of clay type are recognised, namely, montmorillonite, kaolinite and illite.

When used as a soil texture group, such soil contains at least 35 per cent clay and no more than 40 per cent silt. See also **Soil Texture**.

Clay Loam A soil texture group representing a well-graded soil composed of approximately equal parts by weight of clay, silt and sand. See also **Soil Texture**.

Claypan A pan made up of a concentration of dense clays in the subsoil. The term is also used for the impermeable clay surface produced as a result of scalding, although this usage is colloquial.

Clod A large compact and coherent soil aggregate produced artificially, usually by ploughing or digging soils that are either too wet or too dry for normal tillage operations. Cf. **Ped**.

Coffee Rock A type of brownish sand-rock formed where iron oxides and organic matter which have leached through the soil profile are precipitated at, or above, a fluctuating water table. A typical feature of some older coastal sands in which podzols have formed.

Coherence Syn. **Soil Coherence**.

Colloidal Material The finest clay and organic material with a particle size generally less than 10^{-4} mm in diameter. Such material represents the finest particles removed in an erosion event, and as such remains permanently in suspension, unless subject to flocculation.

Colluvial Describes material transported largely by gravity. See also **Colluvium.**

Colluvium Unconsolidated soil and rock material, moved largely by gravity, deposited on lower slopes and/or at the base of a slope. Cf. **Alluvium**.

Compaction The process whereby the density of soils is increased by tillage, stock trampling and/or vehicular traffic, often resulting in the formation of plough-pans. Such compaction gives rise to lower soil permeability and poorer soil aeration with resultant increases in erosion hazard and lowered plant productivity. Deep ripping and conservation tillage are used to alleviate the condition.

Concretion Syn. **Nodule**.

Consistence Syn. **Soil Consistence**.

Conspicuous Bleaching See **Bleaching**.

Corrasion That part of natural erosion processes in which earth and rock materials are worn away due to their abrasion when carried in flows of water, air or ice.

Country Rock Syn. **Bedrock**.

Crabhole Gilgai See **Gilgai**.

Crumb A soft, porous, more or less rounded soil aggregate from 1 to 5 mm in diameter. See also **Crumb Structure, Soil Structure**.

Crumb Structure A soil structural condition in which most of the soil aggregates are soft, porous and more or less rounded units from 1 to 5 mm in diameter. The typical surface condition of medium-textured soils recently cultivated after a period of well-managed pasture.

Cutans A surface modification of a soil aggregate due to the concentration of one of its constituents.

Crusting See **Surface Sealing**.

Cyclic Salt Salt deposited on soils from wind or rainfall. Near the sea or inland salt lakes, where the amounts deposited are likely to be significant, such salt may subsequently be leached into the soil and take part in various chemical processes there.

D Horizon See **Soil Profile**.

Debris Loose and unconsolidated material arising from the distintegration of rocks, soil, vegetation or other material transported and deposited in an erosion event. It is generally superficial and contains a significant proportion of coarse material.

In the classification of mass movement, debris refers to material in which 20 to 80 per cent of the fragments are greater than 2 mm in size and the remainder of the material is less than 2 mm. This distinguishes it from earth type movements, where about 80 per cent or more of the material must be smaller than 2 mm in size. (Reference: Shroder, J.F. (1971), 'Landslides of Utah', *Utah Geological and Mineralogical Survey Bulletin* **90**.) Cf. **Detritus.**

Deflation The removal of fine particles from soil by wind.

Deflocculation The process by which masses of colloidal, or very fine, clay particles or 'flocs' separate in water into their constituent particles which go into suspension. It depends on the balance between exchangeable cations on the clay and in solution, and on the overall ionic strength of the solution. Clays high in sodium deflocculate readily. Ant. **Flocculation**. See also **Sodic Soils, Dispersible Soils**.

Detachability Syn. **Soil Detachability**.

Detritus Loose material arising from the mechanical weathering of rocks and transported and deposited in an erosion event. Cf. **Debris**.

Dispersibility Syn. **Soil Dispersibility**.

Dispersible Soil A structurally unstable soil which readily disperses into its constituent particles (clay, silt, sand) in water. Highly dispersible soils are normally highly erodible and are likely to give problems related to field and earthwork tunnelling. See also **Soil Dispersibility, Sodic Soil**.

Dispersion Percentage A measure of soil dispersibility representing the proportion of clay plus fine silt (< 0.005 mm approx.) in a soil which is dispersible, expressed as a percentage. It is determined in the laboratory by comparing the amount of fine material, in a soil sample, dispersed by a ten-minute shaking in water, to the amount dispersed by a 120-minute shaking in water containing dispersant. Highly dispersible clays have a high dispersion percentage. See also **Soil Dispersion**.

Dump Gypsum See **Gypsum**.

Duplex Soil A soil in which there is a sharp change in soil texture between the A and B horizons (such as loam overlying clay). The soil profile is dominated by the mineral fraction with a texture contrast of 1.5 soil texture groups or greater between the A and B horizons. Horizon boundaries are clear to sharp. The texture change from the bottom of the A horizon to the top of the B horizon occurs over a vertical distance of 10 cm or less. See also **Soil Profile, Primary Profile Form, Gradational Soil, Uniform Soil**.

E.S.P. (ESP) Syn. **Exchangeable Sodium Percentage**.

Earth A general term commonly used to describe a range of soil materials. In pedological terms, it is used to describe great soil groups such as Black Earths and Red Earths. The term also refers to gradational soils with an earthy fabric in their B horizons.
In the classification of mass movement, earth refers to material in which about 80 per cent or more of the particles are smaller than 2 mm in size. This distinguishes it from debris type movements which contain 20 to 80 per cent of fragments greater than 2 mm. (Reference: Shroder, J.F. (1971), 'Landslides of Utah', *Utah Geological and Mineralogical Survey Bulletin* **90**.) As part of the classification, earth can be subdivided on the basis of its constituent separates, ranging from non-plastic sand to highly plastic clay.

Edaphic Pertaining to the soil.

Effective Soil Depth The depth of soil material that plant roots can penetrate readily to obtain water and plant nutrients. It is the depth to a layer that differs sufficiently from the overlying material in physical or chemical properties to prevent or severely retard the growth of roots.

Electrical Conductivity A measure of the conduction of electricity through water or a water extract of soil. It can be used to determine the soluble salts in the extract and hence soil salinity. The unit of electrical conductivity is the siemens and soil salinity is normally expressed as decisiemens per metre at 25°C. (Symbol: E_c, Units: dS/m.)

Conductivity values of 1.5 (for a 1:5 soil:water suspension) or 4.0 (for a saturation extract) indicate the likely occurrence of plant growth restrictions. See also **Saline Soil**.

Elutriator An instrument which lifts particles from a fluidised bed of dry soil in a rising current of air. By changing the rate of lift, particles of different sizes may be separated to enable soil to be characterised by particle size distribution or to determine the proportion of suspension fraction capable of being mobilised during wind erosion. This process is called elutriation.

Eluviation The downward removal of soil material in suspension, or in solution, from a layer or layers of a soil. The loss of material in solution is described by the term leaching. Some of the eluviated materials are typically deposited in lower layers or horizons. Cf. **Illuviation**.

Emerson Aggregate Test A classification of soil aggregates based on their coherence in water. Small dry aggregates are placed in dishes of distilled water and their behaviour observed. The conditions under which they slake, swell and disperse allow the different aggregates to be separated into eight classes. The test is particularly valuable in a soil conservation context as it grades soil aggregates according to their stability in water.

The test uses natural peds, with the first separation being based on slaking. Those aggregates which do not slake are placed in Class 7 if they swell and in Class 8 if they do not.

Of those which do slake, which form the majority of soils, those which show complete soil dispersion are placed in Class 1 and those showing only partial dispersion are placed in Class 2. Those showing no dispersion are remoulded at field capacity and reimmersed in water.

Aggregates which disperse after remoulding are placed in Class 3 and those which do not are further separated by the presence or absence of carbonate or gypsum. Those with carbonate or gypsum fall into Class 4, while those without are made up into a 1:5 suspension aggregates:water. Those soils which then show dispersion are placed in Class 5 and those which show flocculation fall into Class 6.

(Reference: Emerson, W.W. (1967), 'A classification of soil aggregates based on their coherence in water', *Aust. J. Soil Res.* **5**, 47–57.)

In general, the degree of stability of soils increases from Class 1 through to Class 8. In a soil conservation context, it may be useful to further subdivide Classes 2 and 3 according to the degree of dispersion observed.

Entrainment The process by which detached soil particles are drawn into the flow of air or water during an erosion event.

Erodibility Index A quantitative expression of soil erodibility based on measured soil properties. See also **Soil Erodibility**.

Exchange Capacity The total ionic charge of a soil, expressed in centimoles of charge per kilogram of soil. Its numerical value is identical to the value expressed in milliequivalents per 100 grams of soil. See also **Cation Exchange Capacity, Adsorption**.

Exchangeable Sodium Percentage (ESP) The proportion of the cation exchange capacity occupied by sodium ions, expressed as a percentage. Sodic soils are categorised as those with an ESP from 6 to 14 per cent, strongly sodic soils are those with an ESP of 15 per cent or more. Soils with a high ESP are typically unstable and, as a consequence, have high erodibility and often present problems in soil conservation earthworks. See also **Soil Dispersibility, Dispersible Soil**.

Expansive Soil (Swelling Soil) A soil which significantly changes its volume with changes in moisture content. It typically cracks when drying out, and expands on wetting. The shrinking/swelling characteristic is normally due to the presence of montmorillonite type clays in the soil, and is characterised by testing for linear shrinkage.

Extended Principal Profile Form See **Principal Profile Form**.

Fabric Syn. **Soil Fabric**.

Factual Key The alpha-numeric coding system for recognition and classification of Australian soils based on observable soil profile features. It has a hierarchical structure which uses the bifurcating principle to successfully separate out primary profile forms, subdivisions, sections, classes and principal profile forms. The soil profile features used include soil texture, colour, structure, pH, presence of pans, consistence, coherence and other special features related to both A and B horizons. Additional features may be added to give an extended principal profile form.
(Reference: Northcote, K.H. (1979), *A Factual Key for the Recognition of Australian Soils*, Rellim Tech. Publns, SA.)

A description of Australian soils based on the Factual Key is presented in Northcote *et al.* (1975), *A Description of Australian Soils*, Wilke and Co. Ltd, Vic.

Fertility Syn. **Soil Fertility**.

Field Capacity (Water-holding Capacity) The amount of water held in a soil after any excess has drained away following saturation, expressed as a percentage of the oven-dry weight of the soil. As a general rule, soils are considered to be at field capacity after draining for 48 hours. Cf. **Available Water-holding Capacity**. See also **Soil Water Potential**.

Flocculation The process by which colloidal or very fine clay particles, suspended in water, come together into larger masses or loose 'flocs' which eventually settle out of suspension. They are easily redispersed. Flocculation depends on the balance between exchangeable ions on the clay and those in solution, and on the overall ionic strength of the solution. Ant. **Deflocculation**.

Gilgai Surface microrelief associated with some clayey soils, consisting of hummocks and/or hollows of varying size, shape and frequency. This phenomenon is a continuing long-term process due to the shrinking and swelling of deep subsoils with changes in moisture content. It is usually associated with the occurrence of expansive soils.

Normal gilgai are irregularly spaced and have subcircular depressions with vertical intervals usually less than 300 mm and horizontal intervals usually 3 to 10 m. They are associated with flat or very gently sloping terrain. Gilgai that deviate significantly from this pattern include the following main types.

CRABHOLE GILGAI
Small mounds and depressions separated by a more or less continuous shelf and the horizontal interval extends from 3 to 20 m.

LINEAR GILGAI

Long narrow parallel elongate mounds and broader elongate depressions more or less at right angles to the contour, usually in sloping terrain. Vertical interval is usually less than 300 mm and horizontal interval usually 5 to 8 m.

LATTICE GILGAI

Discontinuous elongate mounds and/or elongate depressions more or less at right angles to the contour. Usually in sloping terrain, commonly between linear gilgai on lower slopes and plains.

MELONHOLE GILGAI

Irregularly distributed large depressions, usually greater than 3 m in diameter, subcircular or irregular, and varying from closely spaced in a network of elongate mounds to isolated depressions set in an undulating shelf with occasional small mounds. Vertical interval is usually greater than 300 mm and horizontal interval usually 6 to 50 m.

Gleying   The grey or greenish grey colouration of soils often produced under conditions of poor drainage, which gives rise to chemical reduction of iron and other elements. It typically occurs in the clayey lower B horizons of soils situated in low topographic positions where the water table remains high for much of the year. See also **Soil Colour**.

Gradational Soil   A soil in which there is a gradual change in soil texture between the A and B horizons (for example, loam over clay loam over light clay). The soil is dominated by the mineral fraction and shows more clayey texture grades on passing down the solum of such an order that the texture of each successive horizon changes gradually to that of the one below. Horizon boundaries are usually gradual or diffuse. The texture difference between consecutive horizons is less than 1.5 soil texture groups, while the range of texture throughout the solum exceeds the equivalent span of one texture group. See also **Primary Profile Form, Soil Profile, Duplex Soil, Uniform Soil**.

Grading   Refers to the distribution of particle sizes in a soil. A well-graded soil consists of particles that are distributed over a wide range in size or diameter. Such a soil's density or bearing properties can normally be easily increased by compaction. A poorly graded soil consists mainly of particles nearly the same size, or is deficient in particles of a certain size. Because of this, the soil's density can be increased only slightly by compaction. An assessment of the grading of a soil can be made by particle size analysis and is important to determine the behaviour of a given soil when used in soil conservation earthworks.

Gravel   A mixture of coarse mineral particles larger than 2 mm, but less than 75 mm in equivalent diameter.

Great Soil Group   A soil classification category in which soils are classified according to their mode of formation as reflected in major morphological characteristics and profile form.

The grouping depends on an appraisal and interpretation of features such as the colour, texture, structure and consistence of soil material, the various horizons in the soil profile and the nature of the boundary between horizons. A widely used classification of Australian soils, major developments of which have been set out in *A Manual of Australian Soils* (Stephens, C.G., 1962) and *A Handbook of Australian Soils* (Stace, H.C.T. *et al.*, 1968).

Gypsum   A naturally occurring soft crystalline mineral which is the hydrated form of calcium sulfate, having the formular $CaSO_42H_2)$. Deposits occur mainly in arid inland areas of Australia. contains approximately 23 per cent calcium and 18 per cent sulfur.

This mineral is also a by-product of the manufacture of phosphoric acid. Such by-product gypsum, also called DUMP GYPSUM, is more variable in quality, particularly with respect to moisture content.

Gypsum is normally used as a soil ameliorant to improve soil structure and reduce crusting in hardsetting clayey soils. The applied calcium increases soil aggregation, which results in improved water infiltration, seed germination and root growth. Typical rates used are up to 5 t/ha, with heavier rates being required on highly sodic soils.

Gypsum is also a useful source of nutrient calcium and sulfur, and can also be used for clearing muddy water in dams.

NB It is suggested that any unknown gypsum be tested for flocculating potential before use.

Hardpan Syn. **Pan.**

Hardsetting The condition of a dry surface soil when a compact, hard and apparently apedal condition prevails. Because of this characteristic, such soils tend to give rise to high rates of runoff compared with better structured soils. Clods formed by the tillage of hardsetting soils usually retain the condition until completely broken down by further tillage operations.

Soil on which surface sealing develops may or may not be hardsetting—a surface seal is not a criterion for the hardsetting condition. The majority of soils throughout the wet-dry climatic zones of Australia set hard in the dry season. Soils which do not set hard are pedal in the dry, as well as in the moist, state (clay loams, clays), or are single-grained (sands). Cf. **Self-mulching**.

Horizon A general term used to describe individual layers within a soil. See also **Soil Horizon**.

Hue See **Soil Colour**.

Hydraulic Conductivity The flow of water through soil per unit of energy gradient. For practical purposes, it may be taken as the steady-state percolation rate of a soil when infiltration and internal drainage are equal, measured as depth per unit time.

Hydrophobic Soils Syn. **Water-Repellent Soils.**

Illite Clay material comprising a group of aluminosilicate-mica minerals of indefinite chemical composition with a 2:1 crystal lattice structure. Its properties are generally intermediate between those of kaolinite and montmorillonite.

Illuviation The process of deposition of soil material in the lower horizons of a soil as a result of its removal from upper horizons through eluviation. Materials deposited may include clay, organic matter and iron and aluminium oxides.

Impervious Describes a soil through which water, air or roots penetrate slowly or not at all. No soil is absolutely impervious to water and air all the time.

Indurated Rendered hard. Refers to the hardening of sediments in rock or soil layers into pans by heat and/or pressure and/or cementation.

Infiltrability A general term used to describe the capacity of the soil to take in water at its surface, depending largely on surface texture and structure. See also **Infiltration**.

Infiltration The downward movement of water into the soil. It is largely governed by the structural condition of the soil, the nature of the soil surface including presence of vegetation, and the antecedent moisture content of the soil.

Infiltration Rate The rate at which water moves downward into the soil at any given time. It is measured as volume flux per unit of surface area, in units of mm/h.

Runoff occurs when the rainfall rate exceeds the infiltration rate for a given soil condition. The saturated (or 'steady-state') infiltration rate is the rate which occurs when the soil is saturated and infiltration and drainage are equal. See also **Hydraulic Conductivity**.

Infiltrometer   An apparatus for measuring infiltration.

Internal Drainage (Profile Drainage)   The rate of downward movement of water through the soil governed by both soil and site characteristics. It is assessed in terms of soil water status and the length of time horizons remain wet (soil bolus exudes water when squeezed). It can be difficult to assess in the field and cannot be based solely on soil profile morphology. Vegetation and topography may be useful guides. Soil permeability, groundwater level and seepage are also important. The presence of mottling often, but not always, reflects poor drainage.

Categories are as follows:

Very poorly drained: Free water remains at or near the surface for most of the year. Soils are usually strongly gleyed. Typically a level or depressed site and/or a clayey soil.

Poorly drained: All soil horizons remain wet for several months each year. Soils are usually gleyed, strongly mottled and/or have orange or rusty linings of root channels.

Imperfectly drained: Some soil horizons remain wet for periods of several weeks. Subsoils are often mottled and may have orange or rusty linings of root channels.

Moderately well-drained: Some soil horizons may remain wet for a week after water addition. Soils are often whole coloured, but may be mottled at depth and of medium to clayey texture.

Well-drained: No horizon remains wet for more than a few hours after water addition. Soils are usually of medium texture and not mottled.

Rapidly drained: No horizon remains wet except shortly after water addition. Soils are usually of coarse texture, or shallow, or both, and are not mottled.

Kaolinite   Clay material comprising a group of aluminosilicate minerals with a 1:1 crystal lattice structure. They are generally stable clays with low shrink-swell potential and low cation exchange capacity. See also **Illite, Montmorillonite**.

Lattice Gilgai   See **Gilgai**.

Leachate   Liquid containing dissolved minerals and salts as a result of leaching. Under natural conditions, refers to seepage water containing dissolved minerals and salts.

In a mining context, the minerals and salts are dissolved as water seeps through an ore body or waste material. In some waste treatment processes, chemicals are added to promote leaching.

In a soil context, the leachate may be formed naturally by water seeping through the soil, or artificially in the laboratory in certain soil testing procedures.

In a landfill context, the leachate may result from water passing through waste material such as at a garbage disposal site.

Leaching   The removal in solution of the more soluble minerals and salts by water seeping through a soil, rock, ore body or waste material. See also **Leachate**.

Lime   A naturally occurring calcareous material used to raise the pH of acid soils and/or supply nutrient calcium for plant growth. The term normally refers to ground limestone ($CaCO_3$), but may also include processed forms such as hydrated lime ($Ca(OH)_2$) or burnt lime (CaO). The processed forms are also effective for treating dispersible soils.

The effectiveness of lime application depends on its fineness and subsequent incorporation into the top few centimetres of soil. The finer grades of lime are more effective.

Lime Requirement Test   A laboratory test used to determine the amount of lime required to raise the pH of a soil to a predetermined level, normally in the range 6.0–6.5.

Linear Gilgai   See **Gilgai**.

Linear Shrinkage   The decrease in one dimension of a soil sample when it is oven dried, at 105°C for 24 hours, from the moisture content at its liquid limit, expressed as a percentage of the original dimension. (AS 1289 C4.1 — 1977.)

Liquid Limit   See **Atterberg Limits**.

Lithosol (Skeletal Soil)   A shallow soil showing minimal profile development and dominated by the presence of weathering rock and fragments therefrom. Such soils are typically found on steep slopes, exposed hillcrests and rocky ranges where natural erosion exceeds the formation of new soil material.

Litter   The uppermost layer of organic material in a soil, consisting of freshly fallen or slightly decomposed organic materials which have accumulated at the ground surface.

Loam   A medium-textured soil of approximate composition 10 to 25 per cent clay, 25 to 50 per cent silt, and less than 50 per cent sand. Such a soil is typically well graded. See also **Soil Texture, Grading**.

Lysimeter   A device to measure the quantity or rate of water movement through or from a block of soil, usually undisturbed and *in situ*, or to collect such percolated water for quality analysis.

Massive   Refers to that condition of a soil layer in which the layer appears as a coherent, or solid, mass which is largely devoid of peds, and is more than 6 mm thick. Cf. **Self-mulching**.

Matrix   Finer grained fraction, typically a cementing agent, within a soil or rock in which larger particles are embedded.

Mechanical Analysis   Syn. **Particle Size Analysis**.

Mechanical Stability   The ability of a dry soil to maintain its structure under the influence of mechanical agents, such as tillage or abrasion from windborne materials. It relates to soil coherence and is characterised in the laboratory by repeated dry sieving on a rotary sieve. Cf. **Structural Stability**.

Melonhole Gilgai   See **Gilgai**.

Montmorillonite   Clay material comprising a group of aluminosilicate minerals with a 2:1 expanding crystal lattice structure. They are reactive clays generally with high shrink-swell potential and high cation exchange capacity. See also **Illite, Kaolinite**.

Mottling   The presence of more than one soil colour in the same soil horizon, not including different nodule colours. The subdominant colours normally occur as scattered blobs or blotches, which have definable differences in hue, value or chroma from the dominant colour. Mottling is often indicative of slow internal drainage, but may also be a result of parent material weathering.

Munsell Colour System See **Soil Colour**.

Nitrogen Fixation Generally, the conversion of free nitrogen to nitrogen combined with other elements. Specifically in soils, the assimilation of atmospheric nitrogen from the soil air by soil organisms to produce nitrogen compounds that eventually become available to plants.

Nodule (Concretion) A small segregated mass of material that has accumulated in the soil because of the concentration of one or more particular constituents, usually by chemical or biological action. Nodules vary widely in size, shape, hardness and colour, and may be composed of iron or manganese compounds, calcium carbonate or other materials.

Non-capillary Porosity See **Soil Porosity**.

Non-plastic Describes soil material which shows no plastic behaviour, irrespective of its moisture content. See also **Soil Plasticity**.

O, O_1, O_2 Horizon See **Soil Profile**.

Organic Soil A soil in which soil organic matter dominates the profile. The surface 30 cm should contain 20 per cent or more organic matter if the clay content of the mineral soil is 15 per cent or lower, or 30 per cent or more organic matter if the clay content of the mineral soil is higher than 15 per cent. See also **Primary Profile Form**.

Outcrop The exposure at the surface of rock that is inferred to be continuous with underlying bedrock.

Paleosol Syn. **Buried Soil**.

Pan (Hardpan) A hardened, compacted and/or cemented horizon, or part thereof, in the soil profile. Such pans frequently reduce soil permeability and root penetration, and thus may give rise to plant growth and drainage problems. Deep ripping or chisel ploughing is used to overcome such problems. The hardness is caused by mechanical compaction or cementation of soil particles with organic matter or with materials such as silica, sesquioxides or calcium carbonate. The hardness does not change appreciably with changes in moisture content, and pieces of the hard layer are not subject to slaking. See also **Claypan, Plough-Pan**.

Parent Material The geologic material from which a soil profile develops. It may be bedrock or unconsolidated materials including alluvium, colluvium, aeolian deposits or other sediments.

Particle Size Analysis (Mechanical Analysis) The laboratory determination of the amounts of the different separates in a soil sample such as clay, silt, fine sand, coarse sand and gravel. The amounts are normally expressed as percentages by weight of dry soil and are determined by dispersion, sedimentation, sieving, micrometry or combinations of these techniques.

Ped An individual, natural soil aggregate. Cf. **Clod**.

Pedal Describes a soil in which some or all of the soil material occurs in the form of peds in the moist state. Strongly pedal soils have two-thirds or more of their soil material in the form of peds, and weakly pedal soils have less than one-third of their soil material in the form of peds. Cf. **Apedal**.

Pedology The study of soils, particularly their formation, morphology, distribution and classification.

Penetrability	The ease with which a probe can be pushed into the soil. May be expressed in units of distance, speed, force or work, depending on the type of penetrometer used.
Penetrometer	An instrument used to measure resistance to penetration in soil. Such measurements are important in relation to soil density studies or the location of pans.
Percolation	The downward movement of water through soil, contributing to internal drainage. See also **Soil Permeability**.
Permanent Wilting Point	A laboratory measure of the amount of water held in a soil at the point when foliage wilts and does not recover when placed in a humid atmosphere. Expressed as a percentage of the oven-dry weight of the soil.
	In the field, existing foliage withers and dies as a result of moisture stress. However, in the long term, plants may recover if more water becomes available, due to the production of new shoots. See also **Field Capacity, Available Soil Water, Soil Water Potential**.
Permeability	Syn. **Soil Permeability**.
pH	A measure of the acidity or alkalinity of a soil. A pH of 7.0 denotes neutrality, higher values indicate alkalinity and lower values indicate acidity. Strictly, it represents the negative logarithm of the hydrogen ion concentration in a specified soil/water suspension on a scale of 0 to 14. Soil pH levels generally fall between 5.5 and 8.0 with most plants growing best in this range. See also **Acid Soil**. Soil pH is commonly measured in the field by a colorimetric method using Raupach's indicator. In the laboratory, a number of methods may be used, depending on need. These are commonly based on more accurate electrical techniques generally using 1:2.5 or 1:5 mixtures of soil with water or 0.01M $CaCl_2$ solution. The $CaCl_2$ method gives pH values approximately 0.5 units lower than the water method. In reports quoting pH levels, the method of measurement should be noted.
	(Reference: Raupach, M. and Tucker, B.M. (1959), 'The field determination of soil reaction', *J. Aust. Inst. Ag. Sci.* **25**, 129–33.)
Physical Fertility	See **Soil Fertility**.
Piezometer	A narrow tube, open at each end, inserted down a hole in the ground to the depth of the water table, which enables measurement of its elevation or hydraulic head.
Plant-available Water Capacity	Syn. **Available Water-holding Capacity**.
Plastic	Describes soil material which is in a condition that allows it to undergo permanent deformation without appreciable volume change or elastic rebound, and without rupture. The importance of this property in a soil conservation context relates to the soil's behaviour when used in earthworks or when cultivated. This behaviour is characterised according to the system of Atterberg limits. See also **Soil Plasticity**.
Plastic Limit	See **Atterberg Limits**.
Plasticity Index	See **Atterberg Limits**.
Plough-Pan	A pan made up of a layer of soil compacted by repeated tillage at a constant depth over many years.

Pore Space The fraction of the bulk volume or total space within soils that is not occupied by solid particles. See also **Bulk Density, Soil Porosity**.

Primary Particles The individual mineral particles of which a soil is made up, such as sand, silt and clay particles, in their non-aggregated state.

Primary Profile Form The first division of the Factual Key soil classification system. Four primary profile forms are recognised — organic soils, uniform soils, gradational soils and duplex soils.

Principal Profile Form The end point of the Factual Key soil classification system. A principal profile form code, such as Ug5.16, Gn2.23 or Dy3.41, describes the soil profile to such an extent that it will be possible to make a reasonably concise statement concerning the soils belonging to it.

An EXTENDED PRINCIPAL PROFILE FORM may include additional information describing the surface soil in more detail and/or material below the solum. For example, see Northcote (1979), pp. 121–2.

Profile Drainage Syn. **Internal Drainage**.

Puddling The act of destroying soil structure by compaction or tillage of the soil at high moisture content, thereby reducing its porosity, permeability and aggregation. Stock often damage soil structure in this way when they are left on wet or waterlogged pasture.

R Horizon See **Soil Profile**.

Regolith The layer or mantle of loose, non-cohesive or cohesive rock material, of whatever origin, that nearly everywhere forms the surface of the land and rests on bedrock. It comprises rock waste of all sorts; volcanic ash; glacial drift; alluvium; windblown deposits; accumulations of vegetation, such as peat; and soil.

S.T.P.P. (STPP) Syn. **Sodium Tripolyphosphate**.

Saline Soil A soil which contains sufficient soluble salts to adversely affect plant growth and/or land use. Generally, a level of electrical conductivity of a saturation extract in excess of 4dS/m at 25°C is regarded as a suitable criterion to define such a soil. See also **Soil Salinity**

Sand A soil separate consisting of particles between 0.02 and 2.0 mm in equivalent diameter. Fine sand is defined as particles between 0.02 and 0.2 mm, and coarse sand as those between 0.2 and 2.0 mm.

Saprolite Parent rock undergoing weathering and/or breakdown

Saturation Extract A solution derived by saturating a soil sample with water under standard conditions for a period long enough to dissolve the soluble constituents present. The solution is subsequently extracted by filtration or centrifugation, and used in chemical analysis or measurement related to the soluble constituents and/or other ions present.

Seasonal Cracking A phenomenon in expansive soils which, during a dry period, develop cracks as wide or wider than 6 mm and which penetrate at least 0.3 m into the soil material.

Self-mulching   The condition of a well-aggregated soil in which the surface layer forms a shallow mulch of soil aggregates when dry. Aggregation is maintained largely as a response of the clay minerals present to the natural processes of wetting and drying. Such soils typically have moderate to high clay contents and marked shrink-swell potential. Any tendency to crust and seal under the impact of rain is counteracted by shrinkage and cracking, thus producing a mulch effect as the soil dries out. Tillage when wet may appear to destroy the surface mulch which, however, will re-form upon drying. Cf. **Hardsetting, Massive**.

Shear Failure   A breakdown of the ability of soil material to maintain its aggregated structure such that one body of soil material is caused to move past, or in relation to, the adjacent materials. It is a principal factor in the initiation of landslides and landslips. Shear failure is triggered by:

— factors that contribute to increased shear stress including: removal of lateral support (such as erosion by water and ice, previous mass movement, human activity by cut and fill operations); surcharge (such as weight of precipitation, seepage pressures, construction of fill, weight of buildings); transitory stresses (such as earthquakes, vibrations from blasting); removal of underlying support (such as undercutting by flowing water or waves, weathering, tunnel erosion); and lateral pressure (such as freezing water in cracks)

— factors that contribute to low or reduced shear strength including: initial state of the material giving it inherent low strength through its composition (such as sedimentary clays and shales), texture (such as unconsolidated material) and gross structure and slope geometry (such as faults, bedding planes, inclined strata); changes due to weathering (such as physical disintegration of granular rocks, removal of matrix material); and changes in intergranular forces due to water content and pressure in pores and fractures

— factors that contribute to both increased shear stress and reduced shear strength, such as the addition of water.

Shear Strength   The internal resistance of a soil to shear along a plane. The resistance is caused by intergranular friction and cohesion.

Shear Stress   The force per unit area acting along a given plane and tending to cause shear failure within a soil mass. All landslides involve shear failure of land surface materials under shear stress.

Shrink-Swell Potential   The capacity of soil material to change volume with changes in moisture content, frequently measured by a laboratory assessment of the soil's linear shrinkage. Relates to the soil's content of montmorillonite type clays. High shrink-swell potential in soils, such as cracking clays, can give rise to problems in earth foundations and soil conservation structures. Categories used are:

Shrink-Swell Potential	Linear Shrinkage
Low	0–12 per cent
Medium	12–17 per cent
High	17–22 per cent
Very High	> 22 per cent

In an urban context, shrink-swell potential is particularly relevant in consideration of building and road foundations. It is non-critical when linear shrinkage values are low, marginal when medium, critical when high and very critical when very high. Where marginal or higher values are recognised, structural engineering expertise is required to ensure the construction of stable foundations. See also **Expansive Soil**.

Silt   A soil separate consisting of particles between 0.002 and 0.02 mm in equivalent diameter.

Skeletal Soil   Syn. **Lithosol**.

Slaking   The partial breakdown of soil aggregates in water due to the swelling of clay and the expulsion of air from pore spaces. It does not include the effects of soil dispersion. It is a component, along with soil dispersion and soil detachment, of the process whereby soil structure is broken down in the field. This breakdown results from the action of raindrop impact, raindrop splash, runoff and seepage, and contributes to soil erosion and the failure of earth structures.

Slickensides   Polished surfaces of macro-aggregates in soils dominated by swelling clays, and caused by aggregates sliding against each other during wetting and drying.

Sodic Soil   A soil containing sufficient exchangeable sodium to adversely affect soil stability, plant growth and/or land use. Such soils would typically contain a horizon in which the ESP or amount of exchangeable sodium expressed as a percentage of cation exchange capacity would be six or more. The soils would be dispersible and may be improved by the addition of gypsum. Strongly sodic soils are considered to be those with an ESP of 15 per cent or more. See also **Soil Dispersibility**.

Sodium Tripolyphosphate (STPP)   A manufactured chemical of the general formula $Na_5P_3O_{10}6H_2O$, used for the dispersion of aggregated soils. This may be in the laboratory for particle size analysis, or for sealing leaking water-holding structures built in, or from, aggregated soils.

Soil Aeration   The process by which air in the soil is replenished by air from the atmosphere. In a well-aerated soil, the soil air is similar in composition to the atmosphere above the soil. Poorly aerated soils usually contain a much higher percentage of carbon dioxide and a correspondingly lower percentage of oxygen. The rate of aeration depends largely on the volume and continuity of pores in the soil.

Soil Aggregate   A unit of soil structure consisting of primary soil particles held together by cohesive forces or by secondary soil materials such as iron oxides, silica or organic matter. Aggregates may be natural, such as peds, or formed by tillage, such as crumbs and clods.

Soil Ameliorant (Soil Conditioner)   A substance used to improve the chemical or physical qualities of the soil. For example, the addition of lime to the soil to increase pH to the desired level for optimum plant growth, or the addition of gypsum to improve soil structure.

Soil Amendment   The alteration of the properties of a soil by the addition of substances such as lime, gypsum and sawdust, for the purpose of making the soil more suitable for plant growth. See also **Soil Ameliorant**.

Soil Association   A soil mapping unit in which two or more soil taxonomic units occur together in a characteristic pattern, such as a toposequence. The units are combined because the scale of the map, or the purpose for which it is being made, does not require delineation of individual soils.

The soil association may be named according to the units present, the dominant unit or be given a geographic name based on a locality where the soil association is well developed.

Soil Class   The common taxonomic unit for a group of soils that are characterised by a particular set of morphological features or surface features that are related to soil management. It is commonly used in the mapping of soils for specific purposes and represents a group of soils that respond similarly to a set of management practices.

While no specific taxonomic units can be attributed to a soil class, as their definition depends on the purpose of the soil mapping, soil class often coincides with soil series, soil phase or an extended principal profile form.

Soil Classification   The systematic arrangement of soils into groups or categories on the basis of similarities and differences in their characteristics. Soils can be grouped according to their genesis (taxonomic classification), their morphology (morphological classification), their suitability for different uses (interpretative classification) or according to specific properties.

The purposes of soil classification are as follows:

— generally, as a means of grouping soils into useful categories — so that statements about one particular soil are likely to apply to other soils in the same group;

— with experience, the identification and categorising involved may lead to the inference of other soil properties (apart from those used in the classification);

— a formal system of classification encourages the scientific and logical study of soils;

— the standardisation and objectivity involved are desirable for communication purposes.

See also **Factual Key, Great Soil Group, Soil Survey.**

Soil Coherence   The degree to which soil material is held together at different moisture levels. If two-thirds or more of the soil material, whether composed of peds or not, remain united at the given moisture level, then the soil is described as coherent.

Soil Colour   The colour of soil material is determined by comparison with a standard Munsell soil colour chart (Munsell Color Company, 1975) or its equivalent. The colour designation thus determined specifies the relative degrees of the three variables of colour — HUE, VALUE and CHROMA. Hue represents the spectral colour, for soils normally in terms of red and yellow. Value represents the lightness or darkness of colouration, and chroma its intensity. For example, 5YR 4/6 has a hue of 5YR, a value of four and chroma of six. Equivalent descriptive colour names can be used if desired, using those listed in the Munsell chart.

During soil survey, soil colour is determined on a freshly broken aggregate of both dry and moist soil material, to ensure complete documentation of colour. Since a soil's colour may vary, depending on soil moisture content, the moist soil colour provides a base for comparison with other soil samples.

A whole-coloured soil is one in which less than 10 per cent of the soil mass is affected by mottling. Its colour would be specified as a single colour, whereas mottled soils would be described in terms of the dominant matrix colour and the subdominant colour of the mottles.

Soil colour is particularly important in the identification of bleaching and gleying in a soil profile.

Soil Complex   A soil mapping unit in which two or more soil taxonomic units occur together in an undefined or complex pattern. The soils are intimately mixed and it is undesirable or impractical to delineate them at the scale of the map. The soil complex may be named according to the units present, the dominant unit or be given a geographic name based on a locality where the soil complex is well developed.

Soil Compressibility   The capacity of a soil to decrease in volume on application of loading. Such decrease is equal to the decrease in the volume of the soil's pores. Changes in volume of the particles themselves are considered to be negligible.

Soil compressibility is commonly measured by an oedometer, in which the soil sample is laterally confined in a rigid metal ring but is compressed vertically by mechanical or hydraulic means. The changes in thickness with increasing and decreasing incremental loading over various time periods are plotted and thence used to compute a stiffness modulus (E) which is the expression of the soil's compressibility (meganewtons per square metre or MN/m^2).

Soil Conditioner   Syn. **Soil Ameliorant**.

Soil Consistence   The resistance of soil material to deformation or rupture. Terms used for describing consistence of soil materials at various soil moisture contents and degrees of cementation are: wet — non-sticky, slightly sticky, sticky, very sticky, non-plastic, loose, very friable, friable, firm, very firm, and extremely firm; dry — loose, soft, slightly hard, hard, very hard and extremely hard; cementation — weakly cemented, strongly cemented and indurated.

Soil Crusting   Syn. **Surface Sealing**.

Soil Degradation   Decline in soil quality commonly caused through its improper use by humans. Soil degradation includes physical, chemical and/or biological deterioration. Examples are loss of organic matter, decline in soil fertility, decline in structural condition, erosion, adverse changes in salinity, acidity or alkalinity and the effects of toxic chemicals, pollutants or excessive flooding.

Soil Detachability   The susceptibility of a soil to the removal of transportable fragments by an erosive agent, such as rainfall, running water or wind. Depends largely on soil texture and structure, and is an important component of soil erodibility.

Soil Dispersibility   The characteristic of soils relating to their structural breakdown in water, into individual particles. Usually associated with high levels of exchangeable sodium on the clay fraction, and low levels of soluble salts in the soil. These factors cause clay particles to separate in water. As clay is one of the chief agents holding soil materials together, this leads to collapse of the soil structure and consequent instability.

Qualitative categories of soil dispersibility are high, moderate and low which generally relate to Classes 1, 2 and 3 to 8 respectively of the Emerson Aggregate Test. See also **Dispersible Soil, Deflocculation, Sodic Soil, Slaking**.

Soil Dispersion   The process whereby soil aggregates break down and separate into their constituent particles (clay, silt, sand) in water, due to deflocculation. The process is different from, but often associated with, slaking. See also **Dispersible Soil, Soil Dispersibility**.

Soil Erodibility   The susceptibility of a soil to the detachment and transportation of soil particles by erosive agents. It is a composite expression of those soil properties that affect the behaviour of a soil and is a function of the mechanical, chemical and physical characteristics of the soil. It is independent of the other factors influencing soil erosion such as topography, land use, rainfall intensity and plant cover, but may be changed by management.

The qualitative categories of soil erodibility used are low, moderate, high, very high and extreme. The most highly erodible soils are those that are most easily detached and transported by erosive forces. High soil dispersibility is a good indicator of high soil erodibility.

In the Universal Soil Loss Equation, soil erodibility is represented by the 'K' factor which is defined as the rate of soil loss per erosion index unit as measured on a unit plot. Such a plot is 72.6 feet long with a uniform lengthwise slope of 9 per cent, in continuous fallow, tilled up and down the slope. The K value can also be estimated from a knowledge of soil properties, through the use of a nomograph. (Reference: Wischmeier and Smith, 1978.) See Chapter 12.

Soil Fabric   Describes the appearance of the soil material using a ×10 hand lens. Differences in fabric are associated with the presence or absence of peds; and the lustre, or lack thereof, of the ped surfaces; and the presence, size and arrangement of pores (voids) in the soil mass. Descriptive terms used are earthy, sandy, rough-ped and smooth-ped. (Reference: Northcote, 1979.) See p. 68.

Soil Family	A unit of soil classification and soil mapping intermediate between soil series and great soil group, consisting of or describing groups of similar soil series. Soil properties used to define a soil family include depth of solum, differences in structure, occurrence of mottling or sodium contents. The use of the term is not widespread in Australia.

Soil Fertility The capacity of the soil to provide adequate supplies of nutrients in proper balance for the growth of specified plants, when other growth factors such as light, moisture and temperature are favourable. The more general concept of soil fertility can be divided into three components:

— CHEMICAL FERTILITY refers specifically to the supply of plant nutrients in the soil;

— PHYSICAL FERTILITY refers specifically to soil structure conditions which provide for aeration, water supply and root penetration;

— BIOLOGICAL FERTILITY refers specifically to the population of micro-organisms in the soil, and its activity in recycling organic matter.

Soil Horizon A layer of soil material within the soil profile with distinct characteristics and properties which are produced by soil-forming processes, and which are different from those of the layers below and/or above. Generally, horizons are more or less parallel to the land surface, except that tongues of material from one horizon may penetrate neighbouring horizons.

The boundary between soil horizons defines the nature of the change from one horizon to another. It is specified by the width of the transition zone and the shape as expressed in vertical section. Width of boundary may be expressed as:

— sharp = boundary is less than 2 cm wide

— clear = boundary is 2 to 5 cm wide

— gradual = boundary is 5 to 10 cm wide

— diffuse = boundary is more than 10 cm wide.

Shape of boundary may be expressed as:

— even = boundary is almost a plane surface

— wavy = boundary waves up and down and the pockets so formed are relatively wider than their depths

— irregular = boundary waves up and down and the pockets so formed are relatively deeper than their widths

— broken = boundary is discontinuous.

Soil Landscape An area of land that has recognisable and specifiable topography and soils, that is capable of being presented on maps and of being described by concise statements.

Thus, a soil landscape has a characteristic landform with one or more soil taxonomic units occurring in a defined way. It is often associated with the physiographic features of the landscape and is similar to a soil association but, in a soil landscape, the landform pattern is specifically described. The soil landscape may be named according to the soil taxonomic units present, the dominant unit or be given a geographic name based on a locality where it is well developed.

Soil Moisture Characteristic The graphical relationship between soil water content and soil water potential for a given soil. It may vary widely, depending on the texture, structure and pore size distribution of the soil. The relationship is used to indicate the ease or difficulty of removing water from the soil at different soil water contents and can therefore be important in relation to soil structure and plant growth studies.

Soil Organic Matter   That fraction of the soil including plant and animal residues at various stages of decomposition, cells and tissues of soil organisms, and substances synthesised by them. It is of vital importance as it contributes to the cation exchange and field capacities of the soil and provides a major source of plant nutrients and substances which assist in soil structure maintenance.

Soil Permeability   The characteristic of a soil, soil horizon or soil material which governs the rate at which water moves through it. It is a composite expression of soil properties and depends largely on soil texture, soil structure, the presence of compacted or dense soil horizons and the size and distribution of pores in the soil. The rate varies widely in the field, from more than 3000 mm/day in poorly graded sands and gravels, to less than 0.01 mm/day for some heavy plastic clays. The qualitative categories of permeability for general use are:

slowly permeable — less than 10 mm per day
moderately permeable — 10 mm to 1000 mm per day
highly permeable — more than 1000 mm per day.

When applied to a soil profile, the rate of water transmission is controlled by the least permeable layer in the soil. Cf. **Internal Drainage**.

Soil Phase   A subdivision of a soil taxonomic unit based on characteristics that affect the use and management of the soil, but do not change the classification of the soil. Such characteristics include slope, erosion, depth, stoniness and rockiness, drainage, depositional layers, gilgai or scalding.

Soil Plasticity   The degree to which a soil is plastic. A highly plastic soil has plastic properties over a wide range of moisture contents. A subplastic soil has properties suggesting that less clay is present than is actually the case. Such soil materials may be identified by determining two soil textures. The initial texture obtained by a one to two-minute working of the soil sample will appear to be of a less clayey grade than the texture obtained after a ten-minute working. A non-plastic soil has properties which do not allow plastic behaviour, whatever the moisture content. See also **Atterberg Limits, Soil Texture**.

Soil Porosity   The degree to which the soil mass is permeated with pores or cavities. Porosity can be generally expressed as a percentage of the whole volume of a soil horizon that is unoccupied by solid particles. In addition, the number, sizes, shapes and distribution of the voids are important. Generally, the pore space of surface soil is less than one-half of the soil mass by volume, but in some soils it is more than half. The part of the pore space that consists of small pores that hold water by capillary action is called CAPILLARY POROSITY. The part that consists of larger pores that do not hold water by capillary action is called NON-CAPILLARY POROSITY. See also **Bulk Density**.

Soil Profile   A vertical cross-sectional exposure of a soil, extending downwards from the soil surface to the parent material or, for practical purposes, to a depth of 1 m where the parent material cannot be differentiated. It is generally composed of three major layers designated A, B and C horizons. The A and B horizons are layers that have been modified by weathering and soil development and comprise the solum. The C horizon is weathering parent material which has not, as yet, been significantly altered by biological soil-forming processes. A surface organic (O) horizon and/or a subsolum (D) horizon may also occur.

The boundaries between successive soil horizons are specified by their width and shape.

O HORIZON:
A surface layer of plant materials in varying stages of decomposition not significantly mixed with the mineral soil. Often not present or only poorly developed in Australian soils, except in some forests. When highly developed, it can be divided into two parts:

O_1 HORIZON is the surface layer of undecomposed plant materials.

O_2 HORIZON is the layer beneath the O_1 which is partly decomposed.

A HORIZON:
This is the original top layer of mineral soil. It can be divided into two parts:

A_1 HORIZON is the surface soil and generally referred to as topsoil. Relative to other horizons it has a high content of organic matter, a dark colour and maximum biological activity. This is the most useful part of the soil for revegetation and plant growth. It is typically from 5 to 30 cm thick.

A_2 HORIZON is a layer of soil of similar texture to the A_1 horizon, but is paler in colour, poorer in structure and less fertile. A white or grey colouration, known as bleaching, is often caused by impeded soil drainage and/or eluviation. The A_2 horizon is typically from 5 to 70 cm thick, but does not always occur.

B HORIZON:
The layer of soil below the A horizon. It is usually finer in texture (that is, more clayey), denser and stronger in colour. In most cases it is a poor medium for plant growth. Thickness ranges from 10 cm to over 2 m. It can be divided into two parts:

B_1 HORIZON is a transitional horizon dominated by properties characteristic of the underlying B_2 horizon.

B_2 HORIZON is a horizon of maximum development due to concentration of silicate clay and/or iron, and/or aluminium and/or translocated organic material. Structure and/or consistence are unlike that of the A and C horizons and colour is typically stronger.

C HORIZON:
Layers below the B horizon which may be weathered, consolidated or unconsolidated parent material little affected by biological soil-forming processes. The C horizon is recognised by its lack of pedological development, and by the presence of remnants of geologic organisation. Its thickness is very variable.

D HORIZON:
Layers below the solum which are not C horizon, and are not related to the solum in character or pedologic organisation.

R HORIZON:
Hard rock that is continuous.

Soil Reaction Trend   The change in pH with depth in a soil profile, from surface soil to deep subsoil. Four such trends have been defined: strongly acid, acid, neutral and alkaline.

(Reference: Northcote, K.H. (1979).) See p. 68. See also **pH**.

Soil Resource   The total extent of soil within a given area available as a natural medium for plant growth. It is limited and exhaustible and, thus, its management must aim to avoid degradation to ensure its potential productive capability is maintained or improved.

Soil Salinity   The characteristic of soils relating to their content of water-soluble salts. Such salts predomi-nantly involve sodium chloride, but sulphates, carbonates and magnesium salts occur in some soils. High salinity adversely affects the growth of plants, and therefore increases erosion hazard. Soil salinity is normally characterised by measuring the electrical conductivity of a soil/water saturation extract and is expressed in decisiemens per metre at 25°C (dS/m). See also **Saline Soil**.

Soil Series   A unit of soil classification and soil mapping comprising or describing soils which are alike in all major profile characteristics. Each soil series is developed from a particular parent material, or group of parent materials, under similar environmental conditions. It approximates to the extended principal profile form. The name given to a soil series is geographic in nature, and indicates a locality where the soil series is well developed.

Soil Solution   The water in a soil containing ions dissociated from the surfaces of soil particles, and other soluble substances. See also **Saturation Extract**.

Soil Stabiliser   A substance or material used to improve soil stability, strength or bearing capacity. In a soil conservation context, the primary purpose of a soil stabiliser is to reduce erosion potential. For example, the addition of proprietary chemicals to achieve soil flocculation, or the incorporation of organic materials such as hay or straw into newly topsoiled areas.

Soil Structure   The combination or spatial arrangement of primary soil particles (clay, silt, sand, gravel) into aggregates such as peds or clods and their stability to deformation. Structure may be described in terms of the grade, class and form of the soil aggregates, as follows.

Grade   —   expresses the degree and strength of soil aggregation determined on moist soil. The grades range from structureless, if there is no observable aggregation, to strong, where more than two-thirds of the soil is aggregated.

Class   —   expresses the main size range of the aggregates.

Form   —   expresses the shape of the individual aggregates as crumb, granular, subangular blocky, angular blocky, prismatic, columnar or platy.

Soil structure is an important property with respect to the stability, porosity and infiltration characteristics of the soil. Well-structured soils tend to be more resistant to erosion due to their ability to absorb rainfall more freely and over longer periods, and because of the resistance of their aggregates to detachment and transport by raindrop splash and/or overland flow. They also have good soil/water/air relationships for the growth of plants. Poorly structured soils have unstable aggregates and low infiltration rates. They tend to break down quickly under heavy rainfall which leads to soil detachment and erosion. Under certain conditions, surface sealing occurs and this gives rise to rapid and excessive runoff.

Soil Survey   The systematic examination, description, classification and mapping of soils, with the aim of categorising soil distribution within a defined area. In practice, most soil surveys also include statements on the geology, topography, climate and vegetation of the area concerned. They may be carried out for general use, in which case a wide range of soil properties is examined, or for a particular purpose such as crop irrigation or urban planning, in which case only a limited number of soil properties may be relevant. Soil survey involves the following steps:

1. deciding which properties of the soil are important for the particular purpose;

2. selecting categories for each property relevant to the particular purpose;

3. classifying soils into map units so that soil variation within units is less than that between units;

4. locating and plotting the boundaries of these units on maps; and

5. preparing maps and reports for publication.

Soil Taxonomic Unit   A general term for a grouping of soils based on similarities of the soils within the group, and differences compared with other groups. Note that the general taxonomic units such as great soil group and principal profile form need to be distinguished from the soil survey taxonomic units such as soil series and soil association.

Soil Texture   The coarseness or fineness of soil material as it affects the behaviour of a moist ball of soil when pressed between the thumb and forefinger. It is generally related to the proportion of soil particles of differing sizes (sand, silt, clay and gravel) in a soil, but is also influenced by organic matter content, clay type and degree of structural development of the soil.

Six main soil texture groups are recognised:

Texture Group	Approx. clay content
1. Sands	< 5%
Coarse	
2. Sandy Loams	10–15%
3. Loams	20–25%
Medium	
4. Clay Loams	30%
5. Light Clays	35–40%
Fine	
6. Heavy Clays	> 45%

For field identification of these main groups, take a small handful of soil and knead with water until a homogeneous soil ball or bolus is obtained. Large pieces of grit and organic material should be discarded. Small clay peds should be crushed and worked with the rest of the soil. The feel, behaviour and resistance of the soil to manipulation during this process are important. The bolus should be kept moist so that it just fails to stick to the fingers. The six main texture groups should be apparent as follows.

1. *Sands* — have very little or no coherence and cannot be rolled into a stable ball. Individual sand grains adhere to the fingers.

2. *Sandy Loams* — have some coherence and can be rolled into a stable ball, but not a thread. Sand grains can be felt during manipulation.

3. *Loams* — can be rolled into a thick thread, but this will break up before it is 3 to 4 mm thick. The soil ball is easy to manipulate and has a smooth spongy feel with no obvious sandiness.

4. *Clay Loams* — can be easily rolled to a thread 3 to 4 mm thick, but it will have a number of fractures along its length. Soil becoming plastic, capable of being moulded into a stable shape.

5. *Light Clays* — can be rolled to a thread 3 to 4 mm thick without fracture. Plastic behaviour evident, smooth feel with some resistance to rolling out.

6. *Heavy Clays* — can be rolled to a thread 3 to 4 mm thick and formed into a ring in the palm of the hand without fracture. Smooth and very plastic, with moderate–strong resistance to rolling out.

In a soil conservation context, soil texture is very important as it not only has a major influence on the erodibility of soils, but also on their performance when used in water storage structures. In particular, it largely determines soil permeability.

See also **Grading, Particle Size Analysis**.

Soil Texture Grade   A minor category of soil texture as set out in the table under **Soil Texture Group**.

Soil Texture Group   A major category of soil texture as set out in the following table:

Texture Groups	Texture Grades
Sands	sand; loamy sand; clayey sand;
Sandy Loams	sandy loam; fine sandy loam; light sandy clay loam;
Loams	loam; loam, fine sandy; silt loam; sandy clay loam;
Clay Loams	clay loam; silty clay loam; fine sandy clay loam;
Light Clays	sandy clay; silty clay; light clay; light medium clay;
Medium-Heavy Clays	medium clay; heavy clay.

Soil Type   A general term used to describe a group of soils that can be managed similarly and which exhibit similar morphological features. It is largely a layman's term and now has no formal soil taxonomic meaning.

Soil Variant   A soil having a morphology which is distinct from the surrounding soils, but comprises such a limited geographic area that the delineation of a new map unit, or the naming of a new soil taxonomic unit, is not justified.

Soil Water   Water contained in, or in transit by drainage through, the soil.

Soil Water Potential   In general terms, the amount of work (or suction) which must be applied to remove water from soil. For more detailed explanation, see Beckmann *et al.* (1976), ASSSI publication No. 6.

The water content at a suction of −15 bars is taken to approximate the permanent wilting point, and the water content at a suction of −0.1 bars is taken to approximate field capacity.

Solum   The upper part of a soil profile above the parent material, in which current processes of soil formation are active. The solum consists of either the A and B horizons or the A horizon alone when no B horizon is present. The living roots and other plant and animal life characteristic of the soil are largely confined to the solum.

Solution   That part of natural weathering processes whereby substances in rocks and soil are dissolved in groundwater. Such water usually contains some carbon dioxide and organic compounds from plant breakdown which make it mildly acid, thereby enhancing its ability to dissolve rock materials.

Sporadic Bleaching   See **Bleaching**.

Structural Degradation Hazard   The susceptibility of a soil's structure to breakdown as a result of cultivation. Three categories are recognised:

— *high* — applies to a soil with a structure that readily degrades to a massive condition following cultivation;

— *moderate* — applies to a soil with a structure that degrades to a massive condition after some years of continuous cultivation;

— *low* — applies to a soil with a structure that resists degradation to a massive condition except after many years of continuous cultivation.

Structural Stability   The ability of a soil to maintain its structure under the influence of tillage, rainfall, trampling or other adverse forces, which tend to disintegrate such structure. Sometimes characterised in the laboratory by various wet and/or dry sieving techniques. See also **Soil Structure, Water-stable Aggregation.** Cf **Mechanical Stability**.

Structure   Syn. **Soil Structure**.

Surface Sealing (Soil Crusting)   The orientation and packing of dispersed soil particles in the immediate surface layer of the soil, rendering it relatively impermeable to water. Typically occurs due to the effect of raindrop impact on bare soil and results in a reduction in infiltration. Runoff and the potential for soil erosion are thus increased, and a crust may form on drying out.

However, surface sealing can be an important factor in reducing wind erosion, as the seal tends to resist removal of soil particles by wind action.

Swelling Soil   Syn. **Expansive Soil**.

Texture   See **Soil Texture**.

Texture Contrast Soil   Syn. **Duplex Soil**. For use in the Australian Soil Classification system, a stricter definition has been adopted. Refer Isbell, R.F. (1996), *The Australian Soil Classification*, CSIRO, Melbourne.

Threshold Velocity   The wind velocity at which saltation is initiated.

Toposequence   A repetitive sequence of soils encountered between hillcrests and the valley floor. A catena is a special case of a toposequence in which the parent material is uniform.

Toxicity   The characteristic of a soil relating to its content of elements or minerals which adversely affect plant growth. It is of particular concern in relation to acid soils. Soils with pH less than 5.0 may give rise to manganese and aluminium toxicities, which reduce plant growth and hence ground cover. It is also of concern in the rehabilitation of heavy metal mines, where toxic levels of such elements as copper, zinc and lead in mine tailings create difficulties in their revegetation.

Transportation   That part of erosion processes in which detached soil or rock material is moved from one place to another. This may be accomplished by running water, rainfall, wind, gravity, ice action or subsurface seepage.

Truncated   Describes a soil profile that has been cut down by accelerated erosion or by mechanical means. The profile may have lost part or all of the A horizon and sometimes the B horizon, leaving only the C horizon. Comparison of an eroded soil profile with a virgin profile of the same area, soil type and slope conditions indicates the degree of truncation.

U.S.C.S. (USCS)   Syn. **Unified Soil Classification System**.

Unified Soil Classification System (USCS)

A soil classification system based on the identification of soil materials according to their particle size, grading, plasticity index and liquid limit. These properties have been correlated with the engineering behaviour of soils including soil compressibility and shear strength. (Reference: Casagrande, A. (1947), *Proc. Am. Soc. Civ. Eng.* **73**, 783–810.) The system is used to determine the suitability of soil materials for use in earthworks, optimal conditions for their construction, special precautions which may be needed, such as soil ameliorants, and final batter grades to be used to ensure stability. See also **Atterberg Limits**.

Uniform Soil

A soil in which there is little, if any, change in soil texture between the A and B horizons (for example, loam over loam, sandy clay over silty clay). The soil is dominated by the mineral fraction and shows minimal texture differences throughout, such that no clearly defined texture boundaries are to be found. The range of texture throughout the solum is not more than the equivalent span of one soil texture group. See also **Primary Profile Form, Soil Profile, Duplex Soil, Gradational Soil**.

Value

See **Soil Colour**.

Virgin Soil

A soil that has not been significantly disturbed from its natural environment.

Water Balance

An estimated state of equilibrium within the soil moisture regime based on rainfall, evapotranspiration, runoff, drainage and soil moisture storage.

Soil moisture is assumed to severely limit plant growth in months when rainfall, together with an antecedent moisture content up to a maximum of 100 mm, is less than 40 per cent of pan evaporation. At other times, soil moisture is regarded as being adequate for plant growth. When the assumed field capacity of the soil is reached, any further rainfall is regarded as being lost as runoff. (Reference: Fleck, B.C. (1971), *J. Soil Cons. NSW* **27**, 135–44.)

The water balance provides information as to suitable tillage and sowing times, condition of ground cover and also whether other activities are restricted by wet or dry soil conditions. It also gives useful information concerning periods of major catchment flow and trickle flows. However, being based on monthly averages, short periods of intense rainfall are omitted. This rainfall is important in initiating soil erosion and providing runoff for stock watering purposes.

Water-holding Capacity

Syn. **Field Capacity**.

Waterlogged

The condition of a soil which is saturated with water and in which most or all of the soil air has been replaced. The condition, which is detrimental to most plant growth, may be caused by excessive rainfall, irrigation or seepage, and is exacerbated by inadequate site and/or internal drainage.

Water-repellent Soils (Hydrophobic Soils)

Soils which resist wetting when dry. Drops of water do not spread spontaneously over their surface and into pores. The degree of water repellence may be severe where water drops remain on a flattened surface for some minutes. In other cases, drops appear to be absorbed readily, but quantitative measurements show that the height of capillary rise is diminished.

This characteristic is mainly a feature of some sandy soils (topsoils) and is generally attributed to organic coatings on the sand grains which resist water entry into the soil.

Water-stable Aggregation

An indication of the resistance of soil aggregates to breakdown by water. This is normally measured by the degree to which different size fractions of aggregates are broken up by agitation in water using a wet-sieving procedure.

The degree of water-stable aggregation is used as a general indicator of soil erodibility. However, the procedure is not widely accepted because of problems of standardisation and the paucity of clear evidence relating it objectively to a soil's susceptibility to erosion.

Weathering   The physical and chemical disintegration, alteration and decomposition of rocks and minerals at or near the earth's surface by atmospheric and biological agents.

Index